Biomedical Applications of

Immobilized Enzymes and Proteins

Volume 1

Biomedical Applications of

Immobilized Enzymes and Proteins

Volume 1

EDITED BY

THOMAS MING SWI CHANG, M.D., PH.D., F.R.C.P. (C)

Professor of Physiology, Professor of Medicine
Associate, Medical Research Council of Canada
Director, Artificial Organs Research Unit
McGill University
Montreal, Quebec, Canada

PLENUM PRESS · NEW YORK AND LONDON

Library of Congress Cataloging in Publication Data

Main entry under title:

Biomedical applications of immobilized enzymes and proteins.

Includes bibliographical references and index.
1. Immobilized enzymes—Therapeutic use. 2. Immobilized enzymes. 3. Immobilized proteins. I. Chang, Thomas Ming Swi. [DNLM: 1. Enzymes. 2. Proteins. QU135 B617]
RM666.E55B56 615'.35 76-56231
ISBN 0-306-34311-8

A Division of Plenum Publishing Corporation
227 West 17th Street, New York, N.Y. 10011

Printed in the United States of America

To my wife, Lancy,

my children, Harvey, Victor, Christine, Sandra,

and my parents.

Contributors

WILLIAM H. BARICOS
Department of Biochemistry
Tulane University
New Orleans, Louisiana

PAUL D. BERK
National Institutes of Health
Bethesda, Maryland

FRED R. BERNATH
Department of Chemical and Biochemical Engineering
Rutgers University, New Brunswick, New Jersey

GEORGES B. BROUN
Departement de Genie Biologique
Universite de Technologie
Compiegne, France

CANH TRAN MINH
Ecole Nationale des Mines, Saint Etienne, France

ROBERT P. CHAMBERS
Department of Chemical Engineering
Auburn University
Auburn, Alabama

T. M. S. CHANG
Departments of Physiology and Medicine
McGill University
Montreal, Quebec, Canada

ICHIRO CHIBATA
Research Laboratory of Applied Biochemistry
Tanabe Seiyaku Co. Ltd.
Osaka, Japan

WILLIAM COHEN
Department of Biochemistry
Tulane University
New Orleans, Louisiana

R. J. DESNICK
Mayo Clinic
University of Minnesota Hospitals
Minneapolis, Minnesota

DONALD C. EMMERLING
Battelle, Columbus Laboratories
Columbus, Ohio

RICHARD D. FALB
Battelle, Columbus Laboratories
Columbus, Ohio

M. B. FIDDLER
Department of Pediatrics and the Dight Institute for Human Genetics
University of Minnesota Hospitals
Minneapolis, Minnesota

DAVID L. GARDNER
Battelle, Columbus Laboratories
Columbus, Ohio

ROBERT GLEW
Department of Biochemistry
University of Pittsburgh School of Medicine
Pittsburgh, Pennsylvania

GREGORY GREGORIADIS
Division of Clinical Investigation
Clinical Research Center
Harrow, Middlesex, United Kingdom

L. S. HERSH
Research and Development Laboratories
Corning Glass Works
Corning, New York

GARRET IHLER
Department of Biochemistry
University of Pittsburgh, School of Medicine
Pittsburgh, Pennsylvania

PETER R. KASTL
Department of Chemical Engineering
Auburn University
Auburn, Alabama

ANN-CHRISTIN KOCH-SCHMIDT
Department of Biochemistry
University of Lund
Lund, Sweden

NORMAN N. LI
Corporate Research Laboratories
Exxon Research and Engineering
Linden, New Jersey

CHUNG C. LIU
Department of Surgery and Pharmacology
School of Medicine
University of Pittsburgh
Pittsburgh, Pennsylvania

SHELDON W. MAY
School of Chemistry
Georgia Institute of Technology
Atlanta, Georgia

TAKAO MORI
Research Laboratory of Applied Biochemistry
Tanabe Seiyaku Co. Ltd.
Osaka, Japan

LAWRENCE S. OLANOFF
Department of Chemical and Biochemical Engineering
Rutgers University
New Brunswick, New Jersey

PAUL H. PLOTZ
National Institutes of Health
Bethesda, Maryland

BRUCE F. SCHARSCHMIDT
National Institutes of Health
Bethesda, Maryland

CHARLES R. SCRIVER
The deBelle Laboratory for Biochemical Genetics
McGill University
Montreal Children's Hospital Research Institute
Montreal, Quebec, Canada

E. D. SIU CHONG
Department of Physiology
McGill University
Montreal, Quebec, Canada

L. D. STEGER
Dight Institute for Human Genetics
University of Minnesota Hospitals
Minneapolis, Minnesota

S. R. Thorpe
Department of Pediatrics
University of Minnesota Hospitals
Minneapolis, Minnesota

Tetsuya Tosa
Research Laboratory of Applied Biochemistry
Tanabe Seiyaku Co., Ltd.
Osaka, Japan

Stuart Updike
Department of Medicine
University of Wisconsin
Madison, Wisconsin

Rosa Uy
Department of Biochemistry
University of Minnesota
St. Paul, Minnesota

Wolf R. Vieth
Department of Chemical and Biochemical Engineering
Rutgers University
New Brunswick, New Jersey

H. H. Weetall
Research and Development Laboratories
Corning Glass Works
Corning, New York

Lemuel B. Wingard
Department of Surgery and Pharmacology
School of Medicine
University of Pittsburgh
Pittsburgh, Pennsylvania

Finn Wold
Department of Biochemistry
University of Minnesota
St. Paul, Minnesota

Sidney K. Wolfson
Department of Neurosurgery
School of Medicine
University of Pittsburgh
Pittsburgh, Pennsylvania

Shang J. Yao
Department of Chemical and Petroleum Engineering
School of Engineering
University of Pittsburgh
Pittsburgh, Pennsylvania

Oskar R. Zaborsky
National Science Foundation
Washington, D.C.

Acknowledgments

This book was made possible by all the contributors, who, despite their busy schedules, spent a great deal of time preparing first-class manuscripts and fulfilling the deadline requirements. Mr. Thomas Lanigan, Senior Editor of Plenum, has been most helpful in the initial planning and throughout the preparation up to the last stages of the page proofs. Mr. Eric P. Otto from Plenum has been most patient and efficient in looking after the printing of the manuscript.

The seven chapters from this Unit include the cumulative result of research over the last twenty years. Throughout this time, the participation, collaboration, and help in other ways by a large number of people is gratefully acknowledged, especially: Dr. P. Barre, Mr. D. Cameron, Dr. J. Campbell, Dr. A. Chawla, Dr. E. Chirito, Dr. S. Chung, Dr. J. Coffey, Dr. C. Cole, Dr. J. Cousineau, Prof. J. Dirks, Mrs. P. Douglas, Dr. H. Duff, Dr. A. Gonda, Dr. M. Habib, Mrs. M. Hewish, Mr. K. Holeczek, Miss L. Johnson, Mrs. N. Kunterian, Mrs. T. Lee-Burns, Mr. B. Lessor, Dr. M. Levy, Mr. C. Lister, Dr. K. S. Lo, Prof. F. C. MacIntosh, Mrs. N. Malave, Prof. S. G. Mason, Dr. M. McGoldrick, Miss M. Migchelsen, Mr. P. Nasielski, Dr. M. Poznansky, Dr. B. Reiter, Mrs. E. Resurreccion, Dr. A. Rosenthal, Dr. J. Seely, Dr. E. Siu Chong, Miss A. Stark, Miss J. Toms, Dr. P. Tung, Miss C. Vaitekunas, Mrs. A. Versaza, Dr. B. Watson, and Miss W. Yensen. The typing of the chapters from this Unit and the secretarial help related to this book by Miss Joanne Toms is gratefully acknowledged. The work in this Unit has been encouraged and enhanced throughout the years by many people, especially, Professor D. Bates, Professor J. Beck, Professor A. V. S. Bergen, Professor A. Burton, Professor R. F. P. Cronin, Professor O. Denstedt, Professor W. Kolff, Professor F. C. MacIntosh, Professor S. G. Mason, Professor M. McGregor, Professor R. J. Rossiter, and others. The support of the Medical Research Council of Canada under the Presidency of Professor G. Malcolm Brown is gratefully acknowledged.

I am grateful to my wife, Lancy, for her cheerful acceptance of the extra time I needed for editing this book, especially when it came so shortly after the publication of the monograph *Artificial Cells*. She has carried out the very laborious work of preparing and typing the entire index section and reading over the manuscripts of the entire book.

Contents

Chapter 9

L-Asparaginase as a Model for Enzyme Therapy of Substrate-Dependent Tumors

E. D. SIU CHONG AND T. M. S. CHANG

Chapter 10

A Biomedical View of Enzyme Replacement Strategies in Genetic Disease

CHARLES R. SCRIVER

Chapter 11

Experimental Therapy Using Semipermeable Microcapsules Containing Enzymes and Other Biologically Active Material

T. M. S. CHANG

Chapter 12
Stabilized Urease Microencapsulated

DAVID L. GARDNER AND DONALD C. EMMERLING

Chapter 13
Liquid-Membrane-Encapsulated Enzymes

SHELDON W. MAY AND NORMAN N. LI

Chapter 14
Liposomes as Carriers of Enzymes and Proteins in Medicine

GREGORY GREGORIADIS

Chapter 15
Enzyme-Loaded Erythrocytes

GARRET IHLER AND ROBERT GLEW

Chapter 16

Enzyme Entrapment in Erythrocytes and Liposomes for the Treatment of Lysosomal Storage Diseases

R. J. DESNICK, M. B. FIDDLER, S. R. THORPE, AND L. D. STEGER

Chapter 17

Strategy for Enzyme Therapy: Immobilization in Hypoallergenic Gel Versus Entrapment in Red Blood Cell

STUART UPDIKE

Chapter 18

Immobilized Enzymes for Therapeutic Applications and for Large-Scale Production of Biologically Active Compounds

ICHIRO CHIBATA, TETSUYA TOSA, AND TAKAO MORI

Chapter 19

Artificial Kidney, Artificial Liver, and Detoxifiers Based on Artificial Cells, Immobilized Proteins, and Immobilized Enzymes

T. M. S. CHANG

Contents
of Volume 2

1

Introduction

T. M. S. Chang

In nature, most enzymes and proteins are located inside biological cells. Inside cells, most enzymes and proteins are further subdivided into smaller immobilized systems of intracellular organelles or intracellular membrane systems. Enzyme and protein research and its application have gone through interesting stages (Table 1). Stage I included extensive research into the extraction, purification, and crystallization of enzymes from biological cells and other sources. Enzymes obtained this way were used for detailed studies of enzyme kinetics.

With time, enzymologists began to worry that enzyme kinetics of dilute enzyme solutions might not necessarily represent the kinetics of the enzymes in the natural immobilized condition. This resulted in the initiation of stage II of enzyme research. Originally, stage II involved the insolubilization or attachment of enzymes and proteins to solid support systems in an attempt to study the enzyme kinetics of enzymes attached to membrane systems. This approach has resulted in extensive study of the "microenvironment" of enzyme systems (Goldman *et al.,* 1971).

Stage III of enzyme research involves the study of enzymes and proteins in artificially synthesized "intracellular environments" designed to simulate some of the conditions inside biological cells (Change, 1972). Stage III research was based on attempts to use "artificial cells" to solve problems related to the *in vivo* introduction of enzymes and proteins, since, unlike enzymes and proteins functioning in their normal intracellular environment, pure soluble enzymes or proteins administered into the body cause many problems. Interest in stage III research has been increasing in many laboratories, resulting in much extension, modification, and improvement of techniques.

The increasing interest and resulting proliferation of immobilized enzyme systems has resulted in a classification for immobilized enzymes proposed at an Enzyme Engineering Conference (Sundaram *et al.,* 1972): (1) covalently bound, (2) adsorbed, (3) matrix-entrapped, and (4) microencapsulated (Figure 1). Covalently bound, adsorbed, and matrix-entrapped enzymes would represent the study of enzymes and proteins under stage II, with emphasis on the microenvironment

T. M. S. Chang • Departments of Physiology and Medicine, McGill University, Montreal, Canada

Table 1
Stages of Enzyme Research

Stage	Description
I	"Dilute solution" of enzyme.
II	"Microenvironment," including adsorbed, covalently linked, or matrix-entrapped enzymes as models for membrane-bound enzymes.
III	"Intracellular environment," including artificial cells and other forms of encapsulated enzyme systems as models of intracellularly located enzymes.

(Table 1). Microencapsulated enzymes would represent stage III of enzyme research, with emphasis on the study of intracellular environment (Table 1). An excellent general introductory textbook reviewing all aspects of immobilized enzymes is available (Zaborsky, 1973). Other more specialized and detailed books which have appeared recently include: Goldman *et al.* (1971), Wingard (1972), Chang (1972), Pye and Wingard (1974), Salmona *et al.* (1974), Dunlop (1974), Weetall (1975), Mosbach (1976), and others. However, at present there are no books devoted to a discussion of the biomedical applications of the various types of immobilized enzyme systems.

The present book emphasizes the biomedical applications, not the detailed chemistry, preparative procedures, or kinetics of immobilized enzymes and proteins. However, for those entering this area of research for the first time, the preliminary chapters serve as a brief introduction to the general classification and chemistry of the various types of immobilized enzymes. These are followed by chapters that discuss in detail the therapeutic applications of immobilized enzymes and proteins. The first chapter in the applied section presents a brief discussion of the problems involved with therapeutic applications of enzyme solutions and the rationale and strategy of using immobilized enzymes to solve these problems. This is followed by two chapters which are detailed discussions of the problems encountered in two typical examples of experimental enzyme therapy: (1) asparaginase for tumor suppression, and (2) enzyme replacement in hereditary enzyme-deficiency conditions. The remaining portion of the applied section involves detailed discussions of experimental investigations in animal studies, actual therapeutic applications in patients, and perspectives. The majority of research workers presently carrying out research in the therapeutic applications of immobilized enzymes and proteins are participating in this section. Each chapter in this section, therefore, represents a detailed discussion from each laboratory and its special approach to the area.

In the therapeutic application of enzymes and proteins, a great deal of work involves the use of stage III enzyme immobilization in some form of encapsulated enzymes, including artificial cells, since enzymes and proteins normally function in the body in an intracellular environment. On the other hand, in the diagnostic and public health applications discussed herein, most of the applications involve enzyme systems of stage II immobilized systems: covalent linkage, adsorption, or

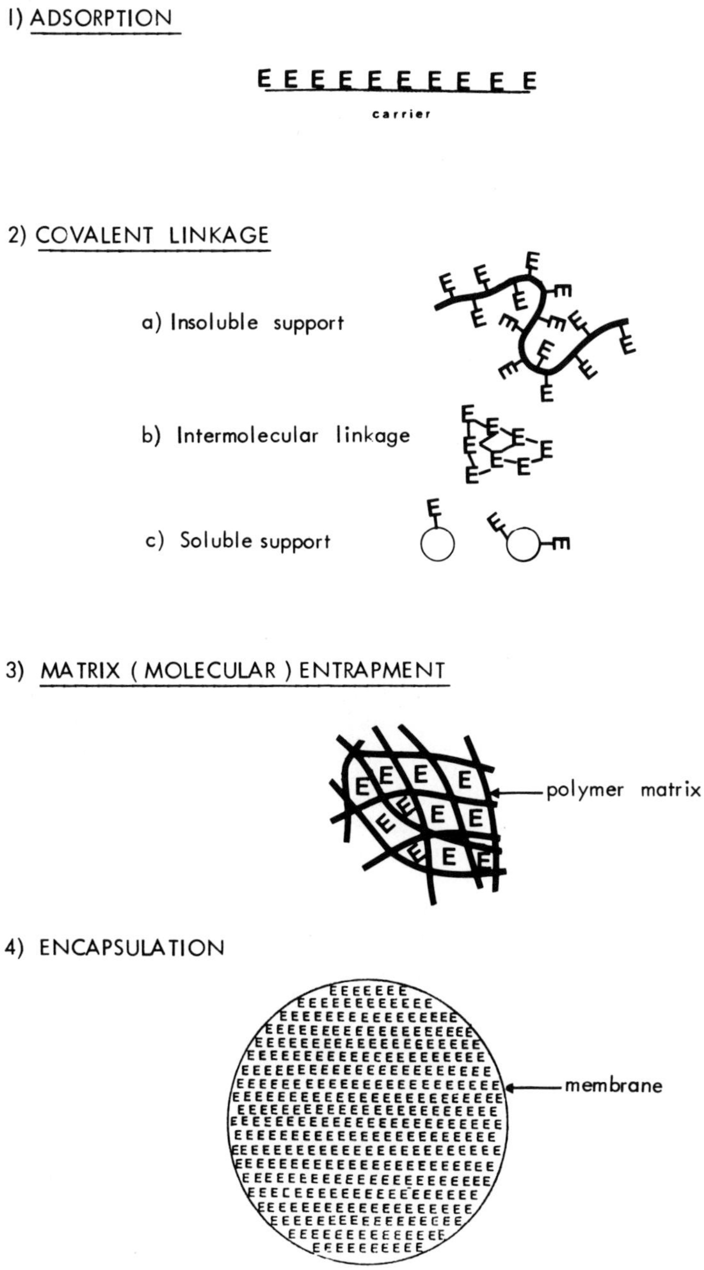

FIGURE 1. Classification of immobilized enzymes. Covalently linked, adsorbed, and matrix-entrapped enzymes represent stage II, research on the microenvironment. Microencapsulation represents stage III, research on the intracellular environment. Further subdivision of microencapsulated enzymes will be found in Chapter 12.

matrix entrapment. In this section, detailed discussions will center on clinical analysis, urine analysis, monitoring of environmental pollution, radioimmune assay, enzyme-linked immunosorbent assay, enzyme electrodes, and other approaches involving immobilized enzymes, antibodies, and antigens.

In the final section, research workers describe and discuss the perspectives of immobilized enzymes and proteins. Here, they speculate on the future potential of possible approaches, even though these may not have been extensively studied or tested at the laboratory stage.

The biomedical applications of enzymes and proteins, especially in the therapeutic area, is in a very early stage of development. Much remains to be explored and studied, and the area is wide open for investigators interested in original research in a new interdisciplinary area.

References

Chang, T. M. S., 1972, *Artificial Cells,* Charles C. Thomas, Publisher, Springfield, Ill.

Dunlop, R. B. (ed.), 1974, *Immobilized Biochemicals and Affinity Chromatography,* Plenum Press, New York.

Goldman, R., Goldstein, L., and Katchalski, E., 1971, 1971, in: *Biochemical Aspects of Reactions on Solid Surfaces* (G. R. Stark, ed.), Academic Press, Inc., New York.

Mosbach, K. (ed.), 1976, *Immobilized Enzymes* (Methods in Enzymology Series), Academic Press, Inc., New York.

Pye, E. K., and Wingard, L. B., Jr. (eds.), 1974, *Enzyme Engineering,* Vol. 2, Plenum Press, New York.

Salmona, M., Saronio, C., and Garanttini, S. (eds.), 1974, *Insolubilized Enzymes,* Raven Press, New York.

Sundaram, P. V., Pye, E. K., Chang, T. M. S., Edwards, V. H., Humphrey, A. E., Kaplan, N. O., Katchalski, E., Levin, Y., Lilly, M. D., Manecke, G., Mosbach, K., Patchornik, A., Porath, J., Weetall, H., and Wingard, L. B., Jr., 1972, Recommendation for standardization of nomenclature in enzyme technology, *Biotechnol. Bioeng.* **14**:15.

Weetall, H. H. (ed.), 1975, *Immobilized Enzymes, Antigens, Antibodies, and Peptides,* Marcel Dekker, Inc., New York.

Wingard, J. B., Jr. (ed.), 1972, *Enzyme Engineering,* Vol. 1, John Wiley & Sons, Inc., New York.

Zaborsky, O. R., 1973, *Immobilized Enzymes,* CRC Press, Cleveland, Ohio.

I. Classification and Chemistries of Immobilized Enzymes

2

Covalent Linkage: I. Enzymes Immobilized by Covalent Linkage on Insolubilized Supports

RICHARD D. FALB

1. *Introduction*

The attachment of enzymes to solids has progressed rapidly in recent years from a laboratory curiosity to a widely expanding field which is useful in many applications. Enzymes immobilized on surfaces now find increasing use in medical, clinical analysis, and synthetic chemistry applications. Immobilized enzymes have provided valuable insight to basic protein interactions. In addition to enzymes, many different types of proteins, including antibodies, enzyme inhibitors, proteinaceous antigens, and peptide hormones, have been attached to polymers and have been shown to possess biological activity. The attachment of an enzyme or other biologically active protein to a solid phase produces an adduct which can have several interesting and useful properties. Since the adducts are water-insoluble, they can be easily recovered from reaction mixtures by centrifugation or filtration. Additionally, enzymes attached to particulate matter can be placed in a column which is used for a continuous flow-through process. In medical applications of enzymes, immobilization can provide advantages because the enzymes do not enter the bloodstream as soluble agents and therefore may have a long lifetime in the body. Enzymes can also be attached to the surfaces of membranes or the inside of tubes or cannulas and can interact with substances flowing past the surface.

Quite often, the attachment of an enzyme to a solid surface results in an increase in stability of the enzyme. The reasons for the increase in stability are not well understood, but from the simplest point of view, immobilization of the enzyme may be viewed as placing the enzyme in a more natural state. In nature, a great many enzymes function not in solution but at an interface or within solid-state assemblages in cells. When an enzyme is isolated and purified, it is removed from the surface to which it was attached, which may destabilize it. Attachment of

RICHARD D. FALB • Battelle, Columbus Laboratories

the enzyme to a solid surface may restore some of the environment in which the enzyme was designed to work.

Enzymes can be placed on solids by a wide variety of methods, including adsorption on the surface, inclusion in gel matrices, cross-linking at solid surfaces, encapsulation, cross-linking with difunctional reagents, and covalent bonding. Immobilization of enzymes by adsorption offers the advantage of simplicity in that the enzyme can usually be attached by simple exposure to the solid particles under very mild conditions. Since these conditions involve no reactive species, the enzyme is not modified chemically during the attachment phase. Adsorbed enzymes, however, can be desorbed under certain conditions of ionic strength or in the presence of substrate. The undesirability of the desorption limits the use of this technique of enzyme immobilization.

Inclusion in a gel phase such as polyacrylamide or starch also offers the advantage of relatively mild reaction conditions and does not significantly alter the protein. The technique has certain disadvantages, such as the relative inaccessibility of the enzyme to its substrate and the possibility of the enzyme leaching out of the gel.

Covalent attachment of an enzyme to a solid surface greatly reduces the loss of the enzyme from the surface during subsequent use. In essence, a covalent bond is formed between a functional group on the enzyme and a reactive group attached to the surface of the solid phase. In the covalent attachment of an enzyme to a solid, several factors must be considered in order to retain the optimum activity of the enzyme. The enzyme must be attached in such a conformation as to allow it interaction with the substrate. The. active site of the enzyme must be accessible to the surrounding milieu and not be buried in a pore or blocked by some other component of the surface. In most covalent attachments, the enzyme is tied down by numerous bonds and the conformational mobility of the enzyme can be restricted. An interesting example of the attachment of an enzyme in an active conformation is provided by the work of Royer and Uy (1973). In this work, they found that the enzyme could be attached and tied down in an active conformation when it was immobilized in the presence of substrate.

A second factor which must be considered in the covalent attachment of enzymes is the possibility of chemically altering the enzyme in such a way that its reactivity is reduced. Since covalent attachment involves the participation of reactive groups on the enzyme, it is possible that groups associated with the active site of the enzyme could be involved in the reaction. This could lead to inactivation of the enzyme and also a blocking of the active site. Conditions used in the attachment reaction could be damaging to the enzyme and activity could be lost in this way.

Several approaches are available to minimize the loss of activity. At present, a very large number of different ways of enzyme attachment are possible, and thus if one mode of attachment does not work well, it is likely that another can be found which will give the desired adduct with good activity. For example, if the active site of the enzyme is known to involve an amino group, an attachment scheme which does not involve reaction with amino groups can be used for the enzyme. In addition, appropriate conditions of pH and ionic strength may be selected so as to maximize the chances for bonding active enzymes. Another ploy which can be taken to protect the enzyme during the attachment is to perform the reaction in the presence of substrates or reversible inhibitors.

For most applications of immobilized enzymes, economics plays a decisive role in determining the ultimate usefulness of the adduct. In selecting a mode of enzyme attachment, one must consider the efficiency with which the enzyme can be attached, the overall reactivity of the adduct, the simplicity of the attachment reaction, and the cost of the support material and reagents used in the process. Since enzymes themselves are usually highly expensive, it is desirable that as much as possible of the enzyme placed in the reaction mixture be recovered in the solid phase. Second, the attachment procedure should be rapid and involve as few steps as possible. Extensive preparation of either the solid or the enzyme can be costly and can result in ultimate loss of the value of any one attachment scheme. In most instances, it is desirable to have a high activity per unit volume of enzyme adduct. In general, this is best accomplished by using a solid material which has a large surface area resulting from very small particle size or the presence of numerous pores throughout the solid phase.

Numerous inexpensive support materials have been developed. Simple methods of obtaining reactive groups at the surface are also available, and these will be discussed below. This paper is not intended to be a comprehensive review of the methods of enzyme attachment. The reader is referred to the excellent book by Oscar Zaborsky (1973) or the more recent book edited by Messing (1975).

2. *Methods of Enzyme Attachment*

As mentioned previously, covalent attachment of enzymes involves reaction of a group at the surface of the solid phase with the reactive group present on the enzyme. Enzymes are polypeptides composed of a variety of different amino acids, and generally several groups are available on any one given enzyme for attachment. One of the most common groups used for attachment is the amino group present either at the N terminus of the enzyme chains or available in side chains of lysine residues. Carboxylic acid groups are available from the C terminus or from aspartic or glutamic acid residues in the chains. Sulfhydryl groups from the cysteine residues in the chain are also available for bonding reactions. It is also possible to attach enzymes by means of the hydroxyl groups present in serine residues. Finally, attachment may be effected through electrophilic substitution of the phenyl moieties present in phenylalanine and tyrosine.

Generally, the attachment reactions are conducted in an aqueous buffer under very mild pH conditions and at ambient temperatures. These requirements somewhat limit the types of reaction which could be performed with the functional groups available on the enzyme. However, a sufficient variety of reactions does exist so that one is usually able to attach an enzyme by a number of different modes.

The solid phase to which the enzyme is attached is frequently called a "carrier," and this term will be utilized for the remainder of this discussion.

Carriers of a wide variety of physical and chemical properties are available, and in the selection of a carrier, several considerations must be made. Mechanical strength and stability are also desirable if one desires the enzyme to be used over a long period. In medical applications, one must select a carrier which does not contain harmful residues and which is stable in the physiological environment. Carriers available at the present time include a whole host of synthetic and natural organic polymers and a number of inorganic supports, such as metals and metal oxides. Table 1 gives a partial list of these carriers.

The inorganic supports include a large number of materials, such as alumina, nickel oxide, stainless steel, titanium, and zirconium. In addition, many clays and ceramic materials are utilized, such as bentonite, brick, fuller's earth, diatomaceous earth, magnetite, and wollastonite.

The most commonly used method for enzyme insolubilization involves the use of cellulose or Sepharose derivatives which are activated through cyanogen bromide. Indeed, any polymer possessing vicinal hydroxy groups can be utilized in this method. Sepharose is a trademark for an agarose gel prepared by Pharmacia. Agarose is a linear polymer consisting of *d*-galactose and 3,6-anhydro *d*-galactose. Sephadex is a cross-linked dextran available from Pharmacia in the form of beads. The reaction of materials of this type with cyanogen bromide was first described by Axèn *et al.* (1967). Cyanogen bromide reacts to form a cyclic reactive imidocarbonate. This group is susceptible to nucleophilic attack by amino groups present in the enzyme. In practice, the attachment is quite simple. The carrier is first reacted with cyanogen bromide for a brief period and then washed to remove unreacted cyanogen bromide. It is then exposed to the enzyme and attachment takes place. The ease with which the reaction can be performed, the mild conditions employed, the number of different supports which can be utilized, and the generally high levels of enzyme attachment account for the popularity of this mode of enzyme immobilization.

Another reaction which works well with cellulose and cellulosic-type polymers

TABLE 1
Enzyme Immobilization Supports

Synthetic polymers
Ethylene maleic anhydride
Polyacrylamides
Polystyrene and derivatives
Polyamino acids
Nylon
Silicone rubber
Natural polymers
Cellulose and derivatives
Agarose and derivatives
Starch and derivatives
Dextran and Sephadex derivatives
Inorganic materials
Silica glass beads
Metals and metal oxides

is the activation with cyanuric chloride or dichloro derivatives, such as 2-amino-4,6-dichloro-*s*-triazine (Kay and Cook, 1967). In this reaction scheme, the polysaccharide is reacted first with the cyanuric chloride derivative and subsequently exposed to the enzyme.

Many polysaccharide derivatives exist which contain carboxyl groups (carboxymethyl cellulose, carboxymethyl Sephadex, etc.). Numerous ways of linking supports of this type to enzymes are available, and these all depend upon the activation of the carboxyl group to make it susceptible to nucleophilic attack from amine groups present in the enzyme. A direct and simple way of activating the carboxyl group consists of a reaction with a water-soluble carbodiimide. After reaction with the carbodiimide, the activated support is washed and subsequently reacted with the enzyme. Carboxyl-containing polymers can also be activated with Woodward's reagent K (Patel *et al.,* 1967). Another reaction for activation of carboxyl groups involves formation of an azide (Mitz and Summaria, 1961). In this reaction, the carboxyl group is first transformed to a hydrazide and subsequently reacted with nitrous acid to form the azide. After washing to remove unreacted constituents, the azide derivative is contacted with the enzyme resulting in a covalent bond.

Several types of supports for enzyme immobilization are commercially available. Included are materials designated as Enzacryl, obtainable from Koch-Light Laboratories, Ltd. These materials consist of beads of polyacrylamides which are derivatized with various functional groups, such as phenylamines, hydrazyls, thiols, and thiolactones. Derivatives containing a phenylamino group can be converted either to a diazonium salt or an isothiocyanato derivative. The hydrazide derivative can be converted to an acid azide by reaction with nitrous acid. Attachment to the thiol groups can be obtained by oxidative coupling with a sulfhydryl-containing protein. The cyclic thiolactone reacts directly with nucleophilic groups in enzymes such as amines and thiol groups.

Another commonly used method of enzyme support is based upon a copolymer of ethylene and maleic anhydride (Lennon *et al.,* 1964). This polymer is commercially available and reacts directly with nucleophilic groups in enzymes to form a covalent bond.

Polystyrene was used in early methods of enzyme attachment, and a frequently used derivative was polyaminostyrene. This polymer, which is available commercially, can be reacted with nitrous acid to form a polydiazonium salt or with thiophosgene to give an isothiocyanato group (Grubhofer and Schleith, 1953). Reactions of this type have been utilized more recently on aminoaryl derivatives of cellulose and glass.

A highly versatile method of enzyme coupling which is capable of attaching enzymes to a wide range of carriers employs the method of activation with metal salts (Emery, 1972). In this reaction scheme, the support, which may consist of materials as diverse as borosilicate glass, soda glass, filter paper, cellulose of all types, and nylon 66, is first reacted with a metal salt, such as $TiCl_4$, $SnCl_4$, $SnCl_2$, $ZrCl_4$, VCl_3, or $FeCl_2$, for a 24-hr period. After washing to remove unreacted metal salts, the support is then placed in a solution of the enzyme for enzyme attachment.

Enzymes can be attached to a dialdehyde starch by a variety of methods. The dialdehyde starch polymer is produced by cleavage of the glucosyl rings of starch by periodic acid and contains two aldehyde groups per glucosyl residue. These aldehyde groups can react directly with amino groups present in the enzyme to form an insoluble derivative. The resulting Schiff base can be reduced with sodium borohydride to give a nonreversible covalent bond between the enzyme and the starch resin. Dialdehyde starch can also be derivatized with methylene dianiline as described by Goldstein *et al.* (1970). This adduct can be diazotized to produce a reactive polymer which readily forms covalent bonds with enzymes. Because of the highly reactive nature of the polymer, the amounts of enzyme attached are usually quite high.

Phenolic resins can be used as enzyme carriers and have the advantages of being low in cost and highly reactive. These phenol formaldehyde-type resins can be based on phenol, resorcinol, catechol, phloroglucinol, *o*-cresol, and other phenolic compounds. Enzymes can be coupled to materials of this type either by reaction through glutaraldehyde or by an oxidative formylation system (Stanley and Palter, 1973). The amounts of enzyme attached are usually quite high, with good retention of activity. Since the polymers can be polymerized conveniently in bead form, they have good properties for column applications.

An extremely versatile and widely used method of enzyme attachment to inorganic supports involves the use of derivatized silane coupling reagents. In essence, the silanizing agent consists of a trialkoxysilane containing an alkyl chain directly attached to the silicon atom. This group is capable of forming covalent bonds with a wide variety of materials, both organic and inorganic. The alkyl chain can be derivatized to contain a number of different functional groups, such as amino, aminophenyl, and hydroxyl groups. Because of their wide reactivity, silanizing agents provide an excellent and versatile means of derivatizing numerous supports. One of the primary supports utilized has been glass (Weetall, 1969). After reaction with aminopropyl triethyoxysilane, the amine groups attached to the glass can be coupled with enzymes by means of a glutaraldehyde cross-linking reagent. In addition, aminoaryl groups can be diazotized and then reacted directly with enzymes to form a covalent bond. Aminoalkyl and aminoaryl derivatives of glass are commercially available from Corning. Glass has a number of advantages as a support material, particularly because of its stability and inertness to microbial attack. To attain high surface area and high enzyme loading, porous glass is commonly used.

All the preceding methods for enzyme immobilization through covalent bond formation have utilized aqueous media in the attachment reaction. It should be emphasized at this point that nonaqueous media can also be utilized in selected cases. An example of nonaqueous media being used is furnished by the work of Bartling *et al.* (1973), who immobilized lysozyme to a derivatized polystyrene in dimethylformamide solvent. Good retention of the enzyme activity and high amounts of enzyme were obtained in this process.

This review has not been intended to be comprehensive in its scope but rather to acquaint the reader with the general methods utilized in enzyme immobilization. The field of enzyme immobilization is advancing rapidly, and several hundred papers in this field appear each year. It is hoped that this brief review will provide the reader with basic knowledge of the approaches as well as the problems involved in enzyme immobilization.

3. *References*

Axèn, R., Porath, J., and Ernback, S., 1967, Chemical coupling of peptides and proteins to polysaccharides by means of cyanogenhalides, *Nature* **214**:1302.

Bartling, G. J., Brown, H. D., and Chattopadahyay, S. K., 1973, Synthesis of a matrix-supported enzyme in non-aqueous conditions, *Nature* **243**:342.

Emery, A. N., 1972, Some applications of solid-phase enzymes in biological engineering, *Chem. Eng.* **71.**

Goldstein, L., Pecht, M., Blunberg, S., Atlas, D., and Levin, Y., 1970, Water-insoluble enzymes: synthesis of a new carrier and its utilization for preparation of insoluble derivatives of papain, trypsin to subtilopeptidase A., *Biochemistry* **9**:2322.

Grubhofer, N., and Schleith, L., 1953, Modified ion-exchanges as specific adsorbents, *Naturwissenschaften* **40**:508.

Kay, G., and Cook, E. M., 1967, Coupling of enzymes to cellulose using chloro-*s*-triazine, *Nature* **216**:514.

Lennon, Y., Hecht, M., Goldstein, L., and Katchalsky, E., 1964, A water insoluble polyanionic derivative of trypsin: I. Preparation and properties, *Biochemistry* **3**:1905.

Messing, R. A. (ed.), 1975, *Immobilized Enzymes for Industrial Reactors,* Academic Press, Inc., New York.

Mitz, M. A., and Summaria, L. J., 1961, Synthesis of biologically active cellulose derivatives of enzymes, *Nature* **189**:576.

Patel, R. P., Lopiekes, D. V., Brown, S. P., and Price, S., 1967, Derivatives of proteins: II. Coupling of α-chymotrypsin to carboxyl containing polymers by use of *N*-ethyl-5-phenylisoxazolium-3[1]-sulfonate, *Biopolymers* **5**:577.

Royer, G., and Uy, R., 1973, Evidence for the introduction of a conformational change of Bovine trypsin by a specific substrate at pH 8, *J. Biol. Chem.* **248**:2627.

Stanley, W. L., and Palter, R., 1973, Lactase immobilization on phenolic resins, *Biotechnol. Bioeng.* **15**:597.

Weetall, H. H., 1969, Trypsin and papain covalently coupled to porous glass: preparation and characterization, *Science* **166**:615.

Zaborsky, O. R., 1973, *Immobilized Enzymes,* CRC Press, Cleveland, Ohio.

3

Covalent Linkage: II. Intramolecular Linkages

Rosa Uy and Finn Wold

1. *Introduction*

In this section on immobilized enzymes a consideration of covalent intramolecular linkages may appear somewhat irrelevant in that very little, if any, immobilization can be achieved by intramolecular cross-linking reactions. However, since most of the techniques and reagents used for intermolecular cross-link formation are the same as those used to form intramolecular bonds, it is safe to conclude that intramolecular linkages are natural side reactions of many of the immobilization reactions. This, in turn, means that in assessing the properties of an immobilized enzyme, one needs to consider the effects of intramolecular cross-links as well as the effects of the immobilization, and such considerations should therefore justify the inclusion of this chapter. A brief review of nature's own intramolecular protein cross-links should provide convincing evidence that the cross-link effects indeed can be quite substantial. The most obvious natural cross-link is the disulfide bond, which in most globular proteins imparts unique permanence to the proper three-dimensional folded structure. Especially in cases where the native molecule is composed of multiple polypeptide chains linked by interchain disulfide bonds is the cross-link an essential feature of the structural integrity of the molecule. In the case of some of the structural proteins, multiple disulfide bonds contribute structural rigidity and inertness to the molecular complexes. Similarly, in the case of the very stable and metabolically inert proteins of connective tissue, collagen and elastin, a whole group of unique cross-links derived from oxidatively deaminated lysines and hydroxylysines contribute to the structural properties and undoubtedly also to the metabolic intractability of these proteins. In the formation of the final product of the blood-clotting cascade, the fibrin network is cross-linked by intra- and intermolecular amide bonds between lysine ϵ-amino groups and glutamic acid γ-carboxyl groups through the action of the enzyme transglutaminase

---

Rosa Uy • Department of Biochemistry, University of Minnesota, St. Paul, Minn. Finn Wold • Department of Biochemistry, University of Minnesota, St. Paul, Minn.

(factor XIII in the series of plasma clotting factors). Again the role of the cross-links appears to be to fix and stabilize specific molecular conformations and associations. Since the transglutaminase enzyme now has been found to be present in other tissues as well as in the plasma, it seems reasonable to conclude that this cross-linking reaction must be important to processes other than just coagulation.

Based on these considerations, it can perhaps be concluded that the major function of the natural protein cross-links is to give structural stability (or rigidity) to both globular and fibrous proteins. Furthermore, since it is well known that "native" proteins in general are quite resistant to protease digestion, the cross-links might be expected to prolong the half-life of proteins exposed to proteolytic activity, and one very direct and immediately relevant effect of protein cross-linking which should be explored much more extensively during the next few years is the possible stabilization of proteins prior to *in vivo* application or in association with protein immobilization experiments. There is a corollary of this general goal of stabilizing proteins which has not been extensively explored, but which should have a good deal of merit in immobilization technology. If an oligomeric protein is to be anchored to a matrix, it is quite reasonable to expect that with a limited number of attachment points, only one subunit will be anchored. During use, the other subunits are therefore liable to be dissociated from the covalently attached one, unless the original multisubunit complex has been cross-linked prior to the attachment to the matrix.

With these practical applications in mind and with the additional consideration of the intramolecular cross-links as common side reactions in immobilization reactions, the design for this chapter is to briefly review the recent advances in intramolecular cross-linking technology and to consider what evidence exists that proteins can be stabilized by cross-linking.

There are two general methods by which intramolecular and intermolecular cross-links can be introduced into proteins: through reaction with bifunctional reagents or through activation and coupling of natural functional groups in the protein with various condensing reagents. In the latter case the condensing agent may be a simple organic reagent or an enzyme. With either of these methods a certain degree of control of the reaction in terms of intra- or intermolecularly linked products would appear to be available through simple manipulations of the reaction conditions (protein concentration, protein dissociation or association through pH or ionic strength adjustment, reagent concentration and specificity), but in practice, with any new protein, the ideal conditions for the production of a given desired derivative will undoubtedly have to be established empirically.

2. *Bifunctional Reagents*

The general classes of bifunctional reagents have been reviewed quite recently with regard to reagents, reaction conditions, and biomedical applications (Wold, 1972; Wold, 1973). Consequently, the emphasis in this section will be on supplementing the previous reviews with the most recent developments and applications. Since intermolecular linkages are treated in another chapter, an attempt has been made to limit the treatment to intramolecular linkage formation, but it should be stated that intersubunit cross-links have been considered to be intramolecular linkages.

It is clear from the summary in Table 1 that the bifunctional imidoesters have become the favorite type of reagent in most laboratories. In spite of the fact that they react rather sluggishly under physiological conditions, they have the very attractive feature of not significantly changing the ionic properties of the protein in the course of the reaction. The positively charged amino groups are converted to positively charged amidine groups, and the probability of recovering full biological activity in the cross-linked protein should therefore be high. Very important additions to this class of compounds are such reagents as dimethyl 4,4′-dithiobisbutyrimidate (Traut *et al.*, 1973; Sun *et al.*, 1974) and dimethyl 3,3′-dithiobispropioimidate (Wang and Richards, 1974; Ruoho *et al.*, 1975). These bifunctional reagents can be readily cleaved by reductive conversion of the disulfide bond to sulfhydryl or by sulfitolysis, and with very mild treatment it is thus possible to disrupt the cross-link without destroying the amidine derivative. This becomes especially important in view of the fact that the reported cleavage of amidines with concentrated ammonia (Hunter and Ludwig, 1962), which appears to work well with amino acids and peptides (Hartman and Wold, 1967), apparently is not readily applicable to proteins.

Among the important applications of diimido esters is the exploration of subunit interactions in multisubunit proteins. From the elegant analytical technique introduced by Davies and Stark (1970) for the determination of the number of subunits in oligomeric proteins, a series of refinements has been developed. Regulated oligomeric enzymes such as phosphofructokinase (Lad and Hammes, 1974; Telford *et al.*, 1975) and isopropylmalate synthetase (Kohlaw and Boatman, 1971) have been cross-linked and investigated with respect to their response to allosteric effectors, and by using the homologous series of reagents of different chain lengths, the architecture of oligomeric enzymes has also been explored. In the case of glutamate dehydrogenase (Hucho and Janda, 1974) it was found that the shorter-chain reagent (adipimidate) gave only trimers, while the longer-chain reagent (suberimidate) gave both trimers and hexamers, thus confirming the model that two units of closely packed trimers make up the hexameric enzyme molecule. Dimethyladipmidate has also been found to be an antisickling agent (Waterman *et al.*, 1975; Lubin *et al.*, 1975).

Glutaraldehyde appears to be the favored reagent as a "fixative" for proteins. It is implied in this assessment that this reagent frequently is used in large excess to accomplish extensive modification. As seen from Table 1, it has also been used for limited cross-linking in soluble proteins, and it is notable that in the case of glycogen phosphorylase (Wang and Tu, 1969), a cross-linked derivative with increased stability toward heat and denaturants was obtained after reaction with glutaraldehyde.

In the group of acylating reagents, relatively few new applications have been reported. An attractive new series of reagents, the bis-hydroxysuccinimide esters, has been added (Lindsay, 1971), and some elegant experiments have been carried out linking the N-terminal end of the insulin A chain with the lysine residue at the

TABLE 1
Bifunctional Reagents

Reagent class	Commercially available[a]	Primary site of reaction	Optimal pH	Proteins reacted (reference)[b]
Diimido esters	Yes	$—NH_2$	8–10	Glyceraldehyde phosphate dehydrogenase (1); aldolase (1,2); tryptophan synthetase (1); aspartate transcarbamylase (1); L-arabinose isomerase (1); sterol dehydrogenase (3); glycerol kinase (4); phosphofructokinase (5,6); isopropylmalate synthetase (7); glutamate dehydrogenase (8); pyruvate kinase(9); cholinergic receptor protein (10); concanavalin A (2); nerve growth factor (11); hemoglobin (12,13); haptoglobulin (14); Qβ RNA replicase (15); hemocyanin (10)
Glutaraldehyde	Yes	$—NH_2$	> 7	Papain (17); glutamate dehydrogenase (18); glycogen phosphorylase (19); trypsin (20)
Aromatic and aliphatic diisocyanates (diisothiocyanates)	Yes	$—NH_2$ (—SH, —OH)	7	α-Galactosidase A (21)
Bis-*p*-nitrophenyl esters of dicarboxylic acids	Yes	$—NH_2$	7	Insulin (22–24)
Bis-hydroxysuccinimide esters of dicarboxylic acids	Yes	$—NH_2$ (—SH, —OH)	7	Insulin (25)
Aromatic disulfonyl chlorides	Yes	$—NH_2$	7	

Bifunctional aryl halides (1,5-difluoro-2,4-dinitrobenzene, *p,p'*-difluoro *m,m'*-dinitrodiphenyl-sulfone)	Yes	—NH_2 (phenolic OH)	7–10	Glyceraldehyde phosphate dehydrogenase (26); lysozyme (27); high-density lipoprotein (28)
Bis-maleimides	Yes	—SH	7–8	
Bis-haloketones	Yes	—NH_2 —SH, R-S-CH_3, imidazole	5–9	Papain (29); bromelain (30)
Bis-diazoketones		COOH	Wide range	Pepsin (31)
Bis-azidobenzene derivatives	Yes	Several	Wide range	(32)

[a] Commercial availability based on recent reports. The reader is referred to the references for the source of the compounds.
[b] References:

1. Davies and Stark (1970)
2. Ruoho *et al.* (1975)
3. Jarabak and Street (1971)
4. Thorner and Paulus (1971)
5. Lad and Hammes (1974)
6. Telford *et al.* (1975)
7. Kohlaw and Boatman (1971)
8. Hucho and Janda (1974)
9. Davies and Kaplan (1972)
10. Hucho and Changeux (1973)
11. Stach and Shooter (1974)
12. Waterman *et al.* (1975)
13. Lubin *et al.* (1975)
14. Lockhart and Smith (1975)
15. Young and Blumenthal (1975)
16. Van Driel and Van Bruggen (1975)
17. Ottesen and Svensson (1971)
18. Josephs *et al.* (1973)
19. Wang and Tu (1969)
20. Beaven and Gratzer (1973)
21. Snyder *et al.* (1974)
22. Brandenburg (1972)
23. Busse and Carpenter (1974)
24. Busse *et al.* (1974)
25. Lindsay (1971)
26. Shaltiel and Tauber-Finkelstein (1971)
27. Matthyssens and Verheulpen (1973)
28. Grow and Fried (1975)
29. Husain and Lowe (1968)
30. Husain and Lowe (1970)
31. Husain *et al.* (1971)
32. Guire (1975)

C-terminal end of the B chain with such a reagent. Similar results have also been obtained with bis-nitrophenyl esters (Brandenburg, 1972). With the same general approach Busse and Carpenter (1974) and Busse *et al.* (1974) succeeded in making a proinsulin analog by cross-linking the same residues with the *p*-nitrophenyl ester of carbonyl bis(L-methionine). By treatment with cyanogen bromide the cross-link was eliminated and the single-chain derivative was converted back to a two-chain structure. In the category of cleavable acylating agents, another family of compounds should be mentioned. Lutter *et al.* (1974) have prepared tartrate derivatives containing diazides as the reactive acylating groups [tartryl diazide, tartryl di(glycylazide), tartryl di(ϵ-amino caproylazide)]. The reagents have in common the glycol of the tartrate moiety and are consequently readily cleaved by periodate oxidation.

In the group of haloketones and diazoketones the major new developments have been to prepare substrate analogs containing these reactive functional groups (Alazard *et al.*, 1973; Bechet *et al.*, 1973). There is an important aspect of this application which has obvious relevance to contemplated enzyme therapy. It would appear very wise to have available specific inhibitors of this kind whenever an enzyme is supplied to a patient from external sources.

The final group of reagents which warrants careful attention is the family of bis-azidobenzene derivatives. These compounds are unique in that they are thermochemically completely stable and only react when activated by light. Although the reaction specificity of the resulting nitrene is low enough to promise some complication in the characterization of cross-linked derivatives, the very effective control of their reaction through the exposure to light suggest that they will be extremely useful reagents.

3. *Condensation Reactions*

Although the intramolecular condensation of the appropriate side chains to form cross-links (carboxyl groups and amino groups to form amides, carboxyl groups and hydroxyl groups to form esters) has not been used extensively in the past, there are some interesting possibilities in this area which warrant attention. A large number of condensing reagents are available and have been used quite extensively in various intermolecular reactions. Among these are various carbodiimides (Means and Feeney, 1971) and *N,N*-carbonylimidazole (Paul and Anderson, 1960). The use of these reagents in coupling reactions has been reviewed (Laursen, 1975). Perhaps the most attractive and mild procedure for such condensations is that based on the use of the enzyme transglutaminase. This enzyme, catalyzing the reaction

$$\underset{\text{glutamine}}{R{-}CONH_2} + \underset{\text{lysine}}{R'{-}NH_2} \rightarrow R{-}CO\overset{H}{N}{-}R' + NH_3$$

has now been purified from serum and guinea pig liver (Folk and Chung, 1973) as well as from red cells (S. C. Brenner, personal communication). Based on the present data the enzyme appears to be quite specific for only a few glutamine residues in certain proteins, and although it does not appear to be very selective with regard to the source of the amino group, it may have only limited use in practice. Nevertheless, a recent report by Dutton and Singer (1975) on successful amine-incorporation into membrane proteins suggests that it may well become an important tool in our cross-linking technology.

The above considerations have, as stated at the outset, been limited to homobifunctional reagents and intramolecular cross-link formation. In closing it must be noted that with this as a starting point, the potential for variation and flexibility of reaction type and desired products is virtually limited only by one's ability to synthesize different reagents. A large group of heterobifunctional reagents has evolved in recent years (Guire, 1975; Olomucki and Diopoh, 1972; Hixson and Hixson, 1975) which, on the basis of the differential reactivity of the two reactive groups, permit the use of controlled, stepwise reactions. These reactions have become an important asset in reactions aiming at intermolecular linkage formation and will be considered in another chapter of this volume.

Why are we interested in intramolecular cross-link technology? One of the major moving forces in the early exploration of bifunctional reagents in protein chemistry was the search for techniques that could give information about the folded three-dimensional structure of globular protein in solution. The basic idea was to use the bifunctional reagent as a yardstick to measure the distance between the two cross-linked residues in the polypeptide chain. In spite of the success of x-ray crystallography in providing more complete and detailed information about three-dimensional structure, this type of information is still sought for proteins in solution, especially as a means of detecting conformational changes during biological action, and the use of bifunctional reagents will undoubtedly continue to contribute important information in this area. This intramolecular-yardstick concept has recently been expanded to include exploration of the molecular architecture of oligomeric proteins, of multiprotein complexes, and of complex structures such as ribosomes and membranes. Examples of all these have been briefly considered in the present discussion or in a previous review (Wold, 1972). In general this topic is more appropriate for the chapter on intermolecular cross-links (chapter 4).

As indicated at the beginning of the chapter, perhaps the most important and relevant aspect of intramolecular cross-links in the context of this book is the prospect of stabilizing biologically active proteins toward various denaturing conditions, including *in vivo* destruction. It is surprising how little information is available on this point. In most of the cross-linked derivatives that have been investigated, there does not appear to have been an attempt made to assess the stability of the product. In some instances, reaction with a given reagent under a given set of conditions did not give increased stability toward heat, urea, and so on. In other instances, however, clear evidence for increased stability toward denaturation and protease action has been obtained [cross-linked serum albumin and ribonuclease (Wold, 1973) and phosphorylase (Wang and Tu, 1969)]. A recent paper reports specifically on the stabilization α-galactosidase A, the defective enzyme in Fabry's disease, by cross-linking with hexamethylene diisocyanate (Snyder *et al.*, 1974). In this case the technology is clearly aimed at enzyme replacement therapy and the significant stabilization observed in the cross-linked derivative should encourage further concerted efforts in this area.

4. *References*

Alazard, R., Bechet, J. J., Dupaix, A., and Yon, J., 1973, Inactivation of α-chymotrypsin by a bifunctional reagent, 2-bromomethyl-3,1-benzoxazin-4-one, *Biochim. Biophys. Acta* **309**:379.

Beaven, G. H., and Gratzer, W. B., 1973, Modification of the enzymatic activity of trypsin by intramolecular cross-links, *Int. J. Peptide Protein Res.* **5**:215.

Bechet, J. J., Dupaix, A., and Yon, J., 1973, Inactivation of α-chymotrypsin by a bifunctional reagent, 3,4-dihydro-3,4-dibromo-6-bromomethylcoumarin, *Eur. J. Biochem.* **35**:527.

Brandenburg, D., 1972, Preparation of N_{α}^{A1}, N_{ϵ}^{B29}-adiopoylinsulin, an intramolecularly cross-linked derivative of beef insulin, *Z. Physiol. Chem.* **353**:869.

Busse, W. D., and Carpenter, F. H., 1974, Carbonyl bis(L-methionine *p*-nitrophenyl ester): a new reagent for the reversible intramolecular cross-linking of insulin, *J. Amer. Chem. Soc.* **96**:5947.

Busse, W. D., Hansen, S. R., and Carpenter, F. H., 1974, Carbonyl bis(L-methionyl) insulin: a proinsulin analog which is convertible to insulin, *J. Amer. Chem. Soc.* **96**:5959.

Davies, G. E., and Kaplan, G. J., 1972, Use of a diimidoester cross-linking reagent to examine the subunit structure of rabbit muscle pyruvate kinase, *Can. J. Biochem. Physiol.* **50**:416.

Davies, G. E., and Stark, G. R., 1970, Use of dimethyl suberimidate, a cross-linking reagent in studying the subunit structure of oligomeric proteins, *Proc. Natl. Acad. Sci. U.S.* **66**:651.

Dutton, A., and Singer, A. J., 1975, Cross-linking and labeling of membrane proteins by transglutaminase catalyzed reactions, *Proc. Natl. Acad. Sci. U.S.* **72**:2568.

Folk, J. E., and Chung, S. I., 1973, Molecular and catalytic properties of transglutaminase, in: *Adv. Enzymol.* **38**:109.

Givol, D., 1969, Inactivation of glyceraldehyde 3-phosphate dehydrogenase by a bifunctional reagent, *FEBS Letters* **5**:153.

Grow, T. E., and Fried, M., 1975, Lipoprotein geometry: I. Spatial relationships of human HDL apoproteins studied with a bifunctional reagent, *Biochem. Biophys. Res. Commun.* **66**:352.

Guire, P. E., 1975, Photoreactive carrier derivatives for immobilization of enzymes and ligands, *Federation Proc.* **34**:690.

Hartman, F. C., and Wold, F., 1967, Cross-linking of bovine pancreatic ribonuclease A with dimethyl adipimidate, *Biochemistry* **6**:2439.

Hixson, S. H., and Hixson, S. S., 1975, *p*-Azidophenyl bromide, a versatile photolabile bifunctional reagent, reaction with glyceraldehyde 3-phosphate dehydrogenase, *Biochemistry* **14**:4251.

Hucho, F., and Changeux, J. P., 1973, Molecular weight and quaternary structure of the cholinergic receptor protein extracted by detergent from electrophorus electricus electric tissue, *FEBS Letters* **38**:11.

Hucho, F., and Janda, M., 1974, Investigation of the quaternary structure of beef liver glutamate dehydrogenase with bifunctional reagents, *Biochem. Biophys. Res. Commun.* **57**:1080.

Hunter, M. J., and Ludwig, M. L., 1962, The reaction of imidoesters with proteins and related small molecules, *J. Amer. Chem. Soc.* **84**:3491.

Husain, S. S., and Lowe, G., 1968, Evidence for histidine in the active site of papain, *Biochem. J.* **108**:855.

Husain, S. S., and Lowe, G., 1970, The amino acid sequence around active site cysteine and histidine residues of stem bromelain, *Biochem. J.* **117**:341.

Husain, S. S., Ferguson, J. B., and Fruton, J. S., 1971, Bifunctional inhibitors of pepsin, *Proc. Natl. Acad. Sci. U.S.* **68**:2765.

Jarabak, J., and Street, M. A., 1971, Studies on the soluble 17β-hydroxysteroid dehydrogenase from human placenta: evidence for a subunit structure, *Biochemistry* **10**:3831.

Josephs, R., Eisenberg, H., and Reisler, E., 1973, Some properties of cross-linked polymers of glutamic dehydrogenase, *Biochemistry* **12**:4060.

Kohlaw, G., and Boatman, G., 1971, Cross-linking of *Salmonella* isopropylmalate synthase with dimethyl suberimidate: evidence for antagonistic effects of leucine and acetyl CoA on the quaternary structure, *Biochem. Biophys. Res. Commun.* **43**:741.

Lad, P. M., and Hammes, G. G., 1974, Physical and chemical properties of rabbit muscle phosphofructokinase cross-linked with dimethyl suberimidate, *Biochemistry* **13**:4530.

Laursen, R. A. (ed.), 1975, *Solid Phase Methods in Protein Sequence Analysis,* Pierce Chemical Company, Rockford, Ill.

Lindsay, D. G., 1971, Intramolecular cross-linked insulin, *FEBS Letters* **21**:105.

Lockhart, W. L., and Smith, D. B., 1975, Cross-linking of hemoglobin, haptoglobin and hemoglobin–haptoglobin complex with bifunctional reagents, *Can. J. Biochem. Physiol.* **53**:861.

Lubin, B. H., Pena, V., Metzer, W. C., Bymun, E., Bradley, T. B., and Packer, L., 1975, Dimethyl adipimidate: a new antisickling agent, *Proc. Natl. Acad. Sci. U.S.* **72**:43.

Lutter, L. C., Ortanderl, F., and Fasold, H., 1974, The use of a new series of cleavable protein-cross-linkers on the *E. coli* Ribosome, *FEBS Letters* **48**:288.

Matthyssens, G. E., and Verheulpen, W., 1973, Preparation characterization and antigenic specificity of a tyrosine-lysine cross-linked hen egg white lysozyme derivative, *FEBS Letters* **35**:239.

Means, G. R., and Feeney, R. E., 1971, *Chemical Modification of Proteins,* Holden-Day, Inc., San Francisco.

Olomucki, M., and Diopoh, J., 1972, Ethyl chloroacetimidate, its properties and its reaction with ribonuclease, *Biochim. Biophys. Acta* **263**:213.

Ottesen, M., and Svensson, B., 1971, Modification of papain by treatment with glutaraldehyde under reducing and non-reducing conditions, *Compt. Rend. Trav. Lab. Carlsberg* **38**:171.

Paul, R., and Anderson, G. W., 1960, *N,N'*-carbonyldiimidazole: a new peptide forming reagent, *J. Amer. Chem. Soc.* **82**:4596.

Ruoho, A., Bartlette, P. A., Dutton, A., and Singer S. J., 1975, A disulfide bridge bifunctional imidoester as a reversible cross-linking reagent, *Biochem. Biophys. Res. Commun.* **63**:417.

Shaltiel, S., and Tauber-Finkelstein, M., 1971, Introduction of an intramolecular cross-link at the active site of glyceraldehyde 3-phosphate dehydrogenase, *Biochem. Biophys. Res. Commun.* **44**:484.

Snyder, P. D., Jr., Wold, F., Bernlohr, R. W., Dullum, C., Desnick, R. J., Krivit, W., and Condie, R. M., 1974, Purified α-galactosidase A: stabilization to heat and protease degradation by complexing with antibody and by chemical modification, *Biochim. Biophys. Acta* **350**:432.

Stach, R. W., and Shooter, E. M., 1974, The biological activity of cross-linked β nerve growth factor protein, *J. Biol. Chem.* **249**:6668.

Sun, T. T., Bollen, A., Kahan, L., and Traut, R. R., 1974, Topography of ribosomal proteins of *E. coli* 30S subunit as studied with the reversible cross-linking reagent methyl 4-mercaptobutyrimidate, *Biochemistry* **13**:2334.

Telford, J. N., Lad, P. M., and Hammes, G. G. 1975, Electron microscope study of native and cross-linked rabbit muscle phosphofructokinase, *Proc. Natl. Acad. Sci. U.S.* **72**:3054.

Thorner, J. W., and Paulus, H., 1971, Composition and subunit structure of glycerol kinase from *E. coli, J. Biol. Chem.* **246**:3885.

Traut, R. R., Bollen, A., Sun, T. T., Hershey, J. W. B., Sunberg, J., and Pierce, L. R., 1973, Methyl 4-mercaptobutyrimidate as a cleavable cross-linking reagent and its application to the *E. coli* ribosome, *Biochemistry* **12**:3266.

Van Driel, R., and Van Bruggen, E. F., 1975, Functional properties of chem. modified hemocyanin: fixation of hemocyanin in the low and high oxygen affinity state by reaction with a bifunctional imido ester, *Biochemistry* **14**:730.

Wang, J. H. C., and Tu, J. I., 1969, Modification of glycogen phosphorylase b by glutaraldehyde: preparation and isolation of enzyme derivatives with enhanced stability, *Biochemistry* **8**:4403.

Wang, K., and Richards, F. M., 1974, An approach to nearest neighbor analysis of membrane proteins: application to the human erythrocyte membrane of a method employing cleavable cross-linkages, *J. Biol. Chem.* **249**:8005.

Waterman, M. R., Yamaoka, K., Chuang, H. A., and Cottam, G. L., 1975, Anti-sickling nature of dimethyl adipimidate, *Biochem. Biophys. Res. Commun.* **63**:580.

Wold, F., 1972, Bifunctional Reagents, in: *Methods in Enzymology* (Vol. 23) (C. H. W. Hirs and S. N. Timasheff, eds.), p. 623, Academic Press, New York.

Wold, F., 1973, Chemical Modification of Enzymes, in: *Enzyme Therapy in Genetic Diseases* (Vol. IX, No. 2) (D. Bergsma, ed.), p. 46, Williams and Wilkins, Baltimore.

Young, R. A., and Blumenthal, T., 1975, Phage Q-B ribonucleic acid replicase subunit relationships determined by intramolecular cross-linking, *J. Biol. Chem.* **250**:1829.

4

Covalent Linkage: III. Immobilization of Enzymes by Intermolecular Cross-Linking

OSKAR R. ZABORSKY

1. Introduction

A very versatile chemical method for immobilizing enzymes and other proteins is intermolecular cross-linking with multifunctional reagents. The method can produce enzyme conjugates of diverse physical character depending on the conditions employed. Thus, conjugates can be prepared that are water-soluble or water-insoluble; granular, gel-like, or filamentous in nature; or that are pure protein. It is reliable and inexpensive.

The underlying principle of the method is the formation of covalent bonds between molecules of the enzyme and the reagent to give a three-dimensional, cross-linked network. Usually the reaction involves a bifunctional reagent such as glutaraldehyde, which has two identical functional groups to react with the protein and gives a spectrum of products ranging in properties (Table 1). Figure 1 is a schematic representation of the reaction of a bifunctional reagent with a protein. Depending on the conditions employed, the reaction can lead to monomers, dimers, trimers, or oligomers which are still water-soluble or it can lead to a "super" polymer that is highly water-insoluble.

In addition to the degree of intermolecular cross-linking, an equally important, mutually dependent consideration, is the extent of *intramolecular* cross-linking in these conjugates. Intramolecular cross-linking of an enzyme with a multifunctional reagent is favored by a low concentration of the enzyme and the reagent, because this enhances the probability that the functional groups of the reagent react with the amino acid residues of only the *same* enzyme molecule. However, as already pointed out by Uy and Wold in chapter 3, the precise control of intra- versus intermolecular cross-linking is difficult to achieve and immobilization of an enzyme (i.e., the preparation of a water-insoluble conjugate) will necessarily involve both types of cross-links.

OSKAR R. ZABORSKY • National Science Foundation, Washington, D. C.

TABLE 1
Multifunctional Reagents Frequently Employed and Mentioned in the Text

Reagent	Structure
Diazobenzidine	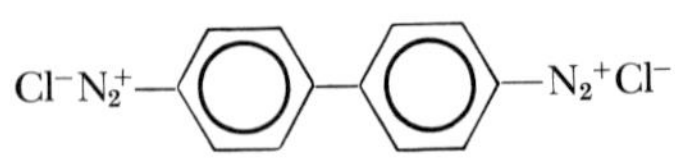
1,5-Difluoro-2,4-dinitrobenzene	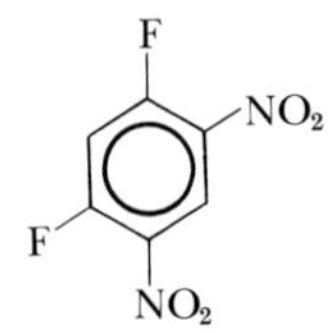
Dimethyl adipimidate	$Cl^- H_2\overset{+}{N}{=}C(OCH_3){-}(CH_2)_4{-}C(OCH_3){=}\overset{+}{N}H_2 Cl^-$
Glutaraldehyde	$OHC{-}(CH_2)_3{-}CHO$[a]
Butene-3,4-oxide	$CH_2{=}CH{-}CH{-}CH_2$[b] (epoxide: CH—CH₂ bridged by O)

[a]For a discussion of the structure of glutaraldehyde in aqueous systems, see Zaborsky (1973).
[b]Reaction with proteins occurs at the epoxide functionality; the unsaturated center is used for copolymerization.

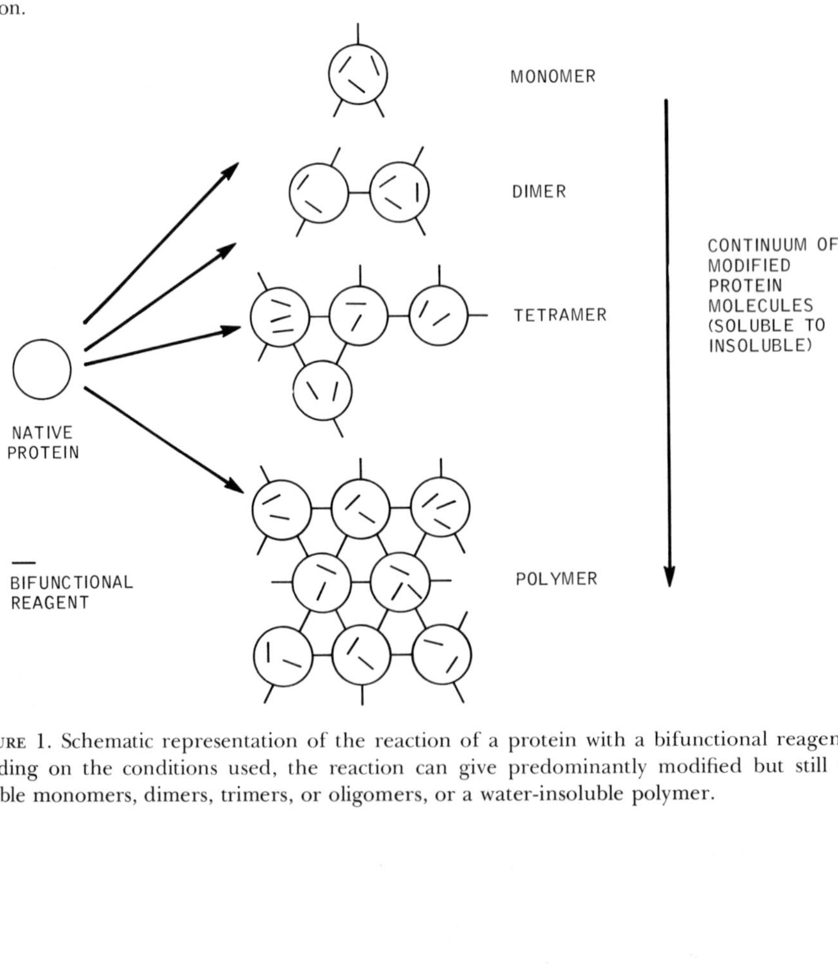

FIGURE 1. Schematic representation of the reaction of a protein with a bifunctional reagent. Depending on the conditions used, the reaction can give predominantly modified but still water-soluble monomers, dimers, trimers, or oligomers, or a water-insoluble polymer.

The reaction of a multifunctional reagent with an enzyme need not necessarily result in the formation of only intra- or intermolecular cross-links. It may involve only one part of the functional group of the reagent; the other part of the molecule can be left unmodified or may be transformed into a nonreactive group through the action of solvent molecules. This monofunctional reaction is also shown in Figure 1.

The preparation of water-insoluble enzyme derivatives using low-molecular-weight multifunctional reagents has been reviewed previously by Zaborsky (1973). Excellent reviews on the utility and scope of multifunctional reagents in protein chemistry have been given by Wold (1972) and Fasold *et al.* (1971). The latter reviews deal with the large number of multifunctional reagents that have been employed to measure interresidue distances in proteins, couple different macromolecules to each other, modify cell membranes, or introduce additional tertiary structure into enzymes with the hope of achieving increased conformational stability.

2. *Methodology*

There are several methods for preparing *water-insoluble* enzyme conjugates with bifunctional reagents. However, before describing these in detail, it may be instructive to discuss some general characteristics and observations of the reactions of enzymes with bifunctional reagents, especially with the most frequently used reagent, glutaraldehyde.

The insolubilization of enzymes through the coupling reaction with bifunctional reagents is dependent on the concentration of the protein and reagent, the pH and ionic strength of the solution, the temperature, and the time of the reaction. For example, Jansen and Olson (1969) observed that the insolubilization of papain with glutaraldehyde increased with an increase in pH of the solution, and that complete insolubilization occurred in 24 hr at 0°C in the pH range 5.2–7.2 and at a glutaraldehyde and protein concentration of 2.3 and 0.2%, respectively. However at pH 4.6, no insolubilization occurred in 24 hr, even though a reaction was evident. In further studies, Jansen *et al.* (1971) reported that the pH of optimum insolubilization (most rapid precipitation) with glutaraldehyde of bovine serum albumin, soybean trypsin inhibitor, lysozyme, and papain were nearly the same as the isoelectric points of these proteins, but that α-chymotrypsin and its zymogen differed. The formation of insoluble and active α-chymotrypsin and α-chymotrypsinogen A was most rapid at pH 6.2 and 8.2, respectively (the isoelectric points being 8.6 and 9.5, respectively). Moreover, only 70% of α-chymotrypsin precipitated at pH 6.4, while no precipitation occurred at pH 6.8. Going below pH 6.2, α-chymotrypsin precipitation decreased sharply (being 68 and 56% at pH 6.0 and 5.9, respectively) with none occurring below pH 5.7. The rate of enzymic inactivation also increased with increasing pH, and the degree of insolubilization was greater at lower ionic strength. It was also observed that the major loss in enzymic activity (60–70%) occurred during the formation of relatively small, water-soluble α-chymotrypsin conjugates (Tomimatsu *et al.*, 1971), a result which indicated that the loss of activity was caused by chemical modification and not by steric or diffusional limitations.

An investigation by Ottesen and Svensson (1971) on the insolubilization of papain with glutaraldehyde revealed an extreme temperature sensitivity. Extensively modified, *water-soluble* conjugates of papain were produced with glutaraldehyde at 0°C, pH 6.0, and at low concentrations. These conjugates were more than 80% modified at their available reactive residues and remained soluble even after the temperature was raised to 25°C. On the other hand, precipitation of papain occurred within 0.5 hr when the same reaction was conducted at room temperature at pH 6.0 even at a 10-fold lower glutaraldehyde concentration. In contrast to the observations with α-chymotrypsin made by Tomimatsu *et al.* (1971), the soluble papain conjugates retained 90% of the initial proteolytic activity even at 80% modification. From this observation, Ottesen and Svensson reasoned that the amino groups of the insoluble papain conjugates could be substituted only to a smaller degree before the activity drastically decreased, and that the major factor responsible for this decrease in the specific activity of the insoluble enzyme derivatives was the formation of the three-dimensional polymeric protein network.

As mentioned previously, multifunctional reagents can produce a variety of different enzyme conjugates, depending on the experimental conditions employed. The reaction of a pure enzyme with a bifunctional reagent such as glutaraldehyde will lead to a conjugate of almost pure protein that can be either water-soluble or water-insoluble. In turn, the oligomeric water-soluble conjugates can be immobilized by ultrafiltration cells or hollow fiber bundles and represent an attractive alternative to water-soluble, enzyme–polymer derivatives prepared from water-soluble activated polysaccharides.

With glutaraldehyde, dimethyl adipimidate, 1,5-difluoro-2,4-dinitrobenzene, and other bifunctional reagents most often employed, the reagent is incorporated into the resulting conjugate. However, several compounds exist that produce water-insoluble conjugates but which are not incorporated into the resulting enzyme conjugate. Two such reagents are ethyl chloroformate (Patramani *et al.*, 1969) and *N*-ethyl-5-phenylisoxazolium-3′-sulfonate, the latter more commonly referred to as Woodward's Reagent K (Patel and Price, 1967). The reactions of these reagents with proteins are given in Eqs. (1) and (2).

$$\mathrm{E{-}CO_2^- + Cl{-}\overset{\overset{\displaystyle O}{\|}}{C}{-}OEt \rightarrow E{-}\overset{\overset{\displaystyle O}{\|}}{C}{-}O{-}\overset{\overset{\displaystyle O}{\|}}{C}{-}OEt \xrightarrow{E{-}NH_2}}$$

$$\mathrm{E{-}\overset{\overset{\displaystyle O}{\|}}{C}{-}NH{-}E \rightarrow\rightarrow \text{water-insoluble enzyme} \quad (1)}$$
$$\mathrm{+\ CO_2 + EtOH}$$

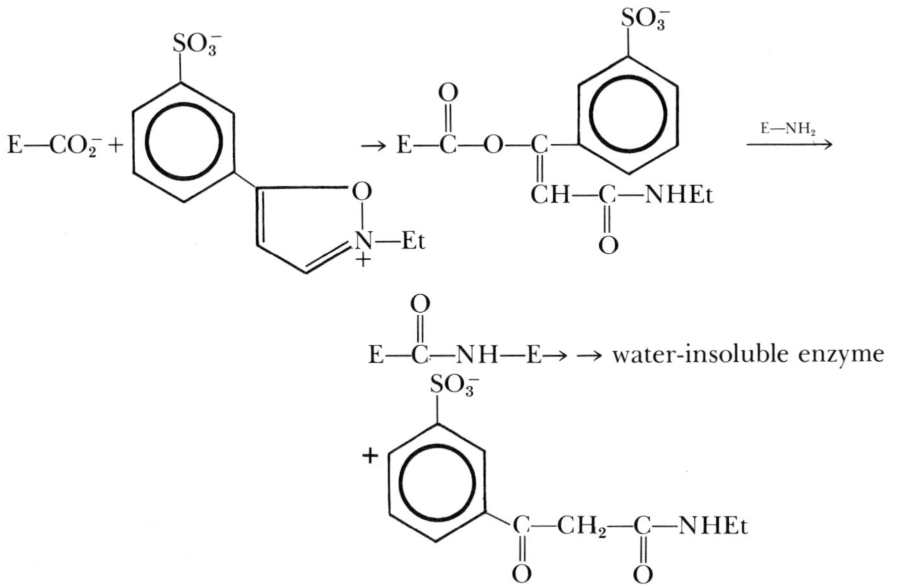

(2)

The preparation of water-insoluble conjugates can also be achieved by treating crystals of an enzyme with a solution of the bifunctional reagent. This has been done with several enzymes, primarily for the purpose of enhancing the mechanical strength of the crystals in order that they could be handled more easily. The treated enzyme crystals retained catalytic activity and were shown to be cross-linked throughout the entire crystal and not just at the surface (Quiocho and Richards, 1964; Ferrier *et al.,* 1972).

The preparation of water-insoluble enzyme conjugates with multifunctional reagents can be achieved even after entrapment of the enzyme in nylon or collodion microcapsules. For example, catalase was first immobilized by entrapment and then insolubilized by treatment of the microcapsules with glutaraldehyde (Chang, 1971).

Water-insoluble enzyme conjugates may also be prepared via the reaction of an enzyme with a second protein and the multifunctional reagent. This method is employed whenever modification of the enzyme leads to extensive inactivation or when the supply of the enzyme is limited. Proteins that have been used for this purpose include bovine serum albumin, human immunoglobulin G, and sheep antibody (Broun *et al.,* 1973).

An interesting variation of this procedure is the preparation of enzyme–protein spheres produced by emulsion polymerization of the enzyme, coprotein, and multifunctional reagent. More specifically, the enzyme and protein are mixed with the bifunctional reagent, and the resulting solution is then dropped rapidly into a vigorously stirred solution of a water-immiscible solvent containing a surfactant. The condensation is allowed to proceed, and the water-insoluble enzyme–protein spheres are transferred to an aqueous solution with the aid of a water-soluble surfactant. In order to prevent further reaction of the spheres in the aqueous medium (before the transfer is made), the bifunctional reagent, if any is present either free or in an active monofunctional form, must be inactivated. With glutaraldehyde, this may be done with $NaBH_4$ treatment.

Enhanced insolubilization of enzymes was reported for trypsin with glutaraldehyde in the presence of ammonium sulfate (Habeeb, 1967) and for subtilopeptidase B with ammonium sulfate or with acetone (Ogata *et al.,* 1968).

Another technique for preparing water-insoluble derivatives with multifunctional reagents consists of adsorbing the protein onto a surface-active, water-insoluble support followed by intermolecular cross-linking to produce more permanent conjugates. In this procedure, the use of nonporous material will result in a surface-layered enzyme conjugate, while the use of a porous support will result in an impregnated, water-insoluble conjugate. Supports that have been used to produce surface-layered enzyme conjugates are colloidal silica (Haynes and Walsh, 1969) and Dow-Corning 500-3 Silastic sheets (Broun *et al.,* 1970). Intermolecularly cross-linked impregnated enzyme conjugates have been prepared from porous collodion membranes (Goldman *et al.,* 1968) and cellophane sheets (Selegny *et al.,* 1971). Goldman *et al.* (1968) reported the preparation of a pure papain membrane that was produced by dissolving the collodion with methanol in a previously synthesized papain–collodion membrane.

Finally, multifunctional reagents can be used to prepare water-insoluble enzyme conjugates through procedures that involve using only one part of the reagent to react with the enzyme. The other part of the multifunctional reagent is used to modify a preformed, water-insoluble support or to react with a monomer that can then initiate a copolymerization. An example of the first method is the glutaraldehyde modification of polyacrylamide (Inman and Dintzis, 1969) or AE-cellulose (Glassmeyer and Ogle, 1971); an example of the second procedure is that reported recently by Jaworek (1974) on the copolymerization of glucose oxidase with the bifunctional reagent butene-3,4-oxide and acrylamide.

3. *Advantages and Disadvantages*

The single most important advantage of using multifunctional reagents to immobilize enzymes is that one reagent can be employed to prepare different kinds of conjugates, such as cross-linked gels, membranes, or spheres; adsorbed monolayered derivatives; and water-insoluble, polymer-bound species. Many bifunctional reagents are commercially available and more-selective ones are being devised constantly (Wetz *et al.,* 1974; Ruoho *et al.,* 1975).

The preparation of water-insoluble conjugates is easy to conduct experimentally and is relatively inexpensive. Enzyme conjugates can be obtained that range from pure enzyme to mostly coprotein or to an inert substance. Immobilized enzymes derivatives of irregular shape may be obtained readily by conducting the cross-linking reaction on the irregular surface that may be part of an apparatus (e.g., the surface of an electrode) or performed in a mold. The cross-linked enzyme gel derivatives can be dispersed in aqueous solution or packed with an inert support material if they are to be used in column operations. The general shortcoming of this approach is the need to control fairly rigidly the experimental conditions of pH, concentration of the reacting components, temperature, and so on, to achieve optimum insolubilization with optimum retainment of activity. Other disadvantages, which may be overcome by changing the techniques slightly, are the need for a large amount of protein or coprotein to produce the insoluble conjugate (circumvented by adsorption followed by cross-linking), and the often unavoidable inactivation of the enzyme (circumvented in part by coprotein immobilization).

The method of preparing water-insoluble enzyme conjugates by adsorbing the protein on a surface-active, water-insoluble support, followed by intermolecular cross-linking with a multifunctional reagent, has the advantage for forming an essentially monolayer of immobilized enzyme molecules. This offers a large proportion of the immobilized enzyme molecules for substrate interaction and transformation to product. However, the disadvantages of this particular approach are that (1) experimental conditions must first be determined to ensure good, if not optimum, adsorption of the enzyme on the support, and (2) the necessity to prevent aggregation of individual colloidal particles. If aggregation does occur, the clusters formed will be difficult, if not impossible, to redisperse.

Low-molecular-weight multifunctional reagents can also be employed advantageously for extending the reactive group from a preformed, water-insoluble polymer and for preparing water-insoluble enzyme derivatives entrapped within semipermeable microcapsules.

4. *Properties*

The properties exhibited by enzyme conjugates prepared by this chemical method are multivariant and often unpredictable—a very similar situation to that observed for other covalently bonded enzyme conjugates. A brief description of some properties will reveal some general trends but also the often unexpected and unexplained result.

4.1. *Activity*

The catalytic activity of water-insoluble enzyme conjugates prepared with multifunctional reagents can vary considerably and has been shown to be dependent on the amount of cross-linking reagent used during the insolubilization as well as on the pH, ionic strength, and temperature. For example, the water-insoluble derivative prepared from 56 mg of papain and 15 mg of diazobenzidine exhibited 19% and 6% relative activity, respectively, with α-*N*-benzoyl-L-arginine ethyl ester (BAEE) and casein as substrates (Silman *et al.*, 1966). If, however, the cross-linking of the same amount of papain was carried out with 25 mg of diazobenzidine, the esterase activity of the enzyme derivative was only 8%. Presumably, the lower activity of the second preparation is due to either a higher degree of modification of the enzyme or to diffusional limitations.

Considerable enzymic activity was even observed for carboxypeptidase A crystals cross-linked with glutaraldehyde (Quiocho and Richards, 1964). The enzymic activity of these crystals for the low-molecular-weight substrate, carbobenzoxyglycyl-L-phenylalanine, varied from 30 to 70% relative to the unmodified crystals.

Jansen *et al.* (1971) described the activity of a "double enzyme" prepared by the reaction of glutaraldehyde with α-chymotrypsin and mercuripapain. After activation of mercuripapain with cysteine, α-chymotrypsin exhibited relative activities of 12% and 0.1%, respectively, toward *N*-acetyl-L-tyrosine ethyl ester (ATEE) and casein, and papain exhibited relative activities of 65% and 8% toward BAEE and casein, respectively. In comparison, α-chymotrypsin cross-linked with glutaraldehyde in the presence of bovine serum albumin showed relative activities of 12% and 0.3% toward ATEE and casein, respectively. An enhancement in the activity of chymotrypsin conjugates by protecting the active site during chemical modification was also noted by these investigators.

4.2. *pH-Activity Behavior*

The pH-activity behavior of these enzyme conjugates has been reported by several workers, most notably by Goldman *et al.* (1968) for enzyme–membrane systems. For example, the pH-activity curve for a three-layer papain–collodion membrane using different substrates (Goldman *et al.*, 1968), consisting of two cross-linked papain layers separated by a collodion layer, differed from one another and from the normal bell-shaped curve of the native enzyme. With BAEE as the substrate, the activity of the enzyme conjugate increased with increasing pH even up to a value of 9.6, where the native enzyme has negligible activity. The enzyme membrane also showed higher activities than the native papain in the pH region 3–6. Equally dramatic departures of the pH-activity curves of the cross-linked papain were observed with *N*-benzoylglycine ethyl ester (BGEE), α-*N*-benzoyl-DL-arginine *p*-nitroanilide (BANA), α-*N*-benzoyl-L-arginine amide (BAA), and acetyl-L-glutamic acid diamide (AGOA) as the substrates. The shapes of the pH-activity curves obtained depended on the nature of the products liberated and on the kinetic parameters determining the rate of hydrolysis; different microenvironments were generated enzymatically by the hydrolysis of the different substrates.

4.3. *Michaelis Constant*

The apparent Michaelis constant, K'_m, of cross-linked conjugates has not been reported too often. A cross-linked glucose oxidase, impregnated in a cellophane membrane and treated with glutaraldehyde, exhibited an identifical value of 13 mM as the native enzyme (Broun *et al.*, 1969). Likewise, an insoluble glucose oxidase film prepared by cross-linking the enzyme with glutaraldehyde in the presence of albumin showed a K'_m value of 13 mM (Selegny *et al.*, 1971).

Considerable theoretical discussions of the K'_m of synthetic membrane-bound enzymes have been presented (Selegny *et al.*, 1971; Goldman and Katchalski, 1971).

4.4. *Specificity*

Changes in the specificity of cross-linked conjugates have been noted. For example, glutamic-aspartic transaminase cross-linked with glutaraldehyde lost its transaminase activity but still retained the ability to form complexes with its antibody (Patramani *et al.*, 1969). Similarly, water-insoluble catalase prepared by cross-linking with glutaraldehyde exhibited a decrease in its peroxidatic activity (Schejter and Bar-Eli, 1970), the magnitude of which was dependent on the particular substrate. The peroxidatic activity decreased 11-fold with methanol and 80-fold with pyrogallol—a difference which was interpreted to be due to diminished accessibility of the substrate to the active site or to a hindrance of conformational changes necessary for peroxidatic activity. The immobilized and native enzyme exhibited identical behavior toward cyanide, hydroxylamine, and aminotriazole.

4.5. *Stability*

The thermal stability of water-insoluble cross-linked enzyme conjugates can vary from being more or less than the native enzyme. For example, glucose oxidase impregated in cellophane sheets and cross-linked with glutaraldehyde exhibited superior thermal stability at 37°C (Broun *et al.*, 1969). On the contrary, papain immobilized as an enzyme–membrane conjugate (the same three-layer papain membrane mentioned previously) showed a decreased thermal stability relative to the native enzyme in the temperature range 65–80°C (Goldman *et al.*, 1968).

The pH-stability behavior of the three-layer papain membrane was also determined. The enzymic activity retained by both the three-layer papain membrane and the native enzyme after 10 min at various pH's was similar and quite high in the pH range 6–10. However, the stability of the three-layer papain membrane was greater at lower pH values. Whereas water-soluble unmodified papain retained only 10% of its initial activity on exposure to pH 2.2, the papain collodion membrane retained about 50% of its initial activity after similar treatment.

The storage stability of several water-insoluble enzyme derivatives has been examined. Papain cross-linked with glutaraldehyde showed no detectable decrease in its esterase activity after storage of 5 weeks at 4°C (Jansen and Olson, 1969). On the contrary, catalase cross-linked by the same reagent showed a decrease of about 20% in its initial activity after 2 weeks of storage at 4°C; after this, no further decrease in activity occurred in 5 months (Schejter and Bar-Eli, 1970). The activity of water-insoluble pepsin prepared by condensation with ethyl chloroformate was observed not to change after 6 months of storage (Rao *et al.*, 1970).

Still other stabilities reported for cross-linked enzyme conjugates are the enhanced mechanical stability of glutaraldehyde-cross-linked carboxypeptidase crystals (Quiocho and Richards, 1964), the enhanced stability toward urea of cross-linked, monolayered trypsin (Walsh *et al.*, 1970), and the stability of trypsin cross-linked with glutaraldehyde during continuous use for casein digestion (Habeeb, 1967).

5. *References*

Broun, G., Selegny, E., Avrameas, S., and Thomas, D., 1969, Enzymatically active membranes: some properties of cellophane membranes supporting cross-linked enzymes, *Biochim. Biophys. Acta* **185**:258.

Broun, G., Selegny, E., Minh, C. T., and Thomas, D., 1970, Facilitated transport of CO_2 across a membrane bearing carbonic anhydrase, *FEBS Letters* **7**:223.

Broun, G., Thomas, D., Gellf, G., Domurado, D., Berjonneau, A. M., and Guillon, C., 1973, New methods for binding enzyme molecules into a water-insoluble matrix: properties after insolubilization, *Biotechnol. Bioeng.* **15**:359.

Chang, T. M. S., 1971, Stabilization of enzymes by microencapsulation with a concentrated protein solution or by microencapsulation followed by cross-linking with glutaraldehyde, *Biochem. Biophys. Res. Commun.* **44**:1531.

Fasold, H., Klappenberger, J., Meyer, C., and Remold, H., 1971, Bifunctional reagents for the cross-linking of proteins, *Angew. Chem. Int. Ed. Engl.* **10**:795.

Ferrier, L. K., Richardson, T., and Olson, N. F., 1972, Crystalline catalase insolubilized with glutaraldehyde, *Enzymologia* **42**:273.

Glassmeyer, C. K., and Ogle, J. D., 1971, Properties of an insoluble form of trypsin, *Biochemistry* **10**:786.

Goldman, R., and Katchalski, E., 1971, Kinetic behavior of a two-enzyme membrane carrying out a consecutive set of reactions, *J. Theoret. Biol.* **32**:243.

Goldman, R., Kedem, O., Silman, I. H., Caplan, S. R., and Katchalski, E., 1968, Papain-collodion membranes: I. Preparation and properties, *Biochemistry* **7**:486.

Habeeb, A. F. S. A., 1967, Preparation of enzymically active, water-insoluble derivatives of trypsin, *Arch. Biochem. Biophys.* **119**:264.

Haynes, R., and Walsh, K. A., 1969, Enzyme envelopes on colloidal particles, *Biochem. Biophys. Res. Commun.* **36**:235.

Inman, J. K., and Dintzis, H. M., 1969, The derivatization of cross-linked polyacrylamide beads: controlled introduction of functional groups for the preparation of special-purpose, biochemical adsorbents, *Biochemistry* **8**:4074.

Jansen, E. F., and Olson, A. C., 1969, Properties and enzymatic activities of papain insolubilized with glutaraldehyde, *Arch. Biochem Biophys.* **129**:221.

Jansen, E. F., Tomimatsu, Y., and Olson, A. C., 1971, Cross-linking of α-chymotrypsin and other proteins by reaction with glutaraldehyde, *Arch. Biochem. Biophys.* **144**:394.

Jaworek, D., 1974, New immobilization techniques and supports, in: *Enzyme Engineering*, Vol. 2 (E. K. Pye and L. B. Wingard, Jr., eds.), pp. 105–114, Plenum Press, New York.

Ogata, K., Ottesen, M., and Svendsen, I., 1968, Preparation of water-insoluble, enzymatically active derivatives of subtilisin type Nova by cross-linking with glutaraldehyde, *Biochim. Biophys. Acta* **159**:403.

Ottesen, M., and Svensson, B., 1971, Modification of papain by treatment with glutaraldehyde under reducing and non-reducing conditions, *Compt. Rend. Trav. Lab. Carlsberg* **38**:171.

Patel, R. P., and Price, S., 1967, Derivatives of proteins: I. Polymerization of α-chymotrypsin by use of *N*-ethyl-5-phenylisoxazolium-3′-sulfonate, *Biopolymers* **5**:583.

Patramani, I., Katsiri, K., Pistevou, E., Kalogerakos, T., Pavlatos, M., and Evangelopoulos, A. E., 1969, Glutamic-aspartic transaminase-antitransaminase interaction: a method for antienzyme purification, *Eur. J. Biochem.* **11**:28.

Quiocho, F. A., and Richards, F. M., 1964, Intermolecular cross-linking of a protein in the crystalline state: carboxypeptidase A, *Proc. Natl. Acad. Sci. U.S.* **52**:833.

Rao, S. S., Patki, V. M., and Kulkarni, A. D., 1970, Preparation of active insoluble pepsin, *Indian J. Biochem.* **7**:210.

Ruoho, A., Bartlett, P. A., Dutton, A., and Singer, S. J., 1975, A disulfide-bridge bifunctional imidoester as a reversible cross-linking reagent, *Biochem. Biophys. Res. Commun.* **63**:417.

Schejter, A., and Bar-Eli, A., 1970, Preparation and properties of cross-linked water-insoluble catalase, *Arch. Biochem. Biophys.* **136**:325.

Selegny, E., Broun, G., and Thomas, D., 1971, Enzymatically active model-membranes: experimental illustrations and calculations on the basis of diffusion reaction kinetics of their functioning, of regulatory effects, of facilitated, retarded and active transports, *Physiol. Vegetale* **9**:25.

Silman, I. H., Albu-Weisenberg, M., and Katchalski, E., 1966, Some water-insoluble papain derivatives, *Biopolymers* **4**:441.

Tomimatsu, Y., Jansen, E. F., Gaffield, and Olson, A. C., 1971, Physical chemical observations on the α-chymotrypsin glutaraldehyde system during formation of an insoluble derivative, *J. Colloid Interface Sci.* **36**:51.

Walsh, K. A., Houston, L. L., and Kenner, R. A., 1970, Chemical modification of bovine trypsinogen and trypsin, in: *Structure–Function Relationships of Proteolytic Enzymes* (P. Desnuelle, H. Neurath, and M. Ottesen, eds.), pp. 56–69, Academic Press, Inc., New York.

Wetz, K., Fasold, H., and Meyer, C., 1974, Synthesis of "long," hydrophilic, protein cross-linking reagents, *Anal. Biochem.* **58**:347.

Wold, F., 1972, Bifunctional reagents, in: *Methods in Enzymology,* Vol. 25B (C. H. W. Hirs and S. N. Timasheff, eds.), pp. 623–651, Academic Press, Inc., New York.

Zaborsky, O. R., 1973, *Immobilized Enzymes,* CRC Press, Cleveland, Ohio.

5

Immobilization of Enzymes by Adsorption

OSKAR R. ZABORSKY

1. *Introduction*

The adsorption of enzymes and other proteins onto water-insoluble, surface-active supports is probably the most economically attractive method of immobilization, for it is simple and inexpensive (McLaren, 1974). Adsorption is a physical phenomenon and, in principle, reversible in that both the enzyme and the support can be recovered unchanged. Adsorption of enzymes onto water-insoluble matrices is also the oldest documented method for immobilizing enzymes. As early as 1916, Griffin and Nelson (1916) immobilized invertase on activated charcoal and gelatinuous aluminum hydroxide. It is equally interesting to note that the first documented case of using an immobilized enzyme for industrial processing, the hydrolysis of *N*-acetyl-DL-amino acids by L-aminoacylases (Tosa *et al.,* 1969a), also employs adsorption as the method of immobilization. Similarly, the latest and most significant commercial enterprise using an immobilized enzyme, glucose isomerase for the isomerization of glucose to fructose, entails adsorption of the catalyst onto the ion-exchange resin, DEAE-cellulose (Schnyder, 1974). At this moment, adsorption as a method of immobilization seems to be regaining wide acceptance.

The adsorption of enzymes and proteins can occur at air–water, water–oil, and water–solid interfaces. Unfortunately, from a process consideration or from an immobilization viewpoint, the main emphasis of most past studies on the adsorption of enzymes has been on the nature of the process or the purification of the protein but not on the changed chemical and physical properties of the adsorbed/immobilized enzyme. Furthermore, even when the activity of the adsorbed enzyme is reported in the early literature, the main interest was not for continuous catalytic operations, and it is extremely difficult to ascertain whether the reported activity of an "adsorbed" enzyme is due to the bound enzyme on the support or due to a desorbed, soluble enzyme. Often little or no information is given about the integrity of the enzyme–adsorbent conjugate. Only within the last two decades, as the incentives for immobilizing enzymes became more apparent

OSKAR R. ZABORSKY • National Science Foundation, Washington, D.C.

and the technology matured, did investigators become more conscious of continuous processing and of reporting the pertinent facts.

This chapter deals with water–solid, interfacially adsorbed enzymes in which the main interest of the investigation was the immobilization of the enzyme for continuous catalytic processing. The adsorption of enzymes at other interfaces (McLaren and Packer, 1970; James and Augenstein, 1966) and for purification purposes (Zittle, 1953) has been reviewed elsewhere. The immobilization of enzymes by adsorption has been reviewed by Messing (1975) and Zaborsky (1973).

2. *Methodology*

This method of preparing water-insoluble enzyme conjugates seems to be the simplest in terms of operational requirements. It consists of contacting an aqueous solution of the enzyme or mixture of enzymes with a surface-active adsorbent and washing the resulting conjugate to remove any nonadsorbed enzyme. To date, many different enzymes and adsorbents have been employed, and the more commonly used adsorbents are listed in Table 1. The adsorbents may be organic or inorganic in nature, and all are commercially available.

The mechanism of the process responsible for adsorption is often multivariant, and it is at times difficult to differentiate clearly between the various possibilities such as charge–charge interactions (formed by charged groups of opposite polarity), hydrophobic interactions (due to localized hydrophobic regions on the protein and the support), and hydrogen bonds. In addition, other modes of immobilization can also occur with or without specific intent. Thus, the immobilization of proteins with controlled-pore glass may be in part due to the physical entrapment of the macromolecules within the appropriately sized pores, and the immobilization of enzymes in flexible matrices such as collagen could be in part

TABLE 1
Adsorbents Employed for the Immobilization of Enzymes

Alumina
Anion-exchange resins
Bentonite
Calcium carbonate
Carbon
Cation-exchange resins
Cellulose
Clays
Collagen
Collodion (cellulose nitrate)
Glass
Hydroxylapatite
Kaolinite
Polyaminostyrene

due to physical containment within the polymeric network upon reduction of the polymer volume that can be caused by a change in the environment. In addition, a combination adsorption/chemical modification method is often used to circumvent the possible problem of enzyme leakage which could occur during actual use. Adsorption of the enzyme followed by cross-linking with bifunctional reagents such as glutaraldehyde or dimethyl adipimidate gives a rather operationally stable, water-insoluble conjugate (Zaborsky, 1973). Treatment of an adsorbed enzyme with a bifunctional reagent results in both inter- and intramolecular cross-linking, both of which are important. Intermolecular cross-linking is essential for producing the linkages that connect the protein molecules to each other in order to provide for the effective envelope or protein matrix, and intramolecular cross-linking may impart additional conformational stability to the adsorbed proteins (Zaborsky, 1974). Unfortunately, the necessary experimental manipulation to achieve the optimum operational and conformational stabilities (intermolecular versus intramolecular cross-linking) is largely still a matter of trial and error.

The reversibility of adsorption is an important theoretical as well as practical concern. In principle, the adsorption of a protein should be a reversible process; a change in pH, ionic strength, temperature, or concentration of a particular species should cause the bound enzyme to desorb from the water-insoluble matrix and be released into the aqueous phase. However, in practice, this may not be attained. Complete or even partial desorption may not be possible because some unforeseen irreversible process has occurred between the carrier and the enzyme, or the appropriate conditions for desorption have not yet been found. Irreversible binding has been reported for urease on kaolinite (Sundaram and Crook, 1971) and for lactate dehydrogenase on collodion (Thang *et al.*, 1968). Of course, irreversible binding is not a problem if the enzyme remains active upon adsorption and if it is to be used in a continuous process. In fact, for continuous processing, it is highly desirable to have a strongly formed association, for weak binding can be a serious problem, leading eventually to depletion of the enzyme, with concomitant contamination of the product.

The degree of adsorption of a protein onto a water-insoluble matrix depends on a delicate balance of factors, which include the ionic strength and pH of the medium, the concentration of the protein and possible other species in the solution (perhaps one being the substrate), the temperature, the time, and the particular characteristics of the enzyme and the support. Conversely, desorption depends on an interplay of these factors, and it is their precise control which leads either to adsorption (immobilization) or to desorption (solubilization) of the enzyme. A good discussion of these variables, the essence of which follows below, was given by Zittle (1953). A more recent but abbreviated description was given by Dillman and Miller (1973).

2.1. *Hydrogen Ion Concentration (pH)*

Changes in pH can influence the adsorption/immobilization of an enzyme because of changes in the charge of the support and/or the protein. In general, maximum adsorption is obtained at the isoelective point of the protein, with a relatively sharp decrease on either side (±1–2 pH units). However, variations in this behavior do exist. For example, albumin on rubber latex also exhibits an increase in adsorption at the extremes of pH (3 and 7) giving a W-shaped adsorption curve. In addition, the adsorption maximum of egg albumin varies with different carbon supports from pH 3 to 6. Furthermore, a change in pH of 3–4 units may have little effect on the adsorption of the protein, as in the case with trypsin on charcoal.

2.2. *Salts*

The influence of salts on adsorption is complex and multivariant. The most commonly accepted belief, that salts prevent adsorption, is certainly valid, and this phenomenon is used for eluting enzymes adsorbed from low-salt solutions. However, examples do exist in the literature in which salts have been shown to enhance the adsorption of proteins or in which the adsorption occurs on materials that are otherwise inert to proteins. The ability of salts to increase the adsorption of proteins has been termed "salting-out" adsorption.

2.3. *Concentration of Proteins*

In general, the adsorption of an enzyme increases with an increase in the concentration of the protein until saturation is reached. Conversely, with an increase in the amount of an adsorbent, the total adsorption increases until the protein is exhausted from solution.

2.4. *Temperature*

Although low-molecular-weight substances usually show a decrease in adsorption with an increase in temperature, proteins show the opposite behavior, an increase in adsorption with an increase in temperature. This effect of temperature suggests that the adsorption of proteins involves some unfolding of the macromolecules on the surface because this behavior is endothermic. Of course, exceptions to this generalization are known.

2.5. *Rate of Adsorption*

The time required for the adsorption of proteins is considerably greater than that required for low-molecular-weight substances. In turn, this slower rate of adsorption is attributed to the lower rate of diffusion of the macromolecular proteins.

2.6. *Support*

Important considerations of a support to the degree of adsorption and activity of the resulting enzyme conjugate are the surface area, porosity (if any

TABLE 2
Relation of Pore Size and Surface Area to Quantity of Adsorbed Stabilized Enzyme[a]

Glass (Corning designation)	Pore diameter (Å)	Enzyme (mol. wt.)	Surface area (m^2/g)	mg-equiv of bound enzyme/g of glass
7740 (test tube)	Not porous	Papain (21,000)	Very low	0
7740 (F-frit)	50,000	Papain	0.2	0.01
X740BXY	900	Papain	20	0.33
7930	92	Papain	60	0.08
7930	68	Papain	118	0.07
X740BXY	900	Glucose oxidase	20	0.12
7930	68	(150,000)	118	0.0001

[a]Taken from Messing (1970).

exists), and possible preconditioning treatments. If the support is truly nonporous, adsorption of the protein is governed by the total surface area of the material. However, if the support is porous, the total adsorption is not always dictated by surface area. For example, the quantity of immobilized papain was both a function of the pore size and the surface area of the glass (Table 2). Thus, even when the support has a high surface area, the availability of that surface for attachment with enzymes may be limited by the size of the macromolecule and the restrictive size of the pores.

Preconditioning a support before immobilization is extremely important with certain enzymes and with certain supports. For example, if an enzyme is acid-labile around pH 4, immobilization on controlled-pore silica could lead to denaturation, because the surface of this support is rather acidic. Denaturation can be prevented by treating the controlled-pore silica with a buffer before immobilizing the enzyme. Preconditioning of a support can also be important with enzymes that require cofactors for full expression of catalysis. Most carriers will bind metal ions quite effectively and hence remove them from the sphere of action of the immobilized enzyme with a concomitant loss of activity. It is again advisable to pretreat the carrier with these substances before immobilizing the enzyme.

3. *Advantages and Disadvantages*

Advantages of this method for immobilizing enzymes are: (1) operational simplicity, (2) large choice of differently charged and shaped carriers, (3) possibility of achieving simultaneous purification and immobilization, (4) reactivation of the enzyme conjugate, and (5) regeneration of the support material.

Many of the adsorbents listed in Table 1 are also used for the purification of enzymes, and it is conceivable that purification and immobilization can be achieved simultaneously with crude enzyme mixtures. This indeed seems to have occurred for Nikolaev and Mardashev (1961) in the purification/immobilization of asparaginase from a crude homogenate adsorbed on CM-cellulose and for Sundaram and Crook (1971) for urease adsorbed on kaolinite. Preferential adsorption/desorption could also be applied to different conformers of the same enzyme (but still active) or to inactive and active forms (Kobamoto *et al.,* 1966).

An example of the reactivation of an enzyme-adsorbent conjugate was reported by Tosa *et al.* (1969a) for the enzyme aminoacylase and DEAE-Sephadex. Reactivation of the column was accomplished simply by addition of more active, soluble enzyme.

Although most of the supports are inexpensive, their regeneration is possible. For example, controlled-pore silica can be regenerated by a heat treatment of the enzyme–conjugate (if this is necessary) or by eluting the adsorbed enzyme with strong base. Likewise, regeneration/reactivation of the aminoacylase–DEAE-Sephadex column described earlier was accomplished by treatment with sodium hydroxide followed by a buffer change and addition of new aminoacylase. Disadvantages of the method exist. Although it is operationally easy to prepare water-insoluble enzyme conjugates by adsorption, the conditions for achieving optimum loading with activity are a matter of trial and error. The correct conditions of pH, ionic strength, temperature, time, and so on, must be determined individually for each enzyme support, and activity and the degree of adsorption may not go hand in hand. Also, weak adsorption between the enzyme and the support will lead to loss of the catalyst, "enzyme leakage," and contamination of the product. Desorption, as adsorption, is dictated by such factors as the concentration of the substrate and the pH and ionic strength of the medium. Examples of substrate-induced desorption, dependent on concentration, are glucose 6-phosphate dehydrogenase adsorbed on collodion (Goldman and Lenhoff, 1971) and aminoacylase on DEAE-cellulose (Tosa *et al.,* 1966). Desorption by a change in the pH of the solution was found for papain adsorbed on glass (Messing, 1970) and catalase on DEAE-cellulose (Mitz, 1956). Catalase adsorbed on DEAE-cellulose exhibited somewhat unusual desorption behavior in that it could be eluted with either a change in pH or ionic strength but not with its substrate. Examples of ionic-strength-induced desorption of enzymes are leucine aminopeptidase adsorbed on calcium phosphate gel (Koelsch *et al.,* 1970) and ATP deaminase adsorbed on DEAE-cellulose (Chung *et al.,* 1968).

4. *Properties of Adsorbed Enzymes*

Immobilization by adsorption is a physical method for localizing enzymes, and hence no permanent change in either a chemical or physical property is expected once the enzymes have been removed from the adsorbent. On the adsorbent, however, considerable changes can be expected to occur because of changes in the intrinsic properties of the enzyme (such as its conformation or charge) or to microenvironmental effects caused by the charge of the carrier or diffusion.

4.1. *Activity*

The activity of adsorbed enzymes can vary from zero (total inactivation) to values approaching those of the soluble enzymes. Representative values are: glucose-6-phosphate dehydrogenase on collodion, 50% (Goldman and Lenhoff, 1971); invertase on bentonite, 1.2% (Monsan and Durand, 1971); leucine aminopeptidase on calcium phosphate gel, 78% (Koelsch *et al.,* 1970); and α-chymotrypsin on kaolinite, 70% (McLaren and Estermann, 1956). DEAE-cellulose immobilized enzymes exhibited similar varied activities; β-fructofuranosidase, 25–31% (Usami *et al.,* 1971) and 30–50% (Suzuki *et al.,* 1967); glucoamylase, 16–58% (Bachler *et al.,* 1970); and catalase, about 70% (Mitz, 1956). The binding capacity of the adsorbents used for immoblizing enzymes varies from micrograms to milligrams per gram of support.

4.2. *pH-Activity Behavior*

The pH-activity behavior of enzyme-adsorbent conjugates is analogous to that expressed by covalently bonded, water-insoluble derivatives. The pH-activity curves of the conjugates, relative to the soluble enzymes, can be displaced either toward more alkaline or acid pH's. On the other hand, the shape and optimum pH of the conjugates may not be changed at all. Displacements in either direction have been attributed to the microenvironment created by the charge of the carrier. Interestingly, McLaren and Estermann (1957) and McLaren (1960) predicted and observed this pH-activity-curve displacement phenomenon for the adsorbed conjugates well before the same behavior was observed for covalently bonded enzyme derivatives. McLaren (1960) also presented a mathematical treatment of the pH difference between the environment surrounding a polyelectrolyte carrier and the bulk phase of the solution.

Shifts toward more alkaline pH's were observed for α-chymotrypsin adsorbed on kaolinite (McLaren, 1957, 1960) and for glucoamylase adsorbed on acid clay (Usami and Shirasaki, 1970). The latter system also exhibited a somewhat narrower pH-activity profile than the native enzyme. Displacements of the pH-activity curve toward more acid pHs were observed for glucoamylase on DEAE-Sephadex A-25 (Usami and Shirasaki, 1970), ATP deaminase on DEAE-cellulose (Chung *et al.,* 1968), aminoacylase on DEAE-cellulose (Tosa *et al.,* 1967), and invertase on DEAE-cellulose (Suzuki *et al.,* 1966). In the case of aminoacylase, the acid pH-activity displacement was noted to be substrate-dependent. For example, the optimum pH for the hydrolysis of *N*-acetyl-DL-methionine was 7.0 and 7.5 for the Sephadex–conjugate and free enzyme, respectively. Other pH optima for the immobilized and native aminoacylase were 6.0 and 7.5 for *N*-acetyl-DL-tryptophan, and 6.0 and 7.0 for *N*-acetyl-DL-valine. However, identical pH optima for the adsorbed and native enzymes were observed for *N*-chloroacetyl derivatives of phenylalanine, tryptophan, and tyrosine. The same optimum was noted for urease adsorbed on kaolinite (Sundaram and Crook, 1971).

4.3. *Michaelis Constant*

The value of the apparent Michaelis constant for adsorbed enzymes has been observed to be higher, lower, or the same as that of the corresponding soluble enzymes. Higher K'_m values were observed for glucoamylase on activated charcoal (Usami and Shirasaki, 1970), aminoacylase on DEAE-Sephadex (Tosa *et al.*, 1969b), and aminoacylase on DEAE-cellulose (Tosa *et al.*, 1967). Lower values were observed for aminoacylase on DEAE-cellulose or DEAE-Sephadex. The apparent difference in the behavior of these same immobilized enzyme systems can be attributed to the use of different substrates. Identical constants (relative to the soluble enzymes) were observed for acid phosphatase and phosphoglucomutase on silica gel (Vorobeva and Poltorak, 1966).

4.4. *Specificity*

Some interesting substrate and inhibitor specificities have been reported for adsorbed enzymes. For example, dextransucrase adsorbed on DEAE-Sephadex (Ogino, 1970) exhibited a changed substrate specificity relative to the soluble enzyme. The polymer formed by the insoluble enzyme had a higher glycogen value and fructose content. ATP deaminase (Chung *et al.*, 1968) showed the same order of reactivity toward ATP, ADP, and AMP as did the native enzyme, but the relative ratio of activities was different. A surprising activation of aminoacylase adsorbed on DEAE-Sephadex or DEAE-cellulose was observed by Tosa *et al.* (1969b, 1971). Incubation of the enzymes for 1 hr in 6 M urea and then assaying for activity in 2 M urea showed enhanced catalysis for these conjugates relative to their initial activities. This activation of the immobilized enzymes by urea was attributed to the conversion of the rigid structures of the conjugates by the denaturant to more flexible and active conformers.

Additional examples of changed behavior of immobilized enzymes are the enhanced resistance of β-fructofuranosidase toward mercuric acetate inhibition (Usami *et al.*, 1971), and the greater inhibition of adsorbed glucoamylase by mercuric acetate (Usami *et al.*, 1967).

4.5. *Stability*

A considerable amount of information is available on the various kinds of stabilities of adsorbed, water-insoluble, enzyme conjugates. As in the case with covalently bonded, water-insoluble enzyme derivatives, a spectrum of stabilities exists.

The thermal stability of adsorbed enzymes can be increased, decreased, or unchanged relative to the soluble enzymes. Enhanced thermal stability was found for aminoacylase on DEAE-Sephadex (Tosa *et al.*, 1969b), NAD pyrophosphorylase on hydroxylapatite (Traub *et al.*, 1969), and aminoacylase on DEAE-cellulose (Tosa *et al.*, 1967). On the contrary, decreased thermal stability was observed for glucose-6-phosphate dehydrogenase adsorbed on collodion (Goldman and Lenhoff, 1971), β-fructofuranosidase on DEAE-cellulose (Usami *et al.*, 1971), and glucoamylase on activated clay (Usami *et al.,* 1967). Essentially similar thermal stability (relative to soluble enzyme) was observed for urease adsorbed on kaolinite (Sundaram and Crook, 1971).

Storage stabilities for adsorbed enzymes have been reported. β-Fructofuranosidase adsorbed on DEAE-cellulose (Usami *et al.,* 1971) lost negligible activity after 15 weeks of storage at 4°C. The stability of urease adsorbed on kaolinite was similar to that of the native enzyme at 2°C (Sundaram and Crook, 1971). Polynucleotide phosphorylase adsorbed on nitrocellulose had a half-life of 1 week at 4°C (Thang *et al.,* 1968).

Other stabilities have also been reported. There is an enhanced stability toward pronase and trypsin hydrolysis for aminoacylase adsorbed on DEAE-Sephadex (Tosa *et al.,* 1969b). The pH stability is diminished in invertase on DEAE-cellulose (Suzuki *et al.,* 1967) and in glucoamylase on activated clay (Usami *et al.,* 1967); but is enhanced in urease on kaolinite (Sundaram and Crook, 1971). The stability of adsorbed enzymes has also been investigated in continuous catalytic processes (Tosa *et al.,* 1969a).

5. *References*

Bachler, M. J., Strandberg, G. W., and Smiley, K. L., 1970, Starch conversion by immobilized glucoamylase, *Biotechnol. Bioeng.* **12**:85.

Chung, S.-T., Hamano, M., Aida, K., and Uemura, T., 1968, Studies on ATP deaminase: III. Water insoluble ATP deaminase, *Agr. Biol. Chem. (Tokyo)* **33**:1287.

Dillman, W. J., Jr., and Miller, I. F., 1973, On the adsorption of serum proteins on polymer membrane surfaces, *J. Colloid Interface Sci.* **44**:221.

Goldman, R., and Lenhoff, H. M., 1971, Glucose-6-phosphate dehydrogenase on collodion membranes, *Biochim. Biophys. Acta* **242**:514.

Griffin, E. G., and Nelson, J. M., 1916, The influence of certain substances on the activity of invertase, *J. Amer. Chem. Soc.* **38**:722.

James, L. K., and Augenstein, L., 1966, Adsorption of enzymes at interfaces: film formation and the effect on activity, *Adv. Enzymol., Relat. Arease Mol. Biol.* **28**:1.

Kobamoto, N., Löfroth, G., Camp, P., Van Amburg, G., and Augenstein, L., 1966, Specificity of trypsin adsorption onto cellulose, glass, and quartz, *Biochem. Biophys. Res. Commun.* **24**:622.

Koelsch, R., Lasch, J., and Hanson, H., 1970, Covalent and noncovalent binding of leucine aminopeptidase to water-insoluble carriers, *Acta Biol. Med. Geriat.* **24**:833.

McLaren, A. D., 1957, Concerning the pH dependence of enzyme reactions on cells, particulates and in solution, *Science* **125**:697.

McLaren, A. D., 1960, Enzyme action in structually restricted systems, *Enzymologia* **21**:356.

McLaren, A. D., 1974, Soil as a system of bound enzymes, *Chem. Ind. (London)* **1974**:316.

McLaren, A. D., and Estermann, E. F., 1956, The adsorption and reactions of enzymes and proteins on kaolinite: III. The isolation of enzyme–substrate complexes, *Arch. Biochem. Biophys.* **61**:158.

McLaren, A. D., and Estermann, E. F., 1957, Influence of pH on the activity of chymotrypsin at a solid–liquid interface, *Arch. Biochem. Biophys.* **68**:157.

McLaren, A. D., and Packer, L., 1970, Some aspects of enzyme reactions in heterogeneous systems, *Adv. Enzymol., Relat. Areas Mol. Biol.* **33**:245.

Messing, R. A., 1970, Relationship of pore size and surface area to quantity of stabilized enzyme bound to glass, *Enzymologia* **38**:39.

Messing, R. A., 1975, Immobilization by adsorption and inorganic bridge formation, in: *Immobilized Enzymes for Industrial Reactors* (R. A. Messing, ed.), pp. 79–98, Academic Press, Inc., New York.

Mitz, M. A., 1956, New insoluble active derivative of an enzyme as a model for study of cellular metabolism, *Science* **125**:1076.

Monsan, P., and Durand, G., 1971, Preparation of insolubilized invertase by adsorption on bentonite, *FEBS Letters* **16**:39.

Nikolaev, A. Y., and Mardashev, S. R., 1961, An insoluble active compound of asparaginase with carboxymethylcellulose, *Biokhimiya* **26**:641.

Ogino, S., 1970, Formation of the fructose-rich polymer by water-insoluble dextransucrase and presence of a glycogen value-lowering factor, *Agr. Biol. Chem. (Tokyo)* **34**:1268.

Schnyder, B. J., 1974, Continuous isomerization of glucose to fructose on a commercial basis, *Die Starke* **26**:409.

Sundaram, P. V., and Crook, E. M., 1971, Preparation and properties of solid-supported urease, *Can. J. Biochem.* **49**:1388.

Suzuki, H., Ozawa, Y., and Maeda, H., 1966, Studies on the water-insoluble enzyme hydrolysis of sucrose by insoluble yeast invertase, *Agr. Biol. Chem. (Tokyo)* **30**:807.

Suzuki, H., Ozawa, Y., Maeda, H., and Tanabe, O., 1967, Water-insoluble enzyme hydrolysis of sucrose by insoluble yeast invertase, *Kogyo Gijutsuin, Hakko Kenkyusho Kenkyu Hokoku* **1967**:11.

Thang, M. N., Graffe, M., and Grunberg-Manago, M., 1968, Observations on the activity of enzymes after filtration on (and through) a nitrocellulose membrane, *Biochem. Biophys. Res. Commun.* **31**:1.

Tosa, T., Mori, T., Fuse, N., and Chibata, I., 1966, Studies on continuous enzyme reactions: II. Preparation of DEAE-cellulose-aminoacylase column and continuous optical resolution of acetyl-DL-methionine, *Enzymologia* **31**:225.

Tosa, T., Mori, T., Fuse, N., and Chibata, I., 1967, Studies on continuous enzyme reactions: III. Enzymatic properties of the DEAE-cellulose–aminoacylase complex, *Enzymologia* **32**:153.

Tosa, T., Mori, T., Fuse, N., and Chibata, I., 1969a, Studies on continuous enzyme reactions: V. Kinetics and industrial application of aminoacylase column for continuous optical resolution of acyl-DL-amino acids, *Agr. Biol. Chem. (Tokyo)* **33**:1047.

Tosa, T., Mori, T., and Chibata, I., 1969b, Studies on continuous enzyme reactions: VI. Enzymatic properties of the DEAE-Sephadex–aminoacylase complex, *Agr. Biol. Chem. (Tokyo)* **33**:1053.

Tosa, T., Mori, T., and Chibata, I., 1971, Studies on continuous enzyme reactions. VII. Activation of water-insoluble aminoacylase by protein denaturing agents, *Enzymologia* **40**:49.

Traub, A., Kaufmann, E., and Teitz, Y., 1969, Synthesis of nicotinamide-adenine dinucleotide by NAD pyrophosphorylase on a column of hydroxylapatite, *Anal. Biochem.* **28**:469.

Usami, S., and Shirasaki, H., 1970, Kinetics of enzyme adsorbed on adsorbent, *Hakko Kogaku Zasshi* **48**:506.

Usami, S., Yamada, T., and Kimura, A., 1967, Activities of enzyme adsorbed on adsorbents: II. Stability of adsorbed enzyme, *Hakko Kyokaishi* **25**:513.

Usami, S., Noda, J., and Goto, K., 1971, Preparation and properties of water-insoluble saccharase, *Hakko Kogaku Zasshi* **49**:598.

Vorobeva, E. S., and Poltorak, O. M., 1966, Comparative study of enzymic activity of acid phosphatase and phosphoglucomutase in adsorption layers of various types, *Vestn Mosk. Univ. Ser. II: Khim.* **21**:17.

Zaborsky, O. R., 1973, *Immobilized Enzymes*, CRC Press, Cleveland, Ohio.

Zaborsky, O. R., 1974, Stabilization and immobilization of enzymes with imidoesters, in: *Enzyme Engineering*, Vol. 2 (E. K. Pye and L. B. Wingard, Jr., eds.), pp. 115–122, Plenum Press, New York.

Zittle, C. A., 1953, Adsorption studies of enzymes and other proteins, *Adv. Enzymol., Relat. Areas Mol. Biol.* **14**:319.

6

Gel-Entrapment of Enzymes

ANN-CHRISTIN KOCH-SCHMIDT

1. *Introduction*

The three-dimensional conformation of an enzyme has an important effect on its catalytic action. Consequently, when immobilizing an enzyme, it is necessary to use such methods and chemicals that the functional tertiary structure will not be affected. Physical entrapment of an enzyme in a gel lattice is an immobilization method in which no modification of the amino acid residues is needed, and which offers the advantage of reaction conditions usually so mild that few significant changes in the enzyme structure occur. A great advantage lies in the fact that the presence of protective and stabilizing agents does not affect the yield of the entrapped material. The method has a broad applicability to most enzymes, purified as well as crude extracts, to whole cells (e.g., microorganisms) and even to culture broths containing the desired enzyme.

The principle behind the entrapping technique is to form a cross-linked polymeric network around the material to be trapped. This is usually performed by mixing the monomers, a cross-linking agent, and the material to be entrapped in a buffered solution and then adding a catalyst system, which initiates the polymerization process.

In the recent past, most enzymes have been entrapped in acrylic polymers, probably because of the great possibilities of varying the monomers used and thereby the chemical and physical properties of the polymer that is formed. Other materials used include starch, silicone rubber, silica gel, fibrin, and collagen. Table 1 lists examples of enzymes entrapped in various polymers.

In the first part of this chapter, different materials used when entrapping enzymes and various entrapping methods will be described. However, most attention will be paid to polymers based on acrylic monomers. In the second part, factors influencing the operational activity and stability of entrapped enzymes will be described, and finally some applications of gel-entrapped enzymes will be given.

ANN-CHRISTIN KOCH-SCHMIDT • Biochemistry II, University of Lund, Lund, Sweden

TABLE 1
Gel-Entrapped Enzymes

Enzyme	Carrier	Reference
Acetylcholinesterase	Silicone rubber	Pennington *et al.*, 1968a, 1968b
Alcohol dehydrogenase (liver)	Polyacrylamide (block)	Wieland *et al.*, 1966
Aldolase	Polyacrylamide (block)	Bernfeld and Wan, 1963; Bernfeld *et al.*, 1968; Brown *et al.*, 1968b
Aminoacylase	Polyacrylamide (block)	Mori *et al.*, 1972; Kawashima and Umeda, 1974
α-Amylase	Polyacrylamide (block)	Bernfeld and Wan, 1963; Kawashima and Umeda, 1974
β-Amylase	Polyacrylamide (block)	Bernfeld and Wan, 1968; Kawashima and Umeda, 1974
Asparaginase	Polyacrylamide (block)	Nadler and Updike, 1974
	Fibrin	Inada *et al.*, 1975
ATP-diphosphohydrolase	Polyacrylamide (block)	Brown *et al.*, 1968b
	Silicone rubber	Brown *et al.*, 1968a
Butyrylcholin esterase	Polyacrylamide (block)	Degani and Miron, 1970
Cholinesterase	Starch	Bauman *et al.*, 1965; Guilbault and Das, 1970
	Polyacrylamide (block)	Guilbault and Das, 1970
α-Chymotrypsin	Polyacrylamide (block)	Bernfeld and Wan, 1963
	Silicone rubber	Brown *et al.*, 1968a
Citrate synthase	Polyacrylamide (block)	Mosbach, 1970; Srere *et al.*, 1973
Depsidase	Polyacrylamide (block)	Mosbach and Mosbach, 1966
β-Galactosidase	Poly(HEMA) (block)	Hinberg *et al.*, 1974b
Glucoamylase	Polyacrylamide (beads)	Beck and Rase, 1973
	Polyacrylamide (block)	Kawashima and Umeda, 1974
Glucose isomerase (from *Streptomyces phaeochromogenes*)	Collagen	Wang and Vieth, 1973
Glucose oxidase	Polyacrylamide (block)	Hicks and Updike, 1966; Gestrelius *et al.*, 1973; Kawashima and Umeda, 1974
	Poly (HEMA-vinylpyrrolidone)	Hinberg *et al.*, 1974a
	Silicone rubber	Pennington *et al.*, 1968b
	Collagen	Wang and Vieth, 1973; Constantinides *et al.*, 1973
Glucose-6-phosphate dehydrogenase	Polyacrylamide (block)	Mosbach and Mattiasson, 1970; Harrison, 1974

(continued)

TABLE 1 (*continued*)

Enzyme	Carrier	Reference
β-Glucosidase	Polyacrylamide (beads)	Johansson and Mosbach, 1974a
Hexokinase	Polyacrylamide (block)	Brown *et al.*, 1968b; Mosbach and Mattiasson, 1970; Gestrelius *et al.*, 1973
11-β-Hydroxylase (from *Curvurlaria lunata*)	Polyacrylamide(block)	Mosbach and Larsson, 1970
Invertase	Collagen	Wang and Vieth, 1973
	Polyacrylamide-starch	Kawashima and Umeda, 1974
	Polyacrylamide (beads)	Köstner and Mandel, in press
Lactase (from *Saccaromyces lactis*)	Polyacrylamide (beads)	Dahlqvist *et al.*, 1973
Lactate dehydrogenase	Polyacrylamide (block)	Hicks and Updike, 1966; Wieland *et al.*, 1966; Srere *et al.*, 1973
Lysozyme	Collagen	Wang and Vieth, 1973
Malate dehydrogenase	Polyacrylamide (block)	Srere *et al.*, 1973
Orcellinic acid decarboxylase	Polyacrylamide (block)	Mosbach and Mosbach, 1966
Papain	Polyacrylamide (block)	Bernfeld and Wan, 1963
Penicillin amidase	Collagen	Wang and Vieth, 1973
	Polyacrylamide (beads)	Köstner and Mandel, in press
Peroxidase	Silicone rubber	Pennington *et al.*, 1968b
Phosphatase, alkaline	Polyacrylamide (block)	v. Duijn *et al.*, 1967
Phosphofructokinase	Polyacrylamide (block)	Brown *et al.*, 1968b
Phosphoglucoisomerase	Polyacrylamide (block)	Brown *et al.*, 1968b
Phosphoglycerate mutase	Polyacrylamide (block)	Bernfeld *et al.*, 1969
Protease, alkaline	Polyacrylamide-starch	Kawashima and Umeda, 1974
Protease, neutral	Polyacrylamide-starch	Kawashima and Umeda, 1974
Ribonuclease A	Polyacrylamide (block)	Bernfeld and Wan, 1963
	Polyacrylamide (beads)	Johansson and Mosbach, 1974a
Steroid Δ^{1-2}dehydrogenase (from *Corynebacterium simplex*)	Polyacrylamide (block)	Mosbach and Larsson, 1970
Trypsin	Polyacrylamide (block)	Bernfeld and Wan, 1963; Mosbach and Mosbach, 1966; Wieland *et al.*, 1966
	Polyacrylamide (beads)	Nilsson *et al.*, 1972; Gestrelius *et al.*, 1973; Johansson and Mosbach, 1974a
	Acrylic acid/acrylamidecopolymer (beads)	Nilsson *et al.*, 1972

(continued)

TABLE 1 (*continued*)

Enzyme	Carrier	Reference
	Poly(HEMA) (block)	O'Driscoll *et al.*, 1972
	Silica gel	Johnson and Whateley, 1971
	Silicone rubber	Brown *et al.*, 1968a
Urate oxidase	Poly(HEMA)	Hinberg and O'Driscoll, 1975
Urease	Polyacrylamide (block)	Guilbault and Das, 1970
	Polyacrylamide (beads)	Johansson and Mosbach, 1974a
	Polyacrylamide-starch	Broun *et al.*, 1973
	Starch	Guilbault and Das, 1970
	Collagen	Karube and Suzuki, 1972; Wang and Vieth, 1973
Uricase	Collagen	Suzuki *et al.*, 1974
Whole cells of		
Umbilicaria pustulata (lichen)	Polyacrylamide (block)	Mosbach and Mosbach, 1966
Curvularia lunata (fungus)	Polyacrylamide (block)	Mosbach and Larsson, 1970
Streptococcus faecalis (bact.)	Polyacrylamide (block)	Franks, 1971
Escherichia coli (bacterium)	Polyacrylamide (block)	Chibata *et al.*, 1974
Streptomyces albus (bacterium)	Polyacrylamide (beads)	Koch-Schmidt and Mosbach, to be published

2. *Materials and Methods Used for Entrapping Enzymes*

2.1. *Acrylic Copolymers*

Among the large number of materials available for the entrapping of enzymes, the synthetically prepared, hydrophilic acrylic copolymers have mostly been used (Table 1). These show good chemical, mechanical, and thermal stability and are inert to microbial degradation (Turková *et al.*, 1973). The most commonly used monomer is acrylamide, but acrylic acid and several of its derivatives (Figure 1) have also been used. Bifunctional monomers, such as *N,N'*-methylenebisacrylamide (BIS) or ethyleneglycol dimethacrylate (EGDMA) are the usual cross-linking agents. The structure of a poly(2-hydroxyethyl methacrylate) gel, cross-linked with BIS, is shown in Figure 2. The chemical composition of an acrylic copolymer can be determined by IR (Ohno and Stahmann, 1971).

In one respect these acrylic copolymers are outstanding, and that is the ease with which various polymers can be prepared "tailor-made" for a particular enzyme or application with regard to stability and/or operational activity (Johansson and Mosbach, 1974b). Parameters that can easily be varied include the degree of porosity and the chemical composition of the matrix. The polymerization is generally carried out in a liquid medium either as a bulk or a bead polymerization.

$CH_2{=}CHCONH_2$	acrylamide
$CH_2{=}CHCOOH$	acrylic acid
$CH_2{=}CHCOOCH_3$	methyl acrylate
$CH_2{=}C(CH_3)COOH$	methacrylic acid
$CH_2{=}C(CH_3)COOCH_3$	methyl methacrylate
$CH_2{=}C(CH_3)COOCH_2CH_2OH$	hydroxyethyl methacrylate
$CH_2{=}C(CH_3)COOC_{18}H_{37}$	stearyl methacrylate
$CH_2{=}C(CH_3)COOCH_2CHCH_2$ (epoxide O)	glycidyl methacrylate
$CH_2{=}CHC{\equiv}N$	acrylonitrile
$CH_2{=}CHCOCl$	acrylyl chloride

FIGURE 1. Acrylic monomers.

The structure and the porosity obtained homogeneously throughout the gel (Fawcett and Morris, 1966) depend on the total weight of monomers per 100 ml of solution, *T*, and the amount of cross-linking agent, *C*, expressed in percentage (w/w) of total amount of monomer (Hjertén, 1962), as well as on the polymerization conditions. Figure 3 illustrates the dependence of the pore size on *T* and *C* in polyacrylamide cross-linked with BIS. A transition point, indicating minimum average pore size, obtained at $C = 5\%$, is a common characteristic of cross-linked polymers (Fawcett and Morris, 1966). As *C* is increased, the polymer becomes opaque, indicating a different kind of organization within the gel. This microprecipitation phenomenom involves the formation of bundles of polymer chains which are highly cross-linked (generating small pores) and which are separated by large pores (Richards and Temple, 1971).

Because of the great number of acrylic monomers available, it is possible to prepare a support with chemical properties suitable for an enzyme and for a particular application. A factor that can be varied in the "tailor-made" approach is the hydrophobicity of the polymer. Liver alcohol dehydrogenase immobilized in a hydrophobic acrylamide–methyl acrylate polymer showed a lower K_m(app) for the

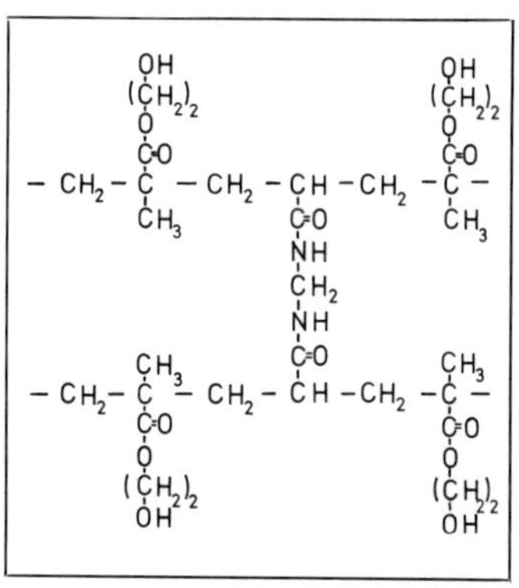

FIGURE 2. Structure of poly(HEMA) cross-linked with BIS.

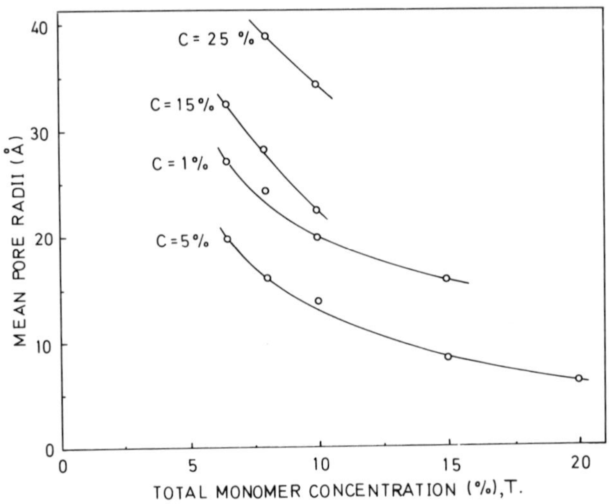

FIGURE 3. Dependence of T and C on the pore size of polyacrylamide cross-linked with BIS.

lipophilic substrate *n*-butanol than when it was bound in a more hydrophilic polyacrylamide gel (Figure 4b) (Johansson and Mosbach, 1974b). In another approach the influences of the negative charge present in an acrylic acid copolymer containing entrapped trypsin on the hydrolysis of the positively charged substrate α-*N*-benzoyl-L-arginine ethyl ester was studied. The catalysis was more effective, especially at low substrate concentrations, when the enzyme was bound to the negatively charged matrix than to a neutral one, probably as a result of attracting forces between the acrylic acid carrier and the substrate (Figure 4a). One more property, the swelling capacity of the gel in water, is summarized for some acrylic copolymers in Table 2.

The polymerization reaction can be brought about by a base-catalyzed oxidation (Mosbach, 1970; Johansson and Mosbach, 1974a), by riboflavin-sensitized photoactivation (Updike and Hicks, 1967a, 1967b), by γ irradiation (Maeda *et al.*,

TABLE 2
Swelling Capacity of Various Acrylic Copolymers Prepared in Bead Form

Acrylic copolymer (bead-formed, 200 μm ϕ) cross-linked with BIS	T/C	Swelling capacity in water (compared to lyophilized)
Acrylamide	10:5	10
Acrylamide/acrylic acid (50:50)	10:5	75
Acrylamide/HEMA (50:50)	10:5	12
Acrylamide	8:30	6
Acrylamide/acrylic acid (50:50)	8:30	31
Acrylamide/HEMA (50:50)	8:30	6.5

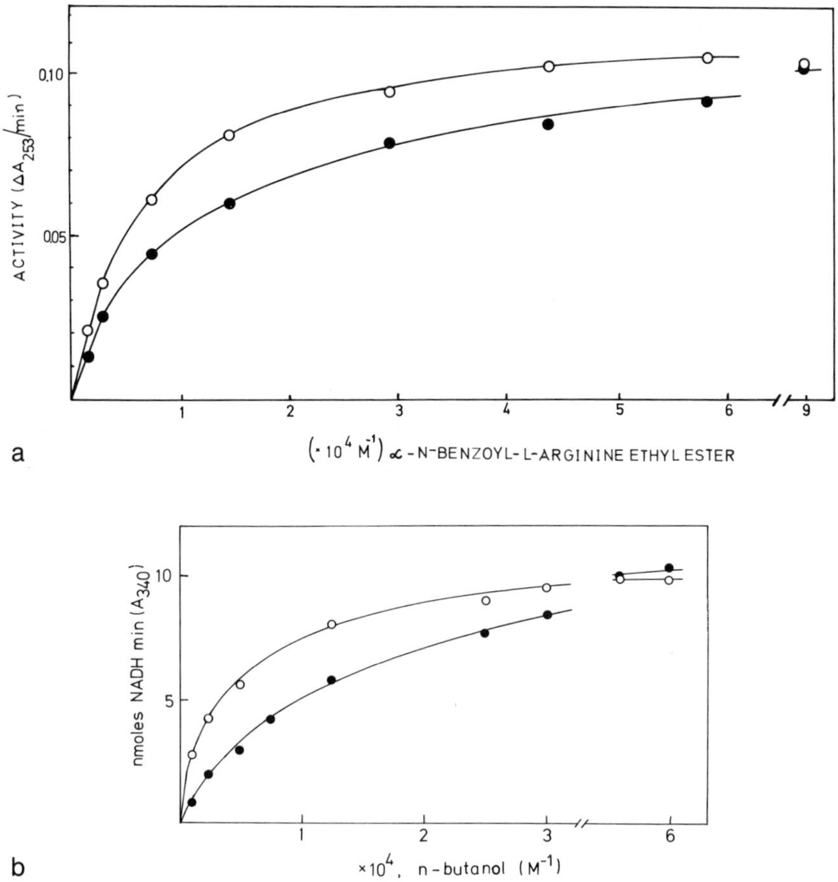

FIGURE 4. The curves shown are Michaelis–Menten plots of (a) trypsin immobilized in (-•-•-) polyacrylamide and (-○-○-) in acrylamide/acrylic acid (50:50) copolymer using benzoyl arginine ethyl ester as substrate at pH 8.1, and (b) liver alcohol dehydrogenase immobilized to polyacrylamide (-•-•-) and to acrylamide–methylacrylate (75:25) copolymer (-○-○-) using *n*-butanol as substrate at pH 8.0. The NAD^+ concentration was 2 mM.

1973, 1975), or, as presented in a recent paper (Kawashima and Umeda, 1974), by aerobic radiopolymerization of frozen (−80°C) acrylamide solutions. Oxygen normally inhibits the polymerization, and thus the process is normally performed under a N_2 atmosphere. As the polymerization reaction is exothermic, the temperature must be carefully controlled (preferably kept at 1–5°C) in order to guard against thermal denaturation of enzymes. The polymerization reaction, independent of the details of the method, takes from some minutes up to several hours, depending on the monomers, the amount of catalysts, temperature, and so on.

Gel formed by bulk polymerization is mechanically fragmentated by cutting it into small slices, which then are pressed through a net (Mosbach, 1970) or by homogenization (Degani and Miron, 1970). The method of bulk polymerization (which is most commonly used) has the disadvantage of giving irregular-shaped, fragile particles showing some enzyme leakage from the ruptured surfaces. By preparing bead-formed polymers, using a bead-polymerization process (Nilsson *et al.*, 1972; Johansson and Mosbach, 1974a), these disadvantages are reduced. An increase in mechanical strength of about 10 times was observed (Köstner and Mandel, in press), and no leakage of enzyme has been found (Johansson and Mosbach, 1974a).

The bead polymer containing entrapped enzyme is formed by dispersing an aqueous buffer (pH 7–8) containing the enzyme with any necessary protecting agents, monomers, cross-linker, and initiators in a hydrophobic medium (containing an emulsifier and having the same density as the water phase) immediately after the polymerization has been initiated. Figure 5a illustrates the apparatus used for this technique, and Figure 5b shows a less sophisticated version, preferably used when preparing smaller quantities of polymer (up to 150 ml of solution).

A method for preparing large amounts of entrapped enzyme in a continuous process has been described by Köstner (1974) and Köstner and Mandel (in press). The apparatus consists of a column which has a hydrophobic inner surface and is filled with deaerated paraffin oil. A solution of monomers and enzyme is mixed with an initiator solution and then immediately dispersed into the oil phase using a vibrating capillary tube. The vibration is transferred to the lower part of the column through an elastic connection. The bead diameter is determined by the vibration frequency, which is checked with a stroboscope. Gel beads in oil continually leave the column, while new oil and monomers are simultaneously added.

The prepared enzyme–polymer should be carefully washed to remove unreacted monomers, initiators, and any loosely trapped enzyme. This step is important, especially in food and medical applications, as the monomers have been shown to be carcinogenic. When entrapping the enzymes β-glucosidase and trypsin, we found that a washing step using the corresponding substrate solution removed weakly bound enzyme in bulk polymerized gels, probably as a result of induction of charge and structural changes in the enzyme. High salt concentrations (e.g., up to 1 M) and some organic solvents (e.g., toluene) seem to give the same effect.

a. Factors Influencing the Yield of Entrapped Enzyme and Its Retained Activity in Acrylic Copolymers. The amount of entrapped enzyme has hitherto been determined by amino acid analysis according to Spackman *et al.*, (1958). By entrapping labeled material, liquid scintillation counting can be used as an alternative. The acrylic polymers are solubilized for some hours in 30% H_2O_2 at 60°C (Young and Fulhorst, 1965) or in Soluene at 65°C for at least 24 hr (Paus, 1971) prior to addition to the scintillation liquid of dioxan and toluene, respectively.

The yield of entrapped enzyme appears to depend on at least two factors: (1) the solubility of the enzyme in the monomer mixture, and (2) the molecular weight of the enzyme. The limitation of maximum loading of an enzyme in a gel can depend on its solubility in the monomer solution (O'Driscoll, in press). However, we found that microorganisms suspended in the monomer solution were well entrapped, that the enzymes studied did not leak out, and that about 30% of the enzyme–polymer (dry weight) consisted of entrapped material in both bulk and

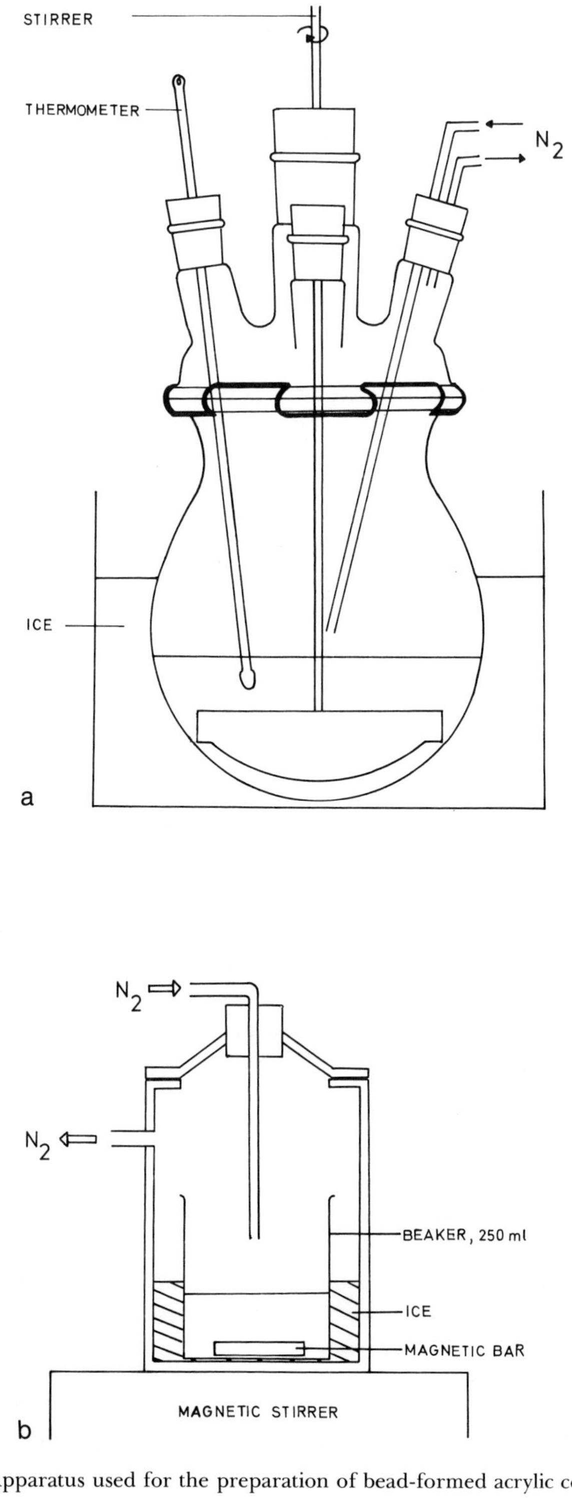

Figure 5. Apparatus used for the preparation of bead-formed acrylic copolymers.

bead polymerization. A positive correlation seems to exist between the molecular weight and the yield of the entrapped substance (Johansson and Mosbach, 1974a). This correlation was also found by Broun *et al.* (1973) when entrapping soluble oligomers of urease (intermolecular cross-linked with glutaraldehyde) in a porous acrylamide/agarose gel. The trend found is probably a result of greater leakage of small molecules than bigger ones from the surface entrapped material. It should be stressed that the yield of entrapped enzyme and its retained activity need not correlate positively; that is rather a question of to what degree the enzyme can endure the polymerization process.

As mentioned above, a transition point at $C = 5\%$ is obtained for acrylic copolymers, indicating minimal pore size (using constant total concentration) at this cross-linking degree. Degani and Miron found the highest retained activity for cholinesterase in polyacrylamide at $C = 5–7\%$, and similar results were found for glucose oxidase when trapped in poly(HEMA) cross-linked with EGDMA (O'Driscoll, in press). These results are probably a consequence of a decrease in leakage of enzyme molecules correlated with a decrease in pore size. They also found that the amount of trapped enzyme increased with the total monomer concentration. However, the retained activity had an optimum at $T = 15\%$, probably owing to two counteracting effects, the amount of entrapped material and the denaturation effects of the monomer.

To guard against leakage, Mosbach (1970) combined the entrapping technique with simultaneous covalent coupling of the enzyme to the polymer. This was performed by having a water-soluble carbodiimide present in the monomer mixture, which bound the protein to the polymer matrix via the protein amino groups and the polymer carboxylic acid groups. Thus, the enzyme was not only entrapped but also covalently bound. Later Jaworek (1974) used a comparable method, by modification of the enzyme with acrylic acid-2,3,-epoxy propylester before the polymerization was started. Recently, however, Harrison (1974) obtained data that indicate formation of a covalent bond directly between the protein and the polyacrylamide matrix, when using BIS as the cross-linker. When the enzyme glucose-6-phosphate dehydrogenase was entrapped in polyacrylamide (the average size was sufficient to allow considerable migration of the enzyme), it did not move when an electric field was applied, in contrast to the behavior of the enzyme when applied to a polyacrylamide gel, prepared in exactly the same way but without enzyme. These results are in accordance with those found by Cavins and Friedman in 1968 when they studied the effect of acrylamide and other acrylic acid derivatives on amino acids and proteins. They found that especially the sulfhydryl groups but also the amino groups were attacked, and that a covalent bond was formed between the β-carbon in the acrylic acid derivative and the sulfur or nitrogen atom, respectively.

Most components of the reaction mixture can cause some denaturation of the entrapped enzyme, thus lowering the activity which is retained during the polymerization. Thus, their concentrations have to be carefully chosen. Degani and Miron (1970) pointed out the denaturing effects of acrylamide on enzymes and draw parallels with urea denaturation. Figure 6 (left part) illustrates how the activity of trypsin disappears after addition to an acrylamide/BIS-solution ($T/C =$

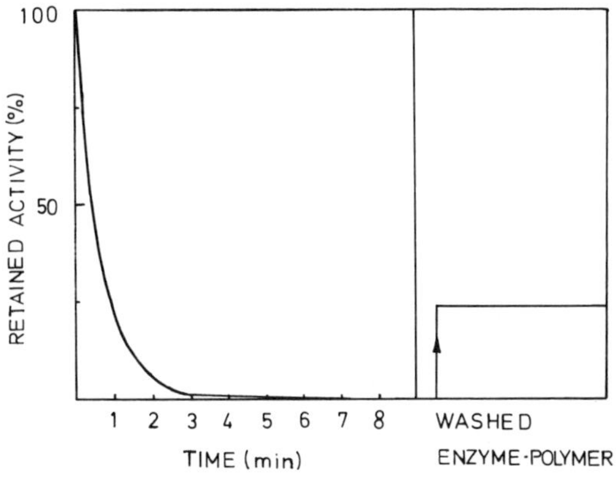

Figure 6. Retained activity of trypsin in a monomer solution of acrylamide and BIS (*T/C* = 8:5) before and after polymerization.

8:5) (Koch-Schmidt, 1976). However, this denaturation seems to have reversible character, as the enzyme activity returns when the monomers polymerize, thereby decreasing the effective acrylamide concentration (right part of Figure 6).

Bernfeld *et al.* (1968) discussed the denaturation caused by the free radicals formed during the polymerization. However, the free radical concentration is usually very low and seldom exceeds 1 μM (O'Driscoll, in press). This type of interaction could, of course, be diminished to some degree by adding other proteins as protecting agents.

A method for avoiding the use of chemical catalysts was recently developed by Kawashima and Umeda (1974) in which a frozen (−80°C) monomer solution containing enzyme, acrylamide, BIS and starch was polymerized by irradiation with 50–70 krad of γ rays. The yield of the polymerized material varied between 25 and 89%, depending on the presence of starch. These figures should be compared to the 80–100% usually obtained with the polymerization method using chemical catalysts.

2.2. *Starch Gel*

For many purposes it would be preferable to exploit naturally occurring polymers as matrices for enzyme immobilization. In 1963 Vesta and Usdin used starch for the immobilization of cholinesterase. Later Bauman *et al.* (1965) prepared entrapped cholinesterase by soaking a urethane foam pad with a warm enzyme starch solution (to get a better air and liquid flow at high pressure of the immobilized enzyme). Addition of glycerine made the pads mechanically more stable and easier to rehydrate. Starch in combination with acrylamide is often used in gel electrophoresis. This type of gel has also been used for the inclusion of enzymes, first by Broun *et al.* (1973) and later by Kawashima and Umeda (1974).

This support has a spongy structure but has the advantage of the mechanical strength imposed by the acrylamide. In order to diminish the enzyme leakage from the macroreticular gel, Broun *et al.* (1973) pre-cross-linked the enzyme molecules (urease) with glutaraldehyde before entrapping. The activity yield had an optimum (50% retained activity) at 0.25% glutaraldehyde concentration in the initial solution.

2.3. *Silicone Rubber*

Pennington *et al.* (1968b) and Brown *et al.* (1968a) entrapped enzymes in a hydrophobic matrix, silicone rubber. The enzyme was mixed with Silastic resin, poly(dimethylsiloxane), $(\text{—Si(CH}_3)_2\text{O—})_n$, and a vulcanizing agent, stannous octoate, was added. The cross-linking was complete after some hours at 5°C. About 8% of the added enzyme (glucose oxidase) remained entrapped and had a specific activity of approximately 50%. Because this polymer is hydrophobic in character, it could find possible medical use for implantation.

2.4. *Inorganic Polymers*

As early as 1955 Dickey entrapped the enzymes catalase, urease, and muscle adenylic acid deaminase in inorganic carriers of silica. Later Johnson and Whateley (1971) trapped trypsin in a silica xerogel. The enzyme solution is added to a silicic acid sol and the ionic strength is adjusted to 0.1. The gel is normally set within 0.5 hr and is allowed to age for additional 12 hr. The hydrogel is lyophilized, ground, and sieved before nonentrapped enzyme is washed off.

The specific activity of trypsin, entrapped as above, was 34% with respect to esterolytic activity, while no proteolytic activity on casein was found. The latter could depend on electrostatic repulsion (both support and substrate are negatively charged at pH 8), steric hindrance, or destroyed proteolytic activity.

2.5. *Natural Fibers of Fibrin*

One material of interest for medical purposes (e.g., enzyme replacement) is fibrin, a natural substance that does not induce antibody formation and is nontoxic. Asparaginase, which is effective in the treatment of children with leukemia, was entrapped in a fibrin polymer by mixing the enzyme with fibrinogen and thrombin in a citrate buffer. The gel particles which formed were stable in 6 M urea, a property that probably depends on small amounts of blood coagulating factor XIII, which causes cross-links between the amino groups of the enzyme and the glutaminyl residues of the fibrin molecules (Inada *et al.*, 1975).

2.6. *Collagen*

Another useful, natural polymer is collagen, a protein which constitutes the structural part of a wide variety of cells and membranes. It has a hydrophilic character and can thus be swollen to a high degree. Owing to its fibrous nature it is easy to prepare in membrane form, which can be made in various reactor configurations. Its protein structure provides a more natural microenvironment for an enzyme compared to some synthetic polymers.

The fundamental unit is a rigid rod of about 300,000 (Constantinides *et al.*, 1973). These macromolecules aggregate spontaneously at physiological pH into bundles whose ends can enter into protein–protein interactions between the enzyme and the collagen. Immobilization of enzymes in collagen is performed through the formation of a network of enzyme and collagen molecules. The stable complexes are the result of a number of different noncovalent bonds, such as salt linkages hydrogen, and van der Waals' bonds, which are acting in a cooperative way.

The collagen membranes can be prepared in different ways. In the so-called macromolecular complexation method the enzyme is added to a collagen suspension which is cast as a membrane (Wang and Vieth, 1973) and then dried, which facilitates complexation between the enzyme and the collagen.

Several enzymes, such as urease (Karube and Suzuki, 1972), glucose oxidase (Constantinides *et al.*, 1973), and uricase (Suzuki *et al.*, 1974), have been entrapped in collagen by an electrochemical method. A dialyzed enzyme solution (salt inhibits this membrane-forming process) is mixed with a collagen fibril solution and exposed to an electrode (anode for a basic electrolyte and a cathode for an acidic electrolyte), around which the enzyme–collagen complex is deposited. Using this technique, Constantinides *et al.* (1973) covered stainless steel helixes with a glucose oxidase–collagen membrane. The collagen–enzyme preparations described show a specific activity varying from 25 to 50%.

3. General Considerations Concerning Immobilization of Catalytic Material

When considering the application and economic advantages of insoluble catalysts (e.g., enzymes), one has to take the following characteristics and interrelated factors into account:

1. The physical and chemical properties of the polymeric material: mechanical stability, size and form of the gel particles, porosity, swelling capacity in different media, and resistance against microbial attack and chemicals.

2. The operational activity of the immobilized enzyme: determined by the catalytic capacity of the matrix, the diffusion resistance for substrates and products to and from the catalytic site, and the chemical qualities of the polymer, which determine the partition coefficient of the substrate and product in the matrix and the microenvironmental conditions of the enzyme molecule.

3. The half-life of an immobilized enzyme preparation: the storage stability, the operational stability, and the heat stability (i.e., the configurational stability of the enzyme as well as of the matrix at higher temperatures).

3.1. Mechanical Stability, Size, Form, and Swelling Capacity of Some Polymers

The mechanical strength and the three-dimensional stability of the polymer are two of the more significant factors determining the reactor design. Inorganic carriers appear to withstand high shear and compression in a column while they function poorly in a stirred tank, because of fragmentation. In contrast, organic material shows greater elasticity and is thus well suited to stirred tanks reactors. A combination of these two positive factors is accomplished by preparing a carrier with an inner core of inorganic material coated with an enzyme-containing polymer (Horvath, 1974). However, several acrylic copolymers, prepared in bead-form, show great mechanical stability (Turková *et al.*, 1973). Nadler and Updike (1974) have prepared polyacrylamide particles of subcellular size for enzyme correction studies by polymerizing the monomers generated as an ultrasonic mist.

When studying enzyme kinetics on β-galactosidase entrapped in HEMA-polymer (Hinberg *et al.*, 1974b) diameters less than 120 μm gave the highest reaction rate possible without any restrictions in diffusion of substrate molecules. However, as shown below, a combination of enzyme loading, turnover number, and substrate diffusion in the gel determines the optimum bead size for a defined purpose. Particles that are too small will plug a column; on the other hand, they show greater mechanical stability than larger ones.

Most frequently, polyacrylamide-entrapped enzymes have been formed by bulk polymerization, after which the gel has to be mechanically dispersed. This usually leads to particles varying widely in size and form. Compared to beads, these particles have relatively low mechanical stability and the enzyme leaks continuously from the ruptured areas. The bead polymerization method is simple, gives bead-formed gel particles with a homogeneous size distribution ($\pm$10%), and is recommended for all water-soluble monomers.

A high water content in the enzyme-polymer is usually coveted in order to create a hydrophilic microenvironment around the enzyme and to facilitate transport of substrates and products to and from the catalyst. Gels based on acrylic acid show great swelling properties, both with pH and variations in ionic strength. Generally, acrylic copolymers contain 50–90% water, depending on T and C and the chemical nature of the monomers (Table 1). Inorganic materials, such as silica gel, show only small variations in size when they are exposed to great changes in pH or ionic strength.

3.2. *Factors Influencing the Operational Activity of Gel-Entrapped Enzymes*

The enzyme loading, which can be defined as the maximal amount of active enzyme molecules per unit volume of support, is governed by a combination of factors such as the solubility of the material to be entrapped in the monomer mixture, the molecular weight of the entrapped enzyme, T and C of the gel, and the denaturation effects induced by the polymerization conditions, as discussed above. Observed effects, such as low specific activity, wide limits in pH or independence of proton concentration on enzyme activity, and increases in thermal or storage stability, are often results of overloading. The effects are due to the fact that under optimal conditions, only a small fraction which is located at or near the surface can act catalytically, since the substrate molecules are consumed before they reach the center of the gel particle. As the conditions are changed to become more unfavorable for the enzyme reaction (e.g., to a pH far from the pH optimum), the catalytic capacity of the enzyme decreases with the result that substrate molecules can diffuse further into the enzyme–gel particle, allowing more enzyme molecules to work catalytically (but less effective) than in the optimal case. Overcapacity also masks the gradual denaturation of an enzyme. If one assumes an identical denaturation pattern for free and entrapped enzyme, the overloading results in an apparent increased stability for the immobilized enzymes as a consequence of the hindered diffusion at optimal conditions. However, in applications where longevity is especially desirable, it can be achieved by overloading.

When studying the transfer of substrate molecules to the active site of an entrapped enzyme, one has to consider two barriers: the Nernst diffusion layer, which always surrounds a particle in a stirred water solution, and the restricted pore diffusion, which is determined by the pore size, the water content of the gel, the chemical nature of the matrix, and any microenvironmental effects generated during the enzymic reaction (e.g., H^+ production). The thickness of the Nernst diffusion layer, and thus the mass-transfer resistance, diminishes when the flow rate (for a packed column) or the stirring rate (for a stirred tank reactor) increases. Thus, at sufficiently high rates the only resistance to be considered is the pore diffusion inside the enzyme gel.

The pore radius is easily varied by changing the cross-linking grade, C, and/or the total concentration, T (see Figure 3). Interesting results were found in a study on the interaction of polyacrylamide-entrapped trypsin (mol. wt. 24,500) and its inhibitor, soybean trypsin inhibitor, STI (mol. wt. 21,000). The enzyme was completely inhibited when gels with T/C of 7.5:1 or 7.5:5 was used, while only around 65% was inhibited when the cross-linking grade was increased to 25%. No enzyme leakage was observed in either case. However, this is not a direct effect of the pore size (Figure 3) but rather of the gel structure. The enzyme molecules are probably more tightly bound in the fibrils ($C = 25\%$), and no complex formation is possible due to steric hindrance. This is one method of avoiding proteolytic digestion of an enzyme while still having large pores.

An interesting technique for increasing the pore size was developed by Beck and Rase in 1973. Together with the enzyme glucoamylase (mol. wt. ~ 50,000) they entrapped the high-molecular-weight substrate amylopectin (mol. wt. ~ 1,000,000). After the polymerization was finished, the gel was treated with soluble glucoamylase, which digested the amylopectin and large pores were developed in the polymer. An increase in the rate of hydrolysis of cornstarch was observed after this treatment.

The particle size has an important effect on the catalytic efficiency of an entrapped enzyme (Hinberg *et al.*, 1974b). β-Galactosidase entrapped in lightly cross-linked poly(HEMA), using a constant substrate concentration, showed identical reaction rates per gel unit for gel particles with a diameter up to 120 μm, and thereafter a decrease in rate followed increasing particle size. A simultaneous increase in K_m(app) by a factor of 1.5 was also found.

When entrapping two or more sequentially working enzymes within the same matrix, the resistance against transport of substrate 2, 3, etc., through the boundary layer is eliminated at the same time as the distance substrate–enzyme is diminished, since these molecules are retained inside the support (Srere *et al.*, 1973).

The chemical nature of the matrix affects the distribution of substrate molecules between the bulk solution and the interior of the gel particle. Partition coefficients of 1 down to 0.2 have been found. A negatively charged matrix attracts positively charged substrates, leading to higher efficiency, especially at low substrate concentrations (Figure 4a), and a hydrophobic substrate is converted more quickly by an enzyme bound in a hydrophobic matrix than by one bound in a hydrophilic one (Figure 4b).

The existence of ionic groups around the enzyme molecules provide an internally buffered milieu, so that the enzyme works in a pH that differs from that in the bulk phase. This is indicated by an apparent change in the pH optimum for the enzyme. Thus by manipulating the charges of the polymer, it is possible to create a favorable pH in the microenvironment around the enzyme while another is maintained in the outer solution.

Another approach to the regulation of activity of immobilized enzyme through pH changes was done by Gestrelius *et al.* in 1973. Glucose oxidase was entrapped in polyacrylamide beads together with either trypsin or urease. The glucose oxidase showed an increased activity in the alkaline region when trypsin (producing protons) was simultaneously working. A reversed effect was measured during urease activity (proton consumption) in that the glucose oxidase activity was stimulated in the acid region but inhibited in the alkaline (compared to the glucose oxidase working alone). The degree of influence of coentrapped enzymes on the glucose oxidase activity is, of course, dependent on the buffer capacity of the system in the microenvironment of the enzyme.

3.3. *Lifetime of an Entrapped Enzyme*

The stability of an immobilized enzyme is an important parameter for its utility. A great advantage of entrapment as an immobilization method lies in the ability to trap crude preparations of relatively unpurified enzyme or even whole cells, in which enzymes normally show greater stability than in the purified state. For the most part the operational stability during time-consuming processes is of interest. Factors influencing this parameter include inactivation due to microbial or proteolytic attack and the presence of inhibiting material and denaturating agents. By lowering the pore size of the gel, microorganisms and some proteolytic enzymes or high-molecular-weight inhibitors are effectively deprived of access to the enzyme.

In cases where high activity is less critical than long-term stability, procedures that increase the Dahmköhler number, β ($\beta = K_{\mathrm{cat}} \cdot [\mathrm{E}]_i \cdot r_0^2/D_s \cdot K'_m$, where $[\mathrm{E}]_i$ is the enzyme concentration in the gel, r_0 is the radius of the particle, and D_s is substrate diffusivity), can be recommended (e.g., increasing the enzyme loading, increasing the particle size or membrane thickness, lowering the diffusivity for the substrate by changing the pore size or the hydration grade of the polymer) (Korus and O'Driscoll, 1975).

The parameter heat stability is defined in several ways. The majority of studies on heat stability have been performed by treating the enzyme at a high temperature for a defined time, then cooling it to room temperature, whereupon the activity is assayed. However, this procedure reflects only the capability of the enzyme to regain its original active conformation and catalytic state after the heat treatment. It is necessary, of course, to use a polymer that itself is stable and does not induce conformational changes in the entrapped material. Acrylic polymers do not show any great conformational changes between 20 and 100°C as measured by differential scanning calorimetry (Koch-Schmidt, 1976).

Around 60°C there is a drastic change in the structure of most proteins. At temperatures over this so-called transition point, the enzymes will generally be in an inactive conformation. Several proteins can, however, regain their original activity when they are cooled below this temperature. By addition of stabilizing factors, such as coenzymes, substrate, buffers, serum albumin, inhibitors, etc., this transition temperature can be raised (Koch-Schmidt and Mosbach, to be published).

3.4. *Medical and Industrial Applications of Entrapped Enzymes*

Enzyme correction therapy has recently been attempted for a number of inborn errors of metabolism. The major problem is the need for continuous administration of the enzyme, which could be solved by developing techniques to stabilize the active enzyme. Wold (1973) outlined potential approaches for stabilizing enzymes which include camouflaging enzymes with a proper glucopeptide for direct transport to certain organs and immobilization (e.g., by inclusion of enzymes in a gel).

The immobilized enzyme could be directly introduced into the bloodstream, injected intraperitoneally, or packed in a shunt chamber connected to the circulatory system. Nadler and Updike (1974) entrapped the antitumor enzyme asparaginase in subcellular gel particles using labeled acrylamide in order to follow their way through the body. These particles were then injected intraperitoneally into rats. The asparaginase concentration was lowered and kept at this state for rather a long period compared to when the free enzyme was injected.

A more useful polymer for injection into the bloodstream (e.g., in treatment of enzyme-deficiency diseases) should be the natural polymer fibrin, which is both hypoallergenic and biodegradable. Asparaginase has been entrapped in such fibers and this enzyme–polymer showed great stability against 6 M urea (Inada *et al.,* 1975). However, *in vivo* studies have not yet been performed.

Another application using gel-entrapped enzymes is the preparation of certain drugs. An example, illustrated in Figure 7, is the preparation of prednisolone, used in the treatment of rheumatoid arthritis, from Compound S. Using conventional chemical methods, this transformation requires several steps, resulting in a relatively low yield and undesirable side reactions. Compound S can be transformed, in one step, to cortisol by the enzyme 11-β-hydroxylase in *Curvularia lunata* mycelium entrapped in polyacrylamide gel. In the second step, the cortisol

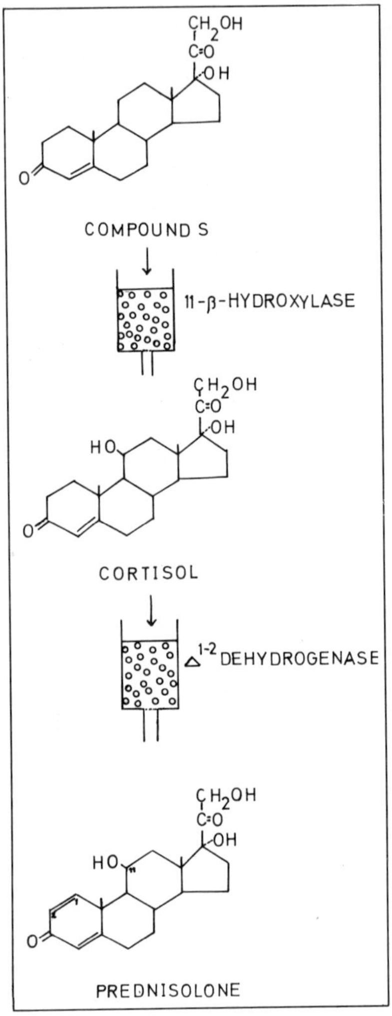

FIGURE 7. Preparation of prednisolone from Compound S by polyacrylamide entrapped enzymes or whole cells (of *Curvularia lunata* in the first step).

is transformed to prednisolone by entrapped Δ^{1-2}-dehydrogenase, which is isolated from *Corynebacterium simplex*.

Biochemical analysis is another area where gel-entrapped enzymes have a great inherent potential. Updike and Hicks (1967a) developed the enzyme electrode, in which the sensitive part was covered with a thin layer of gel-entrapped enzyme. Montalvo and Guilbault (1969) and Guilbault (1971) have applied such electrodes to the measurements of urea in body fluids. Urease was entrapped in a polyacrylamide membrane, which coated the sensitive part of an ammonium-sensitive electrode. Later, Nilsson *et al.* (1973) used common glass electrodes for assaying the glucose concentration in blood by entrapping glucose oxidase in polyacrylamide around the bulb.

Applications for the industrial use of gel-entrapped enzymes have been worked out by Mori *et al.* (1972) and Chibata *et al.* (1974). Chibata *et al.* (1974)

reported on the production of L-aspartic acid using entrapped aspartase-containing *Escherichia coli.* After the entrapment the cells were autolyzed, resulting in greater enzymic activity. They considered these entrapped cells to be advantageous in comparison with purified, entrapped aspartase for the industrial production of L-aspartic acid. In all cases they block-polymerized the gel, which, owing to the irregularity of the particles, unfortunately shows high hydraulic resistance in packed beds and fragmentation in batch processes. The method developed by Köstner *et al.* (1973) for continous production of bead-formed enzyme–polymers on an industrial scale and for industrial use can be recommended.

4. *References*

Bauman, E. K., Goodson, L. H., Guilbault, G. G., and Kramer, D. N., 1965, Preparation of immobilized cholinesterase for use in analytical chemistry, *Anal. Chem.* **37**:1378.

Beck, S. R., and Rase, H. F., 1973, Encapsulated enzyme: a glucoamylase copolymer system, *Ind. Eng. Chem. Prod. Res. Develop.* **12**:260.

Bernfeld, P., and Wan, J., 1963, Antigens and enzymes made insoluble by entrapping them into lattice of synthetic polymers, *Science* **142**:678.

Bernfeld, P., Bieber, R. E., and MacDonnell, P. C., 1968, Water-insoluble enzymes: arrangement of aldolase within an insoluble carrier, *Arch. Biochem. Biophys.* **127**:779.

Bernfeld, P., Bieber, R. E., and Watson, D. M., 1969, Kinetics of water-insoluble phosphoglycerate mutase, *Biochim. Biophys. Acta* **191**:570.

Broun, G., Thomas, D., Gellf, G., Domurado, D., Berjonneau, A. M., and Guillon, C. 1973, New methods for binding enzyme molecules into a water-insoluble matrix: properties after insolubilization, *Biotechnol. Bioeng.* **15**:359.

Brown, H. D., Patel, A. B., and Chattopadhyay, S. K., 1968a, Enzyme entrapment within hydrophobic and hydrophilic matrices, *J. Biomed. Mater. Res.* **2**:231.

Brown, H. D., Patel, A. B., and Chattopadhyay, S. K., 1968b, Lattice entrapment of glycolytic enzymes, *J. Chromatog.* **35**:103.

Cavins, J. F., and Friedman, M., 1968, Specific modification of protein sulfhydryl groups with α,β-unsaturated compounds, *J. Biol. Chem.* **243**:3357.

Chibata, I., Tosa, T., and Sato, T., 1974, Immobilized aspartase-containing microbial cells: preparation and enzymatic properties, *Appl. Microbiol.* **27**:878.

Constantinides, A., Vieth, W. R., and Fernandes, P. M., 1973, Characterization of glucose oxidase immobilized on collagen, *Mol. Cell. Biochem.* **1**:127.

Dahlqvist, A., Mattiasson, B., and Mosbach, K., 1973, Hydrolysis of β-galactosidases using polymer-entrapped lactase: a study towards producing lactose-free milk, *Biotechnol. Bioeng.* **15**:395.

Degani, Y., and Miron, 1970, Immobilization of cholinesterase in cross-linked polyacrylamide, *Biochim. Biophys. Acta* **212**:362.

Dickey, F. H., 1955, Specific adsorption, *J. Phys. Chem.* **59**:695.

v. Duijn, P., Pascoe, E., and van der Ploeg, M., 1967, Theoretical and experimental aspects of enzyme determination in a cytochemical model system of polyacrylamide films containing alkaline phosphatase, *J. Histochem. Cytochem.* **15**:631.

Fawcett, J. S., and Morris, C. J. O. R., 1966, Molecular-sieve chromatography of proteins on granulated polyacrylamide gels, *Separ. Sci.* **1**:9.

Franks, N. E., 1971, Catabolism of L-arginine by entrapped *Streptococcus faecalis* ATCC 8043, *Biochim. Biophys. Acta* **252**:246.

Gestrelius, S., Mattiasson, B., and Mosbach, K., 1973, On the regulation of the activity of immobilized enzyme: microenvironmental effects of enzyme-generated pH-changes, *Eur. J. Biochem.* **36**:89.

Guilbault, G. G., 1971, Enzyme electrode probes, *Pure Appl. Chem.* **25**:727.

Guilbault, G. G., and Das, J., 1970, Immobilization of cholinesterase and urease, *Anal. Biochem.* **33**:341.

Harrison, R. A. P., 1974, The detection of hexokinase, glucosephosphate isomerase and phosphoglucomutase activities in polyacrylamide gels after electrophoresis: a novel method using immobilized glucose-6-phosphate dehydrogenase, *Anal. Biochem.* **61**:500.

Hicks, G. P., and Updike, S. J., 1966, The preparation and characterization of lyophilized polyacrylamide enzyme gels for chemical analysis, *Anal. Chem.* **38**:726.

Hinberg, I., and O'Driscoll, K. F., 1975, Preparation and kinetic properties of gel entrapped urate oxidase, *Biotechnol. Bioeng.* **17**:1435.

Hinberg, I., Kapoulas, A., Korus, R., and O'Driscoll, K. F., 1974a, Gel entrapment of enzymes: kinetic studies of immobilized glucose oxidase, *Biotechnol. Bioeng.* **16**:159.

Hinberg, I., Korus, R., and O'Driscoll, K. F., 1974b, Gel entrapped enzymes: kinetic studies of immobilized β-galactosidase, *Biotechnol. Bioeng.* **16**:943.

Hjertén, S., 1962, "Molecular sieve" chromatography on polyacrylamide gels, prepared according to a simplified method, *Arch. Biochem. Biophys. Suppl.* **1**:147.

Horvath, C., 1974, Pellicular immobilized enzymes, *Biochim. Biophys. Acta* **358**:164.

Inada, Y., Hirose, S., Okada, M., and Mihama, H., 1975, Immobilized L-asparaginase EC 3.5.1.1 embedded in fibrin polymer, *Enzyme* **20**:188.

Jaworek, D., 1974, New immobilization techniques and supports, in: *Enzyme Engineering* (E. K. Pye, and L. B. Wingard, Jr., eds.), Vol. 2, pp. 105–113. Plenum Press, New York.

Johansson, A.-C., and Mosbach, K., 1974a, Acrylic copolymers as matrices for the immobilization of enzymes: I. Covalent binding or entrapping of various enzymes to bead-formed acrylic copolymers, *Biochim. Biophys. Acta* **370**:339.

Johansson, A.-C., and Mosbach, K., 1974b, Acrylic copolymers as matrices for the immobilization of enzymes: II. The effect of a hydrophobic microenvironment on enzyme reactions studied with alcohol dehydrogenase immobilized to different acrylic coploymers, *Biochim. Biophys. Acta* **370**:348.

Johnson, P., and Whateley, T. L., 1971, On the use of polymerizing silica gel systems for the immobilization of trypsin, *J. Colloid Interface Sci.* **37**:557.

Karube, I., and Suzuki, S., 1972, Electrochemical preparation of urease–collagen membrane, *Biochem. Biophys. Res. Commun.* **47**:51.

Kawashima, K., and Umeda, K., 1974, Immobilization of enzymes by radiopolymerization of acrylamide, *Biotechnol. Bioeng.* **16**:609.

Koch-Schmidt, A.-C., 1976, Thesis. University of Lund. Reprocentralen.

Koch-Schmidt and Mosbach, to be published.

Korus, R. A., and O'Driscoll, K. F., 1974, The effects of intraparticle diffusion on the kinetics of gel entrapped enzymes, *Can. J. Chem. Eng.* **52**:775.

Korus, R. A., and O'Driscoll, K. F., 1975, The influence of diffusion of the apparent rate of denaturation of gel entrapped enzymes, *Biotechnol. Bioeng.* **17**:441.

Köstner, A., 1974, Apparatus for Production of Spherical Microbeads. USSR Pat. 458323 (Cl.B 01j 2/06), 15.07.1974, Appl. 26.03.1973; from *Otkytiya, Izobret., Prom. Obraztsy, Tovarnye Znaki* **4,** 7 (1975).

Köstner, A., and Mandel, M., 1977, A method for continuous production of polyacrylamide-entrapped enzyme beads, in: *Methods in Enzymology* (K. Mosbach, ed.), Vol. 44, Academic Press, Inc., New York, in press.

Köstner, A., Kivisilla, K., Mandel, M., and Siimer, E., 1973, Method for Production of Matrix Bound Enzyme, USSR. Pat. 414301 (C1.C. 12d 13/10), 29.05.1973, Appl. 30.10.1971; from *Otkrytiya, Izobret., Prom. Obraztsy, Tovarnye Znaki* **5,** 92 (1974).

Maeda, H., Suzuki, H., and Yamauchi, A., 1973, Preparation of immobilized enzymes by electron-beam irradiation, *Biotechnol. Bioeng.* **15**:827.

Maeda, H., Suzuki, H., Yamauchi, A., and Sakimae, A., 1975, Preparation of immobilized enzymes from acrylic monomers under γ-ray irradiation, *Biotechnol. Bioeng.* **17**:119.

Montalvo, J., Jr., and Guilbault, G. G., 1969, Sensitized cation selective electrode, *Anal. Chem.* **41**:1897.

Mori, T., Sato, T., Tetsuya, T., and Chibata, I., 1972, Studies on immobilized enzymes: X. Preparation and properties of aminoacylase entrapped into acrylamide gel-lattice, *Enzymologia* **43**:213.

Mosbach, K., 1970, Matrix-bound enzymes: I. The use of different acrylic copolymers as matrices, *Acta Chem. Scand.* **24**:2082.

Mosbach, K., and Larsson, P.-O., 1970, Preparation and application of polymer-entrapped enzymes and microorganisms in microbial transformation processes with special reference to steroid 11-β-hydroxylation and Δ^{1-2}-dehydrogenation, *Biotechnol. Bioeng.* **12**:19.
Mosbach, K., and Mattiasson, B., 1970, Matrix-bound enzymes: II. Studies on a matrix-bound two-enzyme system, *Acta Chem. Scand.* **24**:2093.
Mosbach, K., and Mosbach, R., 1966, Entrapment of enzymes and microorganisms in synthetic cross-linked polymers and their application in column techniques, *Acta Chem. Scand.* **20**:2807.
Nadler, H. L., and Updike, S. J., 1974, Gel entrapment of enzymes, *Enzyme* **18**:150.
Nilsson, H., Mosbach, K., and Mosbach, R., 1972, The use of bead polymerization of acrylic monomers for immobilization of enzymes, *Biochim. Biophys. Acta* **268**:253.
Nilsson, H., Åkerlund, A.-C., and Mosbach, K., 1973, Determination of glucose, urea and penicillin using enzyme-pH-electrodes, *Biochim. Biophys. Acta* **320**:529.
O'Driscoll, K. F., Izu, M., and Korus, R., 1972, Gel-entrapment of enzymes, *Biotechnol. Bioeng.* **14**:847.
O'Driscoll, K. F., 1977, Techniques of enzyme entrapment in gels, in: *Methods in Enzymology* (Mosbach, K., ed.), Academic Press, Inc., New York, Vol. 44, in press.
Ohno, Y., and Stahmann, M. A., 1971, Polyacrylamide derivatives of amino acid acylase and trypsin, *Macromolecules* **4**:350.
Paus, P. N., 1971, Solubilization of polyacrylamide gels for liquid scintillation counting, *Anal. Biochem.* **42**:372.
Pennington, S. N., Brown, H. D., Patel, A. B., and Knowles, C. O., 1968a, Properties of matrix supported acetylcholinesterase, *Biochim. Biophys. Acta* **167**:479.
Pennington, S. N., Brown, H. D., Patel, A. B., and Chattopadhyay, S. K., 1968b, Silastic entrapment of glucose oxidase-peroxidase and acetylcholin esterase, *J. Biomed. Mater. Res.* **2**:443.
Richards, E. G., and Temple, C. J., 1971, Some properties of polyacrylamide gels, *Nature Phys. Sci.* **230**:92.
Spackman, D. H., Stein, W. H., and Moore, S., 1958, Automatic recording apparatus for use in the chromatography of amino acids, *Anal. Chem.* **30**:1190.
Srere, P. A., Mattiasson, B., and Mosbach, K., 1973, An immobilized three-enzyme system: a model for microenvironmental compartmentation in mitochondria, *Proc. Natl. Acad. Sci. U.S.* **70**:2534.
Suzuki, S., Sonobe, N., Karube, I., and Aizawa, M., 1974, Electrochemical preparation of uricase–collagen membrane, *Chem. Letters* **1**:9.
Turková, J., Hubálková, O., Křiváková, M., and Čoupek, J., 1973, Affinity chromatography on hydroxyalkyl methacrylate gels: I. Preparation of immobilized chymotrypsin and its use in the isolation of proteolytic inhibitors, *Biochim. Biophys. Acta* **322**:1.
Updike, S. J., and Hicks, G. P., 1967a, The enzyme electrode, *Nature* **214**:986.
Updike, S. J., and Hicks, G. P., 1967b, Reagentless substrate analysis with immobilized enzymes, *Science* **158**:270.
Vesta, B., and Usdin, V., 1963, Melpar Inc., Falls Church, Va., Final Report, Contract DA 18-108-405-CML-828, Section 3.3.4, p. 3.102 (Okt 1963).
Wang, S. S., and Vieth, W. R., 1973, Collagen–enzyme complex membranes and their performance in biocatalytic modules, *Biotechnol. Bioeng.* **15**:93.
Wieland, T., Determann, H., and Bünnig, K., 1966, Über unlösliche, in polyacrylamidegel fixierte enzyme, *Z. Naturforsch.* **21**:1003.
Wold, F., 1973, Chemical modification of enzymes, in: *Enzyme Therapy in Genetic Diseases, Birth Defects* (Original Article Series, Vol. 9), (R. J. Desnick, R. W. Bernlohr, and W. Krivit, eds.) p. 46, The Williams & Wilkins Company, Baltimore.
Young, R. W., and Fulhorst, H. W., 1965, Recovery of S^{35} radioactivity from protein-bearing polyacrylamide gel, *Anal. Biochem.* **11**:389.

7

Encapsulation of Enzymes, Cell Contents, Cells, Vaccines, Antigens, Antiserum, Cofactors, Hormones, and Proteins

T. M. S. CHANG

1. *Principles*

Unlike the other three major methods of enzyme immobilization, which emphasize the "microenvironment," encapsulation of enzymes emphasizes the "intracellular environment" of enzymes and proteins (Figure 1). Here, the enzyme solution or suspension is encapsulated or enveloped within a membrane system in such a way that the membrane creates an intracellular environment for the enzymes, preventing them from leaking out or coming into direct contact with the external environment. Large molecules such as proteins and cells cannot cross the membrane to interact with the enclosed enzymes. Substrates that are permeable can equilibrate rapidly across the membrane to be acted on by the enzymes inside, and the product can diffuse out. Unlike the case of gel entrapment, which involves the entrapping of individual molecules of enzymes in polymer lattices, in encapsulation, any concentration, any volume, and any amount of enzymes can be enclosed within membrane envelopes of different configurations. This principle of allowing any type or concentration of enzymes, cells, or cell extracts to be encapsulated within membrane envelopes allows for an extremely large variation in the membrane composition, configuration, and content. This chapter is a brief review of the general principles and methods of preparation for encapsulation of enzymes and proteins, with emphasis on artificial cells. Only a brief review is made of other forms of encapsulation since they will be dealt with in detail in later chapters.

T. M. S. CHANG • Departments of Physiology and Medicine, McGill University, Montreal, Canada

2. *Artificial Cells*

2.1. *Introduction*

Enzymes and proteins were originally encapsulated in the form of artificial cells (Figure 1) (Chang, 1957, 1964, 1972a; Chang *et al.,* 1963, 1966). Each artificial cell consists of a spherical ultrathin semipermeable membrane of cellular dimensions (Figures 2 and 3). Each artificial cell envelops a solution or suspension of enzymes, cell extracts, cells, or other materials. The enveloping semipermeable membrane separates the contents of the envelope from the external environment. The ultrathin membrane, of about 200 Å thickness, and the very large surface/volume relationship (2.5 cm^2 in 10 ml of 20-μm-diameter microcapsules) allows for an extremely rapid equilibration of substrate, 400 times faster than that of a standard complete dialysis machine used for treatment of patients with chronic renal failure (Chang, 1966, 1972a, 1972b, 1974). In this artificial cell system, the semipermeable membrane with equivalent pore diameters of about 36 Å (Chang,

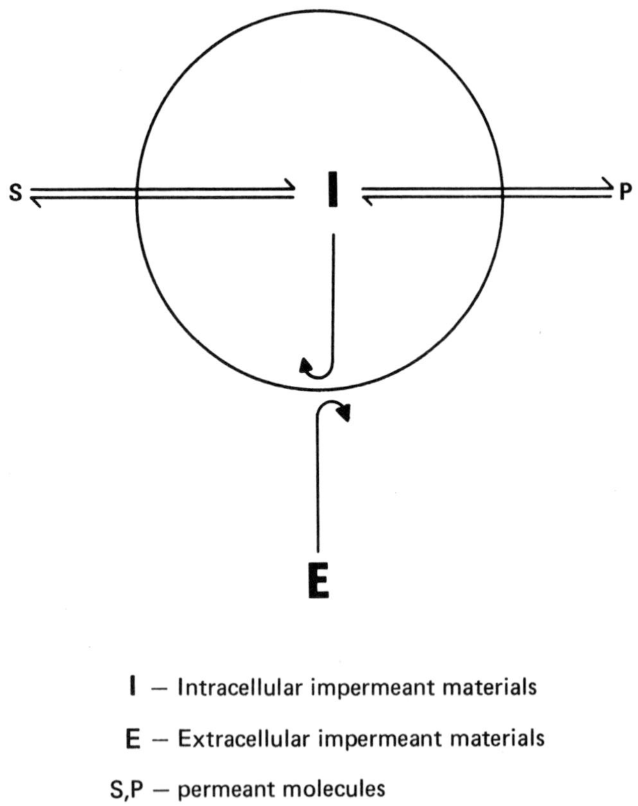

FIGURE 1. Principle of artificial cells. (From Chang, 1972a.)

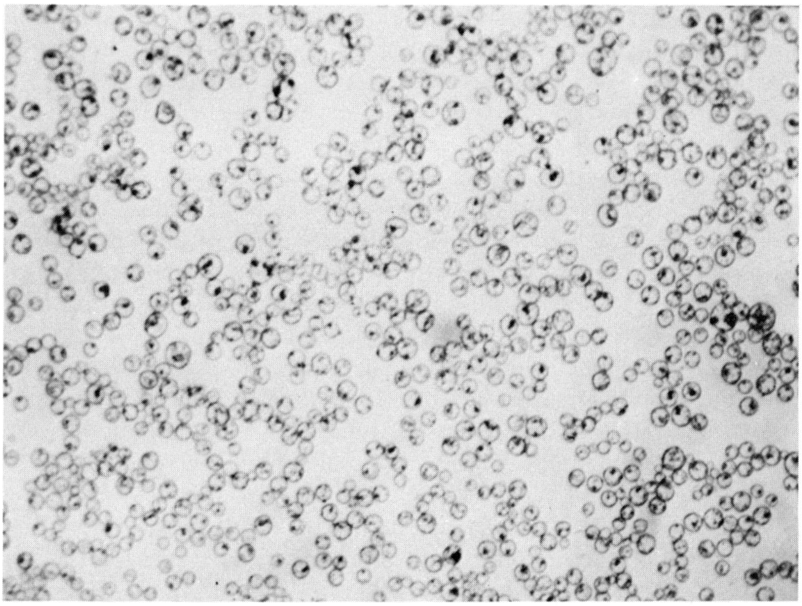

FIGURE 2. Cellulose nitrate membrane artificial cells containing multienzyme systems and proteins with a mean diameter of 5μm. (From Chang, 1972a.)

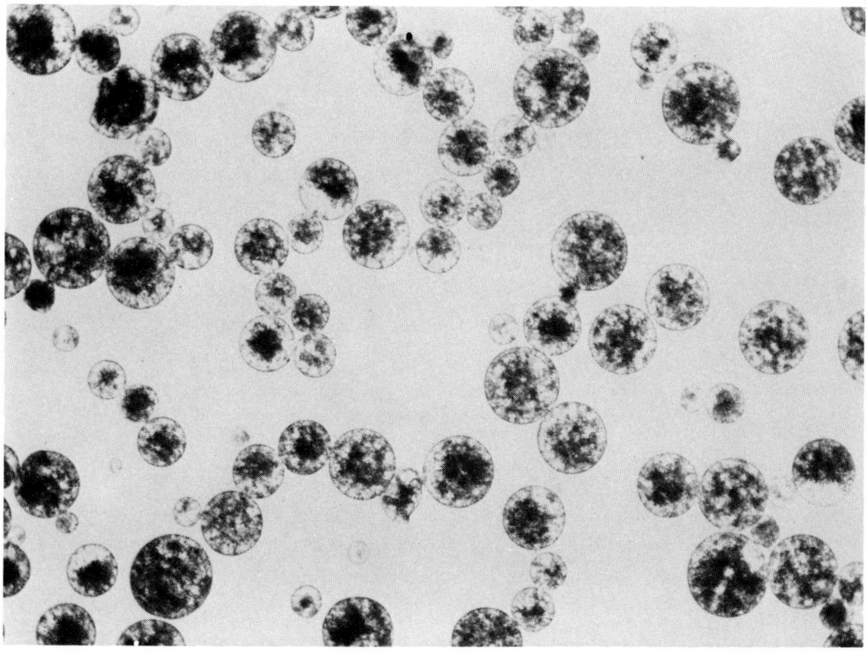

FIGURE 3. Cellulose nitrate membrane artificial cells containing multienzyme systems and proteins with a mean diameter of 80 μm. (From Chang, 1972a.)

1965, 1972a; Chang and Poznansky, 1968) prevents external proteins, antibodies, or cells from entering, but permeant substrates can equilibrate rapidly to come in contact with the enclosed enzymes or proteins. Furthermore, it is very easy to enclose any combination of enzymes, multienzyme systems, cofactors regenerating enzyme systems, cell extracts, whole cells, proteins, adsorbents, magnetic materials, multicompartmental systems, and so on (Figures 4 and 5). Furthermore, the amount and concentration of enzymes encapsulated is unlimited, since the concentration of enzyme depends only on the solubility of the particular enzyme system. Also, with an ultrathin membrane of 200 Å, the polymer membrane/enzyme ratio could be 1:100,000. The conditions fulfilled by encapsulation of enzymes within artificial cells are summarized in Table 1.

The basic principle, the method of preparation, and the biophysical and biochemical properties of artificial cells have already been described in detail elsewhere (Chang, 1972a). However, for the sake of discussion, two simplified examples of the basic methods of preparation of artificial cells by microencapsulation are described here. These methods are reproduced exactly from the book *Artificial Cells* (Chang, 1972a). Further details of the preparation, modification, theory, and extension can be found in the original source (Chang, 1972a), but will be summarized and updated here.

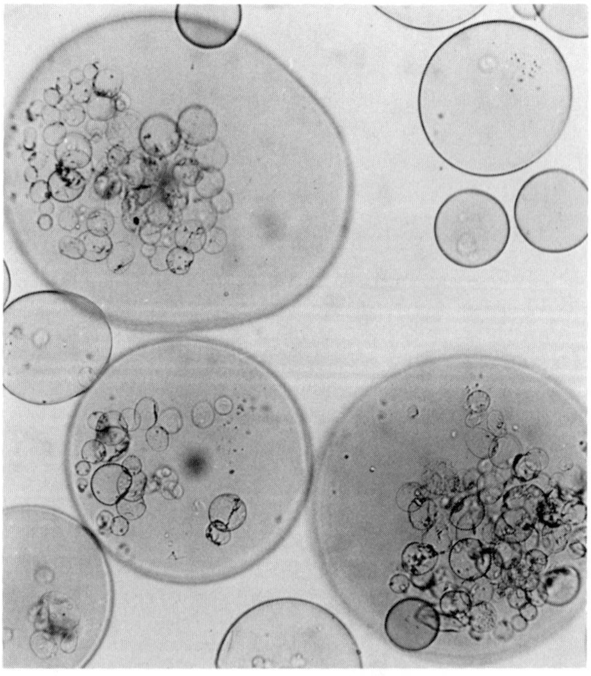

FIGURE 4. Larger microcapsules, each containing many smaller microcapsules with multienzyme systems and proteins. (From Chang, 1965.)

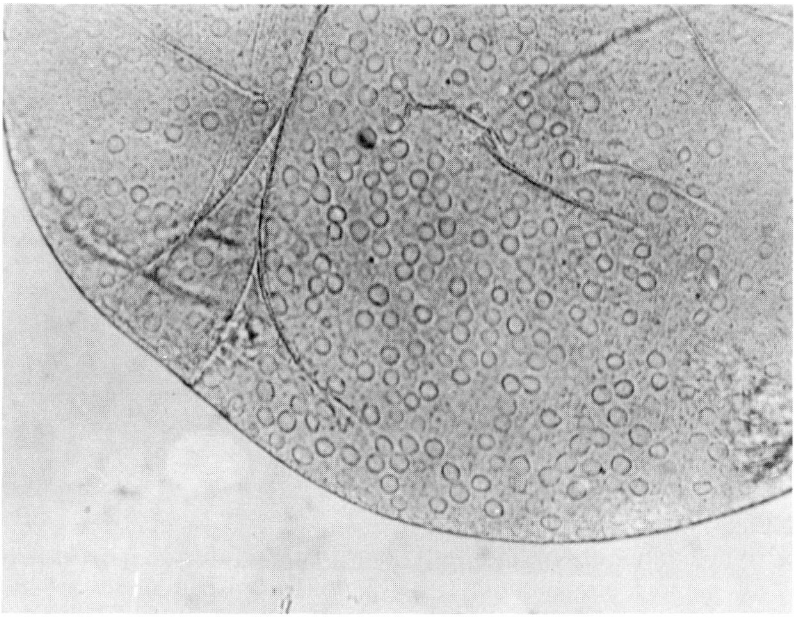

FIGURE 5. Microencapsulation of cells inside cross-linked protein membranes. (From Chang, 1965.)

2.2. *First Basic Method*

The first basic method will give a very high recovery of enzyme activity and is applicable to all enzymes tested so far. The second basic method, which will be described later, is not as good since some enzymes are inactivated by the chemical reaction.

a. Principle. The first method to be described is a simple physical method involving no chemical reactions (Chang, 1957, 1964, 1972a; Chang *et al.,* 1963, 1966). With this method, all enzymes tested have been found to retain their activity. In this procedure, a high concentration of hemoglobin from red blood cells is used to provide an intracellular environment. Enzymes (single, multiple, pure, and crude extracts) are dissolved or suspended in the hemoglobin solution, then enveloped within microscopic, spherical ultrathin membranes by the following procedure. The exact steps, timing, and hemoglobin contents are very important.

b. Material.

1. Hemoglobin solution: 1 g of hemoglobin (hemoglobin substrates, Worthington Company) dissolved in 10 ml of distilled water, then filtered through Whatman No. 42 filter paper. In place of hemoglobin substrate, fresh red blood cell hemolysate can also be used as long as the final concentration of hemoglobin is exactly 10 g%.

TABLE 1
Enzyme Stability as Percentage of Initial Activity

	Catalase		Asparaginase	
	4°C	37°C	4°C	37°C
Free solution	15 days: 50% 35 days: 10%	1 day: 50% 2 days: 10%	20 days: 50% 60 days: 10%	1 day: 50% 2 days: 10%
Microencapsulated with 10 g% hemoglobin	100 days: >90%	1.3 days: 50% 5 days: 10%	100 days: >90%	5 days: 50% 25 days: 10%
Microencapsulated with 10 g% hemoglobin, then treated with glutaraldehyde	100 days: >90%	6 days: 50% 11 days: 10%	100 days: >90%	10 days: 50% 100 days: 25%

2. Enzyme solution: enzymes (single, multiple, insolubilized, cell extract, cells, or combinations of enzymes with other material) dissolved or suspended in the 10 g% hemoglobin solution. If a solution or suspension is added to the hemoglobin solution, suitable adjustment is required to ensure a final concentration of 10 g% hemoglobin. Also, if an acidic enzyme preparation is added to the hemoglobin solution, a minimal pH of 8.5 should be maintained by sufficient Tris buffer.

3. Organic solution: 100 ml of ether (analytical grade) saturated with water by shaking with distilled water in a separating funnel, then discarding the water layer. One milliliter of a water–oil emulsifying agent, Span-85 (Atlas Powder Company, Canada Ltd., Montreal, Quebec, Canada), is added to 100 ml of water-saturated ether just before use.

4. Cellulose nitrate solution: prepared by evaporating collodion (USP, 4 g of cellulose nitrate in a 100-ml mixture of 1 part alcohol and 3 parts ether) to exactly 20% of its original weight, then made up to its original volume by adding ether (analytical grade), and redissolved.

5. *n*-Butyl benzoate–Span-85 solution: 100 ml of *n*-butyl benzoate (Eastman) with 1 ml of Span-85 added just before the procedure.

6. Tween-20 solution: 50% Tween-20 solution prepared by dissolving 50 ml of Tween-20, a water–oil emulsifying agent (Atlas Powder Company) in an equal volume of distilled water. A 1% Tween-20 solution is prepared by dissolving 1 ml of Tween-20 in 99 ml of distilled water.

7. Magnetic stirrer: important to have a magnetic stirrer with sufficient power to give the speed (rpm) stated, especially when stirring the very viscous Tween-20 mixture. The Jumbo Magnetic Stirrer (Fisher Scientific Company, Montreal, Quebec, Canada) is used in the procedures described below. Also needed is a 4-cm magnetic stirring bar and a 150-ml glass beaker with an internal diameter of less than 6 cm.

c. Procedure.

1. To a 150-ml glass beaker containing 2.5 ml of hemoglobin solution, 25 ml of the organic solution is added. The mixture is immediately stirred with the magnetic stirrer at a speed setting of 7 (2600 rpm).

2. After stirring for 5 sec, 25 ml of the cellulose nitrate solution is added and stirring is continued for another 60 sec.

3. The beaker is then covered and allowed to stand unstirred at 4°C for 45 min.

4. If the microcapsules have been completely sedimented by the end of 45 min, all but about 4 ml of the supernatant can be conveniently removed. Otherwise, the suspension will have to be centrifuged at 350*g* for 5 min before the supernatant can be removed. Centrifugation at greater speed or longer duration may adversely affect the formation of the microcapsules.

5. Immediately after the removal of most of the supernatant, 30 ml of the *n*-butyl benzoate–Span-85 solution is added to the beaker and stirred with the suspension on the Jumbo Magnetic Stirrer at speed 5 (1200 rpm) for 30 sec. The beaker is then allowed to stand uncovered and unstirred at 4°C for 30 min.

6. The final step is to transfer the microencapsulated enzyme from the organic liquid phase into an aqueous phase. First, butylbenzoate supernatant should be removed completely. In the case where microcapsules of larger than 50-μm diameter are prepared, the microcapsules would have sedimented completely by 30 min. If the microcapsules have not sedimented completely, centrifugation at 350*g* for 5 min is required.

7. After the removal of supernatant, 25 ml of the 50% Tween-20 solution is added. The suspension is dispersed by stirring with the Jumbo Magnetic Stirrer at a speed setting of 10 for 30 sec. It is important to maintain a speed of 2900 rpm in the presence of the 50% Tween-20 solution. Stirring is then slowed down to speed 5 (1200 rpm) and 25 ml of water is added. After a further 30 sec of stirring, the suspension is further diluted with 200 ml of water. The slightly turbid supernatant may be removed by centrifugation of the suspension at 350*g* for 5 min.

8. The microencapsulated enzymes so obtained are washed repeatedly in 1% Tween-20 solution until no further leakage of hemoglobin takes place and the smell of butyl benzoate is no longer detectable. The removal of *n*-butyl benzoate with 1% Tween-20 solution is very important since unless completely removed, butyl benzoate forms a coating on the microcapsule membrane and may affect the permeability characteristics of the semipermeable microcapsules. Leaving butyl benzoate on the membrane is one way of decreasing the permeability of the membrane. The final preparation is suspended in 0.9 g% sodium chloride solution.

d. Variation. Variation of the diameter of the microcapsules depends on the degree of emulsification (Chang, 1964, 1972a; Chang *et al.,* 1966). For the preparation of microcapsules of less than 10-μm mean diameter, a Virtis homogenizer is required for emulsification. The 10 g% hemoglobin, besides its necessity in the microencapsulation procedure, also stabilizes the microencapsulated enzymes. This is the basic procedure in using cellulose nitrate. Extensions and modifications of other types of membranes may require extensive modifications of the procedure (Chang, 1972a). Secondary emulsion is one example of a simple extension of the first basic method.

2.3. *Second Basic Method*

a. Principles. This is an example of the preparation of microencapsulated enzymes by a chemical method (Chang, 1964, 1972a; Chang *et al.,* 1963, 1966). This method is an extension of the basic method described above. The semipermeable microcapsule membrane is formed from nylon (Poly 610). Although their appearance looks much better than the cellulose nitrate membrane microcapsule, their membranes are not as stable; furthermore, some enzymes, such as catalase, are inactivated by this procedure.

Two examples are given below: one is a drop technique, the other, the emulsification technique. The drop technique is extremely simple and can be used to prepare large microcapsules containing enzymes without the necessity of using the 10 g% hemoglobin solution. However, if a large number of semipermeable microcapsules are required, the drop technique is more time consuming and the diameter is usually larger than 1 mm. The second procedure using emulsification would prepare small semipermeable microcapsules of the order of 50 μm in mean diameter. In the emulsification procedure, a 10 g% hemoglobin solution is required.

b. Material.

1. Hemoglobin solution: 1 g of hemoglobin (hemoglobin substrate, Worthington Company) is dissolved in 10 ml of distilled water. It is then filtered through Whatman No. 42 filter paper. In place of this, fresh red blood cell hemolysate containing 10 g% can be used.

2. Enzyme solution: only enzymes (e.g., urease, asparaginase) which are stable in a high pH condition of the diamine solution can be used with this method. Enzymes (single, multiple, insolubilized, cell extracts, cells, or combinations of enzymes with other material) are either dissolved or suspended in the 10 g% hemoglobin solution. The final solution has a concentration of 10 g% hemoglobin. In the drop method, the enzyme solution can be added directly to the alkaline diamine solution.

3. Alkaline diamine solution: 4.4 g of 1,6-hexamethylenediamine (Eastman), 1.6 g of $NaHCO_3$, and 6.6 g of Na_2CO_3 are dissolved as a 100-ml aqueous solution. This stock solution is filtered and then stored and refrigerated at 4°C but should be discarded after 24 hr.

4. Stock organic–Span-85 liquid: prepared just before use, made up of 40 ml of chloroform and 160 ml of cyclohexane. Span-85 is not required for the drop

technique, but in the emulsification procedure, 2 ml of Span-85 is added to 200 ml of the stock organic liquid.

5. Sebacyl chloride liquid (0.018 M): prepared immediately (within 2 min) before use by adding 0.4 ml of sebacyl chloride (Eastman) to each 100 ml of the stock organic liquid. Failure to prepare satisfactory microcapsules is frequently due to the deterioration of sebacyl chloride after the seal on the bottle has been opened. A new bottle of sebacyl chloride will usually solve the problem.

6. 50% Tween-20 solution: made up by dissolving 50 ml of Tween-20 in 50 ml of distilled water. This is not required for the drop method.

7. Magnetic stirrer, 4-cm magnetic stirring bar, and a 150-ml glass beaker as described for the procedure above.

c. Procedure of Drop Technique (Chang, 1965, 1972a; Chang *et al.,* 1966). First, 150 ml of the sebacyl chloride organic liquid (0.018 M) is prepared as described, but Span-85 is omitted from the organic solution. This sebacyl chloride organic liquid is placed in a 14-cm-diameter flat-bottomed glass petri dish. In this case, enzymes or other material to be microencapsulated are added to 3 ml of the alkaline diamine solution if no hemoglobin is required. If hemoglobin is required, the enzyme could be added to the mixture of 1.5 ml of alkaline diamine and 1.5 ml of hemoglobin. The 3 ml of the aqueous phase containing enzymes is placed in a glass syringe and added dropwise through its attached steel needle placed 1 cm above the surface of the sebacyl chloride solution. It is important to use a glass syringe and a steel needle with no polymer components so as not to be affected by the organic solution. The diameter of the needle would decide the diameter of the droplets and thus the diameter of the microcapsules. The dropwise addition could be done manually or by using a perfusion pump for continuous drop formation. The drop has to be added throughout the surface of the container with no droplets overlapping. After the addition of the droplets, they settle to the bottom of the container and the reaction is allowed to take place for 5 min with gentle shaking. At the end of 5 min, the semipermeable microcapsules are formed. The organic solution can be decanted and any trace amount allowed to evaporate completely before adding saline (0.9 g of NaCl in 100 ml of aqueous solution) to suspend the droplets. It is important to allow the organic solution to evaporate completely before adding this solution; otherwise, the permeability of the membrane may be affected by the organic solution. This is an extremely easy method for forming semipermeable microcapsules, especially as a preliminary step toward the emulsification procedure, which is more complex.

d. Procedure of Emulsification.

1. First, 2.5 ml of the hemoglobin solution containing the enzyme is added to 2.5 ml of the alkaline diamine solution in a 150-ml beaker placed in an ice bath. The two solutions are gently mixed.

2. Within 10 sec after mixing the two solutions, 25 ml of the stock organic–Span-85 liquid is added and then stirred at speed 5 (1200 rpm) on the Jumbo Magnetic Stirrer for 60 sec.

3. While stirring is continued at the same speed, 25 ml of the sebacyl chloride organic–Span-85 liquid is added and stirring is continued for 3 min.

4. Exactly 3 min later, the reaction is quenched by the addition of 50 ml of stock organic–Span-85 liquid; stirring is discontinued and the suspension is centrifuged for 30 sec at 350*g*.

5. All the supernatant is discarded and 50 ml of the 50% Tween-20 solution is added. The suspension is stirred at the Jumbo Magnetic Stirrer speed setting of 10 (2900 rpm) for 30 sec. The speed is then decreased to 5 (1200 rpm), 50 ml of distilled water is added, and stirring is continued for another 30 sec. The suspension is then added to 200 ml of distilled water, and stirring is continued at the same speed for another 30 sec. This suspension is allowed to sediment or centrifuge for 1 min at 350*g* and all the supernatant is discarded. Alternatively, the suspension may be sieved in a mesh screen to remove the supernatant. The microcapsules so prepared are suspended in 0.9 g% sodium chloride solution (Figure 3). Initially, the microcapsules may be "crenated" after being in contact with the hypertonic Tween-20 solution, but they regain sphericity if contact with the strong Tween-20 medium has been kept at a minimum.

e. Variation. Extensive variation is possible in diameter, membrane thickness, content, and innumerable membrane materials (Chang, 1972a).

3. *Enzyme Kinetics and Stability*

3.1. *Stabilization of Microencapsulated Enzymes*

Microencapsulated enzymes in artificial cells are stabilized by microencapsulation with a 10 g% hemoglobin solution. In this way, unlike native enzymes in a dilute solution, the microencapsulated enzyme is at all times retained in an intracellular environment with a high concentration of intracellular protein. Insolubilized enzymes in the form of covalent binding, adsorption, or gel-matrix entrapment could also be microencapsulated in suspension (Chang, 1969a). An easier way to stabilize the enzyme is to directly cross-link the microencapsulated enzyme solution (Chang, 1971). For example, after the 10 g% hemoglobin solution containing catalase, asparaginase, or lactase is microencapsulated within cellulose nitrate membranes, enzyme microencapsulated together with a high concentration of protein is treated with glutaraldehyde by the method as described (Chang, 1971). Here, 10 ml of a solution (0.1 M sodium metaborate and 0.001 M benzamidine HCl) was added to 20 ml of microencapsulated enzyme. Next, 200 μl of 50% glutaraldehyde was added to the suspension, which was kept slightly agitated at 20°C for 1 hr. Following this, 100 ml of 0.05 M sodium borohydride was added to the suspension, which was left at 4°C for 20 min. The supernatant was then removed, and the microcapsules washed twice with 200 ml of 0.9 g% sodium chloride solution. Variation in buffers and reaction conditions may be required for other enzymes.

When asparaginase or catalase are stored at 4°C as a dilute solution, only 50%

of the activity is left after 15–20 days. However, when these enzymes are microencapsulated with a high concentration of protein with or without further cross-linking and stored at 4°C, more than 90% of the original activity is retained after more than 100 days (Chang, 1971, 1973) (Table 1). At a body temperature of 37°C, the microencapsulated enzymes were more stable than the enzyme in free solution (Chang, 1971, 1973) (Table 1). The use of immobilizing enzymes with a high concentration of protein or of cross-linking enzymes with other proteins to stabilize them can be extended to other immobilized enzyme systems.

3.2. *Permeability*

Permeability and transport characteristics of semipermeable microcapsules have been studied in detail (Chang and Poznansky, 1968; Chang, 1972a, 1972b, 1974). The kinetics of microencapsulated enzymes can be found in the literature (Chang, 1964, 1965, 1972a; Boguslaski and Janik, 1971; Mori *et al.,* 1973; Sundaram, 1973; Choi and Fan, 1973; Mogensen and Vieth, 1973). A simple example is shown in Figure 6 (Chang, 1965, 1972a). After urease is microencapsulated with hemoglobin, the activity of 0.25 ml of microcapsules is analyzed. There was no leakage of enzyme, and the reaction was due to the diffusion of urea into the microcapsules, where it is converted to ammonia, which diffuses out. If the microcapsules are homogenized, they contain more than 90% of the urease activity encapsulated originally. No significant activity was detected in the washed microcapsule membrane.

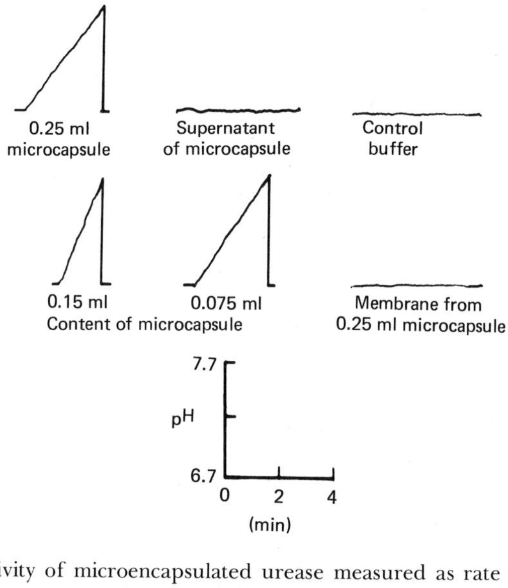

FIGURE 6. Urease activity of microencapsulated urease measured as rate of rise of pH of urea–buffer medium. No measurable enzyme activity is present in the supernatant or washed microcapsule membrane. Enzyme activity is located in the membrane-free homogenate (content of microcapsules). (From Chang, 1972a.)

4. *Variation in Membrane Materials and Configurations Used in the Encapsulation of Enzymes and Proteins*

As emphasized earlier (Chang, 1972a), the basic method used for the preparation of artificial cells are, in fact, physical examples for demonstrating the principle of artificial cells and encapsulation of enzymes. There are many other physical systems that could also be used to demonstrate the same principle. This has now been demonstrated by the increasing numbers of physical model systems both in membrane material and configuration. It would not be possible to describe all these systems in detail. Fortunately, many of the pioneers of these systems are contributing to this book. Their chapters will discuss their approaches in detail. Therefore, only a brief review will be made here to summarize the membrane materials and configurations that are available at present, and references will only be made to the first publication of each system. The different membrane materials and configurations used for encapsulation of enzymes and proteins are summarized in Figure 7. For discussion, one might divide these different systems into two subsections: (1) synthetic membrane systems, and (2) biological or biodegradable systems.

4.1. *Synthetic Membrane Systems (Figure 7)*

a. Spherical Ultrathin Polymer Membranes. The original artificial cells prepared were from cellulose nitrate in the form of a spherical ultrathin membrane of cellular dimensions using a physical method of emulsification followed by interfacial precipitation of the cellulose nitrate on the emulsions of aqueous droplets as described in detail under the first basic method (Figures 2 and 3). Other polymers have also been dissolved in organic solution and then formed into spherical ultrathin membranes by the same procedure (Chang, 1972a). Another way for forming these membranes is the use of interfacial polymerization, as described under the second basic method (e.g., nylon membranes). This approach could also be used to form a combined nylon–protein membrane, or a pure cross-linked protein membrane system; furthermore, different surface properties are possible by variation in membrane materials (Chang, 1964, 1965, 1972a). Having demonstrated the feasibility of preparing spherical ultrathin membranes with interfacial polymerization, the numerous chemical reactions available for interfacial polymerization have been adapted for the same end (Shiba *et al.*, 1970; Mori *et al.*, 1973).

b. Multiple-Compartmental Membrane Systems. A spherical ultrathin membrane system could also be made to contain smaller microencapsulated enzyme systems to compartmentalize different types of microencapsulated enzyme systems for sequential or other special types of reactions. Thus, larger microcapsules made from nylon membrane or protein cross-linked membrane have been used to microencapsulate smaller microcapsules, each containing enzymes and proteins (Figure 4) (Chang, 1965, 1972a). This would have interesting applications, especially in the sequential action of multienzyme systems or in separating different enzyme systems into "intracellular" compartments.

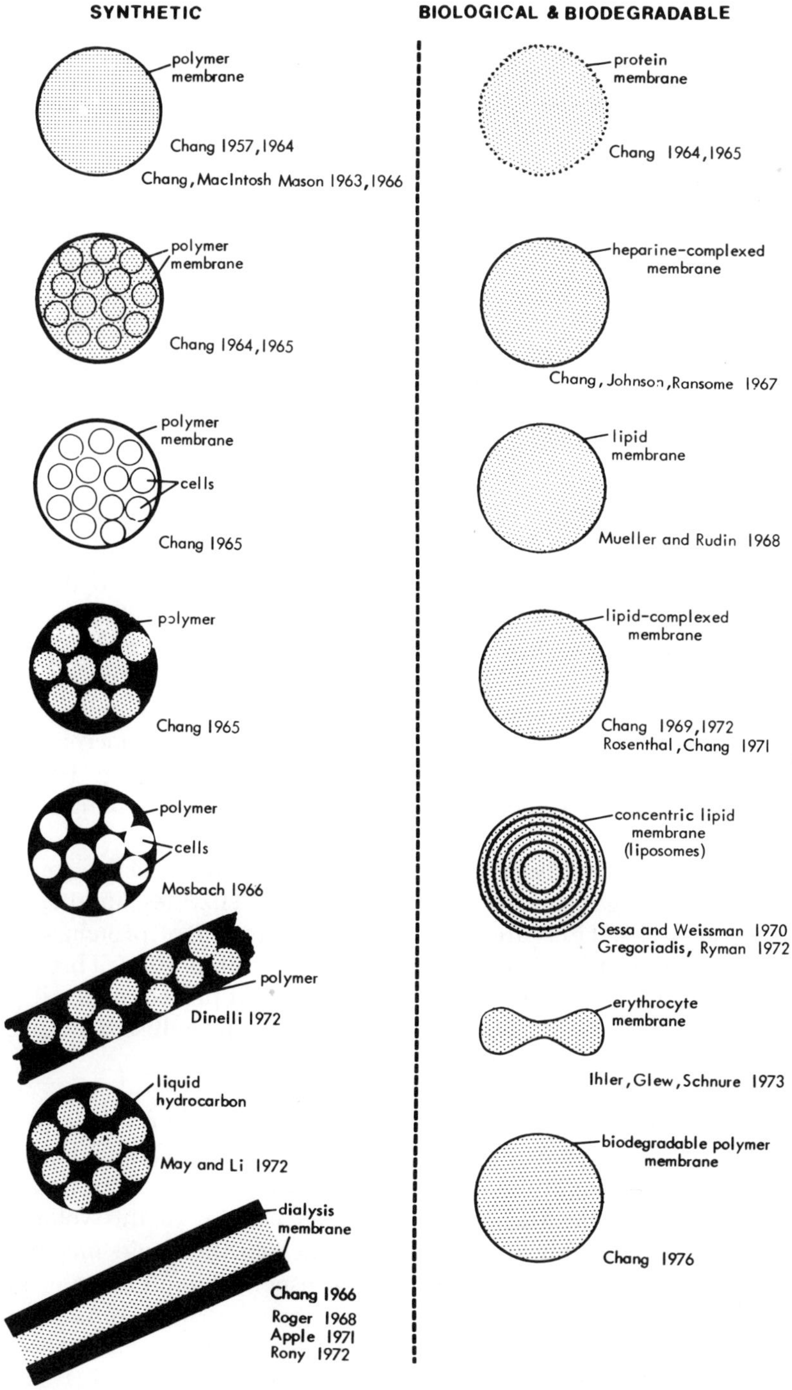

FIGURE 7. Variation of membrane material and configuration for the encapsulation of enzymes and proteins. Dots represent enzyme molecules. Only the reference for the first report of enzyme encapsulation is given.

c. Secondary Emulsion. Another method for forming microcapsules is by the method of secondary emulsion (Chang, 1965, 1966, 1972a). This is a simple extension of the first basic method. After the enzyme or protein solution is emulsified in the polymer solution, instead of continuing the step described, the following is carried out. The emulsion of enzyme solution in the polymer solution is emulsified in an aqueous phase to form a secondary emulsion. Each emulsified droplet in the aqueous phase would now contain a droplet of polymer organic solution inside which are smaller microdroplets of protein and enzyme solution. The secondary emulsion is then allowed to set for the polymer to solidify resulting in a solid polymer sphere, each containing multiple droplets of enzyme solution. This configuration, although not a spherical ultrathin membrane, has the advantage of its ease in preparation. However, it has a thicker membrane and therefore a less effective permeability.

A larger number of polymers have been tested. One type involved the use of silicone rubber (Silastic) to microencapsulate proteins, enzymes, and other materials (Chang, 1966). Very briefly, enzymes, proteins, or other materials to be used are emulsified in two volumes of Silastic (Dow-Corning S-5392). While stirring is continued, 0.025 volume of a catalyst, stannous octoate (Dow-Corning), is added, followed in 2 sec by 20 volumes of 50% Tween-20 solution, and the speed is immediately slowed down and stirring continued for 20 min to allow the silastic to solidify. Other polymers, such as polystyrene or polylactic acid, can be dissolved in benzene. Enzymes or proteins can be emulsified in this polymer solution, then double- emulsified in 50% Tween-20 solution and allowed to stir until the polymer solidifies. This principle of secondary emulsion has been used to microencapsulate enzymes and proteins within cellulose polymer material (Chang, 1965), other polymers (Kitajima *et al.*, 1969), and biodegradable polymers (Chang, 1976).

d. Fiber-Entrapped Enzymes. Instead of forming spherical polymer membranes each containing microdroplets of proteins and enzymes, another configuration is also available. Thus, after emulsifying the enzyme or protein solution in the polymer solution, the emulsion can be extruded into fiber. This results in fibers, each containing microdoplets of the enzyme and protein material (Dinelli, 1972; Marconi *et al.*, 1974). These have advantages in industrial applications, where large flow rate in an external reactor system is used.

e. Secondary Emulsion Using Liquid Membranes. This method is similar to secondary emulsion but, instead of a solid polymer, makes use of a liquid membrane. This approach, using liquid membranes, was originated by Li's group, who will describe this approach in greater detail in their chapter of this volume. Very briefly, the aqueous enzyme solution is emulsified in the liquid membrane, an organic mixture. This water in oil emulsion is dispersed in an aqueous phase to form a double emulsion. Each organic droplet contains many smaller aqueous droplets 1–10 μm in diameter. This way, each spherical hydrocarbon liquid membrane droplet contains enzyme aqueous droplets (May and Li, 1972).

f. Biological Cells. Biological cells have been microencapsulated within nylon, cross-linked protein, and cellulose nitrate membranes (Figure 5) (Chang, 1965,

1972a), including microinjection into preformed microcapsules. Mosbach and Mosbach (1966) have encapsulated microorganisms within solid polymer spheres.

g. Encapsulation of Enzymes in Dialysis Membrane Systems. Microencapsulation of spherical ultrathin membranes in the form of artificial cells are, in fact, microdialysis systems in which the enzymes act on external substrates dialyzing across the enclosing membrane. It has been suggested (Chang, 1966) that many of the advantages of the artificial cells could be mimicked by placing enzymes in the standard dialysis membrane systems, although the main advantage in using spherical ultrathin membranes in the form of artificial cells is the enormous surface/volume relationship, and the ultrathin membrane (10 ml of 20-μm-diameter microcapsules would have 400 times the rate of equilibration as a whole dialysis machine). Nevertheless, in some applications, it has been found to be easier and more advantageous to use dialysis membrane systems containing enzyme solutions (Rogers, 1968; Apple, 1971; Rony, 1971). Making use of standard dialysis membranes, including the capillary hollow fiber system, a number of authors have studied the encapsulation of enzymes within this system. The greatest advantage is in the ease of introducing the enzyme system into this fiber system for studying. However, it would be expected that compared to artificial cells, the rate of equilibration of substrates would be much slower. Also, this system would be more useful for extracorporeal shunt systems, since it would not be as convenient when used for implantation.

4.2. *Biological and Biodegradable Membranes*

If artificial cells or other encapsulated enzyme systems are to be implanted or used in extracorporeal shunt systems in such a way that the encapsulated enzyme systems are required to be permanently encapsulated without being destroyed or released into the patient, synthetic nonbiodegradable membranes will be required. On the other hand, in certain applications one may want to use artificial cells or other encapsulated enzymes in membranes which are partially or completely prepared from biological material or biodegradable material, since in most cases, after the system has completed its function, one may want the polymers to be easily removed, degraded, or metabolized in the body so as to prevent accumulation of these materials. There are a number of systems utilizing biological materials or biodegradable polymers. Some of the examples are as follows.

a. Spherical Ultrathin Cross-Linked Protein Membranes. In the method for preparing artificial cells by interfacial polymerization, if diamine is omitted from the aqueous solution, and if suitable adjustment is made in the preparative procedure in regard to the concentration of the cross-linking agent and the smaller size of the emulsion, a spherical ultrathin membrane of cross-linked protein could be prepared (Chang, 1964, 1965, 1972a; Chang *et al.*, 1966). This spherical ultrathin membrane prepared from protein is easily biodegradable in the body. Different rates of biodegradability could be obtained by mixing different proportions of diamine so as to result in a membrane of cross-linked protein with different proportions of nylon. Spherical ultrathin nylon membranes alone, although much more permanent than cross-linked protein membranes, are also, to some extent, biodegradable.

b. Heparin-Complexed Polymer Membranes. In some cases, where the encapsulated enzymes comes in contact with blood it may be desirable to complex the membrane with heparin so as to prevent coagulation without the need of systemic heparinization. Heparin benzalkonium complex has been applied to the cellulose nitrate or nylon membrane of artificial cells to make them blood-compatible (Chang *et al.,* 1967).

c. Spherical Ultrathin Lipid Membranes. In an attempt to prepare artificial cells with a pure lipid membrane, Mueller and Rudin (1968) used a modification of the original preparative procedure (Chang, 1964) to prepare spherical ultrathin lipid membranes containing red blood cell hemolysate. The membrane consists of a bilayer lipid membrane.

d. Lipid-Complexed Membranes. Spherical ultrathin cross-linked protein membranes have been complexed with lipids to form a lipid–protein membrane containing complex enzyme systems and protein from red blood cell hemolysate (Chang, 1969b, 1972a). In this system, the membrane stability and strength is much greater than the spherical bilayer lipid membrane. Lipids could also be complexed onto spherical ultrathin nylon polymer membranes for the same type of system (Chang, 1969b, 1972a; Rosenthal and Chang, 1971).

e. Liposomes. Liposomes consist of microscopic lipid spheres, each of which consists of concentric onion-skin-like bilayer lipids. Sessa and Weissman (1970) were the first to use this for the entrapment of an enzyme, lysozyme. This method of enzyme encapsulation was seriously extended by Gregoriadis *et al.,* (1971) for the entrapment of enzymes, proteins, and other materials. One of the main advantages of liposomes is their ability to carry protein and enzyme to certain selected intracellular sites, especially in the reticuloendothelial system. This will be discussed later in a chapter by this group of workers.

f. Erythrocyte-Encapsulated Enzymes. The membrane of erythrocytes is normally only permeable to very small molecules unless they are transported by special membrane carrier mechanisms. In a hypotonic solution, red blood cells swell and the stretching of the cell membrane results in its becoming permeable to molecules as large as proteins. As a result, cell proteins leak out and, simultaneously, proteins outside the cells can diffuse into the red blood cells. On replacing the red blood cells in an isotonic solution, the red blood cell membrane returns to its normal shape and permeability, and the enzymes that have entered the cells do not leak out. Ihler *et al.* (1973) made use of this phenomena to encapsulate enzymes inside red blood cells. This way, red blood cells are made to encapsulate different enzyme systems. The advantage of this system is the fact that this is a biological cell and is therefore biocompatible. The amount of enzyme that can be encapsulated this way is limited. The very low permeability of red blood cell membranes to many substrates is such that enzymes, in most cases, would have to be released to carry out their function. One of the important approaches is the use of red blood cell–encapsulated enzymes to deliver and release enzymes intracellularly to the liver and spleen. Further details are available in subsequent chapters.

g. Biodegradable Polymer Membranes. One of the reasons for using biological materials to form the membrane is to ensure that after injection into the body, the membrane could be metabolized and removed. A number of biodegradable synthetic polymer membranes could also be used; for instance, polylactic acid has been used to microencapsulate enzymes and hormones for injection (Chang, 1976).

5. *Variation of Contents in Encapsulated Enzyme–Protein Systems (Figure 8)*

5.1. *Enzymes, Multienzyme Systems, Cell Extracts, Cells, and Other Proteins*

The original method of preparing artificial cells (Chang, 1957, 1964, 1972a; Chang *et al.*, 1963, 1966) involves the microencapsulation of the whole red blood cell hemolysate containing complex multiple enzyme systems of the whole cell content of red blood cells. Any complex enzyme systems, cell extracts, or proteins, including liver cells or liver cell extracts, could easily be microencapsulated into the artificial cell system (Chang, 1965, 1972a, 1976). In fact, this is one of the advantages of this type of artificial cell because of its ease in taking any combination of enzyme systems. Zymase complex and muscle enzyme extracts have also been microencapsulated (Kitajima and Kondo, 1971). The first basic method using interfacial precipitation is more compatible than the second basic method using interfacial polymerization. The chemical method of interfacial polymerization, because of its high pH and use of diamine, inactivates some enzyme systems. However, suitable modification of pH environment and other conditions could overcome this disadvantage (Østergaard and Martiny, 1973).

5.2. *Granules of Enzyme Systems and Proteins*

Instead of solution or suspension, granules of enzymes or proteins could be encapsulated by direct coating with polymer membrane, for instance, cellulose nitrate (Chang, 1957, 1972a).

5.3. *Encapsulation of a Combined Enzyme Adsorbent System*

Urease was encapsulated together with an ammonia adsorbent (Chang, 1966, 1972a; Sparks *et al.*, 1969; Gardner *et al.*, 1971). When the substrate urea is converted by urease to ammonia, the product ammonia could be removed by the ammonia adsorbent. This type of combination has interesting potential for the removal of products of enzymatic reactions. Other combinations of adsorbents

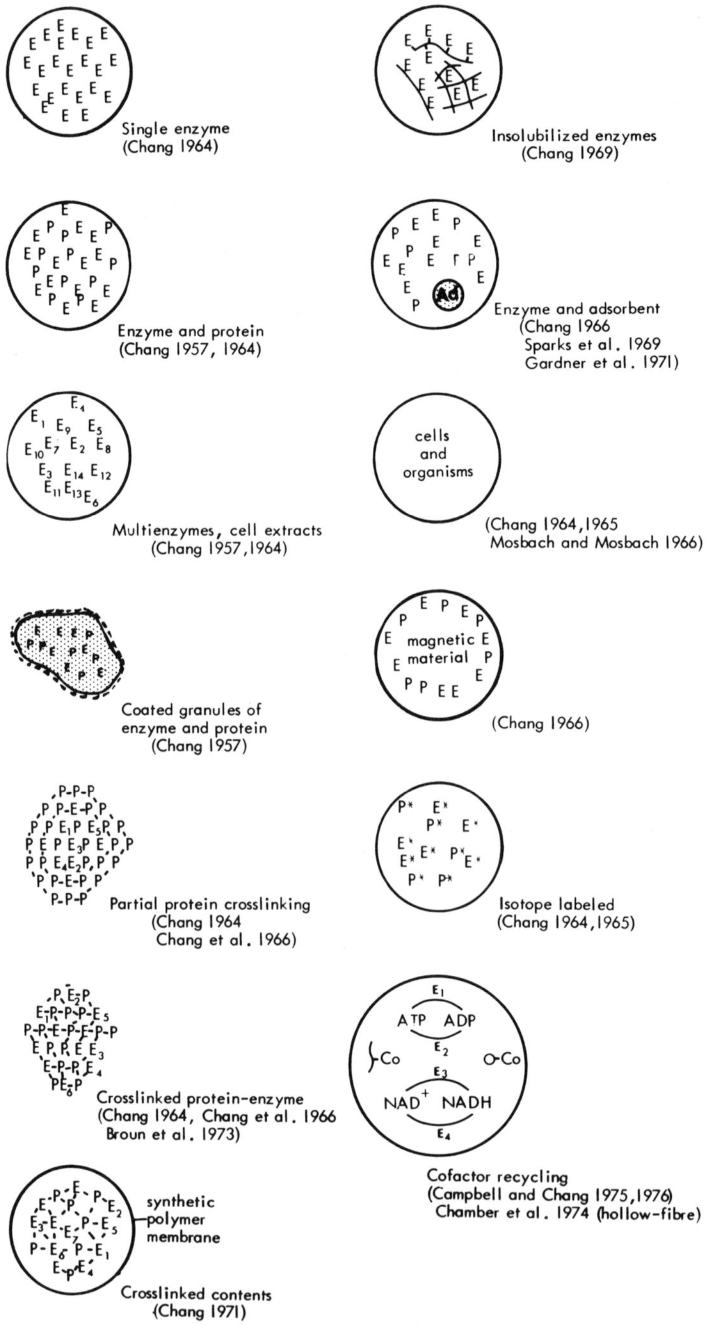

FIGURE 8. Variation in contents of microencapsulated enzymes and proteins. E represents enzyme molecules. P represents protein molecules. Only the reference of the first reports of enzyme encapsulation is shown.

with enzyme systems for other purposes could easily be attempted. In these cases, the adsorbent is suspended in the enzyme solution before encapsulation.

5.4. *Encapsulation of Biological Cells within Spherical Ultrathin Membranes*

Any type of cell can be easily microencapsulated within the spherical ultrathin membranes by one of the basic methods described (Chang, 1965, 1972a; Chang *et al.,* 1966). The enzymes contained in the cell would continue to function after microencapsulation. However, maintaining a viable, living cell culture within this type of spherical ultrathin membrane would require modification of the preparative procedure to allow the cell to survive. Cells have also been introduced by "microinjection" into preformed microcapsules.

5.5. *Encapsulation of Cells within Spherical Solid Particles*

Instead of using a spherical ultrathin membrane, cells could also be dispersed in a polymer solution, which is then cross-linked to form solid particulate, each containing cells in the polymer sphere. This is not unlike the secondary emulsion method. Different types of bacterial cells have been encapsulated this way (Mosbach and Mosbach, 1966).

5.6. *Encapsulation of Spherical Ultrathin Membranes within Larger Ultrathin Spherical Membranes*

As discussed before, any combination of enzyme systems consisting of different multiple-compartmental systems of ultrathin spherical membranes could be encapsulated in one large spherical ultrathin membrane. This allows for an extremely complex compartmental system with different enzyme systems carrying out the functions (Chang, 1965, 1972a; Chang *et al.,* 1966).

5.7. *Encapsulation of Enzymes with Magnetic Materials*

In some cases where it is desirable to locate the encapsulated enzyme, to direct its location, or to maintain it at certain sites, enzymes have been microencapsulated together with magnetic material (Chang, 1966). This way, the microcapsules could be controlled and manipulated by a magnetic field placed externally, thereby directing and maintaining the microcapsules at certain sites.

5.8. *Encapsulation of Radioisotope-Labeled Enzymes and Proteins*

In certain studies, it may be desirable to label the contents of the microcapsule in order to follow the sites of distribution and fate of the injected material. This could easily be done by either labeling the proteins before or after encapsulation (Chang, 1965, 1972a). Any other radioactive-labeled material of macromolecular dimensions could also be encapsulated together.

5.9. *Encapsulation of Insolubilized Enzymes*

For a number of reasons, including increased stability of insolubilized enzymes in the form of covalently bound, adsorbed, or gel-entrapped, they could be encapsulated with spherical ultrathin membranes or other types of encapsulation systems (Chang, 1969a, 1972a). In this form, the advantage of the encapsulation approach is combined with the increased stability of the insolubilized-enzyme systems.

5.10. *Microcapsules of Cross-Linked Proteins and Enzymes*

In the method using interfacial polymerization (Chang, 1964), sebacyl chloride, a difunctional agent, has been used to cross-link microdroplets of protein and enzymes emulsified in a water-immiscible solvent. This way, with very small microdroplets, the entire microdroplet has been cross-linked. If larger microdroplets are used and the time of reaction is short, varying degrees of cross-linking may result. Proteins and enzymes can also be microencapsulated first and then cross-linked using another difunctional reagent, such as a buffered glutaraldehyde solution (Chang, 1971). The glutaraldehyde treatment has been used on microencapsulated catalase (Chang 1971), asparaginase (Chang, 1971, 1973), and lactase, resulting in greatly increased stability. The same general approach has also been used by other workers to form microcapsules of cross-linked protein and enzymes (Broun *et al.,* 1973).

5.11. *Cofactor Recycling Systems*

The first workable system for cofactor regeneration using encapsulated enzymes made use of the dialysis membrane approach (Chambers *et al.,* 1974). Multienzyme systems have also been microencapsulated inside artificial cells so as to allow these systems to carry out cofactor recycling functions (Campbell and Chang, 1975, 1976). In these studies (Campbell and Chang, 1975, 1976) a multienzyme system of hexokinase and pyruvate kinase was microencapsulated into collodion semipermeable microcapsules. This microencapsulated system was shown to recycle ATP and ADP. A multienzyme system of alcohol dehydrogenase and malic dehydrogenase has also been microencapsulated inside collodion microcapsules to recycle both NAD^+ and NADH. Furthermore, cofactors (e.g., NAD^+) have been immobilized to soluble or insoluble polymers, and then encapsulated together with a multienzyme system so as to allow the cofactors to function without leaking out (Campbell and Chang, 1976). Another approach is the use of liquid membrane containing multienzyme systems and cofactors. The liquid membrane, being extremely impermeable, would keep the cofactors inside (May and Landgraff, 1975).

5.12. *Antigens, Antibodies, Vaccines, and Hormones*

There is no limit to the type of material that can be encapsulated. Therefore, this area is completely open for any type of combination of materials for encapsulation for special applications and purposes. For instance, antigens, antibodies, vaccines, and hormones have all been microencapsulated (Chang, 1976). In particular, insulin has been microencapsulated within polylactic acid microcapsules. Rate of release of insulin could easily be adjusted to allow for different hormone levels. This varies from a $T_{1/2}$ of 1 day to a $T_{1/2}$ of greater than 100 days. Polylactic acid in the body is biodegradable and is metabolized into H_2O and CO_2.

6. *References*

Apple, M., 1971, Hemodialysis against enzymes as a method of "gene replacement" in cases of inherited metabolic diseases, *Proc. West. Pharmacol. Soc.* **14**:125.

Boguslaski, R. C., and Janik, A. M., 1971, A kinetic study of microencapsulated bovine carbonic anhydrase, *Biochim. Biophys. Acta* **250**:266.

Broun, G., Thomas, D., Gellf, G., Domurado, D., Berjonnearu, A. M., and Gillon, T., 1973, New method for binding emzyme molecules into a water insoluble matrix—properties after insolubilization, *Biotechnol. Bioeng.* **15**:359.

Campbell, J., and Chang, T. M. S., 1975, Enzymatic recycling of coenzymes by a multi-enzyme system immobilized within semipermeable collodion microcapsules, *Biochim. Biophys. Acta* **397**:101.

Campbell, J., and Chang, T. M. S., 1976, The recycling of NAD^+ (free and immobilized) within semipermeable aqueous microcapsules containing a multi-enzyme system, *Biochem. Biophys. Res. Commun.* **69**:562.

Chambers, R. P., Ford, J. R., Allender, J. H., Baricos, W. H., and Cohen, W., 1974, Continuous processing with cofactors requiring enzymes, coenzyme retention, and regeneration in: *Enzyme Engineering* (E. K. Pye and L. B. Wingard, Jr., eds.), p. 195, Plenum Press, New York.

Chang. T. M. S., 1957, Hemoglogin corpuscles, report of research project for B. Sc. Honours, McGill University.

Chang, T. M. S., 1964, Semipermeable microcapsules, *Science* **146**:524.

Chang, T. M. S., 1965, Semipermeable aqueous microcapsules, Ph.D. thesis, McGill University.

Chang, T. M. S., 1966, Semipermeable aqueous microcapsules ("artificial cells"): with emphasis on experiments in an extracorporeal shunt system, *Trans. Amer. Soc. Artificial Internal Organs* **12**:13.

Chang, T. M. S., 1969a, Clinical potential of enzyme technology, *Science Tools* **16**:33.

Chang, T. M. S., 1969b, Lipid-coated spherical ultrathin membranes of polymer or cross-linked protein as possible cell membrane models, *Federation Proc.* **28**:461.

Chang, T. M. S., 1971, Stabilization of enzymes by microencapsulation with a concentrated protein solution or by microencapsulation followed by cross-linking with glutaraldehyde, *Biochem. Biophys. Res. Commun.* **44**:1531.

Chang, T. M. S., 1972a, *Artificial Cells*, Charles C. Thomas, Publisher, Springfield, Ill.

Chang, T. M. S., 1972b, A new approach to separation using semipermeable microcapsules (artificial cells): combined dialysis, catalysis, and absorption, in: "Recent Development in Separation Science," Vol. I (N. N. Li, ed.), p. 203, CRC Press, Cleveland, Ohio.

Chang, T. M. S., 1973, L-asparaginase immobilized within semipermeable microcapsules: *in-vitro* and *in-vivo* stability, *Enzyme* **14**:95.

Chang, T. M. S., 1974, A comparison of semipermeable microcapsules and standard dialysers for use in separation, *Separ. Purif. Methods* **3**:245.

Chang, T. M. S., 1976, Biodegradable semipermeable microcapsules containing enzymes, hormones, vaccines and biologicals, *J. Bioengineering*, (in press).

Chang, T. M. S., and Poznansky, M. J., 1968, Semipermeable aqueous microcapsules (artificial cells): V. Permeability characteristics, *J. Biomed. Mater. Res.* **2**:187.

Chang, T. M. S., MacIntosh, F. C., and Mason, S. G., 1963, Semipermeable aqueous microcapsules, *Proc. Can. Federation Biol. Soc.* **6**:16.

Chang, T. M. S., MacIntosh, F. C., and Mason, S. G., 1966, Semipermeable aqueous microcapsules: I. Preparation and properties, *Can. J. Physiol. Pharmacol.* **44**:115.

Chang, T. M. S., Johnson, L. J., and Ransome, O., 1967, Semipermeable aqueous microcapsules: IV. Nonthrombogenic microcapsules with heparin-complexed membranes, *Can. J. Physiol. Pharmacol.* **45**:705.

Choi, P. S. K., and Fan, L. T., 1973, Transient behaviour of encapsulated enzyme reactor systems, *J. Appl. Chem. Biotechnol.* **23**:531.

Dinelli, D., 1972, Fibre-entrapped enzymes, *Process Biochem.* **7**:9.

Gardner, D. L., Falb, R. D., Kim, B. C., and Emmerling, D. C., 1971, Possible uremic detoxification via oral-ingested microcapsules, *Trans. Amer. Soc. Artificial Internal Organs* **17**:239.

Gregoriadis, G., Leathwood, P. D., and Ryman, B. E., 1971, Enzyme-entrapment in liposomes, *FEBS Letters* **14**:95.

Ihler, G. M., Glew, R. H., and Schnure, F. W., 1973, Enzyme loading of erythrocytes, *Proc. Natl. Acad. Sci. U.S.* **70**:2663.

Kitajima, M., and Kondo, A., 1971, Fermentation without multiplication of cells using microcapsules that contain zymase complex and muscle enzyme extract, *Bull. Chem. Soc. Japan* **44**:3201.

Kitajima, M., Miyano, S., and Kondo, A., 1969, Studies on enzyme-containing microcapsules, *J. Chem. Soc. Japan* **72**:493.

Marconi, W., Guilinelli, S., and Morisi, F., 1974, Fiber-entrapped enzymes, in: *Insolubilized Enzymes* (M. Salmona, C. Saronio, and S. Garattini, eds.), p. 51, Raven Press, New York.

May, S. W., and Landgraff, L. M., 1975, Cofactor recycling in liquid–membrane–enzyme systems, *Biochem. Biophys. Res. Commun.* **68**:786.

May, S. W., and Li, N. N., 1972, The immobilization of urease using liquid–surfactant membranes, *Biochem. Biophys. Res. Commun.* **47**:1179.

Mogensen, A. O., and Vieth, W. R., 1973, Mass transfer and biochemical reaction with semipermeable microcapsules, *Biotechnol. Bioeng.* **15**:467.

Mori, T., Tosa, T., and Chibata, I., 1973, Enzymatic properties of microcapsules containing asparaginase, *Biochim. Biophys. Acta* **321**:653.

Mosbach, K., and Mosbach, R., 1966, Entrapment of enzymes and micro-organisms in synthetic cross-linked polymers and their applications in volumn techniques, *Acta Chem. Scan.* **20**:2807.

Mueller, P., and Rudin, D. O., 1968, Resting and action potentials in experimental lipid membranes, *J. Theoret. Biol.* **18**:222.

Østergaard, J. C. W., and Martiny, S. C., 1973, Immobilization of β-galactosidase through encapsulation in water insoluble microcapsules, *Biotchnol. Bioeng.* **15**:561.

Rogers, S., 1968, Dialysis against enzymes, *Nature* **220**:1321.

Rony, P. K., 1971, Multiphase catalysts: II. Hollow fibre catalysts, *Biotechnol. Bioeng.* **13**:431.

Rosenthal, A. M., and Chang, T. M. S., 1971, The effect of ialinomycin on the movement of rubidium across lipid coated semipermeable microcapsules, *Proc. Can. Federation Biol. Soc.* **14**:44.

Sessa, G., and Weissman, G., 1970, Incorporation of lysozyme into liposomes, *J. Biol. Chem.* **245**:3295.

Shiba, M., Tomioka, S., Koishi, M., and Kondo, T., 1970, Studies on microcapsules: V. Preparation of polyamide microcapsules containing aqueous protein solution, *Chem. Pharm. Bull. (Tokyo)* **18**:803.

Sparks, R. E., Salemme, R. M., Meier, P. M., Litt, M. H., and Lindan, O., 1969, Removal of waste metabolites in uremia by microencapsulated reactants, *Trans. Amer. Soc. Artificial Internal Organs* **15**:353.

Sundaram, P. V., 1973, The kinetic properties of microencapsulated urease, *Biochim. Biophys. Acta* **321**:319.

II. Experimental Applications in Therapy

8

Rationale and Strategies for the Therapeutic Applications of Immobilized Enzymes

T. M. S. Chang

1. *Introduction*

Except for digestive enzymes and some proteins and enzymes, the majority of enzymes and proteins in the body function in an intracellular environment. The simplest intracellular form is a high concentration of hemoglobin and complex enzyme systems in the red blood cells. The other extreme in the complexity of intracellular organizations are liver cells, each of which contains extremely complex enzyme systems which are further compartmentalized by being immobilized into intracellular membranes and intracellular organelles. Complex enzyme systems immobilized within these intracellular environments of varying complexities carry out their function by acting sequentially on substrates, including those which cross the cell membranes by passive movement or by special transport mechanisms.

What happens if one were to ignore this natural scheme of things and inject an enzyme in solution into the body? Since most of the presently available enzymes are obtained in large amounts from heterogeneous sources, especially from bacterial culture, introduction of heterogeneous enzymes in solution would be considered as foreign proteins by the body. Introduction of foreign proteins in the free form into the bloodstream or other body fluid may result in a number of problems. There may be hypersensitivity reactions with resulting adverse effects on the patients. With repeated injections, there may be production of antibodies resulting in immunological reactions, rejection, and rapid removal and inactivation of the foreign protein. Enzymes have to be highly purified to remove endotoxins and other contaminants. Even if enzymes can be obtained in the crystalline form from homologous sources, their introduction in the soluble-free form into the body would still result in rapid removal by the reticuloendothelial system, destruction by proteolytic enzymes, or, in the case of some smaller enzymes, excretion by the kidney. Furthermore, free enzymes in solution cannot be kept at the sites where the action is desired, since in the free form, they are

T. M. S. Chang • Departments of Physiology and Medicine, McGill University, Montreal, Canada

quickly carried away by the bloodstream and other body fluids, or ingested by phagocytes and then removed. Another problem is that to carry out their therapeutic action, some enzymes may have to be located specifically in certain tissues, organs, or even in selected intracellular sites. Another problem is that enzymes in free solution are not stable, especially at a body temperature of 37°C, and, as a result, the activity would rapidly decrease after injection. Because of these problems, enzyme solutions have not been extensively used in therapeutic applications. The next two chapters will give detailed examples of actual therapeutic administration of free enzymes in solution with emphasis on the types of problems that have been encountered.

Problems related to the therapeutic application of enzyme solutions have resulted in extensive research into the possible therapeutic uses of immobilized enzymes and proteins. The first reported study on the use of immobilized enzymes and proteins for experimental therapy involves the use of artificial cells in the form of microencapsulated enzymes and proteins (Chang, 1957, 1964, 1965, 1972a; Chang and MacIntosh, 1964; Chang and Poznansky, 1968). With these demonstrations of the therapeutic potentials of enzymes immobilized by microencapsulation, a large number of laboratories have, in more recent years, started to seriously investigate the possible therapeutic applications of the different types of immobilized enzymes and proteins. The majority of workers in this area have contributed chapters in this section, and it will not be necessary to review their important research contributions here. However, it would be useful at this stage to summarize: (1) the therapeutic significance of different types of immobilized systems, (2) the examples of the routes of administration, and (3) the examples of the different types of therapeutic conditions.

2. *Biomedical Significance of Different Types of Immobilized Enzymes*

Insofar as the type of immobilized enzymes used, as can be seen in the publications available and also in chapters of this book, the most widely studied approach involves the stage III type of enzyme research using the approach of "intracellular environment" in the form of artificial cells and other encapsulated enzymes. The stage II type of immobilized enzymes, such as covalently linked enzymes and matrix-entrapped enzymes, is also being intensively investigated. A brief discussion will be made of the significant features of different types of immobilized enzyme systems in therapy.

2.1. *Covalent Linkage*

The enzyme/carrier ratio is very variable in covalently bound enzymes. Thus, in the case of enzymes covalently linked to insoluble carriers, such as glass or other polymers, the ratio of enzyme to polymer may be as low as a few micrograms of enzyme per gram of carrier, or as high as 3.5 g of enzyme per gram of carrier. On the other hand, enzymes immobilized by intermolecular linkage using bifunctional agents have an extremely high ratio of enzyme. Enzymes covalently linked to soluble carriers may have some interesting potential advantages, especially in multienzyme systems and cofactor regenerating systems. One of the most important and useful features of covalently bound enzymes in therapeutic applications is their increased stability as compared to the free-enzyme system. As a rule, although not in all cases, covalently bound enzymes usually have increased thermal and storage stabilities. This means that at a body temperature of 37°C, their duration of action would last much longer than the enzyme in free solution. On the other hand, at least part of the covalently bound enzymes are exposed to the surface and would come in direct contact with other proteins and antibodies or cells in the body to result in immunological reactions or initiate immunological response and removal. To circumvent these problems and to make use of their advantages, covalently bound enzymes have been used in combination with encapsulated enzymes (Chang, 1971, 1972a). This way, the stability and duration of action was extensively increased as compared to free enzyme enclosed within artificial cells. In other cases, for instance, in the use of immobilized antibodies for the removal of circulating antigens, the exposure of the protein to external proteins would be a distinct advantage. The covalently linked enzyme system is also extremely useful in industrial processes where immunological reactions are not a problem.

2.2. *Adsorption*

In the case of enzymes immobilized by adsorption, the advantage in this case is the ease of preparation and, in many cases, the increase in enzyme stability. However, the carrier/enzyme ratio is usually high. The adsorbed enzymes are exposed to the external environment. The adsorption of enzymes on carriers is reversible to varying degrees, depending on the adsorbents. Changes in pH, electrolyte concentration, or temperature may increase the desorption of enzymes. In therapeutic applications, adsorbed urease has been microencapsulated to solve some of these problems (Gardner *et al.,* 1971).

2.3. *Matrix Entrapment*

In the case of matrix entrapment, the ratio of enzyme to matrix polymer is usually rather low (for instance, 1 part enzyme to 100 parts polymer). Furthermore, unlike covalent linkage or adsorption, much of the enzymes are located deep inside a solid matrix of polymer limiting the diffusion of substrate. As a result, the ratio of activity of the immobilized enzyme is usually low (about 10% of the enzyme in solution). Some immobilized enzymes of this type have increased stability. This type of enzyme, as will be discussed in the next section, is extremely important in many of the *in vitro* biomedical applications of clinical diagnostic analysis and preventive medicine. If the immobilized enzymes are to be introduced into the body for therapeutic applications, the very large polymer component in comparison to enzyme (1 part enzyme to 100 parts polymer) will result in the introduction of a large amount of exogenous polymers. Furthermore, the solid polymer reduces the rate of diffusion of substrate into the immobilized enzyme. Also, there is always some leakage of enzyme from the polymeric network because of the principle of molecular entrapment (Updike *et al.*, 1973b; Nadler and Updike, 1974). Increasing the cross-linking would restrict the leakage, but this tends to greatly reduce the substrate permeability (Updike *et al.*, 1973b; Nadler and Updike, 1974). However, unlike covalently bound or adsorbed enzymes, matrix-entrapped enzymes are protected from antibodies and immunological reactions. In extracorporeal hemoperfusion systems, the solid polymer material would allow for a more stable system, especially with a high flow rate.

2.4. *Encapsulation*

In the case of artificial cells, the detials have been reviewed elsewhere (Chang, 1972a). Very briefly, the enzyme systems are microencapsulated within spherical ultrathin membranes (200 Å thick), and any concentration of enzyme could be used. As a result, the ratio of enzyme to membrane can be more than 100,000 to 1. The small cellular dimensions of the microencapsulated enzyme allows a very large surface/volume relationship. This, in addition to the ultrathin membrane, allows for extremely fast permeation of substrates. Furthermore, the enzymes are enveloped within a spherical ultrathin membrane and thus do not come in contact with external proteins or cells and do not result in any immunological reactions. The stability of encapsulated enzymes can be greatly increased by microencapsulating enzymes together with a high concentration of another protein, microencapsulating insolubilized enzymes, or cross-linking the microencapsulated enzymes. This way, unlike enzymes in dilute solution, microencapsulated enzymes have an intracellular environment stabilizing and separating them from external environments. Another important feature is that the microcapsule with its 200-Å-thick polymer membrane is a sphere consisting mostly of enzymes with a negligible small amount of polymer membrane surrounding the sphere. This way, the encapsulated enzymes do not leak out but can act on substrates equilibrating rapidly across the membranes. The permeability of the membranes can be modified to have suitable permeability. Furthermore, the membrane material can be made nonthrombogenic, biocompatible, or even biodegradable. Different types of polymer or biological membranes could be made to have different degrees of biodegradability so that their rate of destruction in the body could be varied depending on the type of function that is required. For example, asparaginase has been microencapsulated within polylactic acid, and continued to function effectively. A system of this type could be introduced into the body without accumulation; it would be metabolized in the body into carbon dioxide and water after it has carried out its function.

The original method of artificial cells for encapsulation of enzyme systems has now been extended to a very large number of other systems. Liposomes (Gregoriadis and Ryman, 1972) have the advantage that they can be made extremely small and could be especially useful for intravenous injection for localization at intracellular sites, and lipid carrier material of liposomes could easily be metabolized in the body. The amount of enzymes carried by the liposomes is rather small compared to the carrier material, and the enzymes have to be released to carry out their action, since the carrier is impermeable to substrates. Another very interesting approach is the encapsulation of enzymes within erythrocytes (Ihler *et al.*, 1973). Injection of erythrocyte- encapsulated enzymes this way would allow the enzyme to be removed and located in the reticuloendothelial cells. Since the permeability of the erythrocyte membrane is rather limited, the use of erythrocyte-encapsulated enzymes is more appropriate for those enzymes that require to be released in the reticuloendothelial system. In using dialysis systems (Apple, 1971), it is very easy to enclose any type of enzyme system and to retain it, while allowing substrate to diffuse in from outside to be acted on by the encapsulated enzyme. The ease of preparation using the dialysis system is at the expense of the substrate equilibration rate, which is about 400 times slower than that of artificial cells. The therapeutic application here is toward extracorporeal perfusion. Other systems will be discussed in appropriate chapters.

3. *Routes of Therapeutic Administration*

Ideally, one would like to have immobilized enzymes in a location where they can come readily in contact with a good supply of the intended substrate. The immobilized enzymes should also be able to stay at the required site of action to carry out their function. A number of routes of administration have been studied. They are briefly summarized below.

3.1. *Local Implantation by Injection*

The immobilized enzyme preparation, in the form of a fine suspension, can be injected intramuscularly, subcutaneously, intraperitoneally, or elsewhere. Equilibration of blood-borne substrates is faster in the intramuscular or intraperitoneal sites. Although a very large amount of immobilized enzymes could be introduced into the peritoneal cavity, this way of introduction carries a risk of peritonitis. However, for animal experimentation, the intraperitoneal route is an easy way of administering a large amount of immobilized enzymes into a location where they can come in contact with those substrates which can equilibrate rapidly across the peritoneal cavity. Table 1 gives examples of publications presenting this approach.

3.2. *Intravenous Injection and Subsequent Localization*

A detailed analysis has been carried out to study the fate and sites of distribution of intravenously injected microencapsulated enzymes, other particles, and red blood cells (Chang, 1965, 1972a). In these studies it was found that artificial particulates larger than 2 μm in diameter are filtered out to a great extent by the pulmonary capillaries. Those that passed through the pulmonary capillaries were removed mainly by the reticuloendothelial system of the liver and spleen. Surface properties were found to play an important role in the distribution. By

TABLE 1
Examples of Routes of Administration

Local implantation
Chang, 1964, 1966, 1969a, 1971, 1972a, 1973b, 1975
Chang and MacIntosh, 1964
Chang and Poznansky, 1968
Updike *et al.*, 1973b
Poznansky and Chang, 1974
Ohnuma *et al.*, 1974
Siu Chong and Chang, 1974
Cooney *et al.*, 1975
Intravenous injection and subsequent localization
Chang, 1964, 1966, 1972a
Sekiguchi and Kondo, 1966
Gregoriadis and Ryman, 1972
Updike *et al.*, 1973b
Ihler *et al.*, 1973
Thorne *et al.*, 1975
Extracorporeal systems
Chang, 1966, 1969b, 1972a, 1973a, 1975
Chang *et al.*, 1967
Chang and Poznansky, 1968
Hyden, 1970
Apple, 1971
Sampson *et al.*, 1972
Allison *et al.*, 1972
Horvath *et al.*, 1973
Hersh, 1974
Salmona *et al.*, 1974
Venkatasubramanian *et al.*, 1974
Venter *et al.*, 1975
Administration into the gastrointestinal tract
Chang and Poznansky, 1968
Chang and Loa, 1970
Gardner *et al.*, 1971
Sparks *et al.*, 1971
Chang, 1972a
Asher *et al.*, 1975
Applications to local lesions
Chang, 1972b

variations in the surface properties of microcapsules, one can vary the distribution of microcapsules to different organs. Thus, negatively charged particles of small diameters are less liable to be trapped in the lung capillaries but are removed more by the reticuloendothelial system of the liver and spleen (Chang, 1965, 1972a). Other surface properties studied included the treatment of red blood cells with neuraminidase, and the resulting rapid removal of these treated homologous red blood cells by the reticuloendothelial system of the liver and spleen (Chang, 1965, 1972a). Intravenously injected liposomes are also removed by the liver and spleen, where their contents can enter the cells and even the intracellular oganelles

(Gregoriadis and Ryman, 1972). Erythrocyte-encapsulated enzymes can also be introduced intravenously to be removed by the reticuloendothelial system (Ihler *et al.*, 1973).

3.3. *Use of Immobilized Enzymes for Extracorporeal Shunt Systems*

It was demonstrated that if the substrate to be acted on is in the bloodstream or in the body fluid, one may not have to inject the immobilized enzymes into the body, since immobilized enzymes can be used in an extracorporeal system to act on substrates of blood recirculating through the system (Chang, 1966). An extracorporeal shunt system containing microencapsulated urease was used in this way to remove blood urea (Chang, 1966). The additional use of heparin attached to microcapsule membranes, shunt chamber, and tubings were used to prevent clotting of blood and avoid the necessity of systemic heparinization (Chang *et al.*, 1967). Extracorporeal shunts containing microencapsulated catalase have also been used to recirculate peritoneal fluid for the removal of perborate in acatalasemic mice (Chang and Poznansky, 1968). With these demonstrations of the feasibility of extracorporeal immobilized enzymes, a large number of other extracorporeal immobilized enzyme systems were studied. These include immobilizing enzymes on glass (Hyden, 1971; Venter *et al.*, 1975); polymethemethacrylate (Sampson *et al.*, 1972; Hersh, 1974); nylon tubings (Allison *et al.*, 1972); gel attached to nylon tubings (Horvath *et al.*, 1973); collagen systems (Venkatasubramanian *et al.*, 1974); dialysis systems (Apple, 1971); fibers (Salmona *et al.*, 1974); and combined artificial cell synthetic capillary systems (Chang, 1973a). Albumin immobilized on the surface of microencapsulated adsorbents has resulted in a blood-compatible extracorporeal system being used for the treatment of patients with renal failure, liver failure, and drug intoxication (Chang, 1969b, 1972a, 1975; Chang and Malave, 1970; Chang *et al.*, 1971, 1972, 1973a, 1973b, 1974, 1975). Since the immobilized material in an extracorporeal system does not enter the body at any time, it has become the obvious candidate for early applications in patients.

3.4. *Local Applications*

Immobilized enzymes may be applied directly to local lesions to prevent absorption of the enzyme into the body or to prevent immunological or hypersensitivity reactions. Thus, microencapsulated catalase has been applied directly to oral lesions in mice with hereditary catalase deficiency to replace the enzyme deficiency locally (Chang, 1972b).

3.5. *Administration into the Gastrointestinal Tract*

Some substrates can equilibrate across the gastrointestinal tract and enzymes may be used experimentally in these cases. For example, microencapsulated urease was used to act on urea either by direct introduction into the intestine or by oral administration into animals (Chang and Poznansky, 1968; Chang and Loa, 1970; Chang, 1972a, Gardner *et al.*, 1971; Asher *et al.*, 1975).

4. *Examples of Experimental Therapy*

Immobilized enzymes and proteins have been used in a number of experimental and therapeutic conditions. Some of these are briefly summarized (Table 2), and the details will be available in later chapters.

4.1. *Red Blood Cell Substitutes*

Artificial cells containing red blood cell hemolysate have been tested for use as red blood cell substitutes (Chang, 1957, 1964, 1972a; Sekiguchi and Kondo, 1966).

4.2. *Model Enzyme Systems for Experimental Therapy*

Urease has been used as a model immobilized enzyme system for experimental therapy (Chang and MacIntosh, 1964; Chang, 1964, 1966, 1972a). The basic result obtained paves the way for other types of enzyme replacement therapy.

4.3. *Enzyme Replacement for Hereditary Enzyme-Deficiency Conditions*

The first demonstration of using immobilized enzymes for replacement in hereditary enzyme-deficiency conditions was the use of microencapsulated catalase to effectively replace a hereditary catalase deficiency in acatalasemic mice (Chang and Poznansky, 1968; Chang, 1972b; Poznansky and Chang, 1974). Subsequently, liposome microencapsulated enzymes have been used for replacement in hereditary enzyme-deficiency conditions related to storage diseases (Gregoriadis and Ryman, 1972). More recently, red blood cell entrapped enzymes and liposome-entrapped enzymes have also been tested for possible use in storage diseases (Ihler *et al.,* 1973; Thorne *et al.,* 1975).

4.4. *Enzyme Therapy Using Asparaginase for Substrate-Dependent Tumors*

The most extensive research into the therapeutic applications of immobilized enzymes has been the use of asparaginase for tumor suppression. Having demonstrated the effectiveness of microencapsulated asparaginase for experimental tumor suppression (Chang, 1969a, 1971, 1972a), more detailed studies were carried out on the various aspects of microencapsulated asparaginase (Chang, 1973b; Mori *et al.,* 1972, 1973; Siu Chong and Chang, 1974). Since then, a large amount of work is being carried out by many centers using all available types of immobilized enzymes for injection and extracorporeal shunts. Many of the studies will be discussed in detail in subsequent chapters of this book. These include matrix entrapment of asparaginase in polyacrylamide gel which is fragmentated to fine particles for peritoneal or intravenous injection (Updike *et al.,* 1973b). Another group of investigators (Ohnuma *et al.,* 1974) used matrix-entrapped L-asparaginase in poly-2-hydroxyethylmethacrylate which was frozen, dried, crushed, and sieved through a 38-μm screen, giving a mean particle size of 12 μm for injection intraperitoneally into C3H mice for tumor suppression. A number of extracorporeal shunt approaches to be discussed in subsequent chapters include the immobilization of asparaginase to glass plates by covalent linkage (Hyden, 1971), poly(methyl methacrylate) by covalene linkage (Sampson *et al.,* 1972; Hersh, 1974), nylon tubing by covalent linkage (Allison *et al.,* 1972), polycarboxylic gel layer attached to the inner wall of a nylon tube (Horvath *et al.,* 1973), and others. L-Asparaginase has also been bound covalently to Dacron vascular pros-

TABLE 2
Examples of Immobilized Enzymes and Proteins in Experimental Therapy

Substitutes for erythrocytes	
	Chang, 1957, 1964, 1966, 1972a
	Sekiguchi and Kondo, 1966
	Jay and Edwards, 1969
Basic model for enzyme therapy	
	Chang and MacIntosh, 1964
	Chang, 1964, 1966, 1972a
	Venter *et al.,* 1975
Hereditary enzyme-deficiency conditions	
	Chang and Poznansky, 1968
	Chang, 1972b
	Poznansky and Chang, 1974
	Gregoriadis and Ryman, 1972
	Ihler *et al.,* 1973
	Thorne *et al.,* 1975
Substrate-dependent tumor	
	Chang, 1969a, 1971, 1972a, 1973b
	Hyden, 1970
	Sampson *et al.,* 1972
	Allison *et al.,* 1972
	Horvath, *et al.,* 1973
	Mori *et al.,* 1972, 1973
	Siu Chong and Chang, 1974
	Hersh, 1974
	Marconi *et al.,* 1974
	Venkatasubramanian *et al.,* 1974
	Nadler and Updike, 1974
	Ohnuma *et al.,* 1974
	Cooney *et al.,* 1975
Organ failure and metabolic disorders	
	Chang, 1966, 1969b, 1972a, 1975
	Levine and LaCourse, 1967
	Gordon *et al.,* 1969
	Chang and Loa, 1970
	Broun *et al.,* 1970
	Sparks *et al.,* 1971
	Gardner *et al.,* 1971
	Kusserow *et al.,* 1971, 1973
	May and Li, 1972
	Bessman and Schultz, 1973
	Venter *et al.,* 1975
	Asher *et al.,* 1975

thesis for installation in the inferior vena cava, or as circular Dacron grafts for implantation in the peritoneal cavity (Cooney *et al.,* 1975). Another material investigated for extracorporeal hemoperfusion is the use of asparaginase complexed to collagen wound to form a spiral multipore reactor (Venkatasubramanian *et al.,* 1974). Another approach is to combine artificial cells with capillary dialysis membranes (Chang, 1973a).

4.5. *Use of Immobilized Enzymes and Proteins in Organ Failure*

Table 2 contains examples of this area of research. Thus, immobilized enzymes and proteins have been used for the construction of artificial kidneys, artificial livers, and detoxifiers (Chang, 1966, 1969b, 1972a, 1975; Chang and Malave, 1970; Chang *et al.,* 1971). Other types of immobilized enzymes have been studied for use in oxygenators (Broun *et al.,* 1970; Updike *et al.,* 1973a), artificial pancreas (Bessman and Schultz, 1973), and biocompatible surfaces (Chang, 1969b; Kusserow *et al.,* 1971, 1973). An artificial organ based on albumin immobilized on microencapsulated adsorbent has been used for a number of years for the treatment of patients with renal failure, liver failure, and acute intoxication (Chang, 1969b, 1972a, 1975; Chang and Malave, 1970; Chang *et al.,* 1971, 1972, 1973b, 1974, 1975).

5. *General*

The rationale and strategy of the therapeutic applications of immobilized enzymes briefly summarized in this chapter will be discussed and illustrated in great detail in the subsequent chapters by research workers who are experts in specialized areas.

6. *References*

Allison, J. P., Davidson, L., Gutierrer-Hartman, A., and Kitto, G. B., 1972, Insolubilization of L-asparaginase by covalent attachment to nylon tubing, *Biochem. Biophys. Res. Commun.* **47**:66.

Apple, M., 1971, Hemodialysis against enzymes as a method of "gene replacement" in cases of inherited metabolic diseases, *Proc. West. Pharmacol. Soc.* **14**:125.

Asher, W. J., Bovee, K. C., Frankenfeld, J. W., Hamilton, R. W., Henderson, L. W., Holtzapple, P. G., and Li, N. N., 1975, Liquid membrane system directed toward chronic uremia, *Kidney Int.* **7**:S409.

Bessman, S. P., and Schultz, R. D., 1973, Prototype glucose oxygen sensor for the artificial pancreas, *Trans. Amer. Soc. Artificial Internal Organs* **19**:361.

Broun, G., Selegny, E., Minh, C. T., and Thomas, D., 1970, The use of proteic and enzymatic coatings and/or membranes for oxygenators, *FEBS Letters* **7**:223.

Chang, T. M. S., 1957, Hemoglobin corpuscles, report of research project for B.Sc. Honours, McGill University.

Chang, T. M. S., 1964, Semipermeable microcapsules, *Science* **146**:524.

Chang, T. M. S., 1965, Semipermeable aqueous microcapsules, Ph.D. thesis, McGill University.

Chang, T. M. S., 1966, Semipermeable aqueous microcapsules ("artificial cells"): with emphasis on experiments in an extracorporeal shunt system, *Trans. Amer. Soc. Artificial Internal Organs* **12**:13.

Chang, T. M. S., 1969a, Asparaginase-loaded semipermeable microcapsules for mouse lymphomas, *Proc. Can. Federation Biol. Sci.* **12**:62.

Chang, T. M. S., 1969b, Removal of endogenous and exogenous toxins by a microencapsulated adsorbent, *Can. J. Physiol. Pharmacol.* **47**:1043.

Chang, T. M. S., 1971, The *in-vivo* effects of semipermeable microcapsules containing L-asparaginase on 6C3HED lymphosarcoma, *Nature* **229**:117.

Chang, T. M. S., 1972a, *Artificial Cells*, Charles C. Thomas, Publisher, Springfield, Ill.

Chang, T. M. S., 1972b, Effects of local applications of microencapsulated catalase on the response of oral lesions to hydrogen peroxide in acatalasemia, *J. Dental Res.* **51**:319.

Chang, T. M. S., 1973a, Biomedical applications of artificial cells, *Bio-med. Eng.* **8**:334.

Chang, T. M. S., 1973b, L-Asparaginase immobilized within semipermeable microcapsules: *in-vitro* and *in-vivo* stability, *Enzyme* **14**:95.

Chang, T. M. S., 1975, Microencapsulated adsorbents for uremia, liver failure, and intoxication, *J. Kidney Int.* **7**:5378.

Chang, T. M. S., and Loa, S. K., 1970, Urea removal of urease and ammonia adsorbents in the intestine, *Physiologist* **13**:70.

Chang, T. M. S., and MacIntosh, F. C., 1964, Semipermeable aqueous microcapsules, *Pharmacologist* **6**:198.

Chang, T. M. S., and Malave, N., 1970, The development and first clinical use of semipermeable microcapsules (artificial cells) as a compact artificial kidney, *Trans. Amer. Soc. Artificial Internal Organs* **16**:141.

Chang, T. M. S., and Migchelsen, M., 1973, Characterization of possible "toxic" metabolites in uremia and hepatic coma based on the clearance spectrum for larger molecules by the ACAC microcapsule artificial kidney, *Trans. Amer. Soc. Artificial Internal Organs* **19**:314.

Chang, T. M. S., and Poznansky, M. J., 1968, Semipermeable microcapsules containing catalase for enzyme replacement in acatalasemic mice, *Nature* **218**:243.

Chang, T. M. S., Johnson, L. J., and Ransome, O., 1967, Semipermeable aqueous microcapsules: IV. Nonthrombogenic microcapsules with heparin-complex membranes, *Can. J. Physiol. Pharmacol.* **45**:705.

Chang, T. M. S., Pont, A., Johnson, L. J., and Malave, N., 1968, Response to intermittent extracorporeal perfusion through shunts containing semipermeable microcapsules, *Trans. Amer. Soc. Artificial Internal Organs* **15**:163.

Chang, T. M. S., Gonda, A., Dirks, J. H., and Malave, N., 1971, Clinical evaluation of chronic, intermittent, and short term hemoperfusions in patients with chronic renal failure using semipermeable microcapsules (artificial cells) formed from membrane-coated activated charcoal, *Trans. Amer. Soc. Artificial Internal Organs* **17**:246.

Chang, T. M. S., Gonda, A., Dirks, J. H., Coffey, J. F., and Lee-Burns, T., 1972, ACAC microcapsule artificial kidney for the long term and short term management of eleven patients with chronic renal failure, *Trans. Amer. Soc. Artificial Internal Organs* **18**:465.

Chang, T. M. S., Coffey, J. F., Lister, C., Taroy, E., and Stark, A., 1973a, Methaqualone, methyprylon, and glutethimide clearance by the ACAC microcapsule artificial kidney: *in-vitro* and in patients with acute intoxication, *Trans Amer. Soc. Artificial Internal Organs* **19**:87.

Chang, T. M. S., Coffey, J. F., Barre, P., Gonda, A., Dirks, J. H., Levy, M., and Lister, C., 1973b, Microcapsule artificial kidney: treatment of patients with acute drug intoxication, *Can. Med. Assoc. J.* **108**:429.

Chang, T. M. S., Migchelsen, M., Coffey, J. F., and Stark, A., 1974, Serum middle molecule levels in uremia during long term intermittent hemoperfusions with the ACAC (coated charcoal) microcapsule artificial kidney, *Trans. Amer. Soc. Artificial Internal Organs* **20**:364.

Chang, T. M. S., Chirito, E., Barre, B., Cole, C., and Hewish, M., 1975, Clinical performance characteristics of a new combined system for simultaneous hemoperfusion-hemodialysis-ultrafiltration in series, *Trans. Amer. Soc. Artificial Internal Organs* **21**:502.

Cooney, D. A., Weetall, H. H., and Long, E., 1975, Biochemical and pharmacologic properties of L-asparaginase bonded to dacron vascular prostheses, *Biochem. Pharmacol.* **24**:503.

Gardner, D. L., Falb, R. D., Kim, B. C., and Emmerling, D. C., 1971, Possible uremic detoxification via oral-ingested microcapsules, *Trans. Amer. Soc. Artificial Internal Organs* **17**:239.

Gordon, A., Greenbaum, M. A., Marantz, L. B., McArthur, M. J., and Maxwell, M. H., 1969, Adsorbent based low volume recirculating dialysate, *Trans. Amer. Soc. Artif. Intern. Organs* **15**:347.

Gregoriadis, G., and Ryman, B. E., 1972, Fate of protein-containing liposomes injected into rats: an approach to the treatment of storage diseases, *Eur. J. Biochem.* **24**:485.

Hersh, L. S., 1974, L-Asparaginase from *Escherichia coli* II and *Erwinia carotovora* bound to poly-(methyl methacrylate), *J. Polymer Sci.* **47**:55.

Horvath, C., Sardi, A., and Woods, J. S., 1973, L-Asparaginase tubes: kinetic behavior and application in physiological studies, *J. Appl. Physiol.* **34**:181.

Hyden, H., 1970, An extracorporeal shunt apparatus for blood detoxification, *Arzneimittel-Forschung* **21**:1671.

Ihler, G. M., Glew, R. H., and Schnure, F. W., 1973, Enzyme loading of erythrocytes, *Proc. Natl. Acad. Sci. U.S.* **70**:2663.

Jay, A. W. L., and Edwards, M. A., 1969, Mechanical properties of semipermeable microcapsules, *Can. J. Physiol. Pharmacol.* **46**:731.

Kusserow, B., Larrow, R., and Nichols, J., 1971, Urokinase- heparin bonded synthetic surfaces, *Trans. Amer. Soc. Artificial Internal Organs* **17**:1.

Kusserow, B., Larrow, R., and Nichols, J., 1973, Surface bonded, covalently cross-linked urokinase synthetic surfaces, *Trans. Amer. Soc. Artificial Internal Organs* **19**:8.

Levine, S. N., and LaCourse, W. C., 1967, Materials and design consideration for a compact artificial kidney, *J. Biomed. Mater. Res.* **1**:275.

Marconi, W., Guilinelli, S., and Morisi, F., 1974, Fiber-entrapped enzymes, in *Insolubilized Enzymes* (M. Salmona, C. Saronio, and S. Garattini, eds.), p. 51, Raven Press, New York.

May, S. W., and Li, N. N., 1972, The immobilization of urease using liquid-surfactant membranes, *Biochem. Biophys. Res. Commun.* **47**:1179.

Mori, T., Sato, T., Matuo, Y., Tosa, T., and Chibata, I., 1972, Preparation and characteristics of microcapsules containing asparaginase, *Biotechnol. Bioeng.* **14**:663.

Mori, T., Tosa, T., and Chibata, I., 1973, Enzymatic properties of microcapsules containing asparaginase, *Biochem. Biophys. Acta* **321**:653.

Nadler, H. L., and Updike, S. J., 1974, Gel-entrapment of enzyme: strategy for enzyme correction, *Enzyme* **18**:150.

Ohnuma, T., O'Driscoll, K. F., Korus, R., and Walczak, I. A., 1974, Pharmacological studies and antitumor activity of *E. coli* asparaginase immobilized in 2-hydroxyethylmethacrylate, *Abstr. Papers Amer. Chem. Soc.* **174**:81.

Poznansky, M. J., and Chang, T. M. S., 1974, Comparison of the enzyme kinetics and immunological properties of catalase immobilized by microencapsulation and catalase in free solution for enzyme replacement, *Biochim. Biophys. Acta* **334**:103.

Salmona, M., Saronio, C., Bartosek, I., Vecchi, A., and Mussini, E., 1974, Fibre-entrapped enzymes, in: *Insolubilized Enzymes* (M. Salmona, C. Saronia, and S. Garattini, eds.), p. 189, Raven Press, New York.

Sampson, D., Hersh, L. S., Cooney, D., and Murphy, G. P., 1972, A new technique employing extracorporeal chemotherapy, *Trans. Amer. Soc. Artificial Internal Organs* **18**:54.

Sekiguchi, W., and Kondo, A., 1966, Studies on microencapsulated hemoglobin, *J. Japan. Soc. Blood Transfusion* **13**:153.

Siu Chong, E. D., and Chang, T. M. S., 1974, *In-vivo* effects of intraperitoneally injected L-asparaginase solution and L-asparaginase immobilized within semipermeable nylon microcapsules with emphasis on blood L-asparaginase, "body" L-asparaginase, and plasma L-asparagine levels, *Enzyme* **18**:218.

Sparks, E. R., Mason, N. S., Meier, P. M., Litt, M. H., and Lindan, O., 1971, Removal of uremic waste metabolites from the intestinal tract by encapsulated carbon and oxidized starch, *Trans. Amer. Soc. Artificial Internal Organs* **17**:229.

Thorne, S. R., Fiddler, M. B., and Desnick, R. J., 1975, Enzyme therapy V: *in-vivo* fate of β-glucuronidase deficient mice, *Pediat. Res.* **9**:918.

Updike, S. J., Shults, M. C. N., and Magnuson, J., 1973a, Catalase for oxygenator membranes, *J. Appl. Physiol.* **34**:271.

Updike, S. J., Prieve, C., and Magnuson, J., 1973b, Immobilization in hypoallergenic gel, a method of protecting enzymes from proteolysis and antibody complexing, in: *Enzyme Therapy in Genetic Diseases, Birth Defects* (Original Article Series, Vol. 9), p. 77, The Williams & Wilkins Company, Baltimore.

Venkatasubramanian, K., Vieth, W. R., and Bernath, F. R., 1974, Use of collagen immobilized enzymes in blood treatment, in: *Enzyme Engineering,* Vol. 2 (E. K. Pye and L. B. Wingard, Jr., eds), p. 439, Plenum Press, New York.

Venter, J. C., Venter, B. R., Dixon, J. E., andKaplan, N. O., 1975, Extracorporeal immobilized enzyme reactors, *Biochem. Med.* **12**:79.

9

L-Asparaginase as a Model for Enzyme Therapy of Substrate-Dependent Tumors

E. D. SIU CHONG AND T. M. S. CHANG

1. *Introduction*

A great deal of work is being carried out using L-asparaginase as a model system for the "selective nutritional deprivation" approach to cancer therapy. Certain tumor cells, unlike normal cells, cannot synthesize enough asparagine and depend on extracellular supplies for asparagine. Selective depletion of this extracellular source of asparagine by L-asparaginase can lead to the suppression of asparagine-dependent tumors.

Problems related to the therapeutic applications of asparaginase solution are in many cases similar to problems related to enzyme therapy in general. Furthermore, most centers working on the therapeutic applications of immobilized enzymes have used asparaginase as a model system. A brief review of asparaginase at this point would therefore serve as an introduction to the subsequent chapters on immobilized enzyme therapy. More detailed review articles (Adamson and Fabro, 1968; Broome, 1968b; Boiron and Levy, 1969; Cooney and Handschumacher, 1970; Capizzi *et al.*, 1970; Crowther, 1971; Weinberger, 1971; Wriston and Yellin, 1973; McCredie *et al.*, 1973; Capizzi and Handschumacher, 1973) and proceedings from two international symposia on L-asparaginase (Grundmann and Oettgen, 1970; International Symposium on L-Asparaginase, Paris, 1971) are available in the literature.

2. *Historical Background*

Kidd (1953a, 1953b) observed that injection of normal guinea pig serum caused the complete regression of established Gardner 6C3HED lymphosarcoma and the Lorenz lymphoma II in mice, and delayed the appearance of the Murphy–

E. D. SIU CHONG • Department of Physiology, McGill University, Montreal, Canada T. M. S. CHANG • Departments of Physiology and Medicine, McGill University, Montreal, Canada

Sturm lymphosarcoma in rats. Sera from several other species, namely the horse, rabbit, and human, were without effect. Susceptibility to the effects of guinea pig serum was confined to certain tumor cell lines. It was thought that guinea pig serum acted through its high complement activity, thus augmenting an immunological reaction between the host and the implanted tumor. However, no fraction of complement could be identified with tumor-inhibitory properties, and rabbit antisera, which also contain a high degree of complement activity, did not have any antitumor effect. Broome (1961) began a search to find constituents in guinea pig serum which were not present in other animal sera found to be ineffective against tumors. He came across a paper published by Clementi (1922) showing that L-asparaginase was present in the liver and kidney of many species, but occurred in the blood of only one of the animals he studied, the guinea pig. Broome compared the antitumor properties of guinea pig serum with its L-asparaginase activity, and in a series of papers (Broome, 1961, 1963a, 1963b) convincingly provided evidence that the L-asparaginase activity of guinea pig serum was responsible for its antilymphoma effects. Furthermore, Neuman and McCoy (1956) and Haley *et al.* (1961) showed that cells of the Walker carcinosarcoma 256 and the L5178Y mouse leukemia had an absolute nutritional requirement for growth *in vitro* for L-asparagine. Mashburn and Wriston (1964) discovered that L-asparaginase present in *Escherichia coli* had tumor-inhibitory properties. This made it possible to obtain larger quantities of enzyme for extensive investigation and evaluation. Subsequent studies showed that *E. coli* L-asparaginase was also effective against several other types of murine leukemias and solid tumors (Adamson and Fabro, 1968; Boyse *et al.*, 1967) and spontaneous dog lymphosarcoma (Old *et al.*, 1967).

Before adequate supplies of *E. coli* L-asparaginase became available, De Barros *et al.* (1965) treated one patient with melanoma using crude agouti serum L-asparaginase, and Dolowy *et al.* (1966) treated one patient with acute lymphocytic leukemia using crude guinea pig serum L-asparaginase. When adequate supplies of L-asparaginase from *E. coli* did become available, more extensive clinical trials were carried out by Hill *et al.* (1967) and by Oettgen *et al.* (1967). Since then, intensive research on this enzyme has been carried out both at the basic and clinical levels.

3. *Distribution and Sources of L-Asparaginase*

L-Asparaginase is widely distributed, being found in animal, plant, and microbial sources. It was first found in guinea pig serum by Clementi (1922), who also reported its absence from sera of other common mammals. It has since been found in the sera of four other members of the superfamily Cavioidea to which the guinea pig belongs, namely the capybara, the Patagonian hare, the paca, and the agouti (Holmquist, 1962, 1963; Old *et al.* 1963). With the exception of a report of traces of L-asparaginase activity in rabbit serum (Tower *et al.*, 1963), and its presence in the sera of two species of New World monkeys (Broome, 1968b; Peters *et al.*, 1970), L-asparaginase appears to be absent from the sera of all other mammalian species.

The enzyme is, however, present in the tissues of several animals. It has been found in the liver of the rat (Errera and Greenstein, 1947; Patterson and Orr, 1969; Bonetti *et al.,* 1969; McGee *et al.,* 1971), chicken (Ohnuma *et al.,* 1967), and guinea pig (Krebs, 1950; Suld and Herbut, 1965, 1970), and in the liver and kidney of certain birds (Clementi, 1922). It is also present in the tissues of fish (Adamson and Fabro, 1968). A recent detailed report (Milman and Cooney, 1974) on the distribution of L-asparaginase in the principal organs of representative mammals and birds indicate that L-asparaginase is present in the pancreas, liver, brain, ovary, spleen, lung, testes, and kidney of many of the animals surveyed. L-Asparaginase has also been found in a plant source, barley rootlets (Adamson and Fabro, 1968).

L-Asparaginase is present in many microorganisms, especially bacteria, making these the major sources of today's supply of L-asparaginase. It should be noted that only some of these bacterial L-asparaginases possess antineoplastic activity, most notably *Escherichia coli, Erwinia carotovora,* and *Serratia marcescens.* A review of the various types of microorganisms in which L-asparaginase has been found is available in the literature (Wriston and Yellin, 1973; Roberts *et al.,* 1968; Peterson and Ciegler, 1969; Bilimoria, 1969; Tosa *et al.,* 1971).

4. *Some Properties of E.* coli *L-Asparaginase EC-2 (L-Asparagine Amidohydrolase; EC 3.5.1.1)*

The properties of this enzyme have been described in detail in several papers (Wriston and Yellin, 1973; Cooney and Handschumacher, 1970; Campbell *et al.,* 1967; Whelan and Wriston, 1969; Rauenbusch *et al.,* 1970; Irion and Arens, 1970). The enzyme is quite stable in aqueous solution in the pH range 5–9. At room temperature under sterile conditions, enzyme activity is preserved for weeks. At 60°C, the enzyme loses its catalytic activity within 1 hr. The enzyme is relatively stable in the presence of organic solvents such as acetone, alcohol, and others. The enzyme has an isolectric point of 4.6–5.5. It has a broad pH optimum between pH 5 and 8.5. K_m values for *E. coli* L-asparaginase have been reported to be 1.15×10^{-5} M–1.25×10^{-5} M. Although there is some discrepancy in the literature with respect to the exact molecular weight of *E. coli* L-asparaginase, the generally accepted value is 130,000. The enzyme has a tetrameric structure composed of four identical subunits. Electron microscopy reveals a generally globular appearance for the whole molecule and supports the existence of a tetrameric structure with four approximately spherical subunits.

The substrate specificity of L-asparaginase appears to be restricted to 4 and 5-carbon L- or D- α-amino acids with an ω-nitrogen function. The hydrolysis of L-glutamine by L-asparaginase is an intrinsic property of the enzyme and is not due to the presence of contaminating L-glutaminase. *Escherichia coli* L-asparaginase has 3–4% of the activity toward L-glutamine that it has toward L-asparagine. Most of the known L-asparaginases, with the notable exception of guinea pig serum L-asparaginases, have inherent L-glutaminase activities.

5. *Spectrum of Sensitivity of Tumors to* L*-Asparaginase*

Many animal tumors have been shown to be inhibited by L-asparaginase (Adamson and Fabro, 1968; Broome, 1968b; Cooney and Handschumacher, 1970). L-Asparaginase is effective against over 50 neoplasms in the mouse, including the L5178Y and EARAD1 leukemias, the 6C3HED and P1798 lymphosarcomata, and the Lorenz lymphoma (Kidd, 1953a; Boyse *et al.*, 1967; Horowitz *et al.*, 1968), and the Murphy-Sturm lymphosarcoma, the Walker 256 carcinosarcoma, the ACMC42 and the Jensen sarcomata in the rat (Kidd, 1953a; Jameson *et al.*, 1956; Kwak *et al.*, 1961), and spontaneous lymphosarcoma in the dog (Old *et al.*, 1967; Hill *et al.*, 1967).

Since the initial clinical trials mentioned earlier (De Barros *et al.*, 1965; Dolowy *et al.*, 1966; Hill *et al.*, 1967; Oettgen *et al.*, 1967), more extensive clinical trials with L-asparaginase have been conducted in many centers in well over 1000 cancer patients (Grundmann and Oettgen, 1970; International Symposium on L-Asparaginase, Paris, 1971; Haskell *et al.*, 1969a; Whitecar *et al.*, 1970; Ohnuma *et al.*, 1970, 1971; Clarkson *et al.*, 1970; Tallal *et al.*, 1970; Beard *et al.*, 1970; Capizzi *et al.*, 1971a; Jaffe *et al.*, 1971; Sutow *et al.*, 1971; Oettgen, 1971). These studies have shown that L-asparaginase is most effective in the treatment of acute lymphocytic leukemia, producing complete remissions in about 60% of patients treated, most of whom were resistant to conventional antileukemic drugs. The response rate in patients with acute myelocytic, myelomonocytic, or monocytic leukemia has been much less (10–15%). Occasional responses have been obtained in patients with malignant melanoma, chronic myelocytic leukemia, lymphosarcoma, and reticulum cell sarcoma. No significant responses have been observed in patients with other types of solid tumors.

6. *Mechanism of Antitumor Action*

Although the precise mechanism of the antitumor activity of L-asparaginase is not fully understood, the most widely accepted hypothesis is based on the catalytic action of the enzyme in the hydrolysis of L-asparagine to L-aspartic acid and ammonia. L-Asparaginase-sensitive tumors, in contrast to most normal cells and L-asparaginase-resistant tumors, require an external source of L-asparagine for optimal growth. Thus, administration of L-asparaginase deprives dependent tumor cells of their extracellular source of L-asparagine (Broome, 1963b, 1968c; Boyse *et al.*, 1967; Sobin and Kidd, 1965, 1966). It has been shown that these L-asparagine-dependent cells cannot synthesize sufficient endogenous L-asparagine due to very low levels of the enzyme, L-asparagine synthetase, or the inability of these cells to increase L-asparagine synthetase activity after L-asparaginase administration (Horowitz *et al.*, 1968; Broome and Schwartz, 1967; Prager and Bachynsky, 1968a, 1968b; Patterson and Orr, 1967).

At the present time, the only known pathways of L-asparagine metabolism in animal tissues are conversion to L-aspartate by hydrolysis or to α-ketosuccinamic acid by transamination, and utilization for protein synthesis (Adamson and Fabro, 1968; Meister, 1965). Because no major route of L-asparagine utilization, other than protein synthesis, is known, it is believed that a lack of L-asparagine results in the inhibition of synthesis of certain proteins necessary for the growth of neoplastic cells. On the other hand, it is possible that there may exist other routes of L-asparagine utilization, which though not yet detected, are essential to the survival of tumor cells. Sobin and Kidd (1965, 1966) and Broome and Schwartz (1967) have shown that protein synthesis is inhibited when sensitive cells are deprived of L-asparagine. Bosmann and Kessel (1970) studied the effect of L-asparaginase on glycoprotein synthesis in L5178Y cell suspensions and suggested that hydrolysis of cell membrane glycoproteins, as well as inhibition of their synthesis by the enzyme, cause rapid cell lysis.

Following inhibition of protein synthesis, DNA synthesis, then RNA synthesis are inhibited (Sobin and Kidd, 1966; Ellem *et al.*, 1970). Although the inhibition of DNA synthesis has been regarded as a secondary effect of disturbed protein metabolism, Hirschmann *et al.* (1971) have suggested that L-asparaginase has a direct influence on DNA metabolism, perhaps by inhibition of *de novo* purine synthesis via some unknown mechanism. A preliminary report by Woods and Dixon (1968) also suggests that the C skeleton of L-asparagine may serve as a direct precursor in the biosynthesis of certain pyrimidine nucleotides isolated from lymphosarcoma cells. It has been proposed (Ellem *et al.*, 1970) that L-asparaginase could stop histone synthesis and thereby inhibit DNA replication. Another *in vivo* effect of L-asparaginase treatment that may relate to the mechanism of tumor death is the results by Mashburn and Wriston (1966) and Mashburn and Landin (1968), who found an increase in both acid and alkaline ribonuclease activities after L-asparaginase treatment of the P1798 lymphosarcoma.

In addition to its effects on protein and nucleic acid metabolism, L-asparaginase treatment may also result in other indirect effects which are harmful to neoplastic cells. For example, Broome (1968c) has suggested that the high levels of L-aspartic acid observed after L-asparaginase treatment may be toxic or may significantly alter metabolic control mechanisms in the tumor cells. Miller *et al.* (1969) proposed that the inherent L-glutaminase activity of L-asparaginase may enhance its antitumor effect by lowering the circulating levels of L-glutamine which is the nitrogen donor for the biosynthesis of L-asparagine in mammalian cells. Ryan and Dworak (1970) and Ryan and Sornson (1970) have shown that administration of L-asparaginase depresses the glycine level in susceptible but not resistant 6C3HED tumors, and suggested that the loss of cellular glycine may be more important than loss of L-asparagine because of the requirement of glycine for purine synthesis. The latter findings may be related to preliminary *in vitro* results obtained in a recent study by Chattopadhyay *et al.* (1973), who reported that glycine inhibits the activity of L-asparaginase from guinea pig liver.

The relation of all these possible effects to the tumor-inhibitory properties of L-asparaginase have not yet been fully evaluated, but inhibition of protein synthesis appears to be of the most significance.

7. *Factors That Influence the Tumor-Inhibitory Effectiveness of* L-*Asparaginase*

There are two major factors that determine the tumor-inhibitory properties of a particular L-asparaginase preparation. These are the affinity of the enzyme for substrate, its K_m; and the clearance rate of the enzyme from the host's circulation, its plasma half-life.

A low K_m (i.e., a high affinity for substrate) at physiological pH is an absolute requirement for antitumor activity. This is especially important since L-asparagine concentration must be lowered to at least 10^{-5} M (10 nmoles/ml) in the medium to become rate-limiting in protein synthesis (Broome, 1968a, 1968c; Sobin and Kidd, 1965, Broome and Schwartz, 1967). Indeed, the difference in the tumor-inhibitory activity of two distinct L-asparaginases isolated from *E. coli* (Mashburn and Wriston, 1964; Campbell *et al.,* 1967; Roberts *et al.,* 1966; Schwartz *et al.,* 1966) has been related to their K_m values: EC-1 enzyme, with low affinity, has no tumor-inhibitory action, whereas EC-2 has high affinity and is a potent antitumor agent (Broome, 1968b). Broome (1968a, 1968b) has similarly attributed the relative tumor inhibitory effectiveness of L-asparaginases from *E. coli* and agouti serum to the difference in their K_m values, the more potent *E. coli* enzyme having a lower K_m. It should be noted that *Erwinia carotovora* L-asparaginase, which is clinically active, has a (low) K_m of 1.0×10^{-5} (Howard and Carpenter, 1972).

The more rapidly the injected enzyme is removed from the host's circulation, the lower is the antitumor activity. The clearance rate of the enzyme itself is influenced by the following factors. The source of the enzyme is an important factor. Thus, the lack of tumor-inhibitory activity of L-asparaginases from yeast (Broome, 1965), *Bacillus coagulans* (Law and Wriston, 1971), chicken liver (Ohnuma *et al.,* 1967), and other sources have been related to the short plasma half-lives of these enzymes. Guinea pig (Broome, 1965) and agouti (Broome, 1968a) serum L-asparaginases have long half-lives and are potent antitumor agents. Even different commercial preparations derived from a common source, for example *E. coli,* vary in their plasma half-lives to some extent (Wriston and Yellin 1973; Mashburn and Landin, 1970).

Mashburn and Landin (1970) suggested that the rate of clearance of injected L-asparaginase can be correlated with the isoelectric point of the enzyme. Rutter and Wade (1971) suggested that the isoelectric point of L-asparaginase may be related to the ease of phagocytosis of the enzyme from the plasma. The half-life of the enzyme in the plasma not only depends on the properties of the enzyme itself, but also on the physiological state of the host.

The lactic dehydrogenase-elevating virus (LDH virus) is present as a contaminant in most of the mouse tumors generally employed for L-asparaginase studies. Studies by Riley's group (Riley, 1968; Riley *et al.,* 1970, 1971, 1974) resulted in the proposal that the presence of this virus impairs the capacity of the host to clear protein from the blood, thus prolonging the plasma half-life of the enzyme.

Many studies (Roberts *et al.,* 1966; Vadlamudi *et al.,* 1970, 1971; Prager, 1971; Goldberg *et al.,* 1973; Baechtel and Prager, 1973; Khan and Hill, 1969; Peterson *et al.,* 1971) have demonstrated that the presence of circulating, humoral anti-L-asparaginase antibodies inhibits the catalytic activity of injected enzyme and causes a dramatically accelerated removal of the enzyme from the bloodstream of immunized hosts, thus reducing the therapeutic value of the enzyme.

Finally, estrogen pretreatment of tumor-bearing mice have been shown to result in increased clearance rate of injected L-asparaginase (Broome, 1965). This is consistent with the observation of Kim (1962), later confirmed by Broome (1965), that estrogen treatment decreases the tumor-inhibitory effectiveness of guinea pig serum. It has been suggested that estrogen treatment may cause an increase in the phagocytic activity of the reticuloendothelial system for removing injected substances (Broome, 1965).

8. *Use of L-Asparaginase in Acute Lymphocytic Leukemia Therapy*

The following observations have been made in regard to the use of L-asparaginase for asparagine-dependent acute lymphocytic leukemia (Zubrod, 1970; Crowther, 1971; Grundmann and Oettgen, 1970; International Symposium on L-Asparaginase, Paris, 1971; Ho *et al.,* 1970): It has a good remission rate and a fairly rapid speed of response. It has the highest therapeutic index (1000) among all clinically available antileukemic agents. All other drugs in the treatment of acute leukemia have therapeutic indexes of <10. Its lack of cross-resistance with other antileukemic agents allows asparaginase to be effective at a very late stage of the disease when the patient has become resistant to other chemotherapeutic agents. Its lack of significant bone marrow depression in a situation where both disease and other antileukemic drugs depress hematopoiesis makes it a prime candidate for use in combination with a variety of other antileukemic but myelosuppressive drugs.

9. *Problems Associated with L-Asparaginase Therapy*

Unfortunately, there are major drawbacks related to the clinical application of L-asparaginase. These include a wide spectrum of adverse side effects, and the development of resistance to repeated L-asparaginase therapy. A large variety of untoward side effects have been observed with the clinical use of L-asparaginase (Zubrod, 1970; Capizzi *et al.,* 1970; Capizzi and Handschumacher, 1973; Grundmann and Oettgen, 1970; Oettgen, 1971; Haskell *et al.,* 1969b; Oettgen *et al.,* 1970, Land *et al.,* 1972). Some of the general side effects reported include anorexia, nausea, vomiting, chills, fever, diarrhea, and weight loss. Liver toxicity was manifested by increased levels of serum glutamic-oxaloacetic transaminase, alkaline phosphatase, 5′-nucleotidase, bilirubin, increased retention of bromsulphalein, and decreased levels of serum albumin (causing generalized edema and low serum calcium), fibrinogen, and other coagulation factors. Fatty metamorphosis of the liver has also been observed. Pancreatic dysfunction was manifested as pancreatitis, and by increased levels of serum amylase, lipase, and blood glucose, with decreased levels of serum insulin, causing diabetes to develop in some cases. Renal dysfunction was manifested by increased levels of blood urea nitrogen and acute renal failure in some cases. Hematological changes include granulocytopenia, lymphocytopenia, and thrombocytopenia. Effects on the central nervous system include headache, drowsiness, lethargy, depression, disorientation, and confusion. Immune reactions to L-asparaginase have developed with hypersensitivity reactions occurring in approximately one-third of treated patients. These reactions were usually of the anaphylactic type, characterized by urticaria, abdominal cramps, hypotension, cyanosis, respiratory distress, and facial edema. Coma and even fatal anaphylaxis has also been reported, as a result of immunological complications.

This wide spectrum of effects has been attributed to a number of factors. Thus, L-asparagine depletion may result in inhibition of protein synthesis in the liver, pancreas, and brain. Also, with repeated injections, the enzyme preparation may evoke hypersensitivity reactions and other immunological complications in the host receiving the injection. Several groups have reported the presence of anti-L-asparaginase antibodies in the sera of L-asparaginase-treated patients (Khan and Hill, 1969; Peterson *et al.*, 1971; Oettgen *et al.*, 1970; Reis and Schmidt, 1970; Dohlwitz *et al.*, 1970; Capizzi *et al.*, 1971a; Peterson *et al.*, 1971). The presence of reagin-type antibodies (IgE) has also been detected in patients who experienced anaphylaxis (Capizzi *et al.*, 1971a; Peterson *et al.*, 1971; Dohlwitz *et al.*, 1970). Possible contamination of the enzyme preparation with bacterial substances such as endotoxins may give rise to some of these side effects. Indeed, bacterial endotoxins are known to cause side effects which resemble many of those observed after L-asparaginase therapy: nausea, chills, fever, vomiting, weakness, hypotension, respiratory distress, coagulopathies, hepatic injuries, renal failure, and delayed hypersensitivity reactions (Ohnuma *et al.*, 1970; Loos *et al.*, 1972; Raskova and Vanecek, 1964; Zweifach and Jenoff, 1965).

The other major limitation of L-asparaginase treatment is development of clinical resistance to the antileukemic effect of the enzyme. There have been four major mechanisms proposed to explain the origin of resistance to L-asparaginase therapy. A marked increase in L-asparagine synthetase levels has been shown to occur in tumor cells, probably by a mechanism of derepression of L-asparagine synthetase, so that the tumor cells are no longer dependent on an external supply of L-asparagine (Horowitz *et al.*, 1968; Broome and Schwartz, 1967; Prager and Bachynsky, 1968a, 1968b; Patterson and Orr, 1967; Haskell and Canellos, 1969). In this regard, Gallo *et al.* (1970a,b) and Gallo and Adamson (1971) have shown that L-asparaginase-resistant cells contain reduced levels of an asparaginyl-tRNA species which they proposed normally acts as a corepressor for L-asparagine biosynthesis. Phenotypically stable L-asparagine-independent mutant cells may arise spontaneously, leading to the development of L-asparaginase-resistant cell populations (Summers and Handschumacher, 1973; Uren *et al.*, 1974). Another mechanism of resistance might involve the development of more efficient means by tumor cells of extracting L-asparagine from the plasma, erythrocytes, or other normal host cells (Broome, 1968c; Cooney *et al.*, 1970). Because bacterial L-asparaginase is a large protein which is foreign to the recipient host, it is not surprising that repeated injections of the enzyme will result in the production of humoral antibodies against the enzyme. This has been reported to occur in several studies (Capizzi *et al.*, 1971a; Roberts *et al.*, 1966; Vadlamudi *et al.*, 1970, 1971; Prager, 1971; Goldberg *et al.*, 1973; Baechtel and Prager, 1973; Khan and Hill, 1969; Peterson *et al.*, 1971; Oettgen *et al.*, 1970; Reis and Schmidt, 1970; Dohlwitz *et al.*, 1970). Humoral antibodies have been shown to inhibit the catalytic activity of injected L-asparaginase and cause an accelerated *in vivo* removal of the enzyme from the host's circulation (Roberts *et al.*, 1966; Vadlamudi *et al.*, 1970, 1971; Prager, 1971; Goldberg *et al.*, 1973; Baechtel and Prager, 1973; Khan and Hill, 1969; Peterson *et al.*, 1971), thus reducing the antitumor effectiveness of the enzyme. Of these two factors, the shortened plasma half-life of the enzyme appears to be the more important, especially since the maximum inhibition of catalytic activity by immune serum has been found to be 45–53% (Roberts *et al.*, 1966; Prager, 1971; Baechtel and Prager, 1973; Khan and Hill, 1969; Peterson *et al.*, 1971).

10. Combination Chemotherapy

In attempts to reduce the problem of tumor cell resistance to L-asparaginase while increasing the total tumor cell kill, L-asparaginase has been studied for use in combination with other chemotherapeutic agents. There are two basic concepts which form the rationale for the use of several chemotherapeutic agents in combination against a particular neoplasm. If two or more drugs have different toxicities for normal tissues and are effective against a particular type of neoplasm, the antitumor effects may be potentiated without concomitant additive toxic effects for any given normal tissue. The use of drugs with different mechanisms of action may decrease the development of resistance by reducing the chances for survival of tumor cells resistant to a single agent.

Three properties of L-asparaginase in particular—its high therapeutic index, different expression of host toxicity, and lack of cross-resistance with other antitumor agents—led to extensive investigation of its use in combination with other drugs. Indeed, Burchenal (1971), paraphrasing the well-known statement of Sir Winston Churchill, appropriately stated, "never have so many compounds with so many different mechanisms of action potentiated the antileukemic activity of one agent." Agents that have been shown to have synergistic effects when used in various combinations with L-asparaginase include prednisone, vincristine, daunomycin, cytosine arabinoside, thioguanine, cyclophosphamide, 1,3-bis(2-chloroethyl)-1-nitrosourea (BCNU), azotomycin, 5-fluorouracil and 5-azacytidine (Grundmann and Oettgen, 1970; International Symposium on L-Asparaginase, Paris, 1971; Jacobs *et al.*, 1970; Capizzi *et al.*, 1971b; Mashburn, 1971; Avery and Roberts, 1973; Pratt *et al.*, 1973; Vadlamudi *et al.*, 1973). No attempt will be made to describe the results obtained in these studies. However, two interesting findings which have emerged from these reports deserve attention. First, the addition of one or more antileukemic agents to L-asparaginase therapy does not necessarily result in any therapeutic advantage. Second, therapeutic synergism for a given combination of drugs may be critically schedule-dependent: one example being whether the drugs are administered simultaneously or sequentially.

11. *Future Prospects and Perspectives*

From experience with L-asparaginase, certain basic, important concepts have emerged concerning the use of enzymes for cancer therapy in general. For example, such experience suggests that if other potential antitumor enzymes are to be clinically active, they should possess the prerequisite properties of stability and catalytic activity at physiologic pH, a low K_m value, and an appreciable plasma half-life. Negative results obtained have also been important insofar as it is only when present limitations are clearly defined that future enzymic approaches to cancer therapy could be developed with increased hopes for success. It has been mentioned that the toxicity observed during L-asparaginase therapy may be partly attributed to the antigenic nature of the foreign enzyme preparation and the presence of accompanying contaminants such as bacterial endotoxins. These difficulties represent fundamental problems which limit the therapeutic usefulness of enzymes in general, since most enzymes used in clinical therapy are derived from heterogeneous sources. Moreover, repeated injections of the enzyme may lead to the production of humoral antibodies which results in the inhibition and rapid *in vivo* removal of the enzyme. If some of these difficulties could be overcome or circumvented, the potential use of enzymes in clinical medicine, in general, and cancer chemotherapy, in particular, would be extended.

A major advance in enzyme technology in this direction has come from the initial demonstration of the *in vivo* effectiveness of microencapsulated enzymes (Chang, 1964, 1972, 1973a, 1975), including L-asparaginase (Chang 1969, 1971, 1973b; Siu Chong and Chang, 1974, 1976a, 1976b). With this demonstration, many other types of immobilized L-asparaginase have since been developed and studied. These studies will be described in detail in subsequent chapters. These studies may pave the way to further applications of enzymes in cancer therapy, especially since other enzyme systems are being investigated for use in the selective depletion of specific substrates. These include glutaminase (Greenberg *et al.*, 1964; El-Asmar and Greenberg, 1966; Roberts *et al.*, 1970, 1972; Schrek *et al.*, 1971; Broome and Schenkein, 1971; Riley *et al.*, 1974; Schmid and Roberts, 1974), serine dehydratase (Regan *et al.*, 1969; Pizer and Regan, 1972) arginase (Holley, 1967; Sasada and Terayama, 1969), phenylalanine ammonia-lyase (Abell *et al.*, 1972; Ohnuma *et al.*, 1973), and, most recently, leucine dehydrogenase (Oki *et al.*, 1973).

12. *References*

Abell, C. W., Stith, W. J., and Hodgins, D. S., 1972, The effects of phenylalanine ammonia-lyase on leukemic lymphocytes *in vitro, Cancer Res.* **32**:285.

Adamson, R. H., and Fabro, S., 1968, Antitumor activity and other biologic properties of L-asparaginase (NSC-109229—review, *Cancer Chemotherapy. Rept.* **52**:617.

Avery, T. L., and Roberts, D., 1973, Combination chemotherapy with cytosine arabinoside, L-asparaginase, and 6-azauridine for transplantable murine leukemias, *Cancer Res.* **33**:391.

Baechtel, S., and Prager, M. D., 1973, Basis for loss of therapeutic effectiveness of L-asparaginase in sensitized mice, *Cancer Res.* **33**:1966.

Beard, M. E. J., Crowther, D., Galton, D. A. G., Guyer, R. J., Fairley, G. H., Kay, H. E. M., Knapton, P. J., Malpas, J. S., and Scott, R. B., 1970, L-Asparaginase in treatment of acute leukemia and lymphosarcoma, *Brig. Med. J.* **1**:191.

Bilimoria, M. H., 1969, Conditions for the production of L- asparaginase 2 by coliform bacteria, *Appl. Microbiol.* **18**:1025.

Boiron, M., and Levy, D., 1969, Asparaginase 1969, *Pathol. Biol.* **17**:1129.

Bonetti, E., Abbondanza, A., Della Corte, E., and Stirpe, F., 1969, The regulation of L-asparaginase activity in rats and mice: effects of normal and malignant growth, of sex and of dietary changes, *Biochem. J.* **115**:597.

Bosmann, H. B., and Kessel, D., 1970, Inhibition of glycoprotein synthesis in L5178Y mouse leukemia cells by L-asparaginase *in-vitro, Nature* **226**:850.

Boyse, E. A., Old, L. J., Campbell, H. A., and Mashburn, L. T., 1967, Suppression of murine leukemias by L-asparaginase—incidence of sensitivity among leukemias of various types—comparative inhibitory activities of guinea pig serum L-asparaginase and *Escherichia coli* L-asparaginase, *J. Exptl. Med.* **125**:17.

Broome, J. D., 1961, Evidence that the L-asparaginase activity of guinea pig serum is responsible for its antilymphoma effects, *Nature* **191**:1114.

Broome, J. D., 1963a, Evidence that the L-asparaginase of guinea pig serum is responsible for its antilymphoma effects: I. Properties of the L-asparaginase of guinea pig serum in relation to those of the antilymphoma substance, *J. Exptl. Med.* **118**:99.

Broome, J. D., 1963b, Evidence that the L-asparaginase of guinea pig serum is responsible for its antilymphoma effects: II. Lymphoma 6C3HED cells cultured in a medium devoid of L-asparaginase lose their susceptibility to the effects of guinea pig serum *in-vivo, J. Exptl. Med.* **118**:121.

Broome, J. D., 1965, Antilymphoma activity of L-asparaginase *in-vivo:* clearance rates of enzyme preparations from guinea pig serum and yeast in relation to their effect on tumor growth, *J. Natl. Cancer Inst.* **35**:967.

Broome, J. D., 1968a, Factors which may influence the effectiveness of L-asparaginases as tumor inhibitors, *Brit. J. Cancer* **22**:595.

Broome, J. D., 1968b, L-Asparaginase: the evolution of a new tumor inhibitory agent, *Trans. N.Y. Acad. Sci.* **30**:690.

Broome, J. D., 1968c, Studies on the mechanism of tumor inhibition by L-asparaginase. Effects of the enzyme on asparagine levels in the blood, normal tissues, and 6C3HED lymphomas of mice: differences in asparagine formation and utilization in asparaginase-sensitive and -resistant lymphoma cells, *J. Exptl. Med.* **127**:1055.

Broome, J. D. and Schenkein, I., 1971, in: *La L-asparaginase,* Du Centre National de la Recherche Scientifique, Paris, p. 95.

Broome, J. D., and Schwartz, J. H., 1967, Differences in the production of L-asparagine in asparaginase-sensitive and resistant lymphoma cells, *Biochim. Biophys. Acta* **138**:637.

Burchenal, J. H., 1971, in: *La L-asparaginase,* Du Centre National de la Recherche Scientifique, Paris, p. 243.

Campbell, H. A., Mashburn, L. T., Boyse, E. A., and Old, L. J., 1967, Two L-asparaginase from *Escherichia coli* B: their separation, purification, and antitumor activity, *Biochemistry* **6**:721.

Capizzi, R. L., and Handschumacher, R. E., 1973, Asparaginase, in: *Cancer Medicine,* (J. F. Holland and E. Frei III eds.), p. 850, Lea & Febiger, Philadelphia.

Capizzi, R. L., Bertino, J. R., and Handschumacher, R. E., 1970, L-Asparaginase, *Ann. Rev. Med.* **21**:433.

Capizzi, R. L., Bertino, J. R., Skeel, R. T., Creasey, W. A., Zanes, R., Olayon, C., Peterson, R. G., and Handschumacher, R. E., 1971a, L-Asparaginase: clinical, biochemical, pharmacological, and immunological studies, *Ann. Int. Med.* **74**:893.

Capizzi, R. L., Summers, W. P., and Bertino, J. R., 1971b, L-Asparaginase induced alteration of amethopterin (methotrexate) activity in mouse leukemia L5178Y, *Ann. N.Y. Acad. Sci.* **186**:302.

Chang, T. M. S., 1964, Semipermeable microcapsules, *Science* **146**:524.

Chang, T. M. S., 1969, Asparaginase-loaded semipermeable microcapsules for mouse lymphomas, *Proc. Can. Federation Biol. Sci.* **12**:62.

Chang, T. M. S., 1971, The *in-vivo* effects of semipermeable microcapsules containing L-asparaginase on 6C3HED lymphosarcoma, *Nature* **229**:117.

Chang, T. M. S., 1972, *Artificial Cells,* Charles C. Thomas, Publisher, Springfield, Ill.

Chang, T. M. S., 1973a, Biomedical applications of artificial cells, *Bio-Med. Eng.* **8**:334.

Chang, T. M. S., 1973b, L-Asparaginase immobilized within semipermeable microcapsules: *in-vitro* and *in-vivo* stability, *Enzyme* **14**:95.

Chang, T. M. S., 1975, Immobilized enzymes and their biomedical applications, in: *Immobilized Enzymes, Antigens, Antibodies, and Peptides,* (H. H. Weetall, ed.), p. 245, Marcel Dekker, Inc., New York.

Chattopadhyay, S. K., Matthews, W. S., and Brown, H. D., 1973, Glycine inhibition of L-asparagine amidohydrolase activity, *J. Natl. Cancer Inst.* **50**:797.

Clarkson, B., Krakoff, I., Burchenal, J., Karnofsky, D., Golbey, R., Dowling, M., Oettgen, H., and Lipton, A., 1970, Clinical results of treatment with *E. coli* L-asparaginase in adults with leukemia, lymphoma, and solid tumors, *Cancer* **25**:279.

Clementi, A., 1922, La Désamination enzymatique de l'asparagine chez les différentes espèces animales et la signification physiologique de sa présence dans l'organisme, *Arch. Int. Physiol.* **19**:369.

Cooney, D. A., and Handschumacher, R. E., 1970, L-Asparaginase and L-asparagine metabolism, *Ann. Rev. Pharmacol.* **10**:421.

Cooney, D. A., Capizzi, R. L., and Handschumacher, R. E., 1970, Evaluation of L-asparagine metabolism in animals and man, *Cancer Res.* **30**:929.

Crowther, D., 1971, L-Asparaginase and human malignant disease, *Nature* **229**:168.

De Barros, T., Filho, M. C., De Santana, C. F., Valencia, M., Da Silva, M. P., Guedes, J., and De Carvalho, A. R. L., 1965, Utilizacao de L-asparaginase empaciente humano portador de neoplasia maligna, *An. Fac. Med. Univ. Recife* **25**:21.

Dohlwitz, A., Franźen, S., Holmgren, A., Killander, A., Killander, D., Wide, L., and Ahstrom, L., 1970, in: *Recent Results in Cancer Research* (E. Grundmann and H. F. Oettgen, eds.), **33**:198.

Dolowy, W. C., Henson, D., Cornet, J., and Sellin, H., 1966, Toxic and antineoplastic effects of L-asparaginase: study of mice with lymphoma and normal monkeys and report on a child with leukemia, *Cancer* **19**:116.

El-Asmar, F. A., and Greenberg, D. M., 1966, Studies on the mechanism of inhibition of tumor growth by the enzyme glutaminase, *Cancer Res.* **26**:116.

Ellem, K. A. O., Fabrizio, A. M., and Jackson, L., 1970, The dependence of DNA and RNA synthesis on protein synthesis in asparaginase-treated lymphoma cells, *Cancer Res.* **30**:515.

Errera, M., and Greenstein, J. P., 1947, Desamidation of glutamine and asparagine in normal and neoplastic hepatic tissues, *J. Natl. Cancer Inst.* **7**:285.

Gallo, R. C. and Adamson, R. H., 1971, in: *La L-asparaginase,* Du Centre National de la Recherche Scientifique, Paris, p. 121.

Gallo, R. C., Longmore, J. L., and Adamson, R. H., 1970a, Asparaginyl-tRNA and resistance of murine leukemias to L-asparaginase, *Nature* **227**:1134.

Gallo, R. C., Longmore, J., and Adamson, R. H., 1970b, The mechanism of asparaginase resistance of leukemic cells: correlations with alterations in asparaginyl-transfer RNA, *Proc. Amer. Assoc. Cancer Res.* **11**:29.

Goldberg, A. I., Cooney, D. A., Glynn, J. P., Homan, E. R., Gaston, M. R., and Milman, H. A., 1973, The effects of immunization to L-asparaginase on antitumor and enzymatic activity, *Cancer Res.* **33**:256.

Greenberg, D. M., Blumenthal, G., and Ramadan, M. E., 1964, Effect of administration of the enzyme glutaminase on the growth of cancer cells, *Cancer Res.* **24**:957.

Grundmann, E., and Oettgen, H. F. (eds.), 1970, Experimental and clinical effects of L-asparaginase, *Recent Results in Cancer Res.* **33**.

Haley, E. E., Fischer, G. A., and Welch, A. D., 1961, The requirement for L-asparagine of mouse leukemia cells L5178Y in culture, *Cancer Res.* **21**:532.

Haskell, C. M., and Canellos, G. P., 1969, L-Asparaginase resistance in human leukemia—L-Asparagine synethtase, *Biochem. Pharmacol.* **18**:2578.

Haskell, C. M., Canellos, G. P., Leventhal, B. G., Carbone, P. P., Block, J. B., Serpick, A. A., and Selawry, O. S., 1969a, L-Asparaginase—therapeutic and toxic effects in patients with neoplastic disease, *New Engl. J. Med.* **281**:1028.

Haskell, C. M., Canellos, G. P., Leventhal, B. G., Carbone, P. P., Serpick, A. A., and Hansen, H. H., 1969b, L-Asparaginase toxicity, *Cancer Res.* **29**:974.

Hill, J. M., Roberts, J., Loeb, E., Khan, A., MacLellan, A., and Hill, R. W., 1967, L-Asparaginase therapy for leukemia and other malignant neoplasms, *J. Amer. Med. Assoc.* **202**:882.

Hirschmann, W. D., Seeber, S., Warnecke, P., and Oerkermann, H., 1971, in: *La L-asparaginase,* Du Centre National de la Recherce Scientifique, Paris, p. 133.

Ho, D. H. W., Whitecar, J. P., Jr., Luce, J. K., and Frei, E., III, 1970, L-Asparagine requirement and the effect of L-asparaginase on the normal and leukemic human bone marrow, *Cancer Res.* **30**: 466.

Holley, R. W., 1967, Evidence that a rat liver "inhibitor" of the synthesis of DNA in cultured mammalian cells is arginase, *Biochim. Biophys. Acta* **145**:525.

Holmquist, N. D., 1962, Effect of normal agouti serum on 6C3HED lymphoma *in-vivo, Federation Proc.* **21**:166.

Holmquist, N. D., 1963, Effect of normal sera of several related rodents on 6C3HED lymphoma *in-vivo, Proc. Soc. Exptl. Biol. Med.* **113**:444.

Horowitz, B., Madras, B. K., Meister, A., Old, L. J., Boyse, E. A., and Stockert, E., 1968, Asparagine synthetase activity of mouse leukemias, *Science* **160**:533.

Howard, J. B., and Carpenter, F. H., 1972, L-Asparaginase from *Erwinia carotovora:* substrate specificity and enzymatic properties, *J. Biol. Chem.* **247**:1020.

International Symposium on L-Asparaginase, Paris, Nov. 28–30, 1971. *Colloques Internationaux du Centre National de la Recherche Scientifique No. 197.* Éditions du Centre National de la Recherche Scientifique, Paris.

Irion, E., and Arens, A., 1970, in: *Recent Results in Cancer Research* (E. Grundmann and H. F. Oettgen, eds.), **33**:39.

Jacobs, S. P., Wodinsky, I., Kensler, C. J., and Venditti, J. M., 1970, Combination therapy studies with L-asparaginase (NSC-109229) and other antitumor agents, *Cancer Chemotherapy Rept.* **54**:329.

Jaffe, N., Traggis, D., Das, L., Moloney, W. C., Hann, H. W., Kim, S. C., and Nair, R., 1971, L-asparaginase in treatment of neoplastic diseases in children, *Cancer Res.* **31**:942.

Jameson, E., Ainis, H., and Ryan, R. M., 1956, Action of guinea pig serum and human gamma globulin on the growth of a rat tumor, *Science* **124**:980.

Khan, A., and Hill, J. M., 1969, Neutralizing precipitin in the serum of a patient treated with L-asparaginase, *J. Lab. Clin. Med.* **73**:846.

Kidd, J. G., 1953a, Regression of transplanted lymphomas induced *in-vivo* by means of normal guinea pig serum: I. Cause of transplanted cancers of various kinds in mice and rats given guinea pig serum, horse serum, or rabbit serum, *J. Exptl. Med.* **98**:565.

Kidd, J. G., 1953b, Regression of transplanted lymphomas induced *in-vivo* by means of normal guinea pig serum: II. Studies on the nature of the active serum constituent: histological mechanism of the regression: tests for effects of guinea pig serum on lymphoma cells *in-vitro:* Discussion, *J. Exptl. Med.* **98**:583.

Kim, S. E., 1962, Effect of sex hormones on the growth of transplanted mouse lymphomas in various hosts: enhanced growth of 6C3HED lymphoma cells in mice treated with estrogenic hormones, *New Med. J. (Seoul, Korea)* **5**:73.

Krebs, H. A., 1950, Monometric determination of L-aspartic acid and L-asparagine, *Biochem. J.* **47**:605.

Kwak, K. S., Jameson, E., Ryan, R. M., and Kurtz, H. M., 1961, The effect of varying implant cell numbers of the inhibitory activity of guinea pig serum on Walker carcinosarcoma 256 in the rat, *Cancer Res.* **21**:44.

Land, V. J., Sutow, W. W., Fernbach, D. J., Lane, D. M., Williams, T. E., 1972, Toxicity of L-asparaginase in children with advanced leukemia, *Cancer* **30**:339.

Law, A. S., and Wriston, J. C., Jr., 1971, Purification and properties of *Bacillus coagulans, Arch. Biochem. Biophy.* **147**:744.

Loos, M., Vadlamudi, S., Meltzer, M., Shifrin, S., Borsos, T., and Goldin, A., 1972, Detection of endotoxin in commercial L-asparaginase preparations by complement fixation and separation by chromatography, *Cancer Res.* **32**:2292.

Mashburn, L. T.,1971, Combination chemotherapy with L-asparaginase (EC-2) using P1798 lymphosarcoma, *Cancer* **28**:1321.

Mashburn, L. T., and Landin, L. M., 1968, Changes in ribonuclease activities in P1798 lymphosarcoma after asparaginase treatment, *Arch. Biochem. Biophy.* **125**:721.

Mashburn, L. T. and Landin, L. M., 1970, in: *Recent Results in Cancer Research* (E. Grundmann and H. F. Oettgen, eds.), **33**:48.

Mashburn, L. T., and Wriston, J. C., Jr., 1964, Tumor inhibitory effect of L-asparaginase from *Escherichia coli, Arch. Biochem. Biophy.* **105**:450.

Mashburn, L. T., and Wriston, J. C., Jr., 1966, Change in ribonuclease concentrations in L-asparaginase-treated lymphosarcomata, *Nature* **211**:1403.

McCredie, K. B., Ho, D. H. W., and Freireich, E. J., 1973, L-Asparaginase for the treatment of cancer, *Ca—A Cancer Journal for Clinicians* **23**:220.

McGee, M., Greengard, O., and Knox, W. E., 1971, The developmental formation of asparaginase in liver and its distribution in rat tissues, *Enzyme* **12**:1.

Meister, A. (ed.), 1965, in: *Biochemistry of the Amino Acids,* Vol. 2, 2nd ed., p. 607, Academic Press, Inc., New York.

Miller, H. K., Salser, J. S., and Balis, M. E., 1969, Amino acid levels following L-asparagine amidohydrolase (EC 3.5.1.1) therapy, *Cancer Res.* **29**:183.

Milman, H. A., and Cooney, D. A., 1974, The distribution of L-asparagine synthetase in the principal organs of several mammalian and avian species, *Biochem. J.* **142**:27.

Neuman, R. E., and McCoy, T. A., 1956, Dual requirement of Walker carcinosarcoma 256 *in-vitro* for asparagine and glutamine, *Science* **124**:124.

Oettgen, H. F., 1971, Clinical evaluation of L-asparaginase, in: *Proceedings of the Tenth International Cancer Congress,* Vol. 2, p. 159, Year Book Medical Publishers, Chicago.

Oettgen, H. F., Old, L. J., Boyse, E. A., Campbell, H. A., Philips, F. S., Clarkson, B. D., Tallal, L., Leeper, R. D., Schwartz, M. K., and Kim, J. H., 1967, Inhibition of leukemias in man by L-asparaginase, *Cancer Res.* **27**:2619.

Oettgen, H. F., Stephenson, P. A., Schwartz, M. K., Leeper, R. D., Tallal, L., Tan, C. C., Clarkson, B. D., Golbey, R. B., Krakoff, I. H., Karnofsky, D. A., Murphy, M. L., and Burchenal, J. H ., 1970, Toxicity of *E. coli* L-asparaginase in man, *Cancer* **25**:253.

Ohnuma, T., Bergel, F., and Bray, R. C., 1967, Enzymes in cancer: asparaginase from chicken liver, *Biochem. J.* **103**:238.

Ohnuma, T., Holland, J. F., Freeman, A., and Sinks, L. F., 1970, Biochemical and pharmacological studies with asparaginase in man, *Cancer Res.* **30**:2297.

Ohnuma, T., Rosner, F., Levy, R. N., Cuttner, J., Moon, J. H., Silver, R. T., Blom, J., Falkson, G., Burningham, R., Gildewell, O., and Holland, J. F., 1971, Treatment of adult leukemia with L-asparaginase (NSC-109220), *Cancer Chemotherapy Rept.* **55**:269.

Ohnuma, T., Walczak, I. M., McKee, S. G., Roberts, J., and Havir, E., 1973, Effects of *E. coli* asparaginase (A-ase), *Erwinia carotovora* A-ase, *Acinetobacter* glutaminase (G-ase)/A-ase, *E. coli* G-ase, *Pseudomonas* histidase (H-ase), potato phenylalanine NH_4-lyase (P-ase) and mushroom tyrosinase (T-ase) on human leukemic cells *in-vitro, Proc. Amer. Assoc. Cancer Res.* **14**:106.

Oki, T., Shirai, M., Ohshima, M., Yamamoto, T., and Soda, K., 1973, Antitumor activities of bacterial leucine dehydrogenase and glutaminase A, *FEBS Letters* **33**:286.

Old, L. J., Boyse, E. A., Campbell, H. A., and Daria, G. M., 1963, Leukemia-inhibiting properties and L-asparaginase activity of sera from certain South American rodents, *Nature* **198**:801.

Old, L. J., Boyse, E. A., Campbell, H. A., Brodey, R. S., Fidler, J., and Teller, J. D., 1967, Treatment of lymphosarcoma in the dog with L-asparaginase, *Cancer* **20**:1066.

Patterson, M. K., Jr., and Orr, G. R., 1967, L-Asparagine biosynthesis by nutritional variants of the Jensen sarcoma, *Biochem. Biophys. Res. Commun.* **26**:228.

Patterson, M. K., Jr., and Orr, G. R., 1969, Regeneration, tumor, dietary, and L-asparaginase effects on asparagine biosynthesis in rat liver, *Cancer Res.* **29**:1179.

Peters, J. H., Lin, S. C., Berridge, B. J., Jr., Chao, W. R., and Cummings, J. G., 1970, Circulatory L-asparaginase activity in primates, *Life Sci.* **9**:431.

Peterson, R. E., and Ciegler, A., 1969, L-Asparaginase production by various bacteria, *Appl. Microbiol.* **17**:929.

Peterson, R. G., Handschumacher, R. E., and Mitschell, M. S., 1971, Immunological responses to L-asparaginase, *J. Clin. Invest.* **50**:1080.

Pizer, L. I., and Regan, J. D., 1972, Basis for the serine requirement in leukemic and normal human leukocytes: reduced levels of the enzymes in the phosphorylated pathway, *J. Natl. Cancer Inst.* **48**:1897.

Prager, M. D., 1971, in: *La L-asparaginase,* Du Centre National de la Recherche Scientifique, Paris, p. 179.

Prager, M. D., and Bachynsky, N., 1968a, Asparagine synthetase in asparaginase resistant and susceptible mouse lymphomas, *Biochem. Biophys. Res. Commun.* **31**:43.

Prager, M. D., and Bachynsky, N., 1968b, Asparagine synthetase in normal and malignant tissues: correlation with tumor sensitivity to asparaginase, *Arch. Biochem. Biophys.* **127**:645.

Pratt, C. B., Roberts, D., Shanks, E., Warmath, E. L., and Jackson, R., 1973, Asparaginase in combination chemotherapy for remission induction of childhood acute lymphocytic leukemia, *Cancer Res.* **33**:2020.

Raskova, H., and Vanecek, J., 1964, Pharmacology of bacterial toxins, *Pharmacol. Rev.* **16**:1.

Rauenbusch, E., Bauer, K., Kaufmann, W., and Wagner, O., 1970, Isolation and crystallization of L-asparaginase from *E. coli,* in: *Recent Results in Cancer Research* (E. Grundmann and H. F. Oettgen, eds.), **33**:31.

Regan, J. D., Vodopick, H., Takeda, S., Lee, W. H., and Faulcon, F. M., 1969, Serine requirement in leukemic and normal blood cells, *Science* **163**:1452.

Reis, H. E., and Schmidt, C. G., 1970, The antigenicity of *E. coli* L-asparaginase in: *Recent Results in Cancer Research* (E. Grundman and H. F. Deltgen, eds.), **33**:194.

Riley, V., 1968, Role of the LDH-elevating virus in leukaemia therapy by asparaginase, *Nature* **220**:1245.

Riley, V., Campbell, H. A., and Stock, C. C., 1970, Asparaginase clearance: influence of the LDH-elevating virus, *Proc. Soc. Exptl. Biol. Med.* **133**:38.

Riley, V., Spackman, D. H., and Fitzmaurice, M. A., 1971, in: *La L-asparaginase,* Du Centre National de la Recherche Scientifique, Paris, p. 139.

Riley, V., Spackman, D., Fitzmaurice, M. A., Roberts, J., Holcenberg, J. S., and Dolowy, W. C., 1974, Therapeutic properties of a new glutaminase-asparaginase preparation and the influence of the lactate dehydrogenase-elevating virus, *Cancer Res.* **34**:429.

Roberts, J., Prager, M. D., and Bachynsky, N., 1966, The antitumor activity of *Escherichia coli* L-asparaginase, *Cancer Res.* **26**:2213.

Roberts, J., Burson, G., and Hill, J. M., 1968, New procedures for purification of L-asparaginase with high yield from *Escherichia coli, J. Bacteriol.* **95**:2117.

Roberts, J., Holcenberg, J. S., and Dolowy, W. C., 1970, Antineoplastic activity of highly purified bacterial L-glutaminase, *Nature* **227**:1136.

Roberts, J., Holcenberg, J. S., and Dolowy, W. C., 1972, Isolation, crystallization, and properties of Achromobacteraceae glutaminase-asparaginase with antitumor activity, *J. Biol. Chem.* **247**: 84.

Rutter, D. A., and Wade, H. E., 1971, The influence of the iso-electric point of L-asparaginase upon its persistence in the blood, *Brit. J. Exptl. Pathol.* **52**:610.

Ryan, W. L., and Dworak, J. E., 1970, Amino acids of the 6C3HED lymphosarcoma following treatment with asparaginase, *Cancer Res.* **30**:1206.

Ryan, W. L., and Sornson, H. C., 1970, Glycine inhibition of asparaginase, *Science* **167**:1512.

Sasada, M., and Terayama, H., 1969, The nature of inhibitors of DNA synthesis in rat-liver hepatoma cells, *Biochim. Biophys. Acta* **190**:73.

Schmid, F. A., and Roberts, J., 1974, Antineoplastic and toxic effects of *Acinetobacter* and *Pseudomonas* glutaminase-asparaginases, *Cancer Chemotherapy Rept.* **58**:829.

Schrek, R., Holcenberg, J. S., Roberts, J., and Dolowy, W. C., 1971, *In-vitro* crytocidal effect of L-glutaminase on leukaemic lymphocytes, *Nature* **232**:265.

Schwartz, J. H., Reeves, J. Y., and Broome, J. D., 1966, Two L-asparaginases from *E. coli* and their action against tumors, *Proc. Natl. Acad. Sci. U.S.* **56**:1516.

Siu Chong, E. D., and Chang, T. M. S., 1974. *In-vivo* effects of intraperitoneally injected L-asparaginase solution and L-asparaginase immobilized within semipermeable nylon microcapsules with emphasis on blood L-asparaginase, "body" L-asparaginase, and plasma L-asparagine levels, *Enzyme* **18**:218.

Siu Chong, E. D., and Chang, T. M. S. 1976a, A comparison of the tumor inhibitory effectiveness of intraperitoneally injected L-asparaginase solution and L-asparaginase microcapsules in non-immunized and immunized mice bearing the 6C3HED lymphosarcoma. (in preparation)

Siu Chong, E. D., and Chang, T. M. S. 1976b, Influence of immunization on the *in-vivo* effects of intraperitoneally injected L-asparaginase solution and L-asparaginase microcapsules. (in preparation)

Sobin, L. H., and Kidd, J. G., 1965, A metabolic difference between two lines of lymphoma 6C3HED cells in relation to asparagine, *Proc. Soc. Expt. Biol. Med.* **119**:325.

Sobin, L. H., and Kidd, J. G., 1966, Alterations in protein and nucleic acid metabolism of lymphoma 6C3HED-OG cells in mice given guinea pig serum, *J. Expt. Med.* **123**:55.

Suld, H. M., and Herbut, P. A., 1965, Guinea pig serum and liver L-asparaginases: purification and antitumor activity, *J. Biol. Chem.* **240**:2234.

Suld, H. M., and Herbut, P. A., 1970, Guinea pig serum and liver L-asparaginases: comparison of serum and papain-digested liver L-asparaginases, *J. Biol. Chem.* **245**:2797.

Summers, W. P., and Handschumacher, R. E., 1973, The rate of mutation of L5178Y asparagine-dependent mouse leukemia cells to asparagine independence and its biological consequences, *Cancer Res.* **33**:1775.

Sutow, W. W., Garcia, F., Starling, K. A., Williams, T. E., Lane, D. M., and Gehan, E. A., 1971, L-Asparaginase therapy in children with advanced leukemia, *Cancer* **28**:819.

Tallal, L., Tan, C., Oettgen, H., Wollner, N., McCarthy, M., Helson, L., Burchenal, J., Karnofsky, D., and Murphy, M. L., 1970, *E. coli* L-asparaginase in the treatment of leukemia and solid tumors in 131 children, *Cancer* **25**:306.

Tosa, T., Sano, R., Yamamoto, K., Nakamura, M., Ando, K., and Chibata, I., 1971, L-Asparaginase from *Proteus vulgaris, Appl. Microbiol.* **22**:387.

Tower, D. B., Peters, E. L., and Curtis, W. C., 1963, Guinea pig serum L-asparaginase: properties, purification, and application to determination of asparagine in biological samples, *J. Biol. Chem.* **238**:983.

Uren, J. R., Summers, W. P., and Handschumacher, R. E., 1974, Enzymatic and nutritional evidence for two-state expression of the asparagine synthetase locus in L5178Y murine leukemia mutants, *Cancer Res.* **34**:2940.

Vadlamudi, S., Padarathsingh, M., Waravdekar, V. S., and Goldin, A., 1970, Factors influencing the therapeutic activity of L-asparaginase (NSC 109229) in leukemic (L5178Y) mice, *Cancer Res.* **30**:1467.

Vadlamudi, S., Padarathsingh, M., Bonmassar, E., Waravdekar, V., and Goldin, A., 1971, Studies on neutralization of L-asparaginase activity *in-vitro* and *in-vivo, Cancer* **27**:1321.

Vadlamudi, S., Krishna, B., Reddy, V. V. S., and Goldin, A., 1973, Schedule-dependent therapeutic synergism for L-asparaginase and methotrexate in leukemic (L5178Y) mice, *Cancer Res.* **33**:2014.

Weinberger, S., 1971, L-asparaginase as an antitumor agent, *Enzyme* **12**:143.

Whelan, H. A., and Wriston, J. C., Jr., 1969, Purification and properties of asparaginase from *E. coli* B, *Biochemistry* **8**:2386.

Whitecar, J. P., Jr., Bodey, G. P., Harris, J. E., and Freireich, E. J., 1970, Current concepts—L-asparaginase, *New Engl. J. Med.* **282**:732.

Woods, J. S., and Dixon, R. L., 1968, Uptake of asparagine C-14 into the pyrimidine synthesis pathway of guinea pig serum sensitive 6C3HED lymphosarcoma cells, *Proc. West. Pharmacol. Soc.* **11**:67.

Wriston, J. C., Jr., and Yellin, T. O., 1973, L-Asparaginase: review, *Advan. Enzymol.* **39**:185.

Zubrod, C. G., 1970, The clinical toxicities of L-asparaginase in treatment of leukemia and lymphoma, *Pediatrics* **45**:555.

Zweifach, B. W., and Jenoff, A., 1965, Bacterial endotoxemia, *Ann. Rev. Med.* **16**:201.

10

A Biomedical View of Enzyme Replacement Strategies in Genetic Disease

CHARLES R. SCRIVER

1. *Introduction*

A nursery rhyme, which some of us learned when we were children brought tinker, tailor, soldier, and sailor together (Opie and Opie, 1951), rather like this volume, where its authors have been convened from many disciplines to describe how the diverse indexes of our individuality—our proteins—can be handled for biomedical purposes. My contribution to the mix is a chapter on strategies about enzyme replacement therapy; other authors provide the tactical details upon which the successful campaign for such treatment will depend.

I will be concerned in particular with the problem of enzyme replacement in patients with inborn errors of metabolism. The basic research documented in other chapters could improve the quality of life for those whose genotype otherwise makes life brief, bereft, or intolerable. There will be readers who would prefer that we not spend too much time on the "few" mutants among us when there are "bigger" problems to be solved. However, my work often brings me face to face with families who must adapt to inherited disease day by day (Scriver, 1971). I believe that the high technology of enzyme isolation, stabilization, protection, and delivery could eventually help such families. Moreover, should one of my impatient readers carry a gene which might cause an inborn error of metabolism in one of his offspring, he might then wish mightily for the benefits of enzyme therapy. Science becomes very personal and important when it affects our immortality.

CHARLES R SCRIVER • The deBelle Laboratory for Biochemical Genetics, McGill University–Montreal Children's Hospital Research Institute; and The Medical Research Council Genetics Group, Departments of Pediatrics (Faculty of Medicine) and Biology (Faculty of Science), McGill University, Montreal, Canada

2. *Objective of the Campaign*

We can begin with a statement about the size of the disease burden caused by aberrant gene expression in mankind. McKusick (1975) lists 2336 monogenic diseases, of which 1142 are secure in their assigned inheritance patterns (autosomal recessive, autosomal dominant, and X-linked). The growth rate of enumeration through four editions of McKusick's catalogs has been exponential over the past decade, and it is likely to continue at this pace in the near future. The inborn errors of metabolism comprise about 10% of the catalog entries.

Recent studies of handicapping disease have revised upward all previous estimates of the amount of hereditary disease in human populations (World Health Organization, 1972: Trimble and Doughty, 1974). It is estimated that 94 of every 1000 live-born infants are affected by conditions related to defective gene expression. While this total burden includes single-gene and chromosomal aberrations, congenital malformations, and irregularly inherited conditions with multifactorial cause, at least a small fraction of the total, perhaps as high as 10% (i.e., 1% of all births) comprises monogenic and multigenic illness affecting metabolic events at some time in the subject's lifetime.

The "fallout" of genetic disease is an important cause of admissions to our expensive hospitals. Several studies (Childs *et al.,* 1972; Scriver *et al.,* 1973; Day and Holmes, 1973) indicate that as many as one-third of pediatric admissions are for gene-influenced illness; and that at least 11% of medical care for adults in hospital relates to gene-influenced disease (Childs *et al.,* 1972).

Deaths accountable to genetic and gene-influenced disease now comprise about 40% of childhood mortality (Carter, 1956; Roberts *et al.,* 1970; Childs *et al.,* 1972). If we consider, for example, that atherosclerosis has a multigenic basis as well as several environmental causes (Motulsky and Boman, 1974), we realize that much of the premature mortality in adult life is also the result of gene-influenced disease.

Such observations have been made largely on white populations in the Northern Hemisphere; eventually they will pertain to all races in all countries of the world. In the successive transitions to postindustrial demography (Dumond, 1975) which characterize the usual modes of social evolution, fewer conceptions per parent eventually result, each of which is then relatively more important to the family. Accordingly, the applicability of our knowledge of genetics, and of enzyme technology, will become a matter of increasing concern and relevance to the family of the future.

3. *Origins of Genetic Variation*

Mutation is the ultimate source of variation in DNA composition between individuals. The origin of genes in any individual is a function of selective forces, random genetic drift, and mutagenic events. The continuing but rare occurrence of "new" mutations and the vast accumulation of "old" ones yields the variable genome which characterizes man. The amount of interindividual genetic variation is impressive. Harris and colleagues (Harris, 1968), who pioneered studies of "normal" human genetic variation, did so by examining electrophoretic profiles of enzymes and proteins in blood cells and plasma. In a summary statement of this work (Harris and Hopkinson, 1972), it was reported that at least 1% of persons have inherited variants of proteins specified by their genes at 28% of 71 structural gene loci examined so far. They calculated the average heterozygosity per locus to be at least 0.067, and because noteworthy frequencies of variability were observed for all the proteins examined only by electrophoresis, the authors presumed this frequency to be an underestimate of the true prevalence of polymorphic genetic variation at human gene loci.

Sir Archibald Garrod, who was Harris' scientific ancestor, was among the first to collect chemical impressions of the biological "sports" of mankind. It was Garrod who first developed the great theme of chemical individuality in his extraordinary paper on alcaptonuria (Garrod, 1902), which we now realize was a statement of ideas far ahead of their time:

> If it be indeed the case that in alkaptonuria—we are dealing with [an] individuality of metabolism—the thought naturally presents itself that [it is] a merely extreme example of variation of chemical behaviour—elsewhere present in minor degrees, and that—just as no two individuals of a species are absolutely identical in body structure, neither are their chemical processes carried out on exactly the same lines.

After three-fourths of a century, Garrod's "rare sports"—the patients with inborn errors of metabolism—have become a prominent fraction of the disease burden in modern societies. They are the targets for enzyme replacement strategies. Moreover, Garrod's predictions about chemical individuality have been proved with a vengeance, as Harris has demonstrated so forcibly.

In a sense, it is our misfortune that replacement therapy with "foreign" exogenous enzyme can be so readily perceived by the host. Cellular defenses will be called forth to reject the invader and the defense mechanism may then turn back upon the host and bring about his death, before the primary disease has run its natural course. Accordingly, the subtle as well as the obvious themes of genetic individuality will come to influence the practice of enzyme therapy; the replacement enzyme, if it is foreign to the host, must be masked and its identity protected. It is this requirement which leads us to the techniques of immobilization and packaging of enzymes for purposes of treatment.

4. *Treatment Strategies*

There are a limited number of basic strategies. Genetic disease amenable to treatment requires intervention before harm is done; the time for that may be *in utero* or immediately after birth; or even at some later time in life.

1. *Negative eugenics:* To avert a disease itself not amenable to treatment, it would be necessary to avoid conception or to terminate pregnancy when the fetus is known to be affected. Genetic counseling and prenatal diagnosis through amniocentesis are the respective methods which can now help families to avoid

having children with untreatable genetic disease. These are negative measures which can prevent disease expression but will also prevent life and any gene expression whatsoever.

2. *Gene modifications:* Genetic technologies may eventually permit modification of the mutant gene's expression when treatment is otherwise unavailable—"disinheriting disease" as it was termed at a recent conference on the human uses of molecular genetics (Earley, 1975). *Genetic transformation* through the introduction of additional, whole DNA into the mutant cell, or transduction by specific *gene replacement* through the use of a viral vector to transport the gene, might become modes of gene therapy under certain conditions. However, gargantuan difficulties must be overcome before medically acceptable gene therapy can be anticipated. Our hopes for treatment lie elsewhere for the foreseeable future.

4.1 Environmental (Euphenic) Engineering

The tangible modes of treatment for hereditary disease involve environmental engineering (Scriver, 1967; Scriver, 1969; Hsia, 1975) which can offset the effect(s) of the mutant allele(s) (Figure 1). Alone or in combination, they are, nonetheless, only subterfuges to restore some degree of metabolic equilibrium in the presence of continuing expression of the mutant gene. The practical modes are substrate restriction, product replacement, amplification of residual apoenzyme activity, and enzyme replacement.

4.2 Enzyme Replacement

With respect to enzyme replacement (Scriver, 1974; Desnick *et al.*, 1975a), it is likely that this form of therapy can be adapted only to postnatally expressed genetic illness affecting soluble proteins possessing catalytic activity. Monogenic (autosomal recessive and X-linked) disorders, or multifactorial illness in which a monogenic component is a major determinant in the evolution of illness, are likely to be the only candidates for enzyme replacement. Autosomal dominant handicaps, chromosomal disorders, and multifactorial congenital malformations all involve mutant genes affecting structural proteins during embryogenesis. Once the defective gene has been expressed in such disorders there is no recall to the normal phenotype.

The replacement of normal gene products which possess catalytic or vectorial functions (e.g., enzyme or transport proteins) is an immense challenge. Sufficient amounts of active enzyme must be delivered to the appropriate location (Figure 2). The likelihood of success diminishes as one proceeds from extracellular to intracellular replacement. Extracellular replacement, either intraluminal in the case of digestive enzymes, or intravascular in the case of blood proteins or its cellular components, is feasible and has been widely practiced in the management of such diseases as trypsinogen deficiency, cystic fibrosis, and hemophilia. Intracellular enzyme replacement requires delivery of the enzyme to the appropriate tissue and uptake into the region of the cell where the enzyme acts normally. In either form of replacement (extracellular or intracellular) naked protein will be exposed to the

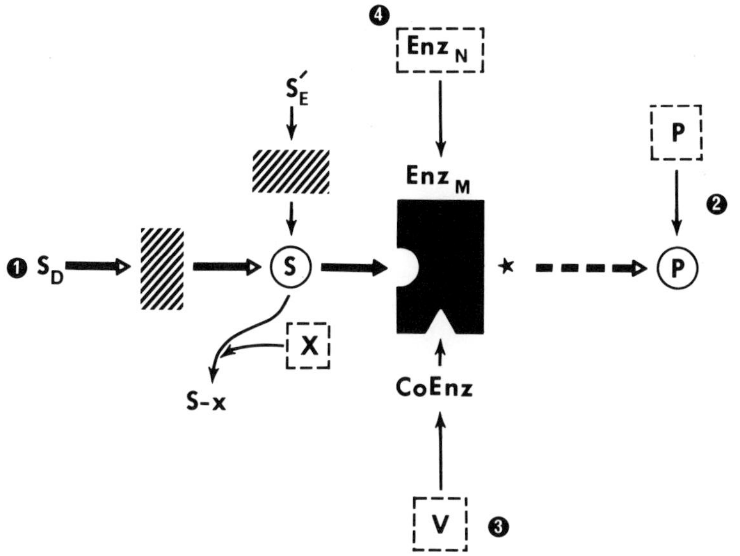

FIGURE 1. Strategies for euthenic (i.e., modification of environment to neutralize effect of allele without modification of genotype) treatment of inborn errors of metabolism. (1) Substrate restriction by removal of substance or precursor (e.g., phenylalanine in PKU and in hereditary tryosinemia) from diet; or by blocking endogenous synthesis of substrate (e.g., with allopurinol to block uric acid synthesis in Lesch-Nyhan syndrome); or by removing offending substance (e.g., with D-penicillamine to remove Cu^{2+} in Wilson's disease; or with glycine to remove isovaleric acid in isovaleric acidemia). (2) Product replacement (e.g., hormone in thyroid hormone biosynthesis defects; or phosphate in X-linked hypophosphatemia). (3) Vitamin supplementation to enhance coenzyme-responsive residual apoenzyme activity. (4) Enzyme replacement, by enzyme administration as insoluble or soluble immobilized protein; or in encapsulated form; or via organ transplantation. Key to symbols: S, substrate of enzyme; S_D, dietary source of S; S'_E, endogenous source of S; S-x or chemical complex of S with X, a therapeutic agent to remove S; P, product of reaction; V, vitamin precursor of coenzyme (CoEnz), Enz_N, normal apoenzyme, Enz_M, mutant apoenzyme.

immunoprotective system of the host and there is a danger of a host defense reaction; both modes require stabile competent protein to achieve replacement. Accordingly, the principles of replacement therapy must recognize both its objectives and its hazards.

a. History of Enzyme (or Protein) Replacement Therapy. The earliest formal use of gene product replacement therapy (excluding the treatment of diabetes, hemophilia, and cystic fibrosis) appears to be an attempt to administer semipurified ceruloplasmin to a patient with Wilson's disease (Bickel *et al.,* 1956); the attempt was unsuccessful. Moreover, we realize now that the wrong protein was probably given in this instance.

A report on the potential of enzyme replacement therapy by Hers's group (Baudhuin *et al.,* 1964) influenced Hug and Schubert (1967a, 1967b) to treat patients with the newly discovered lysosomal disease, type-2 glycogenesis, by repeated intravenous infusion of acid α-glucosidase prepared from *Aspergillus*

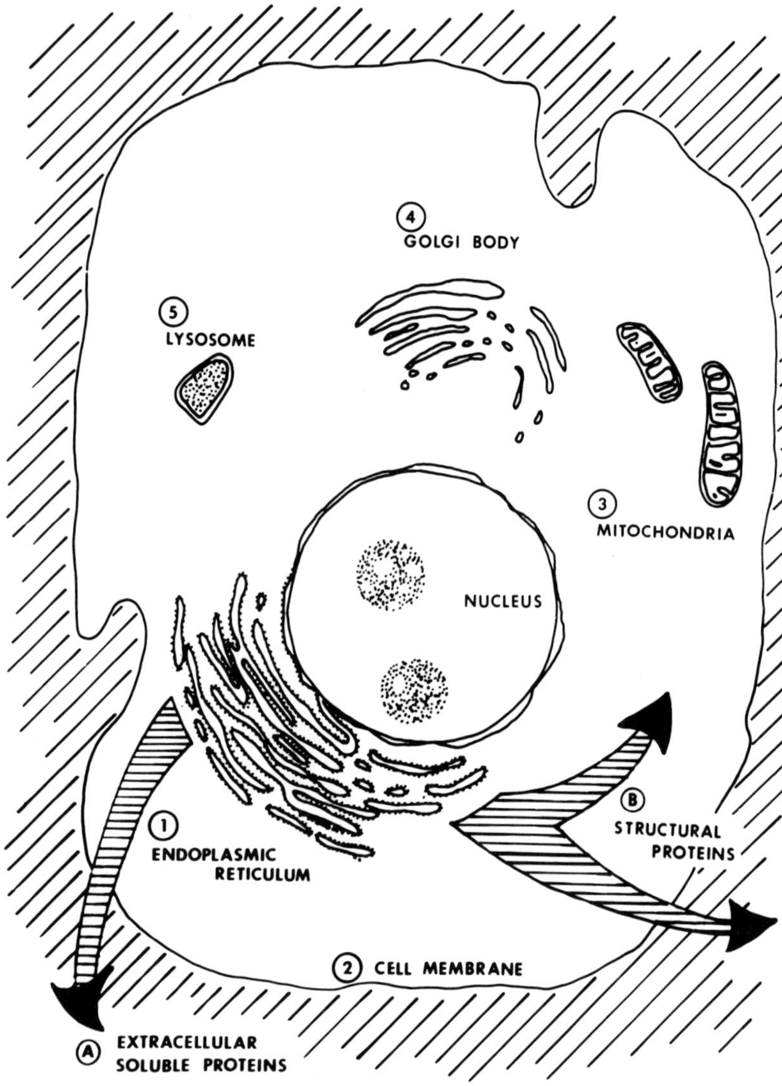

FIGURE 2. Biological expression of inborn errors of metabolism. Gene product deficiency can be expressed in extracellular catalytic proteins (A), extracellular or intracellular structural proteins (B), and intracellular catalytic functions (e.g., in the endoplasmic reticulum, cell membrane mitochondria, Golgi apparatus, and lysosome. Appropriate replacement requires delivery of exogenous protein to a specific location in the cell.

niger. Lysosomal α-glucosidase is deficient in patients with the disease. Hepatic enzyme activity was increased in the patients by the "treatment," and both the glycogen content and the number of abnormal lysosomes were diminished. However, pitfalls were encountered as well. There was insufficient exogenous enzyme for prolonged therapy; and fatal serum sickness followed the therapy in one patient. The next important discovery in this chronology was that normal serum inhibits the development of metachromasia *in vitro,* in cultured fibroblasts obtained from patients with mucopolysaccharidoses (Hors-Cayla *et al.,* 1968); this lead to the demonstration and isolation of "corrective factors" which could be exchanged between normal cells and mucopolysaccharidosis cell lines in coculture (Neufeld and Cantz, 1971). DiFerrante and colleagues (1971) then reported the apparently beneficial *in vivo* use of normal plasma infusions in patients afflicted with one or another form of the mucopolysaccharidoses. Many reports then followed on the preparation and administration of exogenous enzyme for treatment of a variety of inborn errors of metabolism (Table 1). Organ transplantation has also been used as another mode of enzyme replacement; this experience is also summarized in Table 1.

b. Degree of Enzyme Replacement Required. The study of inborn errors of metabolism has revealed ubiquitous genetic heterogeneity (Childs and Der Kaloustian, 1968), which means the occurrence, at the gene locus under study, of more than one *mutant* allele. Different alleles will cause variable expression of the mutant phenotype among individuals who inherit them. The gene product (enzyme protein, transport protein, structural protein, etc.) is the initial determinant of the subsequent phenotypic events. Accordingly, the determination of mutant enzyme activity (or protein structure or amount) in subjects with different forms of any given inborn error of metabolism is one way to evaluate and interpret the adaptive significance of residual gene product activity in mutant phenotypes.

In the example of branched-chain keto acid decarboxylase deficiency (maple syrup urine disease) (Scriver and Rosenberg, 1973), probands with the "classical" form of the disease possess less than 2% of the normal oxidative decarboxylation activity. They become critically ill soon after birth. They can be treated only with special semisynthetic diets which selectively constrain branched-chain amino acid intake to less than 600 mg of leucine per day during infancy and about half that amount for isoleucine and valine; the normal intakes are at least three times greater. Many variants of maple syrup urine disease have now been discovered. In those where the residual enzyme activity is above 2% of normal but less than 8%, the dietary tolerance for leucine is higher (600–1000 mg per day), and it is somewhat easier to manage these patients; and in subjects whose residual enzyme activity approaches 15% of normal, a semisynthetic diet is often quite unnecessary and the subject can usually tolerate a simple low-protein diet. Accordingly, we learn from this model of genetic heterogeneity that if enzyme replacement were feasible and desirable in maple syrup urine disease, it would be necessary to achieve only about 10% replacement to modify the clinical phenotype dramatically. This is an important lesson.

Another example of the corrective potential in small amounts of enzyme activity can be cited. *In vitro* replacement of normal α-*N*-acetyl glucosaminidase in cultured skin fibroblasts with type IIIB mucopolysaccharidosis (Sanfilippo B disease) has been accomplished (O'Brien *et al.,* 1973). As little as 5% of enzyme replacement accomplished 70% correction of the biochemical abnormality in the cultured cells.

TABLE 1
Replacement Therapy in Human Inborn Errors of Metabolism[b]

Disease (enzyme involved and EC number)	Source of replacement enzyme	Route of administration[a]	Tissue uptake by:	Substrate metabolism found in:	Substrate(s) accumulated in disease	Comments	Reference
Wilson's disease (?) (Ceruloplasmin involved)	Human plasma	i.v.			Copper		Bickel *et al.*, 1956
	Human liver	Orthotopic transplantation					
Glycogenoses							
Type II (α-glucosidase; EC 3.2.1.3)	*A. niger*	i.v.	Liver	Liver −	Glycogen		Baudhuin *et al.*, 1964
	A. niger	i.v.	Liver	Liver +		Immunologic reaction	Hug and Schubert, 1967a, 1967b; Hug *et al.*, 1968
	A. niger	i.v.	Liver leukocyte	Liver −			Lauer *et al.*, 1968
	Human placenta	i.v.	Liver	Liver −		Half-life in liver, 4–5 days	deBarsy *et al.*, 1973
Type IV (Branched enzyme; EC 2.4.1.18)	*A. niger*	i.v.	Liver	Liver +	Glycogen	α-Glucosidase monitored	Fernandes and Huijing, 1968; Huijing *et al.*, 1973
Glycosphingolipidoses							
Tay-Sachs disease (β-*N*-acetyhexosaminidase A; EC 3.2.1.52)	Human urine	i.v.			G_{M2} ganglioside, asiola-Gm_2 ganglioside	No hexosaminidase A activity in CNS or CSF	Johnson *et al.*, 1973
	Human urine	i.v.				Rapid clearance from plasma	O'Brien *et al.*, 1973
Sandhoff disease (β-*N*-acetyl hexosaminidase A and B; EC 3.2.1.52)	Plasma concentrate (human)	i.v.		Plasma +	G_{M2} ganglioside, asialo-Gm_2 ganglioside, and globoside		Desnick *et al.*, 1972
Metachromatic leukodystrophy (arylsulfatase A; EC 3.1.6.1)	Bovine brain	i.v./i.t. i.t.	Liver		Galactosyl sulfatide, lactosyl sulfatide	No CNS effect by EM observation; immunological rxn	Greene *et al.*, 1969
	Human urine	i.t.	CSF			Immunolog. rxn	Austin, 1967
Fabry's disease or Ceramidetrihexosidosis (α-galactosidase A; EC 3.2.1.22)	Human plasma	i.v.		Plasma +	-gal-gal-glucosyl ceramide; digalactosyl ceremide		Mape *et al.*, 1970; Sweeley *et al.*, 1972

	Human plasma	i.v.		Plasma −			Brady *et al.*, 1973
	Human leukocytes and platelets	i.v.		Plasma +			Brady *et al.*, 1973
	Human placenta	i.v.	Liver	Plasma +			Brady *et al.*, 1973
	Human kidney	Allo transplant		Plasma +?		No specific α-galactosidase replacement	Desnick *et al.*, 1971; Desnick *et al.*, 1972
	Human kidney	Allo transplant		Plasma +?		No specific α-galactosidase replacement	Philippart *et al.*, 1972
	Human kidney	Allo transplant		Plasma −		Effect on ceramidetrihexoside is nonspecific	Clarke *et al.*, 1972
Gaucher's disease (glucosyl cerebramidase, β-glucocerebrosidase; EC 3.2.1.45)	Human placenta	i.v.	Liver	Plasma +, erythrocyte +, liver +	Glucosylceramide; glucosylsphingosive; sialyl lactosylceramide		Brady *et al.*, 1974
Adult form	Human spleen	Allo transplant					
Infantile form	Human kidney	Allo transplant					
Niemann-Pick disease Type A (sphingomyelin phosphodiesterase; EC 3.1.4.12)	Human liver	Allo transplant		Serum +, leukocytes +/−, tears +, CSF +		Died at 18 months after transplant	Daloze *et al.*, 1976
Mucopolysaccharidoses							
Type I (Hurler) (α-L-iduronidase; EC 3.2.1.76)	Human plasma	i.v.		Urine	Dematan sulfate; heparansulfate		diFerrante *et al.*, 1971
	Human plasma	i.v.		Urine −			DeKaban *et al.*, 1972
Type II (Hunter) (sulfoiduronate sulfatase; EC ?)	Human plasma	i.v.		Urine +	As in type I		DiFerrante *et al.*, 1971
	Human plasma	i.v.		Urine −			Erickson *et al.*, 1972

[a]i.v., intravenous infusion; i.m., intramuscular injection; i.t., intrathecal; +, metabolite effect observed; −, metabolite effect measured and found absent.
[b]adopted from Desnick *et al.*, 1975a.

(continued)

TABLE 1 (*continued*)

Disease (enzyme involved and EC number)	Source of replacement enzyme	Route of administration[a]	Tissue uptake by:	Substrate metabolism found in:	Substrate(s) accumulated in disease	Comments	Reference
	Human leukocytes	i.v.		Urine +		Compatible donor essential	Knudson *et al.*, 1971
	Whole blood (human)	i.v.		Urine −			DeKaban *et al.*, 1972
Type III (Sanfilippo) IIIA; (heparan-*N*-sulfato sulfatase; EC ?)	Human plasma	i.v.		Urine +	Heparan sulfate		Dean *et al.*, 1973
	Human plasma	i.v.		Urine −			Kolodny, 1973; DeKaban *et al.*, 1972
IIIB: (α-*N*-acetyl glucosaminidase; EC 3.2.1.50)	Whole blood (human)	i.v.		Urine −			DeKaban *et al.*, 1972
Type IV (Sly) (β-glucuronidase; EC 3.2.1.31)	Human plasma	i.v.			Chondroitin-4-sulfate; chondroitin-6-sulfate	Blood clearance of enzyme from infused platelets; faster than from infused plasma	Sly *et al.*, 1974
	Human platelets	i.v.					
Severe combined immunodeficiency (adenosine deaminase; EC 3.5.4.4)	Erythrocytes	i.v.		Erythrocyte +		Restored RBC ADA and lymphocyte response to immunological challenge	Polmar *et al.*, 1975
	Fetal hepatocytes	Infused plasma implant	Lymphocyte	Lymphocyte +; erythrocyte +; serum +		Appearance of immunocompetent B and T cells. Death after 1 yr (immunologic)	Keightley *et al.*, 1975

The common message in these two examples, one from *in vivo,* the other from *in vitro* observations, is that a very modest partial replacement is likely to be all that is required. Such a goal is likely to be feasible and it could be effective at the clinical level.

c. Models for Study. It is potentially dangerous to perform replacement experiments in human subjects. Even though one might consider the heroic efforts involved in enzyme replacement to be justified in patients with terminal illness, it is not obligatory that such "experiments" be performed on human patients. There are *in vitro* model systems using cultured cells which may serve a limited objective; and there are nonhuman analogues of human diseases for the *in vivo* study of responses in enzyme replacement (Table 2). In addition to the few animal models cited here there are compendia of nonhuman traits (Lush, 1967; Cornelius, 1969; Jones *et al.,* 1973) which may lead the inquiring investigator to a desirable model. Closer integration of veterinary and human medicine could yield important information on some of the scientific aspects of enzyme replacement (Cass, 1973). Moreover, there would be no ethical conflict regarding human experimentation when the initial research is not performed in human subjects.

The availability of nonhuman, *in vivo* models for long-term studies still remains a problem. Most investigators would have little or no access to such animals; and it is unlikely that investigators possessing a model would be able to service the scientific needs of other, potentially competitive colleagues. A national

TABLE 2
Examples of Nonhuman Vertebrate Models of Inborn Errors of Metabolism in Man[a]

Human form	Host species for animal model	Reference
Hyperuricemia(s)	(Dalmatian) dog	Friedman and Byers (1948)
Acatalasemia	Mouse	Feinstein *et al.* (1966)
Gaucher's disease	Mouse[(b)]	Kanfer *et al.* (1975)
G_{M1} gangliosidosis	Cat	Farrell *et al.* (1973)
Mannosidosis	Cow	Phillips *et al.* (1974)
Batten's disease (neuronal ceroid lipofuscinosis)	Dog	Patel *et al.* (1974)
Krabbe's disease	Dog	Suzuki *et al.* (1972)
C^5-complement deficiency	Mouse B10D2/0	Levy *et al.* (1973)

[a]With one exception,[(b)] these "models" are monogenic in origin in the host species. However, there is no certainty that the mutant alleles are identical in the human and nonhuman diseases.

[b]The Gaucher mouse is nongenetic. Conduritol-β-expoxide is an inhibitor of β-glucosidase activity; the mouse is the host system for this pharmacologic model of Gaucher's disease.

TABLE 3
Principles and Pitfalls of Enzyme Replacement

Criteria	Hazard
1. Availability of enzyme with appropriate activity	1a. Insufficient enzyme for commitment to therapy 1b. Inappropriate specific activity
2. Protection of enzyme	2a. Inactivation *in vivo* 2b. Degradation *in vivo*
3. Administration of enzyme; delivery to normal site of activity and adequate contact with substrate	3a. Failure to reach site of activity 3b. Immunologic intolerance to enzyme
4. Demonstration of desired biological response	4. Inappropriate replacement (nonspecific response, etc.)

or international "repository" of live animal stocks (or cultured cell lines) may be the only feasible and systematic solution*.

5. *Principles and Pitfalls of Replacement Therapy*

To fulfill the expectations of replacement therapy, four criteria must be met, and potential hazards offset (Table 3).

5.1. *Source of Enzyme*

Rationale: Natural or synthetic enzyme, with appropriate activity toward substrate, is required for replacement therapy.

Replacement therapy requires a source of enzyme. Exogenous enzyme in crude or purified form may be administered; alternatively cells or organs which elaborate the required enzyme may be infused, engrafted, or transplanted. In the latter case, the substrate must be able to circulate to the implanted cellular source, or the enzyme must be released from the "foreign" cells and circulate to the substrate for replacement to be achieved.

When an exogenous source of enzyme is used, the protein must first be

*Two models of the "repository" concept exist in Canada: (1) A repository of somatic cell strains comprising normal and mutant skin fibroblasts has been servicing the needs of investigators locally and internationally (Goldman *et al.*, 1974). The available cell lines which could be used for studies of *in vitro* enzyme replacement are listed in Appendix I. (2) The second example is an Index of Vertebrate Stocks for Research in Canada (Canadian Council on Animal Care, 1976). The Index, published in January 1976, provides information on species and stocks, their location in Canada, the curator, the relevant field of interest, the breed and strain, the genetic strains, and additional descriptive material.

isolated for administration. Isolation and purification of enzyme by standard methods may be unsatisfactory. Physical changes brought about by conventional methods of purification may affect the enzyme adversely. If the enzyme has subunits, or if it requires an activator protein or coenzyme, the quatenary structure required for optimal activity may not be maintained. The application of affinity chromatography (Cuatrecasas *et al.,* 1968) may avoid some of these hazards in the isolation and purification of complex enzymes. Refinement in the chemical engineering of enzymes such as the use of lectins to enhance isolation of certain lysosomal glycoproteins (Norden and O'Brien, 1974) may facilitate future work with certain proteins. It seems unlikely, however, that complex multienzyme systems, such as the phenylalanine hydroxylation complex or the branched-chain keto acid decarboxylation complex, will ever be isolated for exogenous replacement therapy. Implantation of donor cells possessing activity appears to be the only method for replacement of such enzymes. Primary synthesis of simple proteins is an attractive alternative when the techniques of solid-phase synthesis (Merrifield, 1967) can be applied. Solid-phase synthesis utilizes reactive residues attached to insoluble polymeric supports as reagents, and an extensive application of these methods has already been developed for polypeptide synthesis (Neckers, 1975). Knowledge of the amino acid sequence of the protein is of course essential; and a resolution of the problems of tertiary structure and intramolecular bridging is necessary for the successful synthesis of an active enzyme.

5.2. *Protection of the Replacement Enzyme*

Rationale: A stabile enzyme with long half-life and adequate activity toward substrate is required for long-term replacement. Exaggerated decay of replacement enzyme is a potential hazard. Accordingly, a method to stabilize the protein and to attenuate its half-life is desirable.

Water-insoluble stabilized proteins have potential clinical use (Cederholm-Williams, 1973). Wold (1973), among others, has recently summarized how this objective can be achieved through covalent bonding, adsorption, microentrapment, or microencapsulation of proteins with inert matrixes or supports. Further details are available elsewhere in this book.

Solid-supported enzymes with "physiological" properties prepared as thin waferlike prostheses for subdermal implantation or as impregnated beads, membranes, or tubing for extracorporeal applications could serve replacement therapy. This strategy requires that the extracellular fluids of the host deliver the substrate to the insoluble enzyme. Moreover, while insoluble enzyme derivatives are catalytically active, their specific activity may be attenuated; this alteration in activity can be offset by the retention of normal substrate-binding kinetics (Bunting and Laidler, 1972) and greatly increased stability of the bonded enzyme.

Stabilization without loss of solubility is necessary if the enzyme is to act in the intravascular space; or if tissue uptake and entry into a cellular compartment is required for replacement. There are methods for stabilization which retain the solubility of proteins (Paillot *et al.,* 1974). Bonding of enzyme to homologous albumin in the presence of glutaraldehyde, when the enzyme is saturated with its substrate, yields a complex whose active site has been protected during the bonding process. The stabilization of unbonded, naked enzyme by the formation of antibody–protein complex or interaction with inorganic bifunctional reagents has also been attempted with some success (Snyder *et al.*, 1974; Desnick *et al.*, 1975b). Further exposure of the protected enzyme to albumin enmeshes the complex so that its antigenicity is masked.

The use of D_2O to stabilize soluble enzymes has apparently received little attention in the present context. Substitution of deuterium for hydrogen stabilizes peptide bonds and increases the half-life of proteins without significant modifications of their catalytic activity (Katz and Crespi, 1971). It is conceivable that isolation and synthesis of enzymes in the presence of D_2O can provide advantages which, if not accompanied by a biological hazard from deuterium exchange *in vivo,* could be utilized for enzyme therapy.

A model system to evaluate the efficacy of soluble immobilized enzyme therapy is suggested in Table 2. Hyperuricemia is a feature of the dalmation coach hound. Stabilized soluble uricase activity could be infused in the dalmation, and if uric acid levels are lowered effectively and without complications, further work in patients with severe hyperuricemic states (e.g., the Lesch–Nyhan syndrome) might be indicated.

5.3. *Delivery of Enzyme*

Rationale: Effective replacement of enzyme requires contact with its substrate. If contact does not normally occur in an extracellular location, uptake of enzyme into the appropriate intracellular compartment is required.

Replacement therapy will be compromised if the enzyme cannot reach its normal site of action. Appropriate delivery of enzyme can also avert the development of undesirable immune responses to the exogenous protein. The immunological aspect of the problem, which has been mentioned several times in this chapter, requires further interpretation.

a. Significance of Cross-Reacting Material. Structural gene mutation which adversely affects catalytic activity may still retain the immunologic identity of the protein. This mutant gene product is potentially cross-reacting material (CRM) and is designated CRM positive (CRM+) if immunologic identity is retained, or CRM negative (CRM−) if it is not. Boyer and collegues (1973) have reviewed the significance of CRM+ and CRM− phenotypes in relation to protein replacement therapy. Their survey of pedigrees in which probands with autosomal recessive inborn errors of metabolism have been evaluated for CRM phenotypes revealed that the majority are CRM+ (Table 4). Replacement of normal human enzyme (or isozyme) unaltered in its antigenic properties by the isolation procedure would be tolerated by CRM+ probands since the replacement enzyme would not be a "foreign" protein. However, the immunological surveillance systems of CRM− subjects would perceive naked homologous enzyme as "foreign" and thus would be intolerant. Accordingly, protective mechanisms which prevent exposure of the

TABLE 4
Proportions of Reported CRM+ and CRM− Mutants in Recessive Disease due to Inherited Protein Abnormality[a,b]

Disease	Implicated protein	CRM+	CRM−
Pseudocholinesterase deficiency: "silent" phenotype (see text)	Acylcholine acylhydrolase EC 3.1.1.8 *,†,‡	7	18
Glycogen storage disease, type III	Amylo-1-6-glucosidase EC 3.2.1.33 ‡	0	5
Metachromatic leucodystrophy (one case each of late infantile, juvenile, and variant types)	Arylsulfatase A EC 3.1.6.1 *	3	0 0
Phosphofructokinase deficiency	ATP: D-fructose-6-phosphate 1-phosphotransferase, EC 2.7.1.11 *,‡	0	1
Glycogen storage disease, type V (McCardle's disease)	α-1, 4-glucan: orthophosphate glycosyl transferase, EC 2.4.1.1 *,‡,§	1	3
Hemolytic anemia (see text)	D-glucose-6-phosphate: N.A.D.P. oxidoreductase, EC 1.1.1.49 ¶	20	0
Glycogen storage disease, type II (Pompe's disease)	α-D-glucoside glucohydrolase *,‡	0	1
Acatalasemia in Swiss	H_2O: H_2O_2, oxidoreductase, EC 1.11.1.6 *	3	0
Sandhoff's disease (see text)	Hexosaminidases A and B †,‡	2	0
Lesch–Nyhan syndrome	Inosine monophosphate: pyrophosphate phosphoribosyl transferase, EC 2.4.2.8 *,‡,§	21	0
Fructose intolerance	Ketose-1-phosphate aldehydelyase, EC 4.1.2.7 *	6	0
Myeloperoxidase deficiency	Myeloperoxidase deficiency *,‡	0	1
Phenylketonuria [one case each of classical (see text) and atypical types]	Phenylalanine-4-Hydroxylase *,‖	1	0
Galactosemia	UDP glucose: α-D-galactose phosphate uridyl transferase, EC 2.7.7.12 *,‡	8	0
Fibrinogen deficiency	Factor I *,§,¶	3	1
Prothrombin dificency	Factor II *,†,§	1	0
Proconvertin deficiency	Factor VII ‡	3	5

(continued)

TABLE 4 (*continued*)

Disease	Implicated protein	CRM+	CRM−
Hemophilia A	Factor VIII *,‡	64	2
Christmas disease	Factor IX *,‡	26	10
Stuart-Prower factor deficiency	Factor X *,‡	3	4
Hageman factor deficiency	Factor XII *,‡	0	14
Fibrin-stabilizing factor deficiency	Factor XIII *,†	14	0
C2 deficiency	Second component of complement *,†	0	5
C4 deficiency	Fourth component of complement *,†	3	0

[a] From Boyer *et al.*, 1973.

[b] Methods of CRM assay used were: *, immunodiffusion; †, immunoelectrophoresis; ‡, antibody consumption (neutralization) after reaction with mutant product; §, immunoprecipitation in liquid; ¶, detection of qualitative enzyme abnormality and thus indirect evidence for CRM+ status; Π, antibody inhibition of enzyme activity. Entries are limited to recessive disorders which cause appreciable disease. It is assumed that CRM status in recessively inherited conditions will usually be uniform among the homozygotes of a kindred. Thus, a kindred should be counted only once in assessing the relative frequency of CRM+ and CRM− mutants. This is reflected in the two columns on the right, where entries refer, except as next noted, to kindreds rather than individuals. In a few instances, relatedness of patients was not specified and entries designate individuals.

Individuals with reduced but still detectable amounts of CRM are scored as CRM+, since the distinction to be made in the table is between those who can and cannot recognize replaced normal protein as nearly like "self"; even reduced CRM should be sufficient to ensure comparative toleration. Omitted from the table are patients with afibrinogenemia in whom no adequate pretreatment CRM assay is described. Growth-hormone deficiency is omitted because of diagnostic uncertainties associated with this condition. For example, in many published accounts, one cannot distinguish acquired from inherited forms of deficiency. The same is true of FSH deficiency, which is also omitted from the table. Although numerous hemophilia A kindreds have been assayed for CRM, included are only those assayed by antibody responsive to interspecies differences.

replacement protein to the host's immunological surveillance system play an important practical role in the strategy of replacement therapy. Moreover, nonhuman enzyme is likely to be antigenic to both CRM+ and CRM− subjects, and replacement of naked enzyme from nonhomologous sources will constitute a hazard to both phenotypes, to be offset by the appropriate delivery strategies.

b. Enzyme Protection and Delivery. Protection of soluble enzymes and proteins was investigated first by Chang (1964), who showed that microencapsulation in semiporous membranes could be accomplished. Gregoriadis *et al.* (1971) were the first to examine the use of biodegradable vesicles as carriers for enzymes *in vivo.*

c. Inert Microcapsules. Collodian membrane microcapsules loaded with crystalline catalase and implanted in the peritoneal cavity of acatalasemic mice will replace catalase activity in the host without stimulation of anticatalase antibodies (Chang and Poznansky, 1968; Poznansky and Chang, 1974).

However, the endogenous use of microcapsules in man is limited by several factors. Synthetic microcapsules possess a high surface charge which causes their rapid removal by the reticuloendothelial system. Anticoagulant coating of the microcapsule surface delays but does not completely eliminate rapid removal from the circulation. Intraperitoneal injection circumvents interaction with the reticuloendothelial system, but microcapsules are still inactivated by mesenteric membranes. An intravascular shunt loaded with microcapsules has been developed to offset these limitations (Chang *et al.*, 1972). Substrate is delivered by the circulation to the encapsulated enzyme; the effluent returns the reaction product to the circulation. The prosthesis is appropriate only if the substrate diffuses readily from tissues into the blood stream; substrate clearance is a function of its equilibrium between tissues and blood, and between blood and the extracorporeal microcapsule system. Furthermore, the physical presence of a prosthesis, even if small, poses a problem in life style in the long-term treatment of an inborn error of metabolism.

d. Biodegradable Vesicles (Liposomes; Erythrocytes). Liposomes will transport active enzyme without evident leakage and will deliver it intact to parenchymal cells and even to subcellular compartments. Stabilization of the soluble enzyme during delivery, and enclosure in liposomes also prolongs the active intracellular life of replacement enzyme (Desnick *et al.*, 1975b).

Variation in the lipid composition of the liposome will determine to some extent whether cellular uptake is achieved by endocytosis or by fusion with the plasma membrane (Papahadjopoulos and Mayhew, 1974). The former permits exogenous enzyme to be taken up primarily into lysosomes while the latter delivers proteins primarily to the cytoplasm. Further studies with liposomes will be necessary to delineate their potential not only for replacement therapy but also for their toxicity, antigenicity, and distribution within the microcirculation.

Erythrocyte entrapment also protects enzyme from rapid intravascular degradation and it prolongs the intracellular life of the enzyme in the target tissue (Desnick *et al.*, 1975b). Enzyme-loaded erythrocytes tend to be taken up preferentially by the reticuloendothelial system and spleen.

The erythrocyte is the preferred vehicle for enzyme (β-glucocerebrosidase) replacement in the spleen and therefore in adult Gaucher's disease, where splenic manifestations are prominent (Ihler *et al.*, 1973). A liposome vehicle sustaining endocytosis would be preferred in Fabry's disease, where renal involvement is prominent and the missing enzyme (α-galactosidase A) is lysosomal in location.

Modification of surface properties of liposomes (or erythrocyte ghosts) by chemical and physical methods may serve to "address" the enzyme carrier to a particular target tissue (Gregoriadis, 1974; Desnick *et al.*, 1975a). Incorporation of specific carbohydrate, glycolipid, glycopeptide, and lipopolysaccharide moities into the surface structure influences the destination of the vesicle.

5.4. *Demonstration of Desired Biological Response*

Rationale: Restoration of appropriate endogenous enzyme activity is required. This response should be measurable by direct assay and confirmed to be effective by evidence for the reversal of the associated biochemical abnormalities and phenotypic aspects of the disease.

Gregoriadis and Buckland (1973) and Desnick *et al.* (1975b) have described "successful" enzyme replacement in model systems *in vivo.* Their animal models permitted total surveillance of replacement enzyme activity *in vivo:* turnover rates and uptake at the desired destination, as well as the level of tissue activity restored. Achievement of such desirable objectives can almost never be documented in man. Three examples can be cited which illustrate the difficulties that face the investigator of enzyme replacement therapy in man.

Successful organ (kidney) transplantation with restoration of α-galactosidase activity in patients with X-linked Fabry's disease (Table 1) is often quoted as the best example of enzyme replacement via donor tissue. In this model it has been assumed that the donor organ synthesizes the appropriate enzyme, which then escapes into the circulating plasma, from where it enters tissues to offset the phenotypic effects of the Fabry allele, causing storage of ceramide trihexoside (galactosylgalactosylglucosylceramide). The replacement enzyme was assumed to be α-galactosidase. Circulating "α-galactosidase" activity was measured in Fabry patients after transplantation of a normal kidney. However, it was not until 1972 that five separate laboratories almost simultaneously reported that α-galactosidase activity is composed of two components (summarized in Scriver, 1974). One component (the A fraction) is heat-stable and has other distinguishing features. Only the A fraction is missing in Fabry's disease. No study has yet confirmed that replacement of α-galactosidase A has been specifically achieved by renal trasplantation in Fabry patients.

There is little question that renal transplantation is justified in Fabry's disease where there is advanced renal failure. The procedure averts the terminal phase of the illness by what amounts to a hydraulic replacement. It had been upheld that renal transplantation, and presumed enzyme replacement, is effective in Fabry's disease because the elevated serum ceramide trihexoside levels, which characterize the illness, decline in the "treated" patient. However, Clarke and colleagues (1972) have warned that the serum trihexoside levels merely seem to parallel those of the precursor tetrahexoside, whose metabolism is not directly influenced by α-galactosidase activity; in fact, they are determined largely by erythrocyte turnover in this disease. Erythrocyte membranes are the major source of serum tetra- and trihexoside; and since repair of renal failure modifies the erythrocyte mass and survival, it is likely that any observed changes in critical metabolite levels in Fabry's disease are only a secondary reflection of this component of the total illness. Thus, one can say that the evidence for successful α-galactosidase A replacement, and for a primary metabolic response to replacement, is still wanting in this example, which seemed initially to be a promising example of enzyme replacement therapy.

A second example, again illustrating the need to evaluate the treatment response, is found in studies of hexosaminidase A infusion in Tay-Sachs disease. In this autosomal recessive illness G_{M2} ganglioside accumulates in the central nervous system (CNS) and in other tissues, because of a specific deficiency of hexosaminidase A activity in tissue lysosomes and body fluids. The question pertaining to replacement therapy in Tay-Sachs disease is whether enzyme can be taken up by the CNS and whether it is effective in G_{M2} cleavage activity if it is. O'Brien (1973) purified human hexosaminidase 10,000-fold from urine. He observed that when this material is infused *in vivo,* it is cleared rapidly from the vascular space. However, evidence is lacking that the enzyme enters the CNS, as it must if it is to alter the clinical course of the illness (Johnson *et al.,* 1973). Schneck and colleagues (1973) have shown that this enzyme is not well taken up *in vitro* by cultured skin fibroblasts from Tay-Sachs' patients, which may indicate that hexosaminidase A is also not able to penetrate brain cells *in vivo.* Furthermore, it is unlikely that postnatal therapy will yield the desired results in the Tay-Sachs patient even if the hexosaminidase A enzyme could reach the affected neurones. O'Brien *et al.,* (1971) observed that Tay-Sachs disease is already well expressed in the midtrimester fetus and that neurones are already storing G_{M2} ganglioside at this age, since the enzyme deficiency is expressed during early embryogenesis. Accordingly, at least 20 weeks of faulty CNS development has already taken place before enzyme could be replaced in the infant after birth. Postnatal enzyme replacement—even if it could be achieved in this illness—would take place in a faulty organ whose later adaptation to the complex pattern of human life would probably be compromised. The only alternative—continuous prenatal enzyme replacement—is quite unattainable. Our second example, like the first, is not promising for successful enzyme replacement therapy.

Our third human example is rather more rewarding. Adenosine deaminase deficiency (Giblett *et al.,* 1972) is an entity which epitomizes many of the criteria that must be met by enzyme replacement therapy, and two recent studies illustrate, in quite favorable terms, what can be learned from the careful clinical evaluation of this treatment strategy.

Adenosine deaminase (ADA) activity is severely deficient in most probands with severe combined immunodeficiency (SCID), an autosomal recessive disorder, in which there is deficient lymphocyte proliferation affecting B (bone marrow) and T (thymic) lymphocytes. ADA is an enzyme which exists in high-molecular-weight and low-molecular-weight forms (isoenzymes), both being effective in the conversion of adenosine to inosine. Adenosine accumulation is known to impair lymphocyte proliferation *in vitro.*

Polmar and colleagues (1975) observed that calf-intestine- and human-erythrocyte-ADA restored the *in vitro* proliferation of lymphocytes obtained from a patient with ADA-deficient SCID. Transfusion of irradiated normal erythrocyte *in vivo* into the same patient caused erythrocyte ADA activity to rise from zero to about 20% of the normal level for 2 weeks, during which time lymphocyte immunocompetence was restored to normal; peripheral blood lymphocytes returned to the circulation and thymic hypertrophy was observed.

Another patient with ADA-deficient SCID (Keightley *et al.,* 1975) has been treated by another form of replacement therapy. This patient received an intraperitoneal injection of human fetal liver cells at 5 months of age. Immunocompetent B and T lymphocytes returned to the circulation and erythrocytes regained 4% of the normal ADA activity within 150 days of the liver cell engraftment. Unfortunately, 1 year after the procedure the patient developed a severe immunologic complication and died.

These two patients illustrate two pitfalls as well as obvious benefits. The form of ADA restored (donor or host) in the host and the hazards of long-term donor erythrocyte ADA replacement have not yet been reported in the first patient. An immunologic hazard either from the disease, which was not corrected by the treatment, or from the treatment itself, occurred in the second. Again, the origin of the replaced ADA activity is unknown in the latter subject. It is unclear whether exogenous ADA activated dormant host enzyme in these two patients or whether the host responded to exogenous active enzyme. The answers are critical for any consideration of long-term therapy. Since solid tissues possess high-molecular-weight isozymes of ADA and erythrocytes possess a low-molecular-weight ADA, it is of interest that the latter can be converted to the former with exposure of the one to the other (Hirschhorn, 1975). Whereas the ADA-deficient SCID patients reported so far appear to have deficiency of a common catalytic component affecting all ADA isozymes, there may yet be discovered a form where a tissue-specific component of ADA is abnormal. Erythrocyte ADA replacement in such patients would probably leave the specific tissue ADA deficiency unaffected. It follows that complete replacement therapy in solid tissues would be impossible in the latter type of ADA deficiency.

From these three modest examples, we see the immense challenge which is ahead for those who seek successful replacement therapy in man.

6. *Comment*

The medical application of enzyme replacement strategies has been largely confined to diseases in which there is a clearly defined enzyme deficiency with associated biochemical abnormalities. Monogenic diseases, as exemplified by the inborn errors of metabolism, provide the simplest models for study in this respect. I would like to think that enzyme replacement *inside* the cell could lead eventually to the means to replace membrane-located proteins and enzymes. Disorders of membrane transport (Scriver *et al.,* 1976) and of membrane-linked events such as those which are believed to occur in Duchenne Muscular dystrophy might then yield, in their turn, to the strategies described here.

Despite the fact that some very handsome technology is now available to facilitate enzyme replacement in the treatment of disease, we are still far from achieving replacement therapy compatible with normal life-style in any chronic illnesses. This important deficit diminishes the state of the art in its medical context.

Alternative or holding strategies therefore merit reflection until such time as enzyme replacement therapy in humans can be perfected. The alternatives include more aggressive use of animal models to perfect the technology, as alluded to earlier in this essay, and the parallel pursuit of chemical therapy as subterfuges for enzyme replacement. The latter can be illustrated by two monogenic diseases, sickle cell anemia and cystinosis.

It is unlikely that replacement of normal hemoglobin ($\alpha_2\beta_2$) or of normal β-polypepide other than by tranfusion of normal erythrocytes is going to be accomplished in patients with sickle cell disease who possess hemoglobin S. On the other hand, stabilization of HbS by chemical means to prevent its crystallization and the formation of sickle cells resulting in the vascular crises and anemia, which characterize the disease, is a practical alternative which might effectively modify its course. Gillette and colleagues (1974) investigated the role of carbamylation and showed that cyanate *in vitro* impeded the sickling of HbS-containing erythrocytes, apparently by modifying HbS oxygen affinity (Jensen *et al.,* 1973). Clinical trials are in progress to determine whether this chemical approach is effective *in vivo* and without hazard. A similar strategy with a different reagent has been suggested by Lubin *et al.* (1975). They have used the bifunctional cross-linking reagent dimethyladipimidate to inhibit sickling of the deoxygenated erythrocyte. Treated cells do not leak potassium or demonstrate increased viscosity under the conditions that precipitate sickling of untreated cells. The reagent appears to influence sickling both by its improvement of oxygen affinity and by a direct effect on erythrocyte membrane or on membrane–hemoglobin interactions. The clinical role of dimethyladipimidate has not yet been investigated.

Infantile nephropatic cystinosis is an inborn error of cystine metabolism associated with cystine accumulation in membrane enclosed regions of the cells believed to be lysosomal in origin (see Scriver and Rosenberg, 1973). The vast excess of cystine is associated with cellular damage and loss of function, particularly in the kidney. The deficient gene product in cystinosis is still unknown. It has been shown that the thiol reagent dithiothreitol (DTT or Cleland's reagent) will deplete cystinotic fibroblasts of their excessive cystine *in vitro* (Goldman *et al.,* 1970); other thiol reagents, in particular, cysteamine, have a similar effect (Thoene *et al.,* 1976). Preliminary trials *in vivo* with these reagents suggests that it may be possible to reverse cystine storage in patients with cystinosis (Goldman *et al.,* 1974; Thoene *et al.,* 1976). It is hoped their use early in the course of the illness may prevent the cellular and functional deterioration that characterizes this disease.

The examples of sickle cell disease and cystinosis illustrate alternative approaches to enzyme (or protein) replacement therapy. They also illustrate how a flexible approach may allow one to deal with some hereditary diseases awaiting therapeutic resolution. We will need many resources to cope with treatment of multifactorial illness and degenerative disease with genetic components. We anticipate that enzyme replacement therapy will not be the preferred answer for many illnesses; but it may be for some. Accordingly, the monogenic diseases of man and animal will remain an important proving ground for the technology of enzyme therapy.

ACKOWLEDGMENTS

I am grateful in particular to Thomas Chang, Robert Desnick, Gregory Gregoriadis, and Peter Rowe for helpful discussion and correspondence. This project was supported by the Medical Research Council of Canada with a grant to the MRC Genetics Group at McGill University; and by the Quebec Network of Genetic Medicine.

7. *References*

Austin, J. H., 1967, Some recent findings in leukodystrophies and in gargoylism, in: *Inborn Disorders of Sphingolipid Metabolism,* (S. M. Aronson and B. W. Volk, eds.) p. 359, Pergamon Press, Inc., Oxford.

Baudhuin, P., Hers, H. G., and Leob, H., 1964, An electron microscopic and biochemical study of type II glycogenosis, *Lab. Invest.* **13**:1139.

Bickel, H., Schultze, H. E., Bruter, W., and Gollner, I., 1956, Versuche zur Coeruloplasminsubstitution bei der hepatocerebralen Degeneration (Wilsonsche Krankheit), *Klin. Wochschr.* **34**:961.

Boyer, S. M., Siggers, D. C., and Krugger, L. J., 1973, Caveat to protein replacement therapy for genetic disease: immunological implications of accurate molecular diagnosis, *Lancet* **2**:654.

Brady, R. O., Tallman, J. F., Johnson, W. G., Gal, A. E., Leahy, W. R., Quirk, J. M., and Dekaban, A. S., 1973, Replacement therapy for inherited enzyme deficiency: use of purified ceramidetrihexosidase in Fabry's disease, *New Engl. J. Med.* **289**:9.

Brady, R. O., Penchev, P. G., Gal, A. E., Hebert, S. R., and Dekaban, A. S., 1974, Replacement therapy for inherited enzyme deficiency: use of purified glucocerebrosidase in Gaucher's disease, *New Engl. J. Med.* **291**:989.

Bunting, P. S., and Laidler, K. J., 1972, Kinetic studies on solid-supported β-galactosidase. *Biochemistry* **11**:4477.

Canadian Council on Animal Care, Animal Resources Committee, 1976, Index of Vertabrate Stocks for Research in Canada.

Carter, C. O., 1956, Changing patterns in the causes of death at the Hospital for Sick Children, *Great Ormond Street Journal* **11**:65.

Cass, J., 1973, One medicine—human and veterinary, *Perspectives Biol. Med.* **16**:418.

Cederholm-Williams, S. A., 1973, Insoluble enzymes—therapeutic potential, in: *Treatment of Inborn Errors of Metabolism* (J. W. T. Seakins, R. A. Saunders, and C. Toothill, eds.), *SSIEM Symposium No. 10,* p. 195, Churchill, Livingston, London.

Chang, T. M. S., 1964, Semipermeable microcapsules, *Science* **146**:524.

Chang, T. M. S., 1972, *Artificial Cells,* Charles C. Thomas Pub., Springfield, Ill.

Chang, T. M. S., and Poznansky, M. J., 1968, Semipermeable microcapsules containing calatase for enzyme replacement in acalatosalmic mice. *Nature* **218**:243.

Chang, T. M. S., Gonda, A., Dirks, J. H., Coffey, J. F., and Lee-Burns, T., 1972, ACAC microcapsule artificial kidney for the long term and short term management of eleven patients with chronic renal failure, *Trans. Amer. Soc. Artificial Internal Organs* **18**:465.

Childs, B., and Der Kaloustian, M., 1968, Genetic heterogeneity, *New Engl. J. Med.* **279**:1205, 1267.

Childs, B., Miller, S. M., and Bearn, A. G., 1972, Gene mutation as a cause of human disease, in: *Mutagenic Effects of Environmental Contamination,* (H. E. Sutton, and M. I. Harris, eds.), p. 3, Academic Press, Inc. New York.

Clarke, J. T. R., Guttman, R. D., Wolfe, L. S., Beaudoin, J. G., and Morehouse, D. D., 1972, Enzyme replacement therapy by renal allotransplantation in Fabry's disease, *New Engl. J. Med.* **287**:1215.

Cornelius, C. E., 1969, Animal models—a neglected medical resource, *N. Engl. J. Med.* **281**:934.

Cuatrecasas, P., Wilchek, M., and Anfinsen, C. B., 1968, Selective enzyme purification by affinity chromatography, *Proc. Natl. Acad. Sci. U. S.* **61**:636.

Daloze, P., Delvin, E. D., Glorieux, F. H., Corman, J., and Bettez, P., 1974, Replacement therapy for inherited enzyme deficiency: liver orthotopic transplantation in Niemann-Pick disease type A, *Clin. Res.* **23**:739A.

Day, N., and Holmes, L. B., 1973, The incidence of genetic disease in a university hospital population, *Amer. J. Human Genet.* **25**:237.

Dean, M. F., Muir, H., and Benson, P. F., 1973, Mobilization of glycosaminoglycans by plasma infusion in mucopolysaccharidosis type III—two types of response, *Nature New Biol.* **243**:143.

deBarsy, T., Jacquemin, P., van Hoof, F., and Hers, H. G., 1973, Enzyme replacement in Pompe disease: an attempt with purified human acid α-glucosidase, in: *Enzyme Therapy in Genetic Diseases, Birth Defects* (Original Article Series, Vol. 9) (R. J. Desnick, R. W. Bernlohr, and W. Krivit, eds.), p. 184, The Williams & Wilkins Company, Baltimore.

DeKaban, A. S., Holden, K. R., and Constantopoulos, G., 1972, Effects of fresh plasma or whole blood transfusions on patients with various types of mucopolysaccharidosis, *Pediatrics* **50**:688.

Desnick, R. J., Allen, K. Y., Simmons, R. L., Majarian, J. S., and Krivit, W., 1971, Treatment of Fabry's disease: correction of the enzymatic deficiency by renal transplantation, *J. Lab. Clin. Med.* **78**:989.

Desnick, R. J., Krivit, W., Synder, P. D., Desnick, S. J., and Sharp, H. L., 1972, Sandhoff's disease: ultrastructural and biochemical studies, in: *Sphingolipidoses and Allied Disorders* (S. M. Aronson and B. W. Volk, eds.), p. 351, Plenum Press, New York.

Desnick, R. J., Simmons, R. L., Allen, K. Y., Woods, J. E., Anderson, C. F., Najarian, J. S., and Krivit, W., 1972, Correction of enzymatic deficiencies by renal transplantation; Fabry's Disease, *Surgery* **72**: 203.

Desnick, R. J., Krivit, W., and Fiddler, M. B., 1975a, Enzyme Therapy in Genetic Diseases: Progress, Principles and Prospects, in: *The Prevention of Genetic Disease and Mental Retardation* (A. Milunsky, ed.), p. 317, W. B. Saunders, Philadelphia.

Desnick, R. J., Thorpe, S. R., Steger, L., Dullum, D., Cumming, D., and Fiddler, M., 1975b, Enzyme replacement therapy: *in vivo* fate of nature, erythrocyte and liposome entrapped β-glucuronidase, *Pediat. Res.* **9**:312.

DiFerrante, N., Nichols, B. L., Donnelly, P. V., Neri, G., Hrgovcic, R., and Berglund, R. K., 1971, Induced degradation of glycosaminoglycans in Hurler's and Hunter's syndromes by plasma infusion, *Proc. Natl. Acad. Sci. U.S.* **68**:303.

Dumond, D. E., 1975, The limitation of human population: a natural history, *Science* **187**:713.

Earley, K., 1975, Disinheriting disease in "The Sciences" *N. Y. Acad. Sci.* **15**(5):19.

Erickson, R. P., Sandman, R., van B. Robertson, W., and Epstein, C. J., 1972, Inefficacy of fresh frozen plasma therapy of mucopolysaccharidosis II, *Pediatrics* **50**:693.

Farrell, D. F., Baker, H. J., Herndon, R. M., Lindsey, J. R., and McKhann, G. M., 1973, Feline GM_1 gangliosidosis: biochemical and ultrastructural comparisons with the disease in man, *J. Neuropathol. Exptl. Neurol.* **32**:1.

Feinstein, R. N., Howard, J. B., Braun, J. T., and Seaholm, J. E., 1966, Acatalasemic and hyposemic mouse mutants, *Genetics* **53**: 923.

Fernandes, J., and Huijing, F., 1968, Branching-enzyme-deficiency glycogenosis: studies in therapy, *Arch. Disease Childhood* **43**:347.

Friedman, M., and Byers, S. O., 1948, Observations concerning the causes of the excess excretion of uric acid in the Dalmatian dog, *J. Biol. Chem.* **175**:727.

Garrod, A. E., 1902, The incidence of alcaptonuria: a study in chemical individuality, *Lancet* **2**: 1616.

Giblett, E. R., Anderson, J. E., Cohen, F., Pollara, F., and Meuwissen, H. J., 1972, Adenosine-deaminase deficiency in two patients with severly impaired cellular immunity, *Lancet* **2**:1067.

Gillette, P. N., Peterson, C. M., Lin, Y. S., and Cerami, A., 1974, Sodium cyanate as a potential treatment for sickle cell disease, *New Engl. J. Med.* **290**:654.

Goldman, H., Scriver, C. R., Aaron, K., and Pinsky, L., 1970, Use of dithiothreitol to correct cystine storage in cultured cystinotic fibroblasts, *Lancet* **1**:811.

Goldman, H., DePape-Brigger, D., Delvin, E., and Scriver, C. R., 1974, Long-term use of oral dithiothreitol (DTT) in nephropathic cystinosis, *Clin. Res.* **22**:740A.

Goldstein, L., and Katchalski, E., 1968, Use of water-insoluble enzyme derivatives in biochemical analysis and separation, *Z. Fur analyt. Chemie* **243**:375.

Greene, H. L., Hug, G., and Schubert, W. K., 1969, Metachromatic leucodystrophy: treatment with arylsulfatase-A, *Arch. Neurol.* **20**:147.

Gregoriadis, G., 1973, Molecular Trojan Horses, *New Scientist,* Dec., p. 890.

Gregoriadis, G., 1974, Structural requirements for the specific uptake of macromolecules and liposomes by target tissues, in: *Enzyme therapy in lysosomal storage disease.* (J. M. Tager, G. J. M. Hooghwichel, and W. T. H. Daems, eds.), p. 131, North Holland Pub. Co., Amsterdam.

Gregoriadis, G., and Ryman, B. E., 1972a, Fate of protein-containing liposomes injected into rats. An approach to the treatment of storage diseases, *Eur. J. Biochem.* **24**:485.

Gregoriadis, G., and Ryman, B. E., 1972b, Lysosomal localization of β-fructofuranosidase containing liposomes injected into rats. Some implications in the treatment of genetic disorders, *Biochem. J.* **129**:123.

Gregoriadis, G., Leatherwood, P. D., and Ryman, B. E., 1971, Enzyme entrapment in liposomes, *FEBS Letters* **14**:95.

Harris, H., 1968, Molecular basis of hereditary disease, *Brit. Med. J.* **1**:135.

Harris, H., and Hopkinson, D. A., 1972, Average heterozygosity per locus in man: an estimate based on the incidence of enzyme polymorphism, *Ann. Human Genet.* **36**:9.

Hirschhorn, R., 1975, Conversion of human erythrocyte-adenosine deaminase activity to different tissue-specific isozymes: evidence for a common catalytic unit, *J. Clin. Invest.* **55**:661.

Hors-Cayla, M. C., Maroteaux, P., and deGrouchy, J., 1968, Fibroblastes en culture au cours de mucopolysaccharidoses: influence du serum sur la métachromasie, *Ann. Genetique* **11**:265.

Hsia, Y. E., 1975, Treatment in genetic diseases, in: *The Prevention of Genetic Disease and Mental Retardation* (A. Milunsky, ed.) p. 277, W. B. Saunders Company, Philadelphia.

Hug, G., and Schubert, W. K., 1967a, Hepatic lysosomes in Pompe's disease: disappearance during glucosodase administration, *J. Clin. Invest.* **46**:1073.

Hug, G., and Schubert, W. K., 1967b, Lysosomes in type II glycogenosis: changes during administration of extract from *Aspergillus niger, J. Cell Biol.* **35**:C1.

Hug, G., Schubert, W. K., and Chuck, G., 1968, Type II glycogenosis: treatment with extract of *Aspergillus niger, Clin. Res.* **16**:345.

Huijing, F., Waltuck, B. L., and Whelan, W. J., 1973, Alpha-glucosidase administration: experiences in two patients with glycogen storage disease compared with animal experiments, in: *Enzyme Therapy in Genetic Diseases, Birth Defects* (Original Article Series, Vol. 9) (R. J. Desnick, R. W. Bernlohr, and W. Krivit, eds.), p. 191, The Williams & Wilkins Company, Baltimore.

Ihler, G. M., Glew, R. M., and Schrure, F. W., 1973, Enzyme loading of erythrocytes, *Proc. Nat. Acad. Sci. Wash.* **70**:2663.

Jensen, M., Bunn, H. F., Halikas, G., Kan, Y. W., and Nathan, D. G., 1973, Effects of cyanate and 2,3-diphosphoglycerate on sickling: relationship to oxygeneration, *J. Clin. Invest.* **52**:2542.

Johnson, W. G., Desnick, R. J., Long, D. M., Sharp, H. L., Krivit, W., Brady, B., and Brady, R. O., 1973, Intravenous injection of purified hexosaminidase A into a patient with Tay–Sachs disease, in: *Enzyme Therapy in Genetic Diseases, Birth Defects:* (Original Article Series, Vol. IX) (R. J. Desnick, R. W. Bernlohr, and W. Krivit, eds.), p. 120, The Williams and Wilkins Co., Baltimore.

Jones, T. C., Hachel, D. B., and Migaki, K., 1973, (eds.), *The Registry of Comparative Pathology; Animal Models of Human Disease,* Baltimore Information Services, Inc., Baltimore.

Kanfer, J. N., Legler, G., Sullivan, J., Raghavan, S. S., and Mumford, R. A., 1975, The Gaucher mouse, *Biochem. Biophys. Res. Commun.* **67**:85.

Katz, J. J., and Crespi, H. L., 1971, Isotope effects in biological systems, in: *Isotope Effects in Chemical Reactions.* (C. J. Collins and N. S. Bowman, eds.), p. 286, Van Nostrand Reinhold, New York.

Keightley, R. G., Lawton, A. R., Cooper, M. D., and Yunis, E. J., 1975, Successful fetal liver transplantation in a child with severe combined immunodeficiency, *Lancet* **2**:850.

Knudson, A. G., Jr., DiFerrante, N., and Curtis, J. E., 1971, Effect of leucocyte transfusion in a child with type II mucopolysaccharidosis, *Proc. Natl. Acad. Sci. U. S.* **68**:1738.

Kolodny, E. H., 1973, commentary in: *Enzyme Therapy in Genetic Diseases, Birth Defects* (Original Article Series, Vol. 9) (R. J. Desnick, R. W. Bernlohr, and W. Krivit, eds.), p. 40, The Williams & Wilkins Company, Baltimore.

Lauer, R. M., Mascarinas, T., Racela, A. S., Diehl, A. M., and Brown, B. I., 1968, Administration of a mixture of fungal glucosidases to a patient with type II glycogenesis (Pompe's disease), *Pediatrics* **42**:672.

Levy, N. L., Synderman, R., Ladda, R. L., and Lieberman, R., 1973, Cytogenetic engineering in vivo: restoration of biologic complement activity to C_5-deficient mice by intravenous inoculation of hybrid cells, *Proc. Natl. Acad. Sci. U. S.* **70**:3125.

Lubin, B. H., Pena, V., Mentzer, W. C., Bymun, E., Bradley, T. B., and Pacher, L., 1975, Dimethyl adipimidate: a new antisickling agent, *Proc. Natl. Acad. Sci. U. S.* **72**:43.

Lush, I. E., 1967, The biochemical genetics of vertebrates except man, in: *Frontiers of Biology,* Vol. 3, (A. Neuberger, and E. L. Tatum, eds.), p. 118, North-Holland Publishing Co., Amsterdam.

Mapes, C. A., Anderson, R. L., Sweeley, C. C., Desnick, R. J., and Krivit, W., 1970, Enzyme replacement in Fabry's disease: an inborn error of metabolism, *Science* **169**:987.

McKusick, V. A., 1975, Mendelian inheritance in man, in: *Catalogs of Autosomal Dominant, Autosomal Recessive and X-Linked Phenotypes,* 4th ed., The Johns Hopkins Press, Baltimore.

Merrifield, R. B., 1967, New approaches to the chemical synthesis of peptides, *Recent Progr. Hormone Res.* **23**:451.

Mosback, K., 1970, Matrix-bound enzymes, Part I: The use of different acrylic copolymers as matrices, *Acta Chem. Scand.* **24**:2084.

Motulsky, A. G., and Boman, H., 1974, Genetics and artherosclerosis, in: *Proc. 3rd Int. Symp. Artherosclerosis,* Oct. 1973 (G. Schettler, and A. Weizel, eds.), Springer-Verlag, Berlin.

Neckers, D. C., 1975, Solid phase synthesis, *J. Chem. Educ.* **52**:695.

Neufeld, E. F., and Cantz, M. J., 1971, Corrective factors for inborn errors of mucopolysaccharide metabolism, *Ann. N. Y. Acad. Sci.* **179**:580.

Norden, A. G. W., and O'Brien, J. S., 1974, Binding of human liver β-galactosidases to plant lectins insolubilized on agarose, *Biochem. Biophys. Res. Commun.* **56**:193.

O'Brien, J. S., 1973, Ganglioside Storage Disease, in: *Advances in Human Genetics* (Vol. 3) (H. Harris and K. Hirschhorn, eds.), p. 39, Plenum Press, New York.

O'Brien, J. S., Okada, S., Fillerup, D. L., Veath, M. L., Adornato, B., Brenner, P. H., and LeRoy, J. G., 1971, Tay–Sachs Disease: Prenatal Diagnosis, *Science* **172**:61.

O'Brien, J. S., Miller, A. L., Loverde, A. W., and Veath, M. L., 1973, Sanfilippo disease type B: enzyme replacement and metabolic correction in cultured fibroblasts, *Science,* **181**:753.

Opie, I., and Opie, P., 1951, *The Oxford Dictionary of Nursery Rhymes,* Oxford University Press, Oxford.

Paillot, B., Remy, N. H., Thomas, D., and Brown, G., 1974, *Path. Biol.* **22**:491.

Papahadjopoulos, G. P., and Mayhew, E., 1974, Cellular uptake of cyclic AMP captured within phospholipid vesicles and effect on cell-growth behaviors, *Biochim. Biophys. Acta* **363**:404.

Patel, V. P., Koppang, N., Patel, B., and Zeman, W., 1974, *p*-phenylenediamine-mediated peroxidase deficiency in English setters with neuronal ceroid-lipofuscinosis, *Lab. Invest.* **30**:366.

Phillips, N. C., Robinson, D., Winchester, B. G., and Jolly, R. D., 1974, Mannosidosis in Angus cattle: the enzymic defect, *Biochem. J.* **137**:363.

Polmar, S. H., Wetzler, E. M., Stern, R. C., and Hirschhorn, R., 1975, Restoration of *in vitro* lymphocyte responses with exogenous adenosine deaminase in a patient with severe combined immunodeficiency, *Lancet* **2**:743.

Philippart, M., Franklin, S. S., and Gordon, A., 1972, Reversal of an inborn sphingolipidosis (Fabry's disease) by kidney transplantation, *Ann. Intern. Med.* **77**:195.

Poznansky, M. J., and Chang, T. M. S., 1974, Comparison of the enzyme kinetics and immunological properties of catalase immobilized by microencapsulation and catalase in free solution for enzyme replacement, *Biochem. Biophys. Acta* **334**:103.

Rifkind, R. A., 1966, Destruction of injured red cells in vivo, *Amer. J. Med.* **41**:711.

Roberts, D. F., Chavez, J., and Court, S. D. M., 1970, The genetic component in child mortality, *Arch. Disease Childhood* **45**:33.

Scriver, C. R., 1967, Treatment in medical genetics, in: *Proc. 3rd Int. Congr. Human Genetics* (Chicago, Sept.) (J. F. Crow, and J. V. Neel, eds.), p. 45, The Johns Hopkins Press, Baltimore.

Scriver, C. R., 1969, Treatment of inherited disease: realized and potential, *Med. Clin. N. Amer.* **53**:941.

Scriver, C. R., 1971, Mutants: consumers with special needs, *Nutr. Rev.* **29**:155.

Scriver, C. R., 1974, Enzyme therapy and induction in genetic disease: pox or pax, *Proc. 4th Int. Birth Defects Conf.* (A. G. Motulsky, and W. Lentz, eds.), p. 114, Excerpta Medica ICS 310, Amsterdam.

Scriver, C. R., and Rosenberg, L. E., 1973, *Amino Acid Metabolism and Its Disorders,* W. B. Saunders Ltd., Philadelphia.

Scriver, C. R., Neal, J. L., Saginur, R., and Clow, A., 1973, The frequency of genetic disease and congenital malformation among patients in a pediatric hospital, *Can. Med. Assoc. Journal* **108**:1111.

Scriver, C. R., Chesney, R. W., and McInnes, R. R., 1976, Genetic aspects of renal tubular transport: diversity and topology of carriers, *Kidney Int.* **9**:149.

Silman, I. H., and Katchalski, E., 1966, Water-insoluble derivatives of enzymes, antigens, and antibodies, *Ann. Rev. Biochem.* **35**:873. (Part II).

Sly, W. S., Glaser, J. G., Roozen, K., Brot, E., and Stahl, P., 1974, Enzyme replacement studies with β-glucuronidase deficiency, in: *Enzyme Therapy in Lysosomal Storage Diseases* (J. M. Tager, G. J. M. Hooghwinkel, and W. Th. Daims, eds.), p. 288, North-Holland Publishing Co., Amsterdam.

Suzuki, K., Suzuki, Y., and Fletcher, T. F., 1972, Further studies on galactocerebroside β-galactosidase in globoid cell leukodystrophy, in: *Sphingolipids, Sphingolipidoses and Allied Disorders* (B. W. Volk and S. M. Aronson, eds.), p. 487, Plenum Press, New York.

Sweeley, C. C., Mapes, C. A., Krivit, W., and Desnick, R. J., 1972, Chemistry and metabolism of glycosphingolipids in Fabry's disease, in: *Sphingolipids, Sphingolipidoses and Allied Disorders* (B. W. Volk and S. M. Aronson, eds.), p. 287, Plenum Press, New York.

Synder, P. D. Jr., Wold, F., Bernlohr, R. W., Dullum, C., Desnick, R. J., Krivit, W., and Condie, R. M., 1974, Enzyme therapy II: Purified human α-galactosidase A. Stabilization to heat and protease aggregation by complexing with antibody and by chemical modification, *Biochem. Biophys. Acta* **350**:432.

Thoene, J. C., Oshima R. G., Crawhall, J. C., Olson D. L., and Schneider, J. A., 1976, Cystinosis intracellular cystine depletion by aminothiols in vitro and in vivo. *J. Clin. Invest.* **58**:180.

Trimble, B. K., and Doughty, J. H., 1974, The amount of hereditary disease in human populations, *Ann. Human Genet. (Lond.)* **38**:199.

Wold, F., 1973, Chemical modification of enzymes, in: *Enzyme Therapy in Genetic Diseases, Birth Defects* (Original Article Series, Vol. 9) (R. J. Desnick, R. W. Bernlohr, and W. Krivit, eds.), p. 46, The Williams & Wilkins Company, Baltimore.

World Health Organization, 1972, *Genetic Disorders: Prevention, Treatment and Rehabilitation, Tech. Rept. Ser. No. 497,* Geneva.

11

Experimental Therapy Using Semipermeable Microcapsules Containing Enzymes and Other Biologically Active Material

T. M. S. Chang

1. *Introduction*

Microencapsulated enzymes and proteins in the form of artificial cells are the first form of immobilized enzymes to be tested in animals for experimental therapy (Chang, 1957, 1964, 1966, 1969b, 1972a; Chang and MacIntosh, 1964; Chang and Poznansky, 1968). This chapter and Chapter 19 will discuss research carried out in this laboratory on the therapeutic applications of artificial cells and immobilized proteins. An increasing number of other centers have become involved in seriously looking into the therapeutic applications of all types of immobilized enzymes. Their research will be discussed in detail in other chapters.

2. *Red Blood Cell Substitutes*

Erythrocytes, with their hemoglobin, carbonic anhydrase, catalase, and complex enzyme systems, efficiently carry out the complex functions of oxygen and CO_2 carriage. There are a number of problems in the use of red blood cells for transfusion. The demand cannot be always met by available donors. Careful and time-consuming blood typing and matching is required because of blood group antigens on the cell membrane. Furthermore, blood stored by the standard methods can only be kept for a few weeks. In an attempt to solve these problems, red blood cell content, with its hemoglobin, carbonic anhydrase, and complex enzyme systems, was microencapsulated to study its feasibility as blood cell substitutes (Chang, 1957, 1964). After microencapsulation within ultrathin spherical cellulose nitrate membranes, a large part of the hemoglobin is changed to methemoglobin, but the unaltered hemoglobin retains its ability to combine reversibly

T. M. S. Chang • Departments of Physiology and Medicine, McGill University, Montreal, Canada

with oxygen. Microencapsulation of additional catalase markedly reduced methemoglobin formation. When silicone rubber (Medical Grade Silastics) was used to form the microcapsules, methemoglobin formation is even further reduced, and the unaltered hemoglobin combines reversibly with oxygen (Chang, 1966). Carbonic anhydrase activity of collodion or nylon microencapsulated erythrocyte hemolysate retained excellent activity in converting carbon dioxide to carbonic acid (Chang, 1964). The catalase activity of the collodion microencapsulated hemolysate is also well maintained (Chang and Poznansky, 1968). Immunological studies carried out using human ABO antisera show that although antisera caused agglutination of incompatible human erythrocytes and heterogenous erythrocytes, the same antisera does not cause any agglutination of microencapsulated red blood cell hemolysate even though the red blood hemolysate is obtained from incompatible human red blood cells or from animal red blood cells (Chang, 1972a). Thus, *in vitro* studies show that microencapsulated red blood cell content continues to carry out the important functions of red blood cells without having problems related to the antigenicity of ABO antisera. Studies carried out in dogs, cats, and rats show that after intravenous injection, these "artificial red blood cells" were removed rapidly from the systemic circulation of the animals. Larger microcapsules are trapped in the pulmonary capillaries. Smaller microcapsules, which can pass through the pulmonary capillaries, are quickly removed by the reticuloendothelial system of the liver and spleen (Chang, 1964, 1965, 1972a). Although the duration of the survival of the artificial red blood cells in the systemic circulation can be increased by changing the surface properties, the time is still extremely brief (Chang, 1964, 1965, 1972a). Thus, further efforts aimed at finding a system that can survive in the systemic circulation would be required. Studies of microencapsulation of red blood cell content have also been carried out in Japan (Sekiguchi and Kondo, 1966) with similar results.

3. *Immobilized Urease as a Basic Model for Experimental Therapy in Vivo*

3.1. *Basic Studies*

In these studies, microencapsulated urease was used as a model system for a feasibility study of the therapeutic applications of immobilized enzymes (Chang, 1964, 1965, 1972a; Chang and MacIntosh, 1964). *In vitro* studies show that microencapsulated urease did not leak out but acted effectively on urea diffusing into the microcapsules converting urea to ammonia. The activity of the artificial cell urease was 37% of the same amount of enzyme in free solution. Furthermore, the enzyme can be stabilized by microencapsulation with a high concentration of hemoglobin in the artificial cells with a $T_{1/2}$ of 1 week at 37°C and 2 weeks at 4°C, as compared to a $T_{1/2}$ of 3 hours with urease not in a high concentration of hemoglobin. Because of the large concentration of urea present in the biological fluid and the easy diffusibility of urea, the study of microencapsulated urease *in vivo* was carried out to see its effect on urea.

3.2. *Implantation by Injection*

In these studies, intraperitoneal injection into dogs of 0.25 ml/kg of controlled microcapsules containing no urease produced no significant changes in the arterial blood level of ammonia. However, intraperitoneal injection of microencapsulated urease (100 Sumner units in 0.25 ml/kg) resulted in a rise of the arterial blood ammonia level as urea in the body is converted enzymatically to ammonia (Figure 1). Since radioisotope-labeled proteins did not leak out of the microcapsules after intraperitoneal injection, urease did not leak out. The microencapsualted urease thus acts on urea diffusing into the peritoneal cavity and then into the microcapsules.

Long-term studies were carried out in an unanesthetized dog. Intraperitoneal injection of 0.5 ml/kg of control artificial cells without urease had no significant effect on the blood ammonia level or the general state of the animal. When artificial cells containing urease were injected (50 Sumner units in 0.5 ml/kg), the ammonia level increased from the control levels from 0.15 μg/ml to a height of 1.70 μg/ml a few hours after injection. The level remained at this height for 24 hr and then fell slowly, reaching 50% 3.5 days after injection. The level is still above the control level 5 days after injection. The high level is maintained despite the fact that the liver is functioning actively in converting ammonia to urea. The $T_{1/2}$ of the microencapsulated urease stored *in vitro* at 37°C is 7 days. The result of these studies show conclusively that when introduced intraperitoneally, the model

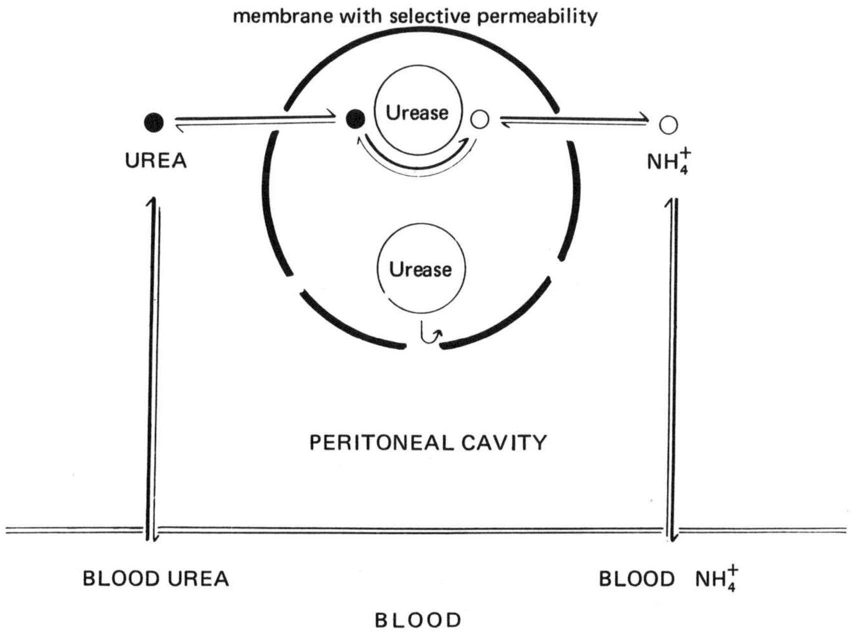

FIGURE 1. Schematic representation of the action of intraperitoneally injected artificial cells loaded with urease. (From Chang, 1965.)

system of microencapsulated urease can act effectively on the model metabolite, urea, converting urea into ammonia. These results lead us to suggest that immobilized enzymes could be used in enzyme replacement therapy since in this form, the enzyme acts on external permeant substrates but does not leak out to become involved in immunological reactions (Chang, 1964, 1965, 1972a; Chang and MacIntosh, 1964).

3.3. *Extracorporeal Perfusion*

In practice, one may not want to inject immobilized enzymes into the body since this may result in the accumulation of foreign material in the body. To avoid this it was suggested and demonstrated that if immobilized enzymes were retained in an extracorporeal system so that blood or other body fluid can recirculate the extracorporeal system, then they can act without being introduced into the body (Chang, 1966). To test this feasibility, microencapsulated urease was again used as a model immobilized enzyme system. Ten milliliters of microencapsulated urease was retained in an extracorporeal shunt chamber. When blood of dogs recirculated through the extracorporeal shunt to come in direct contact with the microencapsulated urease, blood urea was effectively converted into ammonia. This way, the systemic blood urea level fell with a $T_{1/2}$ of 60 min (Chang, 1966). To prevent the need for systemic heparinization, the microencapsulated enzymes, shunt chamber, and connecting tubing were complexed with heparin so that blood could recirculate without clotting (Chang *et al.,* 1967). In this heparinized form, microencapsulated urease still acts effectively in converting urea in the blood to ammonia (Chang *et al.,* 1967).

3.4. *Oral Ingestion*

As described in Chapter in organ replacement therapy, microencapsulated urease also functions effectively after oral administration (Chang and Loa, 1970).

3.5. *Conclusion*

Thus, it was demonstrated that using the model enzyme, urease, immobilized by microencapsulation, can be introduced into the body by injection to act effectively in converting the model metabolite, urea, into its product, ammonia. At the same time, the enzymes remain in the immobilized state without leaking out. Further extensions demonstrated that to avoid introducing enzymes into the body, one could use an extracorporeal shunt system where the blood is allowed to recirculate through the extracorporeal shunt. Thereby, the metabolites in the blood could be acted on by the immobilized enzyme in the extracorporeal shunt. Furthermore, heparinization by complexing heparin on the surface of the immobilized enzyme and the shunt also eliminates the necessity of using systemic heparinization to prevent coagulation. These initial basic studies using the model enzyme, urease, have led to further studies in this and other laboratories in actual experimental enzyme therapy for more specific disorders.

4. *Experimental Enzyme Replacement Therapy for Hereditary Enzyme Deficiency*

One line of research in this laboratory involves the use of microencapsulated enzymes for experimental replacement in congenital enzyme deficiencies (Chang and Poznansky, 1968; Poznansky and Chang, 1974). The first experiment carried out here involved the use of the Feinstein strain of mice with a hereditary deficiency in catalase (acatalasemia) (C_s). Acatalasemia is a condition that is also present in man. These Feinstein acatalasemic mice (C_s^b) have a blood catalase activity approximately 2% of normal and a total body catalase activity approximately 20% of normal.

4.1. *Implantation*

In vivo enzyme-replacement-therapy experiments were carried out using catalase immobilized by microencapsulation with cellulose nitrate membranes (Chang and Poznansky, 1968; Poznansky and Chang, 1974). In one set of experiments, one group of normal control mice (C_s^a) and three groups of acatalasemic mice were used. One group of acatalasemic mice and one group of normal mice received an intraperitoneal injection of 0.15 ml/g of body weight of saline. In the second group of acatalasemic mice, each was given an intraperitoneal injection of 0.15 ml/g of body weight of a 50% suspension of microencapsulated catalase. The third acatalasemic group received, intraperitoneally, 0.75 mg/g of body weight of catalase in solution containing the same assayed enzyme activity as the microencapsulated catalase injected. The five groups of mice were then each given a subcutaneous injection of 0.014 ml/g of body weight of sodium perborate as substrate for catalase. The normal mice remained unaffected, whereas the acatalasemic mice which were given only saline became increasingly immobile and in respiratory distress by the end of 20 min. Acatalasemic mice, protected by the intraperitoneal injections of microencapsulated catalase or catalase solution, remained much more active and were moving about freely at the end of 20 min. Detailed analysis were carried out to quantitatively assess the total body perborate that remained unchanged in each group 20 min after injection. The results showed that after 20 min, the following percentage of injected perborate remained: 2.5% ± 2.5% S.D. in normal mice, 70% ± 8.2% S.D. in acatalasemic mice receiving only saline, 7% ± 3.6% S.D. in acatalasemic mice receiving catalase solution, and 16% ± 3.5% S.D. in acatalasemic mice receiving microencapsulated catalase (Chang and Poznansky, 1968). Further detailed analyses (Poznansky and Chang, 1974) were carried out to analyze the kinetic changes in the level by analyzing the total body perborate level at 5-min intervals. These studies show a much more rapid decrease of the total body perborate in normal mice as compared to acatalasemic mice receiving no enzymes. In acatalasemic mice injected with catalase solution or catalase microcapsules, the rate of decrease of total body perborate was greatly increased, approaching the level of normal mice at the end of 20 min (Figure 2). The difference between the injection of immobilized catalase and catalase in solution was that the enzyme in solution after injection rapidly

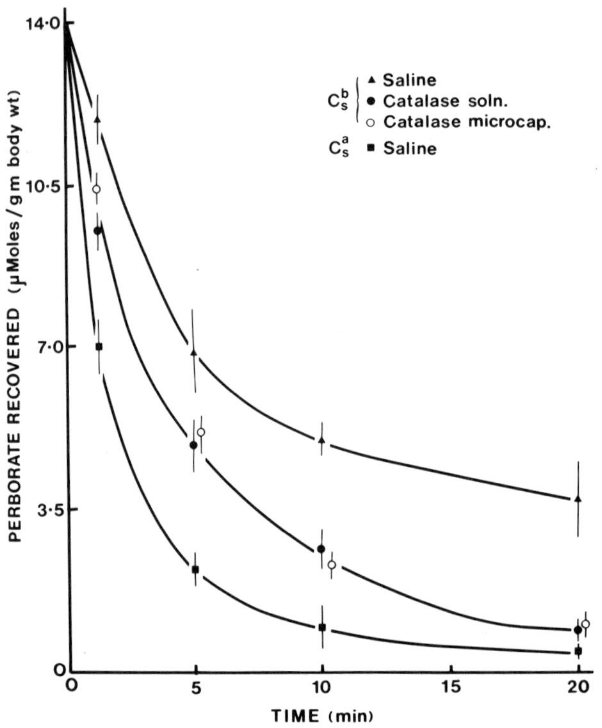

FIGURE 2. Time course of fate of subcutaneously injected $NaBO_3$ in acatalasemic (C_s^b) mice and normal (C_s^a) mice. Immediately before the perborate injections, each animal received one of the following intraperitoneal injections: saline, catalase solution, or microencapsulated catalase. After varying time intervals, the mice were sacrificed and assayed for total $NaBO_3$ remaining in the body. Each point and bar represents the mean and standard error from five animals. (From Poznansky and Chang, 1974.)

entered the circulation and was distributed throughout the body, whereas the microencapsulated catalase after intraperitoneal injection remained at the site of injection and acted on substrates diffusing into the semipermeable microcapsules (Figure 3). In further detailed studies, it was demonstrated that although repeated injections of beef liver catalase resulted in the production of antibodies, repeated injections of microencapsulated catalase did not (Poznansky and Chang, 1974). Further study demonstrated that in acatalasemic mice previously immunized to beef catalase, injections of large doses of beef catalase solution resulted in an anaphylactic type of reaction and the injected enzyme did not protect the animal from perborate injection. On the other hand, microencapsulated catalase injected into these immunized animals continued to act effectively in replacing the deficient catalase and to protect the immunized animals from the injected perborate without any problem of hypersensitivity reactions (Poznansky and Chang, 1974) (Figure 4). Thus, implanted microencapsulated catalase could be used effectively for the replacement of catalase deficiency in acatalasemic mice. Unlike catalase in free solution, immobilzed catalase did not cause immunological and hypersensitivity reactions.

4.2. *Sites of Action*

In practice, however, one would have to look at each individual situation in enzyme replacement for hereditary enzyme deficiency, since in some situations one may not want to inject immobilized enzymes into the body, and in other cases immobilized enzymes may have to act not just on body fluid but may have to be located at specific sites. The following cases are typical examples dealing with specific requirements for practical applications.

4.3. *Extracorporeal Perfusion*

The first example involves a study carried out to extend the enzyme replacement therapy using an extracorporeal shunt containing microencapsulated catalase (Chang and Poznansky, 1968). In these studies, microencapsulated catalase was retained in a small extracorporeal shunt chamber. A small volume of fluid was injected intraperitoneally and the fluid allowed to recirculate through the shunt.

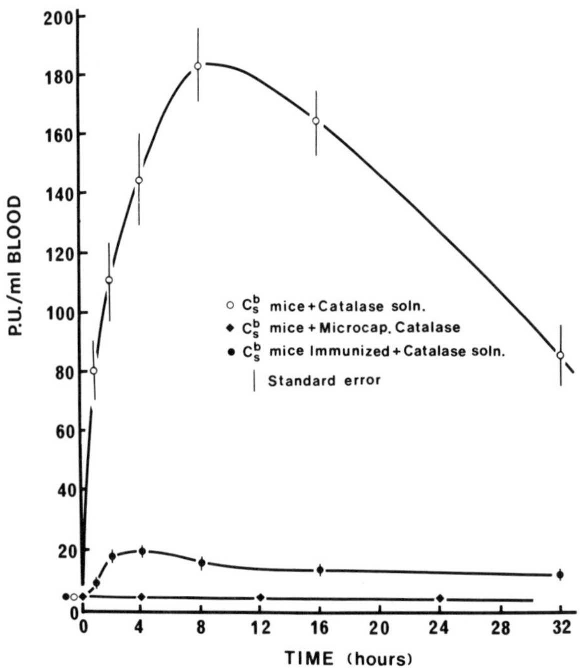

FIGURE 3. Time course of appearance of intraperitoneally injected catalase in the blood stream of acatalasemic C_s^b mice and acatalasemic mice immunized to catalase. To avoid severe anaphylactic reactions, only one-third the usual dose of catalase was used in this experiment. Note that intraperitoneally injected catalase solution appeared in the bloodstream; whereas intraperitoneally injected microencapsulated catalase did not enter the bloodstream. In C_s^b mice immunized to catalase, the intraperitoneally injected catalase which enters the bloodstream is removed much more rapidly from the circulation. (From Poznansky and Chang, 1974.)

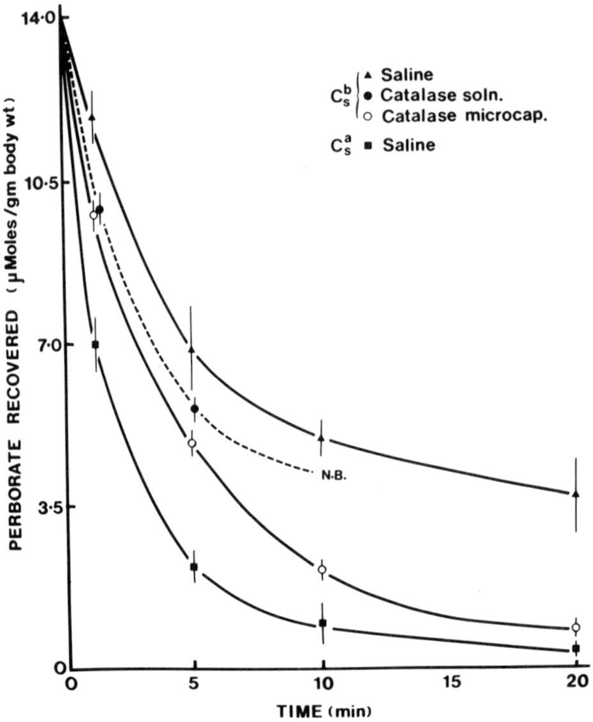

FIGURE 4. Time course of removal of injected $NaBO_3$ in immunized normal (C_s^a) mice and in immunized acatalasemic(C_s^b) mice. N.B., death of immunized mice from anaphylaxis after injection of catalase solution. (From Poznansky and Chang, 1974.)

This way, the extracorporeal shunt chamber containing the microencapsulated catalase was effective in lowering the perborate injected into the body and replacing the deficient catalase in protecting the animal against perborate.

4.4. *Local Applications*

The second example demonstrates another specific way microencapsulated enzymes can be applied to specific sites by local application. In human acatalasemia, the major problem is related to oral lesions as a result of hydrogen peroxide producing bacteria, which causes local anoxia and gangrene formation. Repeated direct application of a catalase solution on an oral lesion may result in absorption and allergic immunological reactions. Furthermore, the enzyme may not be sufficiently stable to last for a sufficient length of time. Catalase immobilized by microencapsulation has been tested experimentally in acatalasemic and normal mice (Chang, 1972b). In these experiments, microencapsulated catalase was applied to oral lesions in mice and was found to act effectively on H_2O_2. Experiments were also carried out in a human, where microencapsulated catalase paste was applied locally to different areas of the gingiva (Chang, 1972b). There was no local irritation or other observable adverse effects and the locally applied microencapsulated catalase continued to act effectively on H_2O_2.

4.5. *Specific Intracellular Locations*

In certain enzyme-deficiency conditions (e.g., storage diseases) the enzymes have to be located at specific intracellular sites. For this, special approaches are required. Thus, liposomes (Gregoriadis and Ryman, 1972) and erythrocyte-entrapped enzymes (Ihler *et al.*, 1973; Thorne *et al.*, 1975) have been used. These and other studies will be described in detail in subsequent chapters.

5. L-*Asparaginase for Substrate-Dependent Tumors*

The first *in vivo* experiment using immobilized asparaginase was carried out in which microencapsulated asparaginase injected intraperitoneally into mice was shown to act effectively in removing asparaginase (Chang *et al.*, 1968). Following this, experiments were carried out here to study the use of immobilized asparaginase in experimental suppression of substrate-dependent tumors in animals (Chang, 1969a, 1971). In these studies, L-asparaginase was microencapsulated within semipermeable microcapsules. Experimental treatment of 6C3HED lymphosarcoma in C3HHEJ mice (Jackson Laboratory) was studied in these experiments. A total of 500,000 lymphosarcoma cells were implanted into each groin. Each mouse was then given one of the following intraperitoneal injections: 0.05 ml/g of body weight of saline as a control; 0.05 ml/g of body weight of a 50% suspension of controlled microcapsules containing no asparaginase; 0.05 ml/g of body weight of microcapsules containing L-asparaginase; and 0.05 ml/g of body weight of L-asparaginase solution with the same assayed activity as that of the microencapsulated asparaginase. The animals were observed from the time the tumor first appeared. The results show that L-asparaginase solution given in the amounts stated significantly delayed the appearance of lymphosarcoma. Microencapsulated L-asparaginase suppressed the growth of implanted lymphosarcoma more effectively than L-asparaginase solution, even though the assay activity of the enzyme in solution and in the microencapsulated form are the same. The times the implanted tumors appeared in the C3H mice were: injection of saline, 9.0 ± 0.9 days; injection of controlled microcapsules, 8.7 ± 1.6 days; injection of L-asparaginase solution, 14 ± 4.5 days; and injection of L-asparaginase microcapsules, >69 days (Figure 5).

Further detailed studies were carried out here on the microencapsulated asparaginase system. It was demonstrated that the asparaginase microencapsulated with hemoglobin was much more stable than the asparaginase in free solution at 4°C and at a body temperature of 37°C. The stability at 37°C could be further enhanced by treating the microencapsulated asparaginase with glutaraldehyde (Chang, 1973b) (Figures 6 and 7). Furthermore, injection of the microencapsulated form maintained a zero asparagine level 3 times longer than the same amount of enzyme in solution (Chang, 1973b) (Figure 8). Further details of the *in*

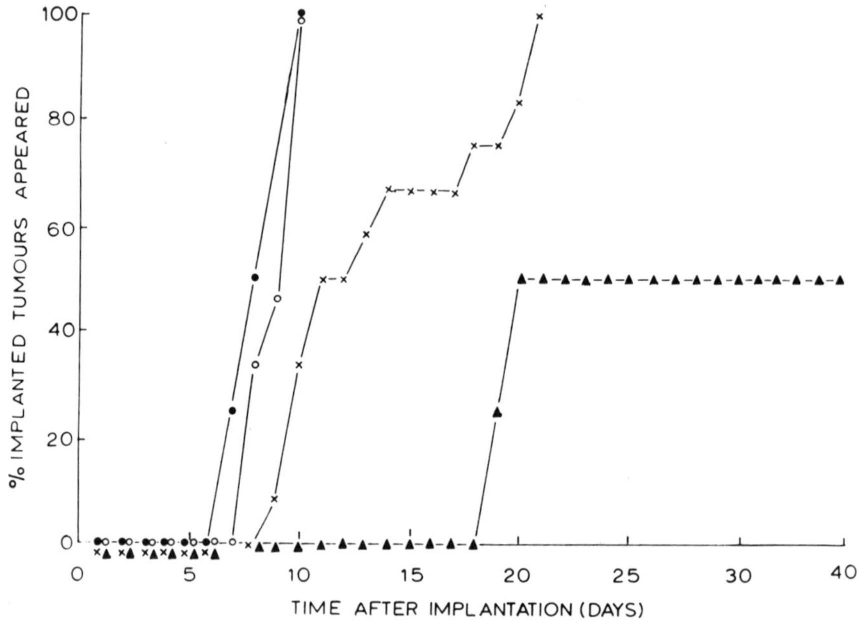

FIGURE 5. Appearance of implanted 6C3HED lymphosarcoma in mice. ●, saline as control; ○, control artificial cells; ×, asparaginase solution, ▲, artificial cells containing asparaginase. (From Chang, 1972a.)

vivo analysis of the effects of intraperitoneally injected L-asparaginase solution and L-asparaginase immobilized within semipermeable microcapsules were studied (Sui Chong and Chang, 1974). It was found in these studies that after injection of L-asparaginase solution, enzyme activity appeared in the blood very rapidly, with the highest concentration occurring after 4 hr. The enzyme was then cleared from the circulation with a $T_{1/2}$ of 4.4 hr. In marked contrast, when microencapsulated L-asparaginase was injected intraperitoneally, no significant L-asparaginase activity appeared in the blood for the entire duration of the study. This demonstrated that microencapsulated L-asparaginase did not leak out to any detectable level. Body L-asparaginase declined very rapidly, with a half-life of 2 hr after injection of L-asparaginase solution, whereas it took 60–72 hr for the body L-asparaginase to decrease to 50% of the original activity after injection of microencapsulated L-asparaginase. The microencapsulated L-asparaginase in the body still retained about 20% of its original activity 16 days after injection. Plasma L-asparagine was maintained at zero concentration 4 days after injection of the solution, compared to 8 days after injection of the microencapsulated enzyme. In these studies, microencapsulated asparaginase was capable of causing regression of the tumor in the advanced, well-established stage (Siu Chong and Chang, 1974).

As will be discussed in a later chapter, asparaginase microencapsulated by interfacial polymerization has also been carried out by another group (Mori *et al.*, 1972, 1973). They found that the membranes of the microcapsule were resistant to both mechanical shock and chymotrypsin. No leakage of microencapsulated asparaginase was detected by the Ouchterlony technique. It was found that although there was reaction from the antibodies of asparaginase and asparaginase solution, there was no precipitant line detected between the antibodies of asparaginase and microencapsulated asparaginase. Studies using trypsin, chymotrypsin, and pronase-*p* show that these did not affect microencapsulated asparaginase; however, they lowered the asparaginase in free solution to 28%, 4%, and 12%, respectively.

Another approach is the use of fiber-encapsulated asparaginase (Salmona *et al.*, 1974). L-Asparaginase was microencapsulated within cellulose triacetate in fiber form. Recirculation of perfusate through the fiber-entrapped asparaginase resulted in the lowering of the L-asparagine level and the increase of the L-aspartic acid level. The L-asparagine level is lowered from 100 nmoles/ml to less than 10 nmoles/ml. *In vitro* studies showed that anti-L-asparaginase serum did not reduce the activity of the fiber-entrapped asparaginase, whereas free enzyme was inhibited to 25% of its original value.

Another encapsulation approach is the use of dialysis against asparaginase (Apple, 1971). Here, the asparaginase is placed outside in the dialysate compartment and blood allowed to pass through the blood compartment of the hemodialysis system. The system used was the capillary hemodialysis system. It was found that, by this approach, L-asparagine could be selectively removed at a clearance of 80 ml/min with a flow rate of about 180 ml/min.

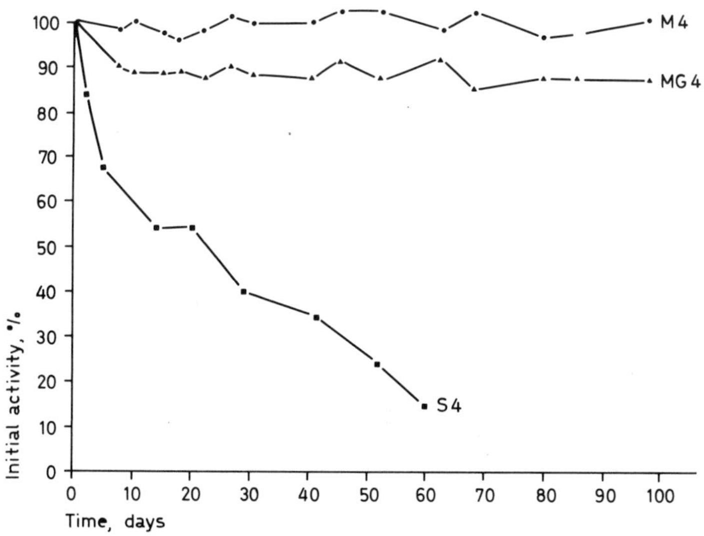

FIGURE 6. Stability of asparaginase stored at 4°C. S4, asparaginase solution kept at a concentration of 10 IU/ml; M4, microencapsulated asparaginase kept at a concentration of 10 IU/ml suspension; MG4, microencapsulated asparaginase which has been cross-linked with glutaraldehyde and kept at a concentration of 10 IU/ml suspension. (From Chang, 1973.)

Since the enzyme in solution outside the dialysate is rather unstable, a new proposal was to combine artificial cells in the form of microencapsulated asparaginase with the artificial capillary system (Chang, 1973a). In these studies, blood passes through the capillary system, and microencapsulated asparaginase in the dialysate compartment was found to act effectively on L-asparagine. By using glutaraldehyde-treated microencapsulated asparaginase, the stability of the microencapsulated L-asparaginase is maintained for months. The combination of the heparinized capillary systems and this long stability may allow for long-term implantation for blood hemoperfusion.

Although the feasibility of using immobilized asparaginase for tumor suppression has been demonstrated (Chang, 1969a, 1971), a large amount of work is still required to develop and make this system practicable for clinical uses. In addition to the encapsulation of asparaginase, a very large amount of work is also being carried out using other types of immobilized asparaginase. Most of this will be discussed in subsequent chapters. In the intermittent extracorporeal approach using asparaginase, although the serum asparagine is lowered quickly to near zero during hemoperfusion, with termination of hemoperfusion the serum asparagine level rapidly increases to normal. In this particular condition, where it is necessary to suppress the L-asparagine for a long period of time, the intermittent use of the extracorporeal approach may not be suitable. It will require a system that can act for long periods of time, as in the case of injected microencapsulated asparagine (Chang, 1969a, 1971), or long-acting extracorporeal devices. Microcapsulated

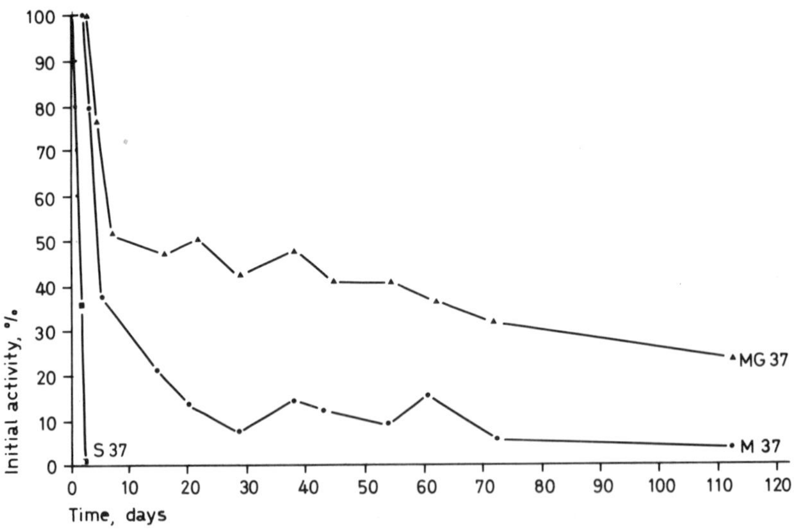

FIGURE 7. Stability of asparaginase stored at 37°C. S37, asparaginase solution kept at a concentration of 10 IU/ml: M37, microencapsulated asparaginase kept at a concentration of 10 IU/ml suspension; MG 37, microencapsulated asparaginase which has been cross-linked with glutaraldehyde and kept at a concentration of 10 IU/ml suspension. (From Chang, 1973b.)

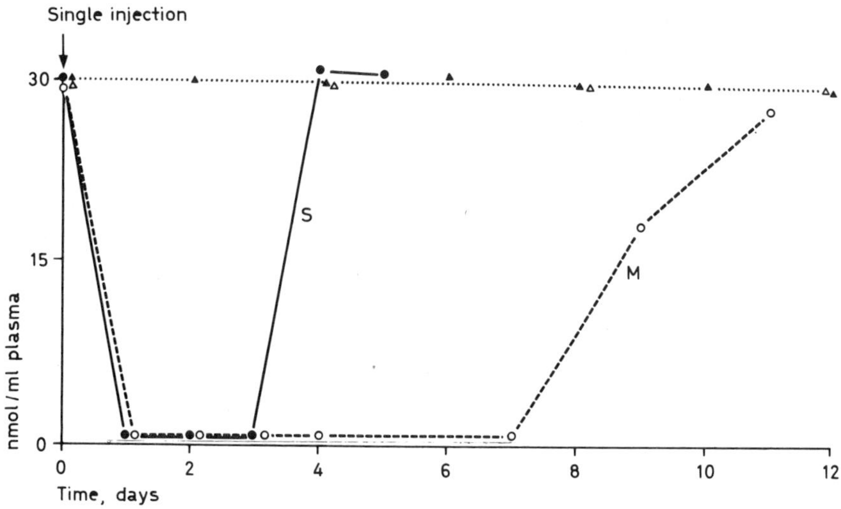

FIGURE 8. Plasma asparagine level in C3HHeJ mice. ▲, mice injected with control saline; △, mice injected with control microcapsules; •, mice injected with asparaginase solution; ○, mice injected with microencapsulated asparaginase. (From Chang, 1973b.)

asparaginase has been prepared using polylactic acid, a biodegradable polymer (Chang, 1976a, 1976b). The asparaginase retained excellent activity on external asparagine diffusing into the polylactic acid microcapsule. Polylactic acid is a biodegradable polymer that is converted into H_2O and CO_2 in the body.

6. *"One-Shot" Vaccine*

Vaccination has protected millions against smallpox, cholera, typhus–yellow fever, plague, and other infectious diseases. In the case of vaccines prepared in the form of killed suspensions of bacteria or virus, or in the form of modified toxins, repeated injection(s) at precise interval(s) are required in order for the vaccination to be effective. These intervals may range from weeks to months. In some areas, the ratio of physicians, nurses, or other medical personnel to population may be extremely low, with a resulting lack of personnel to carry out adequate immunization programs. Even more serious is the fact that there are some regions where either because of geography or other social economic conditions there is no easily accessible medical facilities. Here, a single injection carried out by a team of visiting medical personnel rather than repeated injections would be important. In other cases, the population may come for one vaccination but, because of difficulty in transportation, lack of interest, or other factors, fail to return for subsequent necessary injections. These problems are magnified many times in mass vaccination compaigns, in which an attempt is made to cover the whole of a susceptible population in one intensive effort or in the rapid and effective vaccination for the control of epidemic outbreaks of infectious diseases. With a "one-shot" vaccination requiring only one injection, then, the problem related to the inadequate health personnel and the difficulty of having patients returning at the exact intervals would have much lessened. In order to make this possible, the one-shot vaccination has to be developed in such a way that it contains the standard vaccines plus another system that can release vaccine at the required interval(s). It was proposed that microencapsulated vaccines could be the basis of this approach (Chang, 1975). Vaccines have already been successfully microencapsulated (Chang, 1976a, 1976b). Microcapsules have also been formed with membranes of different degrees of biodegradability. One typical example includes the range of cross-linked protein membrane to membrane formed entirely from nylon. Cross-linked protein membrane is much more easily biodegradable than nylon. It is also possible to have membranes with different proportions of cross-linked protein and nylon resulting in varying degrees of biodegradability. Biodegradable polymers such as polylactic acid have been used successfully here for microencapsulation. Here, the polymer can be metabolized into H_2O and CO_2.

7. *Multienzyme Systems and Requirements for Cofactors*

A large number of the conditions that require enzyme replacement therapy are those systems which depend on a multienzyme system or a multienzyme system that requires a continuous supply of cofactors. Even in the beginning, when artificial cells were first prepared by microencapsulation of red blood cell content (Chang, 1957), the complex enzyme systems of the cell content were easily microencapsulated. Thus, as long as the enzymes are available, there are no problems in microencapsulating multienzyme systems or combinations of multienzyme systems (Chang, 1972a).

Some multienzyme systems require the availability of cofactors. The problems involved have been discussed at some length (Chang, 1965, 1972a). One solution to this is the use of microencapsulated multienzyme systems for the recycling of cofactors, for example, the recycling of ATP $\rightleftharpoons$ ADP (Campbell and Chang, 1975) or the recycling of $NAD^+ \rightleftharpoons NADH$ (Campbell and Chang, 1976). Since cofactors are small molecules that can equilibrate across the microcapsule membrane, a continuous supply of cofactors would be required. To solve this problem, cofactors have been immobilized first to soluble or insoluble macromolecular carriers, and then microencapsulated together with the multienzyme systems (Campbell and Chang, 1976). This way, the cofactor remains inside the microcapsule and can be recycled and used repeatedly (Campbell and Chang, 1976).

A recent study was carried out here to make use of the multienzyme systems and recycling of cofactors for enzyme replacement in enzyme-deficiency conditions. This involved the use of a multienzyme system for galactokinase deficiency in galactosemia. In this study a multienzyme system of galactokinase and pyruvate kinase was microencapsulated inside collodion microcapsules. This way, the following reaction was successfully carried out:

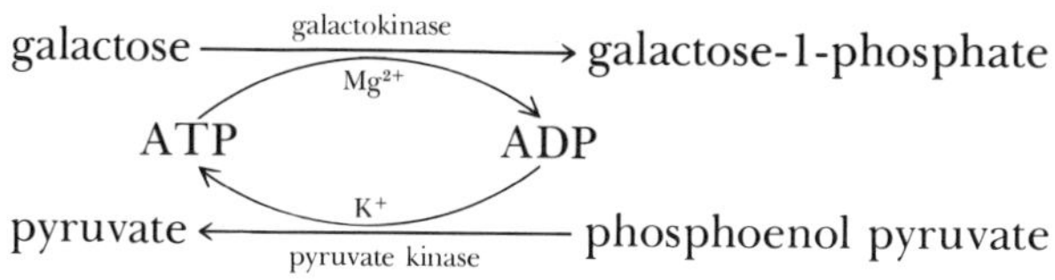

8. *Immobilized Enzymes and Proteins for Artificial Kidney, Artificial Liver, and Detoxifier*

Immobilized enzymes and proteins have been used successfully here and elsewhere for the construction of artificial kidney, artificial liver, and detoxifier. There will be discussed in detail in subsequent chapters.

9. *References*

Apple, M., 1971, Hemodialysis against enzymes as a method of "gene replacement" in cases of inherited metabolic diseases, *Proc. West. Pharmacol. Soc.* **14**:125.

Campbell, J., and Chang, T. M. S., 1975, Enzymatic recycling of coenzymes by a multi-enzyme system immobilized within semipermeable collodion microcapsules, *Biochim. Biophys. Acta* **397**:101.

Campbell, J., and Chang, T. M. S., 1976, The recycling of NAD^+ (free and immobilized) within semipermeable aqueous microcapsules containing a multi-enzyme system, *Biochem. Biophys Res. Commun.* **62**:562.

Chang, T. M. S., 1957, Hemoglobin corpuscles, Report of research project for B.Sc. Honors, McGill University.

Chang, T. M. S., 1964, Semipermeable microcapsules, *Science* **146**:524.

Chang, T. M. S., 1965, Semipermeable aqueous microcapsules, Ph.D. thesis, McGill University.

Chang, T. M. S., 1966, Semipermeable aqueous microcapsules ("artificial cells"): with emphasis on experiments in an extracorporeal shunt system, *Trans. Amer. Soc. Artificial Internal Organs* **12**:13.

Chang, T. M. S., 1969a, Asparaginase-loaded semipermeable microcapsules for mouse lymphomas, *Proc. Can. Federation Biol. Sci.* **12**:62.

Chang, T. M. S., 1969b, Clinical potential of enzyme technology, *Sci. Tools* **16**:33.

Chang, T. M. S., 1971, the in-vivo effects of semipermeable microcapsules containing L-asparaginase on 6C3HED lymphosarcoma, *Nature* **229**:117.

Chang, T. M. S., 1972a, *Artificial Cells,* Charles C. Thomas, Publisher, Springfield, Ill.

Chang, T.M.S., 1972b, Effects of local applications of microencapsulated catalase on the response of oral lesions to hydrogen peroxide in acatalasemia, *J. Dental Res.* **51**:319.

Chang, T. M. S., 1973a, Biomedical applications of artificial cells, *Bio-Med. Eng.* **8**:334.

Chang, T. M. S., 1973b, L-Asparaginase immobilized within semipermeable microcapsules: *in-vitro* and *in-vivo* stability, *Enzyme* **14**:95.

Chang, T. M. S., 1975, The one-shot vaccine, in: *Socio-economic and Ethical Implications of Enzyme Engineering* (C.-G. Heden, ed.), International Federation of Institutes for Advanced Studies, Stockholm, Sweden.

Chang, T. M. S., 1976a, Methods in experimental therapy using immobilized enzymes, in: *Immobilized Enzymes* (K. Mosbach, ed.), from Methods in Enzymology Series, Academic Press, Inc., New York.

Chang, T. M. S., 1976b, Biodegradable semipermeable microcapsules containing enzymes, hormones, vaccines and biologicals, *J. Bioengineering* (in press).

Chang, T. M. S., and Loa, S. K., 1970, Urea removal by urease and ammonia adsorbents in the intestine, *Physiologist* **13**:70.

Chang, T. M. S., and MacIntosh, F. C., 1964, Semipermeable aqueous microcapsules, *Pharmacologist* **6**:198.
Chang, T. M. S., and Poznansky, M. J., 1968, Semipermeable microcapsules containing catalase for enzyme replacement in acatalasemic mice, *Nature* **218**:243.
Chang, T. M. S., Johnson, L. J., and Ransome, O., 1967, Semipermeable aqueous microcapsules: IV. Nonthrombogenic microcapsules with heparin-complexed membranes, *Can. J. Physiol. Pharmacol.* **45**:705.
Chang, T. M. S., Pont, A., Johnson, L. J., and Malave, N., 1968, Response to intermittent extracorporeal perfusion through shunts containing semipermeable microcapsules, *Trans Amer. Soc. Artificial Internal Organs* **15**:163
Gregoriadis, G., and Ryman, B. E., 1972, Fate of protein-containing liposomes injected into rats: an approach to the treatment of storage diseases, *Eur. J. Biochem.* **24**:485.
Ihler, G. M., Glew, R. H., and Schnure, F. W., 1973, Enzyme loading of erythrocytes, *Proc. Natl. Acad. Sci. U.S.* **70**:2663.
Mori, T., Sata, T., Matuo, Y., Tosa, T., and Chibata, I., 1972, Preparation and characteristics of microcapsules containing asparaginase, *Biotechnol. Bioeng.* **14**:663.
Mori, T., Tosa, T., and Chibata, I., 1973, Enzymatic properties of microcapsules containing asparaginase, *Biochim. Biophys. Acta* **321**:653.
Poznansky, M. J., and Chang, T. M. S., 1974, Comparison of the enzyme kinetics and immunological properties of catalase immobilized by microencapsulation and catalase in free solution for enzyme replacement, *Biochim. Biophys. Acta* **334**:103.
Sekiguchi, W. and Kondo, A., 1966, Studies on microencapsulated hemoglobin, *J. Japan. Soc. Blood Transfusion* **13**:153.
Slamona, M., Saronio, C., Bartosek, I., Vecchi, A., and Mussini, E., 1974, Fibre-entrapped enzymes, in: *Insolubilized Enzymes* p. 189, (M. Salmona, C. Saronio, and S. Garattini, eds.), Raven Press, New York.
Siu Chong, E. D., and Chang, T. M. S., 1974, *In-vivo* effects of intraperitoneally injected L-asparaginase solution and L-asparaginase immobilized within semipermeable nylon microcapsules with emphasis on blood L-asparagine, "body" L-asparaginase, and plasma L-asparagine levels, *Enzyme* **18**:218.
Thorne, S. R., Fiddler, M. B., and Desnick, R. J., 1975, Enzyme therapy V: *in-vivo* fate of β-glucuronidase deficient mice, *Pediat. Res.* **9**:918.

12

Stabilized Urease Microencapsulated

DAVID L. GARDNER AND DONALD C. EMMERLING

1. *Introduction*

The possibility of using enzymes for the treatment of enzyme-deficiency diseases or organ dysfunction has generated enthusiasm over the last 10 years. Part of this enthusiasm has stemmed from technological developments for enzyme stabilization to prolong enzyme activity and for microencapsulation processes to form a protective barrier around the soluble enzyme or stabilized enzyme adduct. The microencapsulated enzyme (adduct) product might then be placed in conjunction with the circulatory system (hemoperfusion), injected intramuscularly, subcutaneously, or intraperitoneally, or administered orally as an ingestible product to carry out its intended functions.

Our interest in using the preceding technologies has been centered on an approach which employs the oral ingestion of adsorbents for removal of nitrogenous waste products found in the gastrointestinal tract in patients suffering from kidney dysfunction (uremia). A major metabolic waste product from the standpoint of quantity is urea. However, since it is extremely difficult to chemically bind urea to adsorbents at physiological pH levels, we decided to convert the urea to ammonia and carbon dioxide via the urease enzyme. In addition, Chang and Poznansky (1968) have noted that "microencapsulated soluble urease in the mouse intestinal tract acted efficiently in lowering blood urea." Thus, the routine ingestion of microencapsulated stabilized urease might remove urea from the gastrointestinal tract and thereby maintain an equilibrium of normal urea concentration levels in all body fluids. This paper describes the stabilization of the enzyme urease, the microencapsulation of the stabilized adduct, and preliminary experiments with the encapsulated product.

DAVID L. GARDNER • Battelle, Columbus Laboratories, 505 King Ave, Columbus, Ohio 43201
DONALD C. EMMERLING • Battelle, Columbus Laboratories, 505 King Ave, Columbus, Ohio 43201

2. *Urease Stabilization*

Insoluble forms of enzymes can be obtained by attaching the enzyme molecule to a solid surface by adsorption or a covalent bond. Silman and Katchalski (1966) and Goldstein and Katchalski (1968) have shown that adducts formed on insoluble polymeric matrixes are more stable than the unattached soluble enzyme. Although the mechanism of enhanced stability is difficult to interpret, the improvement appears to be quite general, as reported by Weetall (1970) in his enzyme-stability studies involving a variety of enzymes attached to organic polymers and porous glass particles. Chang (1964) and Sparks *et al.* (1969) have confirmed that urease stabilization would be desirable by showing that soluble urease, even when protected with a urea-permeable microcapsule membrane, is unstable in an aqueous solution at body temperature. Thus, ingested microcapsules containing soluble urease would begin deactivating while traversing the gastrointestinal tract.

We chose to stabilize urease essentially by the procedure of Haynes and Walsh (1969) in which sorption forces initially bind the enzyme to the surfaces of silica particles (0.01 μm). This is a practical advantage for possible large-scale future operations, since the surface preparation is simple and less expensive than that required for covalent binding. Following the sorption step, the adsorbed urease was cross-linked with glutaraldehyde to yield an active and insoluble urease layer on each silica particle.

The protein content of these urease-silica adduct particles was analyzed by the method of Lowry *et al.* (1951) and was found to vary between 2 and 6% (weight basis). The enzymatic activity of the urease-silica adduct (SE) particles was 0.56 mg of NH_3/min/mg of SE (16,500 IU/g) when tested in a 100 mg% buffered (pH 7.3) urea solution at 37°C. When compared to the soluble form, the activity represented 80–90% of soluble activity for an equivalent quantity of protein.

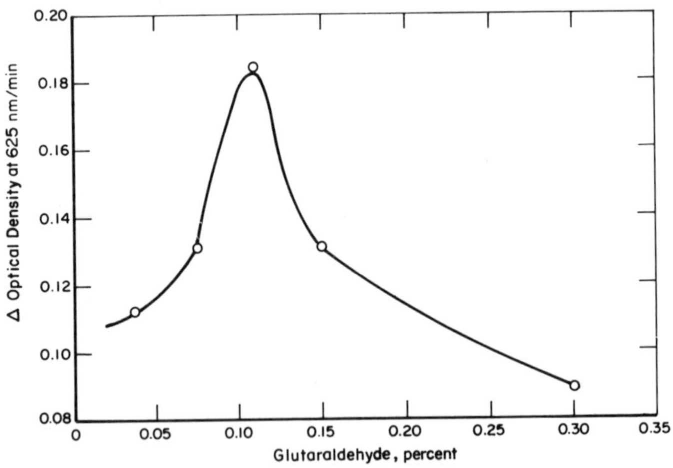

FIGURE 1. Relationship between activity and percent glutaraldehyde.

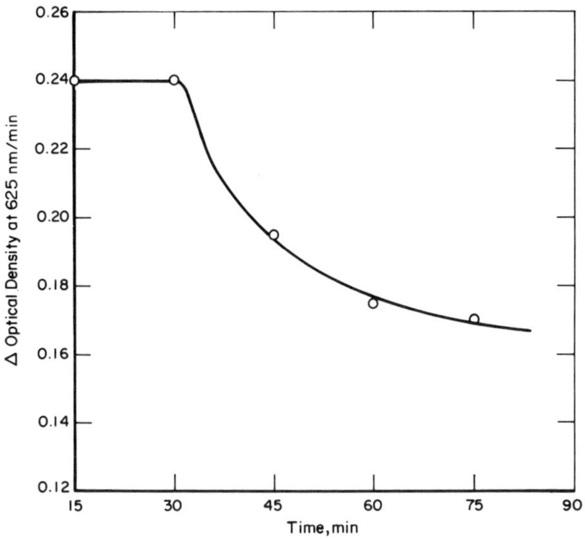

FIGURE 2. Relationship between activity and cross-linking time.

The effect of glutaraldehyde concentration and cross-linking time was investigated for optimizing the silica-enzyme activity. Figure 1 indicates the highest specific activity was attained by the adsorbed cross-linked silica-enzyme adduct when 0.11% glutaraldehyde concentration was used in the cross-linking step. The arbitrary unit of activity shown in Figure 1 is the rate of increase of ammonia in the urea solution expressed as the increased OD/min of the test solution at 625 nm using an ammonia assay by Weatherburn (1967).

Figure 2 indicates that an optimum cross-linking time of no more than 30 min should be employed when using the optimal glutaraldehyde concentration above.

A comparison of soluble urease and urease-silica adduct stability at 22 and 38°C is summarized in Table 1. The greater stability of the urease-silica adduct is evident at both temperatures. At 38°C, the urease-silica adduct retained 42% of its original activity after 15 days, while the soluble form of the enzyme had lost essentially all activity. Of more significance from the standpoint of shelf life, the

TABLE 1
Stability of Soluble Urease versus Urease-Silica Particles

	Activity retained at 38°C (%)		Activity retained at 22°C (%)	
Days	Soluble urease	Urease-silica particle	Soluble urease	Urease-silica particle
3	70	85	—[a]	—
15	10	42	47	100
39	—	—	46	100

[a]Did not run.

urease-silica adduct at room temperature retained all its original activity for 39 days while the soluble form of the enzyme had lost 54% of its original activity. The degree of stability at 38°C indicates a useful therapeutic lifetime. However, other requirements must be met for an acceptable clinical application, and these dictate protection of the adduct particles by microencapsulation.

3. Microencapsulation Technique

An active enzyme intended for ingestion should be protected against inactivation by the multiple action of the digestive processes, including chemical attack (pH), bacterial action, and proteolytic enzymes. Most important is protection of the subject against hypersensitivity and immunological reactions that often can be induced by chronic exposure to foreign proteins, especially if the soluble enzyme were injected subcutaneously or intramuscularly. Microencapsulation is the procedure of choice, even though our experience to date indicates a substantial part of the initial urease-silica adduct activity is unavoidably lost during the processing steps of microencapsulation. Microcapsules are typically spherical in shape, 10–500 μm in diameter, with a thin, semipermeable membrane which retains the enclosed urease-silica adduct but passes the substrate (i.e., urea). They offer the advantages of rapid diffusion and reaction of substrate because of their high surface area and small volume, as well as the necessary protection outlined above.

Stabilized urease-silica particles were encapsulated by a nonaqueous phase-separation process in which a polymeric material is precipitated out of solution by slowly adding a miscible nonsolvent as described by Gardner *et al* (1971). The polymer solution initially contains aqueous droplets of the urease-silica adduct in suspension. As the nonsolvent is added, a precepitate forms at the interface of the droplets and coalesces into a membrane around each droplet.

Exploratory studies suggested cellulose acetate butyrate as the membrane material in preference to several polymers, since almost perfect spherical microcapsules were easily obtained and manipulated with this polymer. In addition, we preferred a cellulosic-type membrane, since it would be relatively resistant to degradation upon passing through the intestinal tract. Suitable solvents for the cellulose acetate butyrate polymer include dichloromethane and chloroform; nonsolvents include toluene, cyclohexane, and *n*-hexane. After the capsule membrane has been precipitated, it remains highly solvated, semifluid, and fragile. The membrane may be hardened by further repeated washing with nonsolvent until the capsules are ready to be filtered and vacuum-dried.

Drying in this manner results in a completely dry and crenated capsule. Crenation, however, poses no problem since the microcapsule will regain its sphericity when reimmersed in aqueous environment. In fact, this phenomenon may be used to advantage to remove excess metabolic water. In addition, drying the capsule should improve the shelf life of the microencapsulated product, since any organic solvent remaining in the capsule would tend to reduce enzyme activity.

During the encapsulation of the stabilized adduct, approximately 70–90% of the enzyme activity before encapsulation is lost. The loss in enzyme activity agrees with the 80% decrease reported by Chang *et al.* (1967) for encapsulated soluble urease. Presumably this decrease represents a diffusion barrier to urea and denaturation of the active enzyme sites by the organic solvents, which inevitably have trace solubility in the aqueous phase. However, even with the loss occurring during encapsulation, the enzyme capsules contain sufficient residual activity for urea conversion. In addition, results show no loss of postencapsulation activity (capsules stored in a sealed jar at room temperature) for periods in excess of 2 weeks, followed by a decrease thereafter to about 25% activity after 6 months.

4. *Experimental Procedure*

The activity of the urease-silica adduct and the integrity of the capsule membrane following crenation and reswelling was demonstrated by the following experiment. Microcapsules containing urease-silica particles were placed in a 100 mg% urea–phosphate–saline buffer solution (pH 7.3). Measurements of the cumulative increase in ammonia produced by the decomposition of urea showed that the capsules were the source of urease activity since ammonia accumulation was interrupted when the capsules were temporarily removed from the solution (Figure 3). Thus, the urease-silica adduct was effectively contained within the microcapsules, demonstrating the flexibility of cellulose acetate butyrate as a capsule membrane.

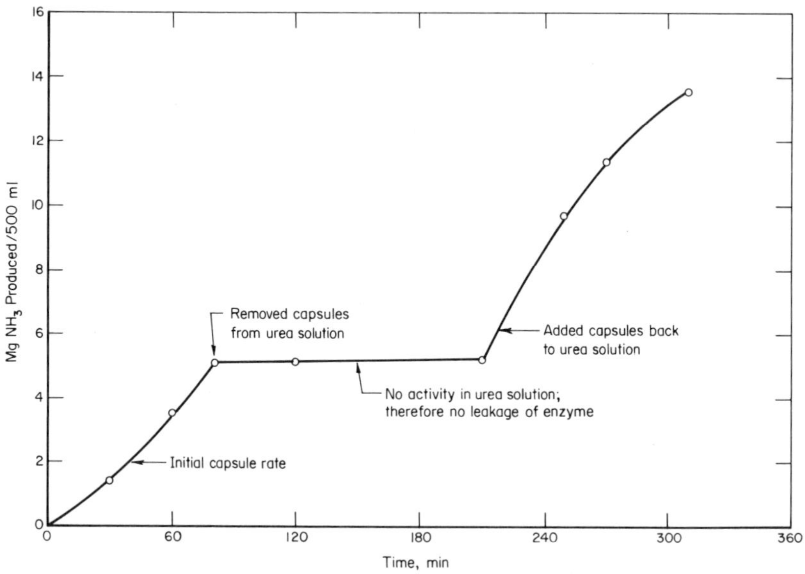

FIGURE 3. Stabilized urease residing within 100 to 400-μm microcapsules.

Gardner *et al.* (1971) have shown that urea can be converted by encapsulated urease-silica particles at rates of 20–50 mg of urea/hr/g of crenated capsules, depending upon the average size and specific activity of the capsules prepared. Based upon these conversion rates, a small daily dosage (10–25 g) of microcapsules would be adequate to hydrolyze the 12 g of urea produced each day in a chronic uremic patient, provided the patient were on a low-protein diet and urea is present in the gastrointestinal tract.

To determine if urea was present in the intestinal tract, dogs were rendered azotemic via a modification of Bricker's (1964) technique, which involves a sequential arterial ligation of the left kidney followed by nephrectomy of the second kidney 2 weeks later. The dogs were then stabilized on a controlled protein and water diet. Gut fluid was obtained from various regions of the uremic dog intestinal tract via a percutaneous catheter which had been implanted at the time the contralateral kidney had been removed.

The results of the studies by Gardner *et al.* (1972) indicated that there was almost a 1:1 relationship between serum urea nitrogen levels and gut urea nitrogen levels, irrespective of the intestinal region aspirated. In addition, further studies involving triple-lumen catheters and standard perfusion techniques in the uremic dog showed there was indeed a favorable flux rate of urea from the blood to the gut lumen, approximating 60 mg/hr/20-cm test segment at a blood urea nitrogen level of 60 mg%. Thus, based upon the aspiration and perfusion studies, urea removal via a microencapsulated enzyme product should be feasible.

To test the above hypothesis, Gardner *et al.* (1973, 1975) performed preliminary studies on the feeding of a combination microcapsule system (stabilized urease plus zirconium phosphate) to stabilized azotemic dogs. The combination capsule system was developed to eliminate the necessity of two separate capsule systems (i.e., a stabilized-urease capsule system and an ammonia binder capsule system) and has advantages over a two-capsule system. These include (1) higher concentrations of ammonium ion inside the capsule, which favors adsorption to the zirconium phosphate, and (2) the wall membrane of the capsule tends to reject ions. This rejection would keep the NH_4^+ within the capsule while repelling the Ca^{2+}, a strong competitive binder to the zirconium phosphate.

The results from these feeding studies indicated that the combination capsule

TABLE 2
Feeding or Instilling the Combination Capsule to Azotemic Dogs

Feeding study No.	Prefeeding BUN (mg%)	Postfeeding BUN (mg%)	Urea removed (g)
1	66	54	3.5
2	55	47	2.4
3	76	65	3.3
4	118	93	7.3

[a]Assuming complete equilibration of urea between body water compartments. Total body water of dog was 13.86 liters. One gram of combination capsule can bind approximately 20 mg of NH_4.

system functioned as intended, i.e., blood urea nitrogen levels decreased in proportion to the calculated theoretical capacity of the zirconium phosphate, the rate-limiting factor in the combination capsule system. In the first three feeding studies, 90 g of crenated capsules were placed in the dog's diet, while in the fourth study, 290 g of crenated capsules were instilled through a percutaneous catheter ($\approx$ 30 g/hr). The results of these studies are shown in Table 2.

5. *Summary*

We have developed a stabilized urease microcapsule system which, when orally administered to dogs, reduced serum urea nitrogen levels. Its potential application to patients suffering from chronic uremia remains to be evaluated. However, the feasibility of using ingestible adsorbents for treating kidney dysfunction appears to be enhanced by these studies.

Acknowledgments

The authors acknowledge the financial support of the National Institute of Arthritis, Metabolism, and Digestive Diseases. In addition, special credit is due Dr. Craig R. Hassler, Mr. Kenneth D. Williamson, and Mr. William C. Baytos for their supportive roles.

6. *References*

Bricker, N. S., 1964, The functional adaptation of the diseased kidney: I. Glomerular filtration rate, *J. Clin. Invest.* **43**:1915.

Chang, T. M. S., 1964, Semipermeable aqueous microcapsules, *Science* **146**:524.

Chang, T. M. S., and Poznansky, M. J., 1968, Semipermeable microcapsules containing catalase for enzyme replacement in acatalosemic mice, *Nature* **218**:243.

Chang, T. M. S., Johnson, L. J., and Ransom, O., 1967, Semipermeable aqueous microcapsules, *Can. J. Physiol. Pharmacol.* **45**:705.

Gardner, D. L., Falb, R. D., Kim, B. C., and Emmerling, D. C., 1971, Possible uremic detoxification via oral-ingested microcapsules, *Trans. Amer. Soc. Artificial Internal Organs* **17**:239.

Gardner, D. L., Emmerling, D. C., Hassler, C. R., and Kim, B. C., 1972, Characterization of nitrogenous metabolites in intestinal fluid of stabilized uremic dogs, *Trans. Amer. Soc. Artificial Internal Organs* **18**(1):18.

Gardner, D. L., Emmerling, D.C., Williamson, K. D., Hassler, C. R., and Baytos, W. C., 1973, Combination microcapsule for controlling chronic uremia urea levels, *Ann. Conf. Eng. Med. Biol.* **26**:267.

Gardner, D. L., Emmerling, D. C., Williamson, K. D., Baytos, W. C., and Hassler, C. R., 1975, Encapsulated adsorbents for removal of nitrogenous metabolites via oral ingestion, *Kidney Intern.* **7**:S-393.

Goldstein, L., and Katchalski, E., 1968, Use of water-insoluble enzyme derivatives in biochemical analysis and separation, Fresenius, *J. Anal. Chem.* **243**:375.

Haynes, R., and Walsh, K. E., 1969, Enzyme envelopes on colloidal particles, *Biochem. Biophys. Res. Commun.* **36**:235.

Lowry, D. H., Rosebrough, N. J., Farr, A. L., and Randall, R. J., 1951, Protein measurement with the folin phenol reagent, *J. Biol. Chem.* **193**:265.

Silman, I. H., and Katchalski, E., 1966, Water insoluble derivatives of enzymes, antigens, and antibodies, *Ann. Rev. Biochem.* **35**:873.

Sparks, R. E., Salemme, R. M., Meier, P. M., Litt, M. H., and Lindan, O., 1969, Removal of waste metabolites in uremia by microencapsulated reactants, *Trans. Amer. Soc. Artificial Internal Organs* **15**:353.

Weatherburn, N. W., 1967, Phenol-hypochlorite reaction for determination of ammonia, *Anal. Chem.* **39**(8):971.

Weetall, H. H., 1970, Storage stability of water insoluble enzymes covalently coupled to organic and inorganic carriers, *Biochim. Biophys. Acta* **212**:1.

13

Liquid-Membrane-Encapsulated Enzymes

SHELDON W. MAY AND NORMAN N. LI

1. Liquid Membrane System

A liquid membrane is a membrane which uses a specifically formulated liquid to form small-diameter spherical shells separating two miscible phases. The potential range of liquid membrane applications is quite wide, and includes separation of hydrocarbons (Li, 1971a, 1971b), removal of contaminants from wastewater (Li and Shrier, 1972; Matulevicius and Li, 1975), oxygenation of blood (Li and Asher, 1973), removal of toxins from blood (Asher *et al.,* 1975), and encapsulation of enzymes and/or microorganisms for reaction and/or separation (May and Li, 1972, 1974; May and Landgraff, 1975; Li *et al.,* 1973; Mohan and Li, 1974, 1975).

Liquid membranes, in general, are formed by first making an emulsion of two immiscible phases and then dispersing the emulsion in a third phase (continuous phase). The liquid membrane phase refers to the phase between the encapsulated phase in the emulsion and the continuous phase. Usually the encapsulated phase and the continuous phase are miscible, but they are not miscible with the membrane phase. The emulsion can be either oil in water or water in oil. By definition, the liquid membrane then will be of the water type in the former case and of the oil type in the latter case. The liquid membrane phase usually contains surfactants, additives, and a base material which is a solvent for all the other ingredients. The surfactants and additives are used to control the stability, permeability, and selectivity of the membrane. For specific applications, liquid membranes can be tailor-made by using the appropriate ingredients in specified combination ratios. The two membrane surfaces facing the encapsulated phase and the continuous phase are interfaces between an oil phase and a water phase. They are, therefore, covered by the molecules of the surfactant and surface-active additives (Edwards, 1972).

When the emulsion is dispersed by agitation in a continuous phase (the third phase), many small globules or balls of emulsion are formed. Their size depends strongly on the system temperature, the nature and concentration of the surfac-

SHELDON W. MAY (to whom correspondence should be addressed) • School of Chemistry, Georgia Institute of Technology, Atlanta, Ga. 30332 NORMAN N. LI • Corporate Research Laboratories, Exxon Research and Engineering, Linden, N.J. 07036

tants in the emulsion, the nature of the encapsulated and the continuous phases, the emulsion viscosity, and the mode and intensity of mixing. In laboratory studies, conditions are usually such that the globules have diameters of about 1 mm. Each emulsion globule contains many tiny encapsulated droplets, typically 1–10 μm in diameter. A large number of globules of emulsion can easily be formed to produce a correspondingly large membrane surface area for rapid mass transfer from either the continuous phase to the encapsulated phase or vice versa. After a desired degree of separation has been achieved, mixing is stopped and the globules of emulsion quickly coalesce and form a layer of emulsion. The emulsion phase can be lighter or heavier than the continuous phase and can be easily separated from the continuous phase. The contacting operation can be either batchwise or continuous, cocurrent or countercurrent (Li and Shrier, 1972; Matulevicius and Li, 1975).

When properly formulated, very stable liquid membranes can be made. To illustrate the kind of high membrane stability that can be achieved, a certain type of liquid membrane was used to encapsulate an aqeuous solution of sodium cyanide at 10 times the lethal dosage for rats. Tests show that all the rats fed with such an emulsion were alive and active, whereas those in the control group, which received the same components but with sodium cyanide not encapsulated by the liquid membranes, were killed almost instantly (Asher *et al.,* 1975). Liquid membrane emulsions can generally be stored for a long time, such as for a period of over 1 year. Even those that show some demulsification after storage can, in general, regain their initial characteristics by intense mixing for a few minutes.

It is obvious that a primary separation mechanism operative in liquid membrane systems is simple solubility in the liquid membrane phase. Thus, materials that are totally insoluble in the liquid membrane phase are confined to either the external or internal liquids. However, anything that can dissolve in the liquid membrane can be transferred across the membrane, and the types of materials most soluble in the liquid membrane, all else being equal, will be transferred most rapidly. Thus, the preferential transfer of a given material can be achieved by "tailor-making" the liquid membrane phase. This type of selectivity has been demonstrated with both aqueous and oil types of liquid membranes. For example, several types of hydrocarbons have been separated from each other where similar characteristics such as size and boiling points make separation by other methods difficult (Li, 1971a, 1971b; Matulevicius and Li, 1975).

Transport of materials through properly formulated liquid membranes can, in general, be much more rapid than transport commonly associated with conventional membrane processes which use polymeric membranes. The diffusion thickness of the liquid membrane phase around an emulsion globule is usually about 0.1–1.0 μm, which is much thinner than the presently available commercial types of polymeric membranes (Li and Long, 1973).

In addition to the separation based on selective dissolution and/or diffusion through the liquid membrane, a "facilitated transport" mechanism may be used to enhance the mass transfer. This is achieved usually by incorporating a reagent in the encapsulated phase which can react with the permeating compound, or by incorporating a compound, such as a complexing agent or an ion-exchange compound, in the membrane phase which increases the solubility of the permeating compound in the membrane phase and thereby increases its rate of permeation. Separation can also be achieved by extraction and selective adsorption by the membrane phase.

The liquid membrane technique is applicable to both liquid and gaseous systems. This wide range of applicability, together with the many types of separation mechanisms which are operative either singly or in various combinations, attests to the versatility of this system as a separation tool.

2. *Preparation of Encapsulated Enzymes*

The steps involved in preparing enzyme-containing liquid membrane systems are analogous to those used for the encapsulation of simple water-soluble substances by this technique. An emulsion is first formed by dropping an aqueous enzyme solution into a rapidly stirring mixture of hydrocarbons, which we refer to as the membrane-forming solution. Once this emulsion is formed, the enzymatic reaction of interest is carried out by dispersing the emulsion in an aqueous solution of substrate. As mentioned previously, under the shear of stirring, the emulsion forms globules which have diameters of about 1 mm, each of which contains droplets with diameters of about 1–10 μm. One such globule of an enzyme-containing preparation is shown schematically in Figure 1 and photographically in Figure 2. The enzyme is undoubtedly associated with the oil–water interface, in addition to being dissolved within the internal aqueous pools, and this association might be expected to be reflected in the chemical and physical properties of the enzyme.

In a typical experiment, the emulsion is prepared as follows. Fifteen milliliters of an aqueous solution approximately 1 mg/ml in enzyme is added dropwise to a baffled mixing vessel containing 18 g of a mixture of 2% Span-80, 3% ENJ-3029, and 95% S100N. S100N is a high-molecular-weight isoparaffin with an average molecular weight of 386.5, a cloud point of 93°F, a pour point of 90°F, and a density of 0.836 g/cm^3. ENJ-3029 is an amine of average MW 2000 manufactured by ENJAY Additives Laboratories, and Span-80 is sorbitan monooleate, a nonionic surfactant. During the addition, mechanical stirring is maintained in excess of 350 rpm and controlled by means of a photometric rpm monitor. The protein concentration in the aqueous enzyme solution can be varied over a wide range depending on the particular experiment, and we have used solutions varying from less than 0.1 mg/ml to more than 5 mg/ml in protein. Although we have routinely kept the weight ratio of aqueous to hydrocarbon components at about 0.8, the total volume of emulsion prepared can be increased essentially without limit, but care must be taken to ensure sufficiently vigorous agitation for complete emulsification.

In principle, the entire emulsification process could be carried out in reverse (i.e., the enzyme incorporated into the hydrocarbon component and water-soluble surfactants and additives used). If the hydrocarbon phase were added dropwise into an excess of the aqueous phase, the result would be an oil-in-water emulsion

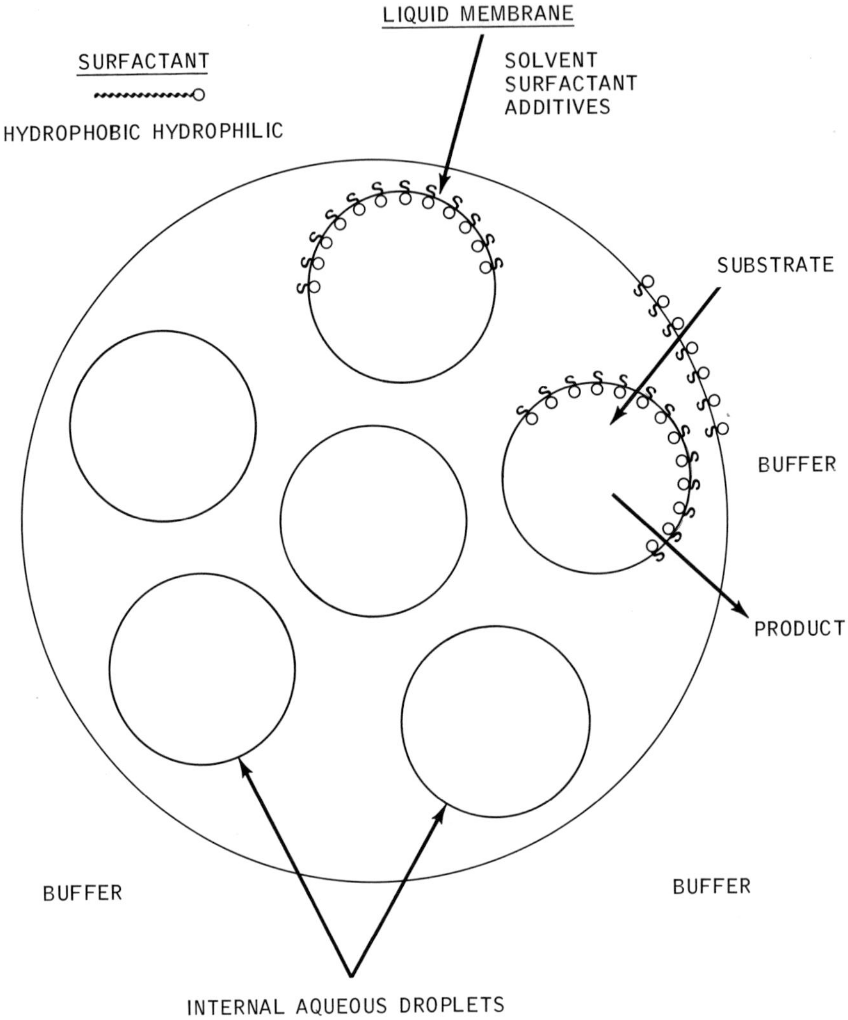

FIGURE 1. Schematic diagram of liquid membrane globule.

which could then be dispersed in an external nonaqueous solution. On the other hand, if the emulsion were prepared in the usual manner, the result would be a normal water-in-oil emulsion, but the enzyme would be contained within the liquid membrane phase when the emulsion was dispersed in an aqueous substrate solution. Although such formulations are especially attractive for the study of membrane-related processes (e.g., chemiosmotic ATP generation), very little work has been done to date on such "reverse" liquid membrane–enzyme systems.

The experimental conditions employed in carrying out reactions with liquid-membrane-encapsulated enzymes can be modified to suit many different types of reactor configurations. For laboratory work, we commonly use small batch reactors in which a given amount of emulsion is added to an excess of aqueous

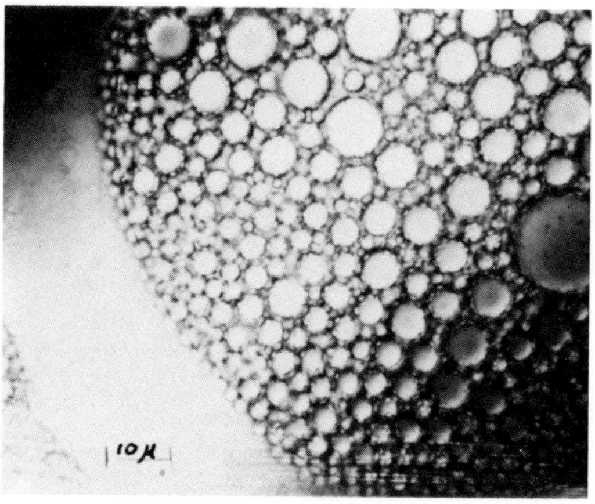

FIGURE 2. Photomicrograph of phenolase-containing liquid membrane emulsion (prepared by S. May).

substrate solution and the heterogeneous mixture is kept well dispersed by carefully adjusted magnetic stirring. The reaction course is monitored by periodically withdrawing samples of the external solution, gravity filtering them to remove any residual emulsion, and carrying out whatever spectrophotometric, colorimetric, or enzymatic assays are appropriate. Continuous monitoring of the external solution has also been successfully employed (Mohan and Li, 1975) using a cellulose-based filtration device, and this could easily be coupled to a flow cell for continuous spectrophotometric recording. Column techniques could also be employed in flow reactor configurations for continuous conversions of substrate to product. Although these would probably be attractive for large-scale commercial applications, it is likely that severe clogging problems would develop with conventional membranes or filters.

3. *Catalytic Activity of Encapsulated Enzymes*

Our initial studies with liquid-membrane-encapsulated enzymes were designed to answer two simple questions: (1) Does the immobilized enzyme retain substantial catalytic activity? (2) How extensively does the enzyme leak out from the emulsion to the external aqueous solution? In view of the well-known sensitivity of enzyme secondary, tertiary, and quaternary structure toward many organic solvents and surfactants, we would not have been surprised to find that the emulsification process caused total inactivation of the enzyme. We chose to work first with urease, since its substrate would be expected to diffuse readily through the membrane phase, its activity could be conveniently followed titrimetrically, and it had potential relevance in the development of artificial kidney systems.

The results of a typical experiment with urease, in which the reaction was followed using an automatic pH stat, are shown in Figure 3. In this experiment,

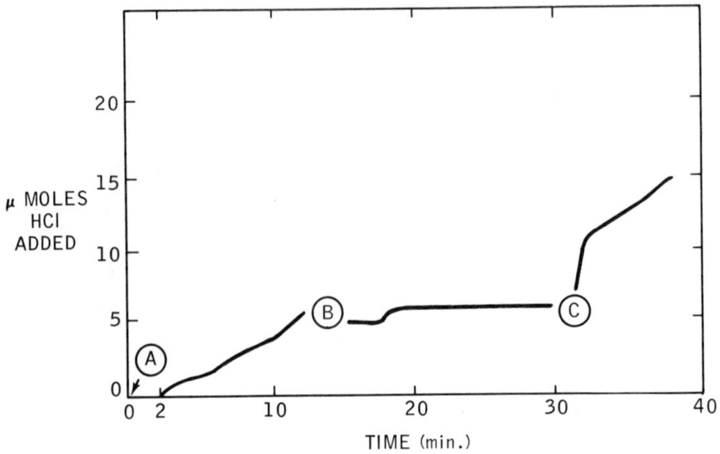

FIGURE 3. Hydrolysis of urea catalyzed by immobilized urease, followed by pH-stat method. (From May and Li, 1972.)

the emulsion was prepared from a 0.065% aqueous urease solution and the standard membrane-forming solution. The reaction was carried out as follows. Thirty milliliters of urea solution in 0.1 M NaCl, 1 ml of 0.025 M phosphate buffer, pH 7.0, and 0.1 μmole dithiothreitol were added to the automatic titration vessel and the pH of the mixture checked (initial conditions: [urea] = 0.43 M; pH 6.7 ± 0.05). Stirring was begun, and at point *A* 2 ml of the enzyme-containing emulsion was added, and the reaction monitored by recording the rate of addition of 0.1 N HCl required to maintain the pH at 6.7. At point *B*, recording and stirring were interrupted, the reaction solution removed from the vessel, and the aqueous and emulsion layers allowed to separate. About 80% of the aqueous layer was transferred to a clean titration vessel and, after adjusting its pH to 6.7, stirring and recording were resumed. At point *C*, recording was interrupted, the emulsion and residual aqueous layers readded to the vessel and recording was resumed. It is evident from the figure that after an initial rapid addition of HCl to neutralize NH_3 produced while the emulsion had been standing in contact with the residual aqueous phase, the reaction resumed its normal course. These results were the first clear indication that the immobilized urease retains substantial enzymatic activity and that no detectable leakage of enzyme into the external aqueous phase occurs during the reaction. In control experiments, where enzyme was contained in the external aqueous phase rather than in the emulsion, removal of the emulsion at point *B* did not terminate the reaction.

In order to eliminate the possibility that the liquid membrane emulsion was responsible for artifacts in the response of the pH-stat electrodes (see May and Li, 1972), the enzymatic hydrolysis of urea was monitored spectrophotometrically using phenolphthalein as an indicator. The results of these experiments confirm those obtained in the pH-stat experiments. The absorbances at 550 nm of the aqueous layers of samples withdrawn from the reaction mixture after 500, 2000, and 2800 sec were 0.036, 0.178, and 0.383, respectively. Moreover, the absorbance of each of these samples was stable for several minutes, after which it slowly decreased, owing to the uptake of carbon dioxide. Since the addition of 1 unit of soluble urease to these aqueous samples caused an immediate increase in absorbance, at an initial rate of 0.04 per min, it is apparent that no detectable leakage of the immobilized urease into the aqueous layer has occurred. Thus, the results of both the pH-stat and spectrophotometric experiments clearly demonstrated for the first time that the liquid membrane technique is a viable method for the immobilization of enzymes.

A distinguishing characteristic of catalytic systems involving immobilized enzymes is the fact that the kinetically determined value of the Michaelis constant, K_m(app), is generally greater than that obtained for the soluble enzyme under similar conditions. This can be viewed as a reflection of the effects of restricted diffusion—the actual substrate concentration that the enzyme "sees" is less than the stoichiometric substrate concentration—and the amount by which K_m(app) is increased is a measure of the extent of diffusion restriction for a given substrate. Figure 4 shows a Lineweaver–Burk plot and the corresponding saturation curve, for the immobilzed-urease-catalyzed hydrolysis of urea at pH 6.7 and 25°C. The value of K_m(app) obtained from this plot is 0.18 M, which is approximately 50 times as large as the value of 3.4×10^{-3} M which we obtained for soluble urease under similar conditions. Although there are certain complications in the interpretation of such plots, which we have discussed elsewhere (May and Li, 1972), this observation is clearly in line with our expectation that a certain amount of diffusional restriction is imposed on the enzyme by the presence of the hydrophobic liquid membrane phase.

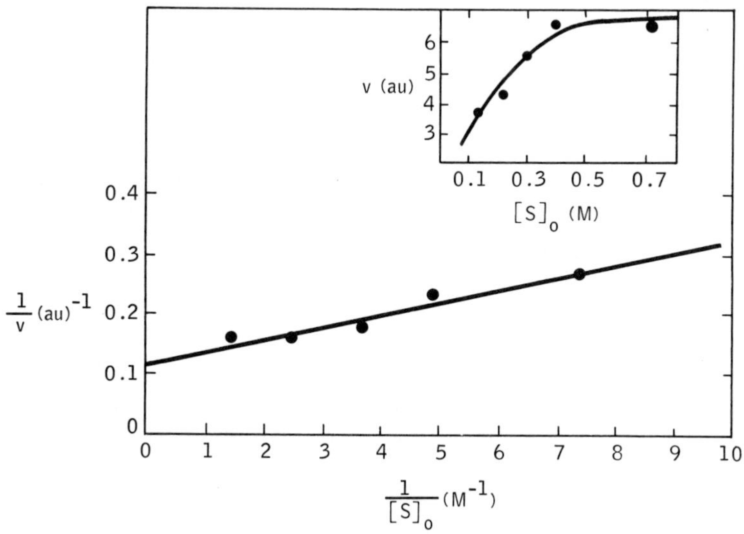

FIGURE 4. Saturation curve and Lineweaver–Burk plot for the hydrolysis of urea by immobilized urease, pH 6.7. The plot gives K_m(app) = 0.18 M.(From May and Li, 1972.)

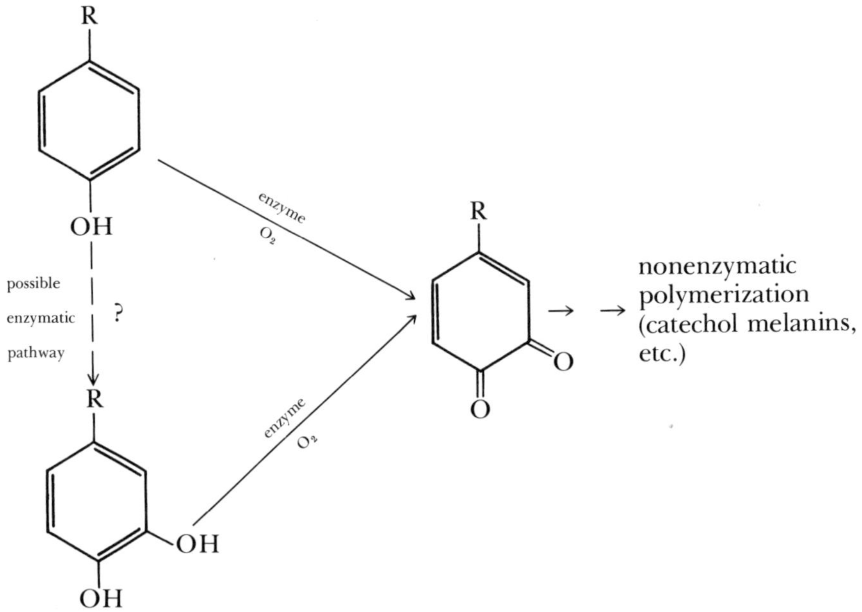

The effects of the permeability characteristics of the liquid membrane on the course of an enzymatic reaction are further illustrated by our results with the enzyme phenolase. This enzyme catalyzes the oxygen-dependent oxidation of a variety of monophenolic and diphenolic compounds as shown. Typical data which are obtained when a phenolase-containing emulsion is dispersed in an aqueous solution of phenol are shown in Figure 5. The amount of phenol in the external solution, as measured by the absorbance at 270 nm, markedly decreases during the course of the reaction, but the concentration of oxidation products in the external solution, as measured by the absorbance at 390 nm, increases only slightly over the same time period. Nevertheless, the oxidation of phenol certainly is proceeding during this period and the accumulation of colored oxidation products can be seen in the hydrocarbon phase, which turns a deep pink-brown. Thus, phenol successfully diffuses through the liquid membrane, contacts the enzyme, and is converted to products that do not diffuse back out to the external aqueous solution.

Just as the permeability characteristics of the liquid membrane can affect the diffusion of enzymatic products back out into the external phase, they can also affect the diffusion of substrate from the external phase to the enzyme in the first place. Thus, these permeability characteristics superimpose additional substrate specificity on that normally exhibited by the particular enzyme in question. For example, protocatechuic acid (PCA) is a substrate of phenolase which is converted by the soluble enzyme to the substituted quinone. In contrast, we have found that phenolase *immobilized* by the liquid membrane technique does not react with PCA. Although phenol and PCA are of similar molecular weight, they differ markedly in that the latter contains a negatively charged carboxylic acid group.

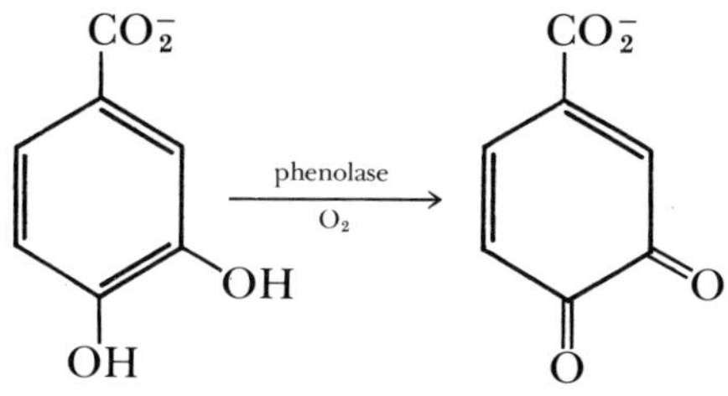

PCA

Data for a phenolase-containing emulsion dispersed in an aqueous solution of PCA are shown in Figure 6. Over the time period indicated, no reaction of PCA has occurred. Not only is there no loss of PCA from the external aqueous solution, as evidenced by the absorbance at 287 nm, but correspondingly there is no accumulation of the oxidation product in the external aqueous solution, as shown by the lack of absorbance at 350 nm. PCA, like most charged molecules, is extremely insoluble in hydrocarbon solutions and is thus unable to permeate the hydrocarbon phase of the LM emulsion and contact the enzyme. As shown in Figure 6, the later addition of 100 μg of soluble phenolase to the external solution caused an immediate reaction to take place. From the rate of this reaction, we calculate that leakage of only about 1% of the immobilized enzyme into the

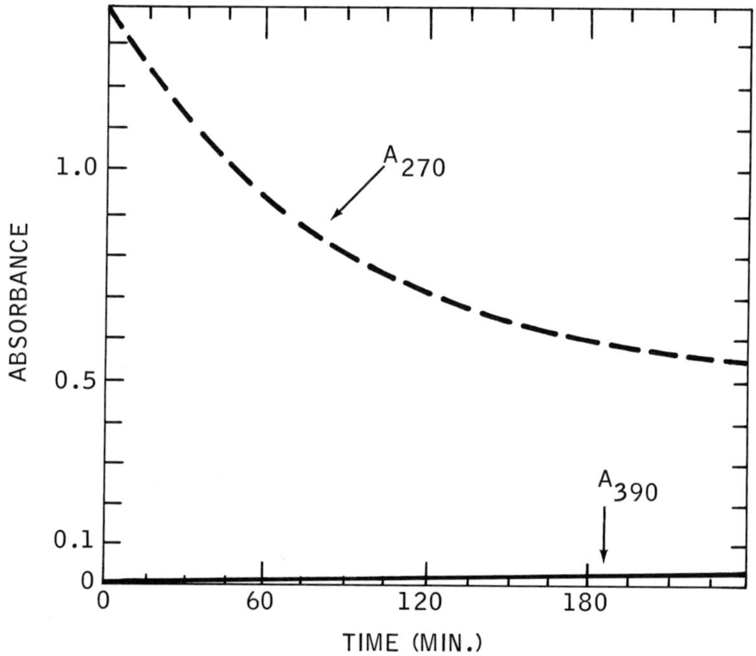

FIGURE 5. Oxidation of phenol by a phenolase-containing emulsion. (From May and Li, 1974.)

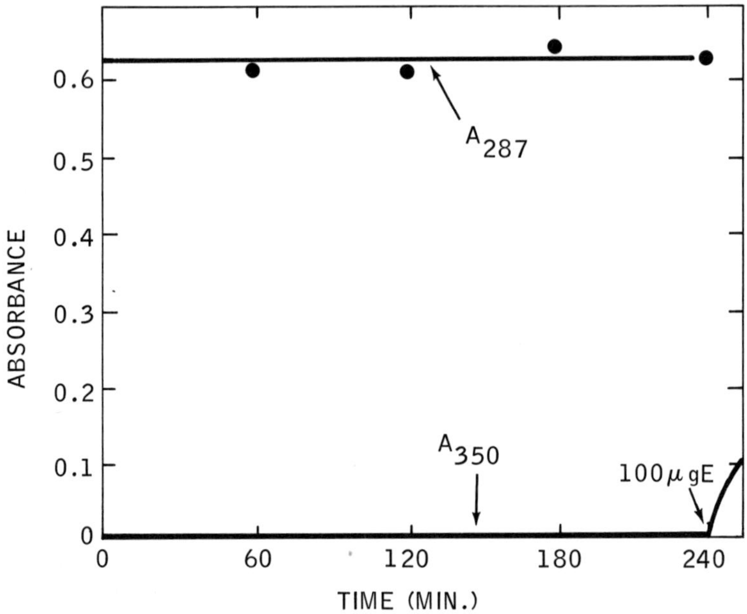

FIGURE 6. Phenolase-containing emulsion dispersed in aqueous solution of protocatechuic acid. (From May and Li, 1974.)

external solution over the time period of the experiment could easily have been detected.

The results of a *direct* leakage test using a very highly loaded phenolase-containing emulsion are shown in Table 1. In this experiment the emulsion contained 200 mg of enzyme in 150 g of hydrocarbon, which is approximately 10 times as concentrated as the emulsions used in the previously described experi-

TABLE 1.
Phenolase Leakage Test[a,b]

	A_{390} of aqueous-layer test		
	30 min	60 min	120 min
Contact time (min)			
30	0.01	0.00	0.01
60	0.00	0.00	0.00
90	0.01	0.01	0.04
Soluble enzyme (5 μg/ml) (2% leakage)	0.01	0.04	0.12

[a] Data from May and Li, 1974.
[b] Thirty milliliters of an emulsion containing 1.33 mg of phenolase/ml were dispersed in 150 ml of deionized H_2O. After the indicated contact time, samples of the external aqueous solution were removed, incubated with phenol for the indicated time period, and examined for absorbance at 390 nm.

ments. The results indicate that after 90 min of contact time, some enzyme had leaked out into the external solution. However, the enzymatic activity of the external solution was considerably less than that of a 5 μg/ml solution of soluble phenolase, which would correspond to leakage of 2% of the total encapsulated enzyme. Clearly, although some enzyme leakage does occur, it is very small.

To ascertain whether enzyme leakage accompanied by inactivation was occurring in our system, we utilized trinitrobenzene sulfonic acid (TNBS), which reacts with the amino groups of proteins to give a trinitrophenylated derivative having a characteristic absorption maximum at 420 nm with a high extinction coefficient (Fields, 1972). Thus, TNBS-modified enzymes can easily be detected in aqueous solutions spectrophotometrically. In our experiments we modified phenolase with TNBS, purified the product using ultrafiltration, and repeated the leakage experiment outlined in Table 1. Once again we observed very little protein leakage into the external aqueous solution, with the results being virtually identical with those obtained from activity measurements.

Recently, May and Landgraff (1975) have carried out a number of studies with encapsulated yeast alcohol dehydrogenase (ADH) as a model for employing the liquid membrane system for cofactor-requiring enzymes. The reaction catalyzed by ADH is

$$CH_3CH_2OH + NAD^+ \underset{}{\overset{\text{ADH}}{\rightleftharpoons}} CH_3CHO + NADH + H^+$$

and the yeast enzyme exhibits a strong specificity for ethanol, longer-chain alcohols reacting much more slowly. In order to utilize ADH in a liquid membrane system it was first necessary to establish whether or not it was possible to retain the requisite cofactors within the emulsion phase. Table 2 presents typical data obtained in direct leakage tests with NAD^+ and NADH, and it is apparent that very little leakage of these cofactors into the external solution occurs over the time scale of the experiment. This property distinguishes liquid membranes from virtually all other immobilization methods based on physical entrapment (e.g.,

TABLE 2
NAD^+ and NADH Leakage Tests[a]

Time (min)	Total leakage NAD^+ (%)	Total leakage NADH (%)
0	0	0
10	0	0
25	0	0
45	1.3	1.4
65	0.1	0
125	0.5	0

[a] In separate experiments 10 ml of an emulsion containing 6 mg of either NAD^+ or NADH was dispersed in 50 ml of Tris buffer, pH 7, and at the times indicated, samples of the external solution were withdrawn and examined spectrophotometrically.

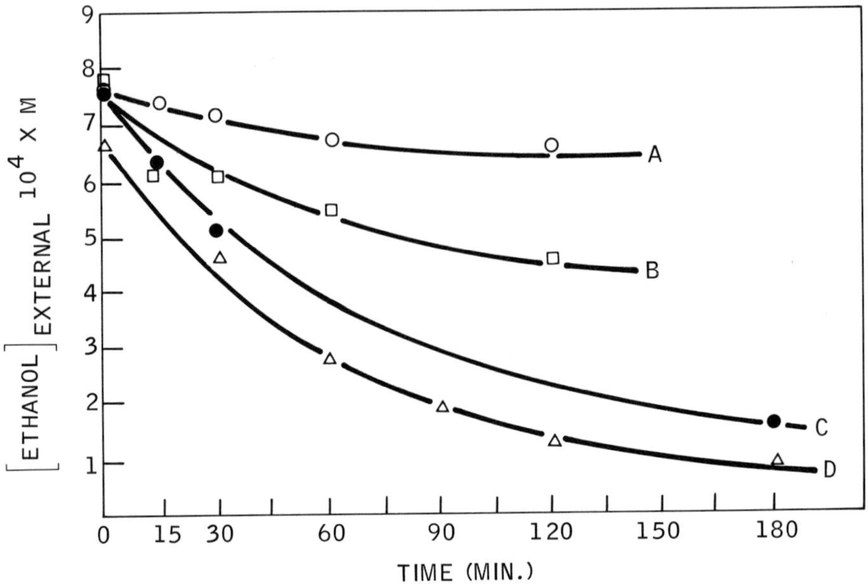

FIGURE 7. Ethanol consumption in ADH–NAD⁺ systems. The emulsion phase contained (per ml): A, 0.125 mg of ADH, 0.65 mg of NAD^+; B, 0.5 mg of ADH, 2 mg of NAD^+; C, 0.475 mg of ADH, 1.5 mg of NAD^+, 0.75 mg of ferricyanide, 25 μg of diaphorase; D, 0.49 mg of ADH, 0.6 mg of NAD, 0.8 mg of ferricyanide, 25 μg of diaphorase; D, 0.49 mg of ADH, 0.6 mg of NAD, 0.8 mg of ferricyanide, 25 μg of diaphorase. Ten milliliters (expt. A) or 20 ml (expts. B, C, D) of emulsion was dispersed in 50 ml of ethanol solution to run the reaction.

hollow fibers, permanent microcapsules or ultrafiltration cells), which do not retain these relatively low molecular weight molecules. Accordingly, cofactor-requiring enzymes can be directly incorporated into liquid membrane systems without resort to techniques such as attachment to macromolecular carriers, which always lower the activity of the cofactor.

Data obtained when an emulsion containing various amounts of ADH and NAD^+ were dispersed in aqueous solutions of ethanol are shown in curves A and B of Figure 7. It is evident that both the rate of ethanol loss from the external solution and the total amount of ethanol consumed increase with the amount of ADH and NAD contained in the emulsion. In the absence of ADH and NAD^+, the extent of ethanol loss from the external aqueous solution is very small. Clearly, the permeability of ethanol through the liquid membrane is inherently low, but is enhanced in the presence of the enzymatic system, which "traps" the internal ethanol by converting it to acetaldehyde, thereby shifting the diffusion equilibrium and allowing more ethanol to diffuse in from the external solution.* This result represents a clear example of enzyme-mediated "facilitated transport," a

*In the experiments with ADH, the emulsion contained enzyme and cofactor in 0.5 M Tris buffer, pH 8.8, which serves to drive the ADH reaction to completion.

process which is utilized in virtually every membrane system within the living cell. Thus, another area in which model studies with liquid membrane systems may find application has emerged from these studies.

4. *Recovery of Enzymes and Denaturation Effects*

Since the liquid membrane technique provides a nonpermanent immobilization system, in principle one should be able to prepare an enzyme-containing emulsion, carry out some chemical conversion to obtain a desired product, and then recover the enzyme at the completion of the process. In this regard, liquid membrane encapsulation differs from the microencapsulation technique of Chang (1972), which employs permanent membranes formed by interfacial polymerization or coacervation. Reversibility allows great flexibility in the design of experiments. For example, active-site-directed chemical modification reactions or active-site titrations can be carried on an enzyme in the *immobilized* state and the modified enzyme recovered and examined. Such experiments can provide valuable information on the effect of microenvironment on enzyme structure and function and cannot be carried out readily with other immobilization techniques. From a practical point of view, reversibility may make processes using scarce or expensive enzymes more attractive, especially if it is desired to recover the enzyme when one of the reactants or cofactors is spent. However, it must be recognized that reversibility has its price in that the long-term stability of emulsions is inherently less than that of permanent microcapsules. Thus, for long-term applications, emulsion-strengthening agents must be developed or cyclical processes employed.

If an enzyme is to be recovered from liquid membrane emulsions and reused, a critical question is: How active is the recovered enzyme? Or, how much irreversible denaturation occurs during the preparation and breakage of the emulsion? Since active-site titration provides a method to answer this question, we have prepared trypsin-containing emulsions, broken the emulsions by high-speed centrifugation, recovered the enzyme by extraction and ultrafiltration, and compared the results of active site titrations of the starting and recovered enzyme. The active-site titrant used in these experiments was *p*-nitrophenyl-*p*′-guanidinobenzoate hydrochloride (Chase and Shaw, 1970). In three separate experiments with different trypsin preparations, the number of "active sites per protein molecule" (based on $E_{1\%}^{280} = 14.3$) for the starting and recovered trypsin were, respectively, 0.74 and 0.62; 0.36 and 0.27; 0.34 and 0.29 (May and Landgraff, 1975). It is evident that only slight decreases in the number of active sites per molecule were observed, and certainly no gross denaturation of the enzyme had occurred during the process.

It is interesting to note that we have found that in experiments such as the above, the actual percentage of the total starting enzyme which can be recovered varies greatly with the particular enzyme used, the pH of the enzyme solution, the nature of the buffer, and the exact conditions used for breaking the emulsion. This undoubtedly reflects the strong association of the enzyme with the liquid membrane phase and the aqueous–hydrocarbon interface. Although recoveries of only 25–40% are perfectly acceptable for active-site titration or chemical modification experiments, it is clear that for commercial applications the development of more efficient enzyme recovery procedures is an important research goal.

There is no doubt that the association of an enzyme with a hydrophobic microenvironment, such as that in the liquid membrane system, is capable of greatly altering the conformation of the enzyme, thereby affecting its stability, activity, and catalytic properties. Accordingly, it would be highly desirable to have in hand a method for titrating the enzyme *in the immobilized state,* to provide quantitative information about the effect of microenvironment factors on the integrity of the active-site region. Moreover, it is a general problem with immobilized enzyme systems to quantitate the concentration of *active* enzyme molecules per gram of carrier or milliliter of gel, and although such information could be obtained from active-site titration experiments, there has been a dearth of such procedures developed which are suitable for use with immobilized enzymes. It is our contention that such procedures can be readily developed for liquid-membrane-encapsulated enzymes, since many of the active-site titrants are readily soluble in hydrocarbons, and a number of them form covalent enzyme derivatives which are sufficiently stable to be recovered from the emulsion and quantitated. It is along these lines that much of our current work is directed.

5. *Encapsulation of Multienzyme Systems and Whole Cells*

The liquid membrane system involves immobilization by physical entrapment and thus would be expected to be eminently suitable for use with multienzyme systems. Such systems, which involve the *concerted* action of more than one protein component to catalyze a given transformation, are exceedingly common, and cannot be immobilized successfully by methods which involve the attachment of protein molecules to inert matrices. Among the particular advantages of the liquid membrane system are ease of preparation and the possibility of recovering the enzyme at the completion of the process. Accordingly, a number of complex enzyme systems, as well as whole bacterial cells, have been encapsulated using liquid membranes.

A coupled enzyme system involving ADH and diaphorase and the cofactors NAD and NADH is outlined in Figure 8. In this system ADH will catalyze the oxidation of ethanol to acetaldehyde, concurrently reducing NAD^+ to NADH. The second enzyme, diaphorase, then reacts with NADH in the presence of a suitable electron acceptor, A_{ox}, to regenerate NAD^+ and produce A_{red}. Thus, this coupled enzyme system provides a method for cofactor recycling and allows the conversion of large amounts of ethanol to acetaldehyde in the presence of a limited amount of NAD^+.

Data obtained by May and Landgraff (1975) when this coupled enzyme system, using ferricyanide as the electron acceptor, was encapsulated in a liquid membrane emulsion and dispersed in an aqueous solution of ethanol are shown in curves C and D of Figure 7. Comparison with curves A and B reveals that cofactor

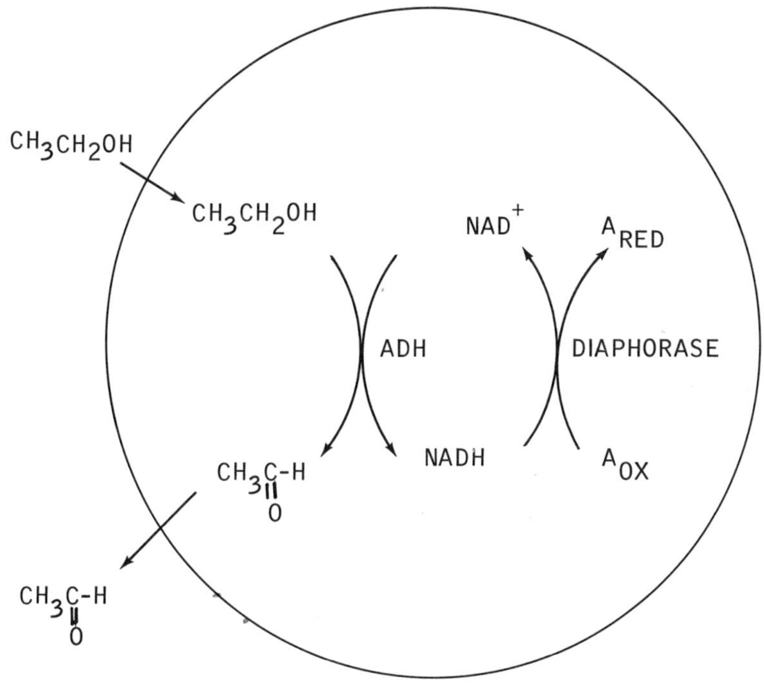

FIGURE 8. Schematic of liquid-membrane-encapsulated coupled ADH–diaphorase enzyme system.

recycling has greatly enhanced both the rate and extent of ethanol consumption, even though the amount of NAD^+ present in the emulsions was considerably less than in that of experiment B. A detailed analysis of the external solution showed that the acetaldehyde formed in the ADH reaction permeates the membrane and accumulates in the external phase (May and Landgraff, 1975). Taken together, the data establish that the coupled process outlined in Figure 8, which involves cofactor recycling, is indeed occurring in the liquid membrane system. The significance of this result lies in the fact that it demonstrates the feasibility of coupling cofactor-recycling reactions to cofactor-requiring enzymatic transformations within a single liquid membrane emulsion. With many other entrapment techniques (e.g., hollow fibers, microcapsules, ultrafiltration cells) extensive leakage of the relatively low molecular weight cofactors occurs, and such *in situ* regeneration systems cannot be readily used. Retention of dissociable cofactors within conventional entrapment systems generally requires either the use of "tight membranes" which severely restrict diffusion, or prior attachment to macromolecular "carriers" which results in drastic losses in activity (reviewed in Baricos *et al.*, 1975).

Mohan and Li (1974) have carried out a detailed study on the activity of a liquid-membrane-encapsulated cell-free preparation (from a strain tentatively identified as *Micrococcos denitrificans*) which reduces NO_3^- and NO_2^-. The encapsulated preparation exhibits substantial enzymatic activity which does not "leak" out into the external aqueous phase. This result clearly demonstrates that highly complex multienzyme systems can be readily accommodated in the liquid membrane system. Although a detailed kinetic analysis was not carried out, the authors suggest that the decreased activity of the immobilized preparation (as compared to the soluble system) is attributable to diffusional restrictions on the permeability of nitrate and nitrite through the membrane. When both enzyme-containing and buffer-containing emulsions were dispersed in aqueous solutions, loss of nitrate from the external solution after 1 hr was substantial in the first case but negligible in the second. This result thus provides another example of permeability enhancement via enzyme-dependent facilitated transport.

Recently, Mohan and Li (1975) have extended their work to an examination of the nitrate and nitrite reduction activity of liquid-membrane-encapsulated whole cells. When whole cells of the same organism were incorporated into a liquid membrane emulsion, they remained visibly intact for several days, and retained over 90% of their nitrite-reductase activity after storage for 72 hr. Unbound cells of the same organism lost their activity completely after 16 hr. The presence of the liquid membrane protected the cells against inhibition by $HgCl_2$, which readily destroys the reductase activity of unbound cells or soluble nitrate and nitrite reductases. Incorporation of Amberlite LA-2 into the liquid membrane formulation enhanced the nitrite-reductase activity of encapsulated whole cells, presumably by facilitating the transport of NO_2^- through the membrane phase. A prototype continuous reactor and coupled analytical system was constructed and used for the reduction of nitrite with encapsulated whole cells. Under the experimental conditions used, a steady-state reduction rate of 0.65 ppm of NO_2^-/min was achieved with this reactor, but optimization would almost certainly lead to higher nitrite-reduction capacities.

Taken together, these results are highly significant in that they underscore one of the primary advantages of liquid membrane systems. The encapsulation of highly complex multienzyme systems, whole cells, and cofactor regeneration systems can be achieved using an effective, nondestructive technique and gentle reaction conditions. Chemical cross-linking or polymerization reactions, which can often lead to reactions at functional groups of enzymes or membrane constituents, are unnecessary, and thus the possibility of adversely affecting enzyme reactivity or cell viability avoided. (An example of a drastic effect of chemical cross- linking on cell viability and enzyme activity is presented by Zaborsky and Schwartz, 1974.) Finally, the method is nonpermanent in nature, and, although this may cause some losses in physical stability, the possibility of recovering the enzymes will often present an advantage which far outweighs any processing problems that must be overcome.

6. *Potential Applications*

The potential applications of liquid-membrane-encapsulated enzymes and whole cells are many. For example, in the area of wastewater treatment, various enzymes and whole cells may be encapsulated to remove specific contaminants.

The use of encapsulated whole cells may have a strong advantage over the conventional bioxy process if the wastewater to be treated contains compounds toxic to the bacteria. In this case it is possible to protect the bacteria by employing liquid membranes which are nonpermeable to the toxins (Mohan and Li, 1975).

Another potential application is the use of encapsulated enzymes in an "artificial kidney" scheme (Li *et al.,* 1973; Asher *et al.,* 1975). This scheme would involve having the kidney patient orally ingest a preparation which is actually a mixture of two types of liquid membrane emulsions, one containing an organic acid and the other containing urease. The emulsion is administered when urea in the patient's blood accumulates in the intestines. When urea permeates into droplets containing the enzyme, it will decompose into ammonia and carbon dioxide. The diffusion of urea into the droplets containing organic acid will stop as soon as the urea concentration is equalized inside and outside of the droplets. Both carbon dioxide and ammonia produced from the enzymatic reaction of urease will diffuse out of the droplets into the gut. Carbon dioxide can be readily eliminated through the lung, whereas the ammonia will diffuse into the acid-containing droplets where the high concentration of acid shifts the ammonia–ammonium ion equilibrium to convert ammonia to the ammonium ion. The liquid membrane formulation is selected to be an ion barrier so that the ammonium ion formed cannot diffuse back out of the encapsulated phase and, therefore, become trapped. Other toxins, such as phosphates, can also be readily removed by the liquid membrane emulsion when proper transforming agent and encapsulated reagent are used.

The liquid membrane phase (the continuous phase of the emulsions) used in this artificial kidney application would be dominantly isoparaffinic oils. Encapsulation in this type of true liquid phase allows the use of active phase materials that would otherwise be inappropriate. For example, an encapsulated enzyme might be protected from a hostile environment, such as the stomach contents, or a concentrated acid phase which would normally cause damage might be acceptable when encapsulated. In addition, species formed by appropriate chemical reactions can be effectively trapped within the encapsulating liquid membrane, and thus kept separated from the patient's system. It is hoped that the use of liquid membrane emulsions to remove some of the uremia toxins through the gastrointestinal tract can be either a valuable adjunct treatment to dialysis or, eventually, a substitute to dialysis.

Another application of liquid membranes is in the area of blood oxygenation (*Medical News*, 1971; Li and Asher, 1973; Wallace *et al.,* 1973). Although no enzymes or whole cells are involved, this is nevertheless a potentially important application in the biomedical field and is, therefore, briefly reviewed here. In this application the blood contacts fluorocarbon liquid membranes which encapsulate oxygen bubbles. The blood may be shunted from the body and passed through a contacting device before being sent back to the body. Among the many types of contacting devices which can be used is a column in which the blood and the encapsulated oxygen bubbles flow countercurrently. The blood serves as the continuous phase, and the method of introduction used plus the normal column flow ensure a uniform dispersion of the bubbles. The liquid membranes encompass the oxygen bubbles completely, thus preventing the formation of a direct blood–oxygen interface while allowing oxygen to diffuse into the blood and carbon dioxide to diffuse from the blood to the gas bubbles.

In the blood oxygenator application, the liquid membranes are composed of biocompatible fluorocarbon-type compounds to minimize both blood hemolysis and protein denaturation. They are typically quite thin—usually 1 μm or less—which is substantially thinner than polymeric membranes. The thinness of the membranes, coupled with the high solubility of both O_2 and CO_2 in fluorocarbons, would be expected to lead to minimal resistance to transport of these substances in this system. It is interesting to note that the use of gas encapsulated within liquid membranes in this application mimics the normal mechanism of the body, which involves gas entrapment within the alveoli of the lungs.

A consideration of the presently available blood oxygenators reveals that they can be divided into two categories. One involves the direct contact of blood with oxygen; the other involves contacting blood with oxygen that has permeated through a solid polymeric membrane. Blood oxygenation in either type of oxygenator usually results in some degree of damage to blood proteins and red cells. There is no apparent way in which this can be avoided in the first method. For the second method, industrial and academic research has been carried out to modify or treat the polymeric surface (e.g., with chemically bound heparin) to make it more compatible with the blood. However, it has not yet been established that polymeric oxygenators can operate over prolonged times with truly negligible blood damage. The liquid membrane technique has the advantages of presenting a blood interface which minimizes hemolysis and protein denaturation and of using deformable liquid membranes.

The general advantages of liquid surfactant membranes over solid polymeric membranes may be summarized as follows: (1) greater surface area for permeation and separation because of high surface area of liquid membranes per unit volume; (2) elimination of pinholes in imperfect solid membranes, which allow all components of a mixture to pass through, just as a broken liquid membrane would; (3) no membrane-life problem, since the liquid membranes are recovered after each pass, and no deposit (e.g., thrombus) accumulation with time to decrease performance; (4) a faster rate of permeation and separation, since liquid membranes are much thinner than solid ones; (5) no need for a mechanical support, such as the frames that hold solid membranes in the mixtures being separated; and (6) no minute flow channels to become blocked, in time, with solids.

Continuing advances in the field of cardiovascular surgery and organ transplantation require a constant search for more efficient and less traumatic means of oxygenating blood. In addition, it is becoming increasingly apparent that there is a definite need for oxygenators to support respiratory functions during pulmonary failure due to reversible pathologic processes, thereby allowing the lungs to fully recover from the underlying disease. Because of this continued need for a better oxygenator, the possible utilization of liquid membranes made of fluorochemicals has been investigated, and this technique has been shown to be a feasible method of oxygenating blood.

In view of the versatility and potential technical and economic advantages of liquid membrane systems, many future biomedical, analytical, and environmental applications may be anticipated.

Acknowledgment

We gratefully acknowledge the support provided for work carried out at the Georgia Institute of Technology by Research Corporation, the National Science Foundation (BMS74-20830) and the Biomedical Sciences Support Program of the National Institutes of Health.

7. *References*

Asher, W. J., Bovee, K. C., Frankenfeld, J. W., Hamilton, R. W., Henderson, L. W., Holtzapple, P. G., and Li, N. N., 1975, Liquid membrane system directed toward chronic uremia, *Kidney Int.* **7**:5–409.

Baricos, W. H., Chambers, R. P., and Cohen, W., 1975, Considerations on the use of cofactor-requiring enzymes in enzyme engineering, *Enzyme Technol. Digest* **4**:39–53.

Chang, T. M. S., 1972, *Artificial Cells,* Charles C. Thomas, Publisher, Springfield, Ill.

Chase, T., and Shaw, E., 1970, Titration of trypsin, plasmin and thrombin with *p*-nitrophenyl-*p'*-guanidinobenzoate HC1, *Methods Enzymol.* **19**:20–27.

Edwards, V. H., 1972, Origins of separation systems, in: *Recent Developments in Separation Science,* Vol. 2, p. 269, CRC Press, Cleveland, Ohio.

Fields, R., 1972, The rapid determination of amino groups with TNBS, *Methods Enzymol.* **25**:464–468.

Li, N. N., 1971a, Permeation through liquid surfactant membranes, *J. Amer. Inst. Chem. Eng.* **17**:459–463.

Li, N. N., 1971b, Separation of hydrocarbons by liquid membrane permeation, *Ind. Eng. Chem. Process Design Develop.* **10**:215–221.

Li, N. N., and Asher, W. J., 1973, *Chemical Engineering in Medicine,* p. 1, American Chemical Society, Washington, D.C.

Li, N. N., and Long, R. B., 1973, Membrane processes, in: *Perry's Chemical Engineer's Handbook,* 5th ed., pp. 17–34, McGraw-Hill Book Company, New York.

Li, N. N., and Shrier, A. L., 1972, Liquid membrane water treating, in: *Recent Developments in Separation Science,* Vol. 1, p. 163, CRC Press, Cleveland, Ohio.

Li, N. N., Mohan, R. R., and Brusca, D. R., 1973, Process for the Reaction and Separation of Components Utilizing a Liquid Surfactant Membrane and an Enzyme Catalyst, U.S. Pat. 3,740,315.

Matulevicius, E. S., and Li, N. N., 1975, Facilitated transport through liquid membranes, Separ. Purif. Methods, **4**:73–96.

May, S. W., and Landgraff, L. M., 1975, Cofactor recycling in liquid membrane enzyme systems, *Biochem. Biophys. Res. Commun.* **68**:786–792.

May, S. W. and Li, N. N., 1972, The immobilization of urease using liquid-surfactant membranes, *Biochem. Biophys. Res. Commun.* **47**:1179–1185.

May, S. W., and Li, N. N., 1974, Encapsulation of enzymes in liquid membrane emulsions, *Enzyme Eng.,* **2**:77–82.

Medical News (Journal of the American Medical Association), 1971, **218**(3):325.

Mohan, R. R., and Li, N. N., 1974, Reductions and separation of nitrate and nitrite by liquid membrane-encapsulated enzymes, *Biotechnol. Bioeng.* **16**:513–523.

Mohan, R. R., and Li, N. N., 1975, Nitrate and nitrite reduction by liquid membrane-encapsulated whole cells, *Biotechnol. Bioeng.* **17**:1137–1156.

Wallace, H. W., Asher, W. J., and Li, N. N., 1973, Liquid–liquid oxygenation: a new approach, *Trans. Amer. Soc. Artificial Internal Organs* **19**:80.

Zaborsky, O. R., and Schwartz, R. D., 1974, The effect of imidoesters on the protocatechuate-3,4-dioxygenase activity of *Acinetobacter calcoaceticus, FEBS Letters* **46**:236–239.

14

Liposomes as Carriers of Enzymes and Proteins in Medicine

Gregory Gregoriadis

1. Problems Associated with the Direct Administration of Proteins

Successful application of proteins in therapy can be hampered by a multitude of difficulties. For instance, use, in the treatment of inherited metabolic disorders, of enzymes that are foreign to the body can promote immune reactions and occasionally fatal serum sickness. Such enzymes might even fail to penetrate, and act in, target areas either because of premature inactivation or because of target inaccessibility. Glucohydrolases, which are potentially useful in the treatment of most lysosomal storage diseases can, en route to their destination, act on their substrates located on cell surfaces or blood glycoproteins and thus upset normal metabolism (Gregoriadis *et al.,* 1974a). Treatment of some forms of human leukemia with asparaginase can be jeopardized by allergic and other toxic reactions (Capizzi *et al.,* 1970), and similar complications arise with the use of uricase in the treatment of gout. Indeed, it is perhaps correct to assume that most of these problems originate from our inability to direct therapeutic agents specifically to diseased areas. It follows that any approach enabling enzymes to reach their target selectively and in a controlled fashion would contribute to a safe and effective enzyme therapy.

Another aspect of protein use in medicine is immunization, and the important task of achieving high antibody titers can be carried out in experimental animals by the administration of the antigen (usually a protein) in conjunction with an adjuvant such as water-in-oil emulsion with or without killed tubercle bacilli (Freund's complete or incomplete adjuvant). However, because of the nonbiodegradable oil they contain and because of the unacceptable granulomas they form, such adjuvants cannot be applied in man. Obviously, a safe and effective adjuvant for use in human immunization programs would be of great value. It should, for instance, reduce the amount of antigen needed for immunization, with resulting economies especially relevant to the developing countries.

Gregory Gregoriadis • Division of Clinical Investigation, Clinical Research Centre, Harrow, Middlesex, England

Such an adjuvant would also reduce risks involved in the use of live virus vaccines (e.g., potential initiation of persistent infections, malignancy by the use of live virus vaccines, or the induction of allergic or even autoallergic reactions from the use of adjuvants consisting of immunogenic materials). On the other hand, allergic reactions induced either because of repeated administration to boost immunity or because of previous natural sensitization of the recipients could be avoided if the antigens were to be presented to the body in a manner that prevents contact of the antigens with their preformed antibodies in the circulation.

2. *Need for a Protein Carrier*

Elimination of problems inherent in the conventional methods of protein administration could be brought about by any method enabling a protein to reach its target selectively. One such approach has been the development of a nontoxic and biodegradable carrier capable of containing a wide variety of materials of biological interest which the carrier could, upon contact with a living entity, transport to the area where they are needed, and subsequently allow to perform their task. The carrier itself should be eventually degraded or otherwise eliminated from the entity.

While such properties of an ideal carrier seem commendable, these can be inadequate when dealing with specific cases, and it is therefore desirable that the carrier be amenable to manipulations in regard to composition, size, surface charge, biodegradability, and, most importantly, target selectivity. Additional advantages of practical and commercial interest should include simplicity in the preparation of the carrier and abundance of the raw materials needed.

Of all proposed carriers, proteins have been the most popular (de Duve *et al.*, 1974). In particular, antibodies raised against target cells have been coupled with enzymes and their ability to deliver these to target areas investigated both *in vivo* and *in vitro*. Other proteins possessing a specific affinity for certain cells have also been employed as carriers. For instance, desialylated fetuin, which, along with other desialylated glycoproteins, binds to parenchymal cells of the liver (Gregoriadis, 1975), was successfully used for the transport of lysozyme and albumin *in vivo* to these cells (Rogers and Kornfeld, 1971). Among carrier macromolecules other than proteins, deoxyribonucleic acid is a notable example, but its use is limited to the transport of substances which are capable of combining with this biopolymer (de Duve *et al.*, 1974).

All such polymer carriers have the advantage of preventing the presence of free therapeutic agents in the extracellular fluids, and their size is compatible with penetration of capillary, endothelial, and other anatomical barriers. However, these advantages can be counteracted by certain limitations. For instance, with most carriers of this nature there is a need for covalent interaction with the agent to be transported. Furthermore, when at the site of action, active agents need to be liberated from the carrier, and this requires the specific action of an enzyme or other factors. Covalent binding of the active agent to the carrier can often abolish or mask the affinity of the latter (e.g., antibody) for its target. Although binding of a protein to a macromolecular carrier can to a certain extent control the fate of the protein, this can still be subject to inactivation by circulating enzymes, inhibitors, or other agents. The bound protein, when foreign, can elicit antibody response and, owing to its availability to the circulating antibodies, promote dangerous allergic reactions upon repeated injection. Unfortunately, the list of problems does not end here, and it is obvious that some protection of the transported proteins from the biological environment is required.

In addition to the "linear" polymers discussed, there have been proposed alternatives in which the transported material is incorporated into the internal spaces of a three-dimensional carrier. As most of these systems are not biodegradable, entrapped agents such as enzymes are expected to act on substrates capable of diffusing into the carrier. The properties and applications of such carriers are discussed elsewhere in this volume. I should like to present research efforts of the last few years which have produced considerable evidence (Gregoriadis, 1976a) that oral or parenteral administration of proteins and other agents in a variety of situations in pharmacology and medicine could be possible via a universal biodegradable carrier, namely, the liposome.

3. *Liposome as Carrier Candidate*

3.1. *Composition and Structure*

Formation of liposomes occurs upon the confrontation of water-insoluble polar lipids, such as phospholipids, with water (Bangham *et al.*, 1974). The highly ordered assemblages which emerge persist in the presence of excess water and are finally arranged to a system of concentric bimolecular sheets of lipid molecules (Figure 1). These multilamellar liposomes, which can measure up to several micrometers in diameter, can be converted upon sonication to smaller structures of a minimum diameter of 25 nm.

The usefulness of liposomes as vehicles derives from the fact that as lipids rearrange to form the concentric spheres of Figure 1, there is unrestricted entry of water and solutes (e.g., proteins) in between the lipid bilayers. In this way any water-soluble substance can be entrapped in the aqueous compartments provided that its size is compatible with the dimensions of the space between the planes of the hydrophilic head groups of lipids (about 7.5 nm width) or the aqueous space of monolamellar liposomes (about 8.5 nm diameter). A much larger space of water exists in the central core of multilamellar liposomes (about 0.15 μm in Figure 1). An additional requirement for entrapment is that the substance in question does not interact with lipids in a way that upsets liposome formation.

Entrapped water-soluble substances such as proteins by no means exist entirely within the aqueous phase of liposomes. Indeed, it is expected that hydrophobic regions of protein molecules will penetrate the lipid phase in a fashion similar to that seen in natural membranes. In addition, proteins can be adsorbed onto the liposomal surface or interact electrostatically with lipids. Such events suggest that for most enzymes or proteins, exclusive existence in the

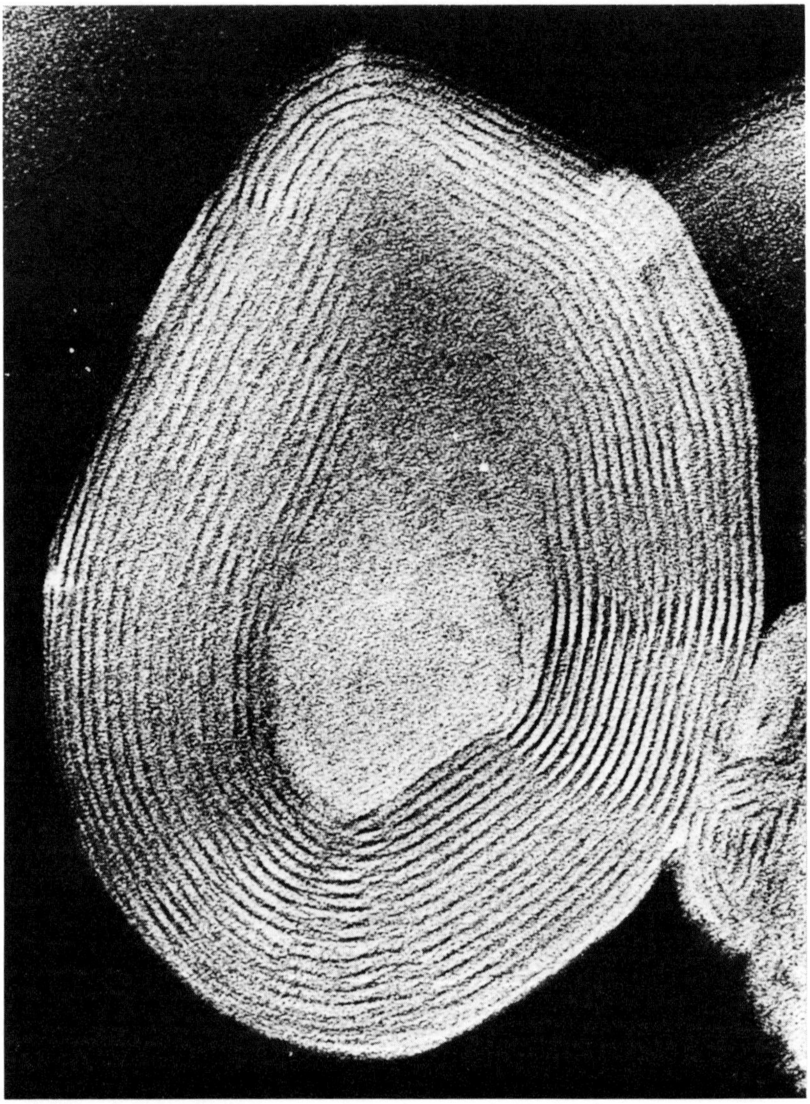

FIGURE 1. Electron micrograph of a multilamellar liposome showing the lipid bilayers separated by electron-opaque aqueous layers. Sodium silicotungstate. 272,000×. (From Segal *et al.*, 1974.)

aqueous milieu of liposomes is unlikely. This does not, however, diminish the value of liposomes as carriers. For instance, if the objective is the transport of an enzyme to a particular target, then, providing that the enzyme follows the carrier to its destination, its exact position in the liposome should be irrelevant. On the other hand, when expression of the enzymatic or immunologic activity of the enzyme is unacceptable in areas other than that of the target, the enzyme, or at least its active sites, should be conveniently accommodated beyond the surface of the carrier.

Phospholipids such as lecithins are the main lipid components of liposomes. While both unsaturated and saturated lecithins can be used, the choice is often dependent on practical as well as on theoretical considerations (Bangham *et al.,* 1974). For instance, since preparation of liposomes is best carried out above the liquid crystalline transition temperature of the phospholipid used, when temperature-labile enzymes are to be entrapped, egg lecithin and other unsaturated lecithins which have low transition points should be the lipids of choice. Unsaturated lecithins should also be used for the preparation of liposomes intended for fusion with cells (Papahadjopoulos *et al.,* 1974a,b). On the other hand, the presence of saturated lecithins (e.g., dipalmitoyl lecithin) in the liposome structure can decrease the permeability of the lipid bilayers to small ions. Beef brain sphingomyelin and some synthetic phosphonyl and phosphinyl analogs of lecithins (but not bacterial phosphatidyl ethanolamine) can replace lecithin as the sole phospholipid.

Formation of liposomes can be improved by the addition of other lipid components (Gregoriadis, 1976b). Thus, the presence of a sterol (e.g., cholesterol, ergosterol) can promote the stability of lipid bilayers, especially those composed of synthetic lecithins. Further, the inclusion of a charged amphiphile will produce electrostatic repulsion between any two charged bilayers either side of an aqueous channel, and this can greatly increase the volume of entrapped enzyme solution, and hence the absolute amount of entrapped enzyme. In addition, the presence of a charged component will confer a positive or negative charge on the surface of liposomes, an event that can decisively influence their biological properties. However, in deciding on the choice of the charged amphiphile, consideration should be given to the nature of the enzyme. For instance, it is preferable to use amphiphiles and enzymes of like charge to avoid electrostatic attraction and formation of precipitable enzyme–lipid complexes. When there is no alternative, adjustment of the pH should temporarily alter the net charge of the protein. It has been possible, for instance, to entrap albumin in positively charged liposomes without precipitate formation at pH 4.2, which is below the isoelectric point of the protein. After separation from nonentrapped albumin, the pH of albumin-containing liposomes can be adjusted to neutrality (Gregoriadis, 1976). Finally, in regard to the relative concentrations of the lipids used for the formation of liposomes, these can vary widely according to the requirements of the preparation in terms of stability, surface charge, and its desired behavior within the biological milieu.

3.2. Entrapment of Proteins in Liposomes

Methods for the formation of liposomes in the presence of materials of which entrapment is desirable have been described (Bangham *et al.,* 1974), and while these differ in details, the basic event is simply the disruption of a dry thin lipid layer with water containing the material to be entrapped. In the procedure that has been adopted in this laboratory (Gregoriadis, 1976b), appropriate lipids are dissolved in an organic solvent (e.g., chloroform) which is subsequently eliminated under reduced argon pressure, leaving a thin lipid layer on the walls of a round-bottomed flask the internal surface of which is approximately proportional to the amount of lipid used. The layer is then flushed with oxygen-free argon to remove traces of chloroform and subsequently dispersed with the protein solution. A few glass beads may be added to facilitate removal of lipids from the glass surface. This part of the procedure leads to the spontaneous formation of large liposomes which can be reduced in size by sonication, preferably at 4°C, when temperature-labile enzymes are used. Separation of liposomes containing some of the enzyme from the nonentrapped enzyme can be carried out by centrifugation or by molecular sieve chromatography.

The extent of entrapment per unit weight of lipid is dependent on a number of parameters, among which the volume of water employed, the presence of charged amphiphiles, and the nature of the protein used are of great importance. For instance, it would be theoretically possible to achieve total entrapment if one could use water (containing the protein) in amounts just sufficient to form liposomes. Proteins that do not interact with lipids normally used for liposome formation, such as albumin and certain enzymes, are expected to be entrapped in relation to the volume of the solution captured within liposomes. Other proteins, such as diphtheria toxoid (Gregoriadis and Allison, 1974), asparaginase (Neerunjun and Gregoriadis, 1976), insulin (Dapergolas *et al.*, 1976), and cerebroside:β-glucosidase (Braidman and Gregoriadis, 1976) seem to form complexes with certain lipids and, provided that this does not upset liposome formation, much of the protein used (often to the extent of 50–90%) can be recovered with liposomes. Obviously, this is of importance when proteins (e.g., antigens used for immunization, or certain enzymes) are either expensive or available only in small quantities.

Enzymes are rendered largely latent upon entrapment, and measurement of their activity requires previous rupture of the lipid envelope with a detergent (e.g., Triton X-100, 0.1% final concentration) (Gregoriadis, 1976b). The latency of enzymes acting on substrates of low molecular weight capable of penetrating lipid bilayers is difficult to assess and, in such cases, a measure of real entrapment, as opposed to extraliposomal adsorption or to partial penetration of enzymically active sites into the extraliposomal space, can be obtained by protease treatment of the extraliposomal radioisotope-labeled material.

As shown in Table 1, a multitude of proteins, many of which are enzymes, have been entrapped in various liposomal preparations, principally for two reasons: in the first, which precedes historically, liposomes are used as models for the study of cell membrane biophysics, and in the second, liposomes have served as convenient carriers of enzymes and other materials in biomedical applications. It is this latter aspect of liposomes which will be discussed here in terms of both their behavior *in vivo* and *in vitro* and their potential role in pharmacology.

4. *Interaction of Protein-Containing Liposomes with the Biological Environment*

4.1. *In Vivo Studies*

Using a variety of liposome-associated proteins and other substances incapable of diffusing out of the carrier in the presence of blood, the behavior of liposomes and their contents *in vivo* and *in vitro* has been studied with a number of preparations of diverse lipid composition, size, and charge (Table 1). The first indication of the carrier potential of liposomes in enzyme therapy and other applications in medicine came from experiments in which the latency of injected liposome-entrapped enzymes was retained in the blood circulation and their fate was shown to be dictated by the liposome carrier (Gregoriadis,1974). It was found, for instance, that the rate of removal from the blood of liposomes and their contents is directly related to the size of the carrier (Juliano and Stamp, 1975), while that of mixed-size multilamellar liposomes could be approximately described as biphasic, with the first phase being rapid and the second slow. Proteins entrapped in positively charged liposomes were shown to leave the circulation at a rate considerably lower than that observed with neutral or negatively charged liposomes and, indeed, even among negative liposomes the nature of the charged amphiphile could influence the rate of elimination (Gregoriadis and Neerunjun, 1974).

Contact of liposomes with blood plasma is followed by changes in their surface charge. Electrophoretic mobility studies have shown that positive and neutral liposomes acquire a net negative charge while that of negative liposomes remains unaltered. This is apparently due to the incorporation of α_2-macroglobulin (man) or α_1-macroglobulin (rat) onto the liposomal surface (Black and Gregoriadis, 1976). α_2-Macroglobulin has been recently shown to promote particle phagocytosis (Allen, 1974), and it is perhaps because of the association of liposomes with this plasma protein that tissues of the reticuloendothelial system are principally involved in the uptake of injected liposomal proteins. However, in addition to the fixed macrophages of the liver and spleen, radioautographic evidence has also implicated the hepatic parenchymal cells (Gregoriadis and Ryman, 1972a). Indeed, this has now been substantiated by electron microscopic observations, which clearly demonstrate participation in liposome uptake by all liver cells, including endothelial (Wisse and Gregoriadis, 1975). It seems that very soon after injection of rats with liposome-entrapped horseradish peroxidase there is an initial involvement of Kupffer and also of endothelial cells followed by the parenchymal cells. The latter event, which coincides with the second phase of slow elimination from the blood, may occur following the penetration of small liposomes through the fenestrations (0.1 μm in diameter; Wisse, 1970) of the sinusoidal lining.

A small but measurable proportion of injected liposomes (and their contents) reaches the lungs, kidney, brain, and other tissues (Gregoriadis and Ryman, 1972a; Kimelberg *et al.,* 1975), and it appears that uptake by these tissues can be influenced considerably by the size and lipid composition of liposomes (Jonah *et al.,* 1975). In view of the major participation of the liver and spleen it should be appreciated that uptake of liposome-entrapped materials by other tissues, which is deemed to be negligible, could be easily misjudged. This is especially true for entrapped proteins the activity of which, enzymatic or otherwise, could be rapidly decreased by catabolic events in the cell's interior (Gregoriadis and Ryman, 1972a). Tissue participation can be accurately ascertained by a method which not only precludes any association of liposome-entrapped markers with cells, other than that mediated through the carrier, but also ensures that markers transported into cells via liposomes remain unaltered. We have found (Dapergolas *et al.,* 1976) that the radiolabel of ^{111}In-labeled bleomycin remains firmly associated with

TABLE 1
Liposome-Associated Proteins of Biological Interest

Substance		Liposomal lipids[a]	Reference
Lysodeictus lysozyme; EC 3.2.1.17	7:1:2	PC, CHOL, DCP	Sessa and Weissmann, 1970
	7:1:2	PC, CHOL, SA	Sessa and Weissmann, 1970
Aspergilus niger amyloglucosidase[b]; EC 3.2.1.3	7:2:1	PC, CHOL, DCP	Gregoriadis and Ryman, 1972a (rat); Scherphof *et al.*, 1975 (fibroblast)
Yeast invertase[b]; EC 3.2.1.26	7:2:1	PC, CHOL, PA	Gregoriadis and Ryman, 1972b (rat); Gregoriadis and Buckland, 1973 (macrophage, fibroblast)
Clostridium perfringens neuraminidase[b]; EC 3.2.1.18	7:2:1	PC, CHOL, PA	Gregoriadis *et al.*, 1974a (rat)
Dextranase[b]; EC 3.2.1.11	7:2:1	PC, CHOL, PA	Colley and Ryman, 1974 (rat)
Jack-bean α-mannosidase[b]; EC 3.2.1.24	7:2:1	PC, CHOL, PA	Patel and Ryman, 1974 (rat)
Bacillus subtilis α-amylase[b]; EC 3.2.1.1		DPL	Batzri and Korn, 1975 (ameba)
Horseradish peroxidase[b]; EC 1.11.1.7	7:2:1	PC, CHOL, PA	Wisse and Gregoriadis, 1975 (rat)
	3:1.1:1	SPH, CHOL, SA	Magee *et al.*, 1974 (HeLa)
	7:1:2	PC, CHOL, DCP	Weissmann *et al.*, 1975 (phagocyte)
Influenza virus hemagglutinin and neuraminidase[b]; EC 3.2.1.18	9:1	PC, DCP	Almeida *et al.*, 1975
Erwinia asparaginase[b]; EC 3.5.1.1	7:2:1	PC, CHOL, PA	Neerunjun and Gregoriadis, 1976 (mouse)
	7:2:1	PC, CHOL, SA	Neerunjun and Gregoriadis, 1976 (mouse)
E. coli asparaginase[b]; EC 3.5.1.1	10:2:1	PC, CHOL, DCP	Fishman and Citri, 1975
Glucose oxidase[b]; EC 1.1.3.4		PI	Gregoriadis *et al.*, 1976 (rat)
	7:2:1	PC, CHOL, PA	Gregoriadis *et al.*, 1976 (rat)
	7:2:1	PC, CHOL, SA	Gregoriadis *et al.*, 1976 (rat)
Human glucocerebroside: β-glucosidase[b]; EC 3.2.1.21	7:2:1	PC, CHOL, PA	Braidman and Gregoriadis, 1976
Cytochrome oxidase; Ec 1.9.3.1	1:1:0.15:0.25	PE, PC, CL, LL	Eytan *et al.*, 1975
Insulin[b]		PI	Gregoriadis *et al.*, 1976 (rat)
	7:2	DPL, CHOL	Dapergolas and Gregoriadis, 1976 (rat)
	10:2:1	PC, CHOL, DCP	Patel and Ryman, 1976 (rat)
	7:2:1	PC, CHOL, PA	Gregoriadis *et al.*, 1976 (rat)
	7:2:1	PC, CHOL, SA	Gregoriadis *et al.*, 1976 (rat)
Cytochrome C	1:1	PC, PE	Papahadjopoulos and Miller, 1967
Albumin	1:1	PC, PE	Papahadjopoulos and Miller, 1967
Albumin[b]	7:2	PC, CHOL	Gregoriadis and Neerunjun, 1974 (rat)

	7:2:1	PC, CHOL, DCP	Gregoriadis and Ryman, 1972b (rat); Gregoriadis and Neerunjun, 1974 (rat); Gregoriadis *et al.*, 1971
	7:2:1	PC, CHOL, PA	Gregoriadis and Ryman, 1972b (rat); Gregoriadis and Neerunjun, 1974 (rat); Segal *et al.*, 1975 (rat)
	7:2:1	PC, CHOL, G	Gregoriadis and Neerunjun, 1974 (rat)
	7:2:1	PC, CHOL, SA	Gregoriadis and Neerunjun, 1974 (rat); Segal *et al.*, 1975 (rat); Gregoriadis, 1973a (rat)
Diphtheria toxoid[b]	7:2:1	DPL, CHOL, PA	Gregoriadis *et al.*, 1974b (man)
	7:2:1	PC, CHOL, DCP	Gregoriadis and Allison, 1974 (mouse)
	7:2:1	PC, CHOL, PA	Gregoriadis and Allison, 1974 (mouse); Allison and Gregoriadis, 1974 (mouse)
	7:2:1	PC, CHOL, SA	Gregoriadis and Allison, 1974 (mouse); Allison and Gregoriadis, 1974 (mouse)
Fetuin,[b] desialylated fetuin[b]	7:2:1	PC, CHOL, PA	Gregoriadis, 1974 (rat); Gregoriadis and Neerunjun, 1975 (rat)
Ricinus communis lectin	2:1:0.04-0.2	DPL, CHOL, G	Surolia *et al.*, 1975
Thyroglobulin	7:1:2	PC, CHOL, DCP	Kawada *et al.*, 1974
Staphylococcal α-toxin	7:1:2	PC, CHOL, DCP	Freer *et al.*, 1968
IgG immunoglobulin			
(antifibroblast, anti-HeLa cell, anti-AKR-A cell)[b]	7:2:1	PC, CHOL, PA	Gregoriadis and Neerunjun, 1975 (HeLa, AKR-A, fibroblast)
(antithyroglobulin)[b]	11:3.7:1	PC, PE, G	Dunnick *et al.*, 1975 (mouse)
(anticoxsackie A-21 virus)[b]	3:1.5:0.66	SPH, CHOL, DCP	Magee and Miller, 1972 (ML cell)
	3:1.5:0.66	SPH, CHOL, SA	Magee and Miller, 1972 (ML cell)
(anti-α-glucosidase)[b]	7:2:1	PC, CHOL, SA	de Barsy *et al.*, 1975 (rat)
nonspecific	7:1:2	PC, CHOL, DCP	Weissmann *et al.*, 1974
	7:1:2	PC, CHOL, SA	Weissmann *et al.*, 1974
IgM immunoglobulin[b]	7:1:2	PC, CHOL, DCP	Weissmann *et al.*, 1975 (phagocyte)

[a]Lipids, with molar ratios preceding, are PC, phosphatidyl choline; CHOL, cholesterol; DCP, dicetyl phosphate; SA, stearylamine; PA, phosphatidic acid; DPL, dipalmitoyl lecithin; SPH, sphingomyelin; PI, phosphatidyl inositol; CL, cardiolipin; LL, lysolecithin; PE, phosphatidyl ethanolamine; G, gangliosides.

[b]Substances were associated with liposomes for the study of the carrier potential of liposomes in pharmacology and medicine. Historically, this is preceded by studies (unmarked substances) in which liposomes were used as models of the cell membrane (Bangham *et al.*, 1974). Association of substances with liposomes was carried out either during the preparation of the latter or after mixing with preformed liposomes (modified from Gregoriadis, 1976a).

liposomes while the latter circulate in the blood, and it stays and decays in the site to which bleomycin is eventually transported for at least 48 hr. Examination of tissues at a time when there is no radioactivity in the blood and transcapillary passage has been already accomplished (usually 24 hr after intravenous injection) can then reveal the extent to which a given tissue has participated in liposome uptake.

The fate of liposome-associated proteins and other materials has also been investigated following administration in animals by routes other than the intravenous. For instance, intraperitoneal injection of rats with liposomal ^{125}I-labeled polyvinylpyrrolidone, a substance which by itself penetrates liver cells poorly, was followed by its uptake by this tissue presumably after the passage of intact liposomes through the peritoneal vascular barrier into the portal vein (Dapergolas *et al.,* 1976). Interestingly, even large unsonicated liposomes were capable of reaching the liver just as efficiently as small sonicated ones. In other studies, large liposomes injected into the rat testicle were slowly disrupted in the site of injection, releasing the entrapped protein, while smaller liposomes and their contents were rapidly removed from the site of injection to enter the blood and the lymphatics. Some of the entrapped protein was partially recovered in the liver and spleen and in the lymph nodes draining the injected testicle (Segal *et al.,* 1975).

In attempts (Patel and Ryman, 1976; Dapergolas and Gregoriadis, 1976) to introduce liposome-associated proteins into the blood by the oral route it was possible to decrease blood glucose levels in normal and diabetic rats over a period of at least 24 hr by the intragastric administration of insulin entrapped in dipalmitoyl lecithin liposomes (Dapergolas and Gregoriadis, 1976). The effect of insulin was less pronounced with egg lecithin liposomes and this was attributed to the increased resistance of dipalmitoyl lecithin liposomes to the action of gut phospholipase. Such liposomes were shown to transport insulin in its entrapped form into the circulation (Dapergolas and Gregoriadis, 1976). A similar reduction (but of shorter duration) in blood glucose levels was obtained by the intragastric administration into normal rats of glucose oxidase entrapped in phosphatidylinositol liposomes (Dapergolas *et al.,* 1976).

Work at the subcellular level has demonstrated beyond any doubt that injected liposome-entrapped enzymes are transported largely into the lysosomes of cells that participate in liposome uptake. For instance, it was shown that at all time intervals after injeciton of liposome-entrapped amyloglucosidase or invertase, enzyme activity in the subfractionated rat liver was recovered mainly in the lysosomes (Gregoriadis, 1974). The possibility of associaiton of liposomes with lysosomes during the homogenization of the tissue was discarded after it was found that very little activity was associated with lysosomes by homogenizing the liver of intact rats in the presence of the injected material (Gregoriadis and Ryman, 1972b).

Lysosomal localization was confirmed in electron microscopic studies of the liver of rats injected with entrapped horseradish peroxidase (Wisse *et al.,* 1976). Indeed, it was observed that all lysosomes of the Kupffer, endothelial, and parenchymal cells contained peroxidase activity for at least up to 48 hr (Figure 2). Liposomal structure, however, was lost soon (3–10 min) after uptake and myelin-like figures seen in the cells after this time were peroxidase-negative. Morphological observations (Wisse *et al.,* 1976) also revealed that lysosomal localization of liposomal proteins occurs through endocytosis of the carrier. Fusion of liposomes with the cell membranes, an event that would have led to the ejection of peroxidase into the cytoplasm, did not appear to occur.

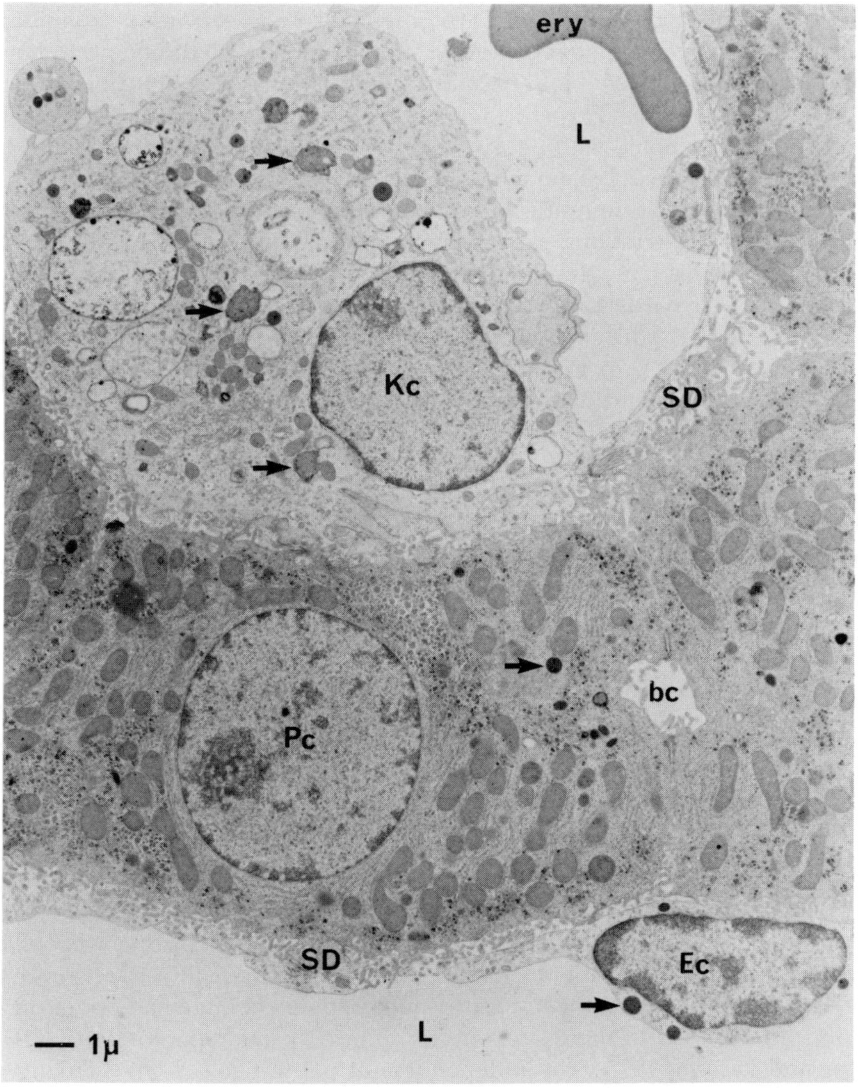

Figure 2. Survey of normal perfusion-fixed rat liver, showing three cell types, which take up HRP-liposomes: Kc, Kupffer cell; Ec, endothelial cell; and Pc, parenchymal cell. Arrows indicate dense bodies (lysosomes), which all became labeled with peroxidatic activity after i.v. HRP-liposome injections. L, lumen of the sinusoid; ery, red blood cell, SD, space of Disse; bc, bile capillary. (From Wisse *et al.,* 1976.)

Although information related to the fate of injected liposome-associated materials is relatively abundant, very little is known about the interaction of the carrier itself with the living animal and the possible changes on normal metabolism which the carrier could effect. It seems that soon after intravenous injection of liposomes, there is some loss of liposomal cholesterol to the low- and high-density plasma lipoproteins, but it has not yet been established that a similar amount of lipoprotein cholesterol attaches to liposomes (Black and Gregoriadis, 1976). Liposomes of certain lipid compositions are potentially toxic. For instance, liposomes composed of egg lecithin and cholesterol supplemented with dicetyl phosphate or stearylamine, when injected into mice intracerebrally, cause epileptic seizures and cerebral tissue necrosis. These do not occur, however, with similar egg lecithin liposomes supplemented with phosphatidic acid or with dipalmitoyl lecithin liposomes (D. H. Adams, J. Greta, V. J. Richardson, B. E. Ryman, and H. M. Wisniewski, personal communication). Liposomes were also found to provoke specific metabolic alterations. For instance, sonicated liposomes made of bovine brain phospholipids can induce both modification of glucose distribution (Bruni *et al.*, 1976) and increased catecholamine metabolism in the brain of mice (Leon and Toffano, 1976) or produce release of acetylcholine from the brain cortex of the rat (Mantovani *et al.*, 1976). Phosphatidyl serine was the active ingredient, while phosphatidyl choline and phosphatidyl inositol were inactive.

The suitability of lecithin and phosphatidic acid as components of innocuous liposomal preparations has also been supported by experiments in which rats were treated chronically with sonicated liposomes composed of egg lecithin, cholesterol, and phosphatidic acid (molar ratio 7:2:1, 0.7 g of lipid injected intravenously during a 6-month period). Rats tolerated such treatment well, and histochemical examination of a wide variety of tissues, including brain, showed no lesions (C. D. V. Black and G. Gregoriadis, unpublished observations). Likewise, clinical experience with a small number of cancer patients injected with similar liposomes repeatedly has also been uneventful (Gregoriadis *et al.*, 1974b).

4.2. *In Vitro Studies*

The observation that liposomes can transport hydrolytic enzymes into cultured cells and influence intracellular catabolism (Gregoriadis and Buckland, 1973) has been followed by an ever-increasing number of studies in which a variety of mammalian and other cells were capable of interacting with diverse liposomal preparations (Table 1). Recent work has shown that for the types of cells examined, uptake of liposomes is temperature-dependent (Papahadjopoulos *et al.*, 1974b), proportional to the lipid concentration (Papahadjopoulos *et al.*, 1974b; Huang and Pagano, 1975), and independent of pH (range 6.8–8.7) (Huang and Pagano, 1975) and lipid composition (Papahadjopoulos *et al.*, 1974b). In addition, in a few cases uptake was found to be more pronounced with positively charged liposomes (Magee and Miller, 1972; Batzri and Korn, 1975; Magee *et al.*, 1974).

Under the controlled conditions of *in vitro* experiments, association of liposomes with cells appears to be dictated to a significant extent by the identity of lipids in both liposomes and cell membrane (Papahadjopoulos *et al.*, 1974a; Batzri and Korn, 1975; Pagano and Huang, 1975; Grant and McConnell, 1973). Liposomes composed of unsaturated lecithins, for instance, are thought to fuse with cells and deliver their contents into the cell's cytoplasm rather than being endocytosed (Papahadjopoulos *et al.*, 1974a). Evidence (Papahadjopoulos *et al.*, 1974b; Pagano and Huang, 1974) against endocytosis includes the absence of significant effect by inhibitors of cellular energy metabolism or by endocytosis stimulators on uptake by cells, while supporting fusion were electron microscopic data (Papahadjopoulos *et al.*, 1974b), the incorporation of liposomal lipid (Papahadjopoulos *et al.*, 1974b; Inbar and Shinitzky, 1974) and enzyme contents (Magee *et al.*, 1974) into the cellular membranes, and also the inhibitory effect on cell growth of cyclic AMP entrapped in unsaturated liposomes and believed to enter the cytoplasm after fusion (Papahadjopoulos *et al.*, 1974a, 1974b). Such an effect was not observed with cyclic AMP entrapped in liposomes composed of dipalmitoyl lecithin (Papahadjopoulos *et al.*, 1974b). These saturated liposomes are endocytosed by cells and end up in the lysosomes, where cyclic AMP is probably inactivated. Paradoxically, ameba (Batzri and Korn, 1975) and mycoplasma (Grant and McConnell, 1973) cells appear to fuse with saturated liposomes and ameba cells (Batzri and Korn, 1975) to endocytose unsaturated ones.

While theories (Batzri and Korn, 1975; Papahadjopoulos *et al.*, 1974a; Pagano and Huang, 1975) pertaining to the mechanism of fusion between appropriate liposomes and cells have been propagated in considerable detail (Pagano and Huang, 1975), unequivocal evidence for fusion has yet to be obtained, and indeed, whatever its extent, fusion of unsaturated liposomes with cells could occur in parallel with endocytosis. Indeed, in experiments with cultured cells, liposomes composed of unsaturated lecithin and containing invertase (Gregoriadis and Buckland, 1973) or amyloglucosidase (Scherphof *et al.*, 1975) were capable of effecting hydrolysis of intralysosomal sucrose and glycogen, respectively, and horseradish peroxidase entrapped in similar liposomes could be visualized within lysosomes (Magee *et al.*, 1974; Weissmann *et al.*, 1975) and endocytic vacuoles (Magee *et al.*, 1974) in the cytoplasm.

4.3. *Mode of Action of Liposome-Associated Proteins*

There are at least two major pathways for the introduction of liposome-associated biologically active proteins into target intracellular sites (Figure 3). In the first, liposomes can transport their contents into the lysosomes following endocytosis (Gregoriadis *et al.*, 1974b), and in the second, liposomal contents are expelled into the cell's cytoplasm after fusion of the outer lipid bilayer with the cell membrane (Papahadjopoulos *et al.*, 1974a; Batzri and Korn, 1975). Regardless of the method employed for the introduction of liposome-associated agents into the appropriate cell compartment, a biologically active protein could be transported near the target intracellular site according to the requirements of the particular case. For instance, with a number of microbial infections or with lysosomal storage diseases in both of which the target is intralysosomal, active proteins (e.g., antibacterial antibodies or hydrolases) should be administered via liposomes that can be endocytosed (Figure 3A). Such proteins should be capable, or at least be rendered capable, of surviving in the lysosomal milieu for a period of time sufficient for effective action. Experience with liposome-entrapped hydrolases employed to "correct" model lysosomal storage conditions (Gregoriadis and Buckland, 1973; Colley and Ryman, 1974; Scherphof *et al.,* 1975) or with antibodies made against specific lysosomal enzymes (de Barsy *et al.,* 1975) shows that an operationally useful survival of some proteins within the lysosomes is achieved. Introduction of proteins into the cytoplasm without the participation of lysosomes could be carried out by fusion of cells with monolamellar liposomes of the appropriate lipid composition (Figure 3B). In this way, cellular organelles such as endoplasmic reticulum Golgi apparatus or even mitochondria and the nucleus could become accessible to enzymes and other proteins. Fusion would also allow the insertion of a lipid-soluble protein (carried in the liposomal bilayer) within the framework of the plasma membrane (Figure 3B). Finally, multilamellar liposomes with the ability to fuse could serve as donors of smaller vesicles which upon their ejection into the cytoplasm could become slow-release depots of proteins (Figure 3C).

Action of liposome-associated protein in the extracellular space is also possible. This reverse action of the liposomal carrier, in which, for instance, substrates

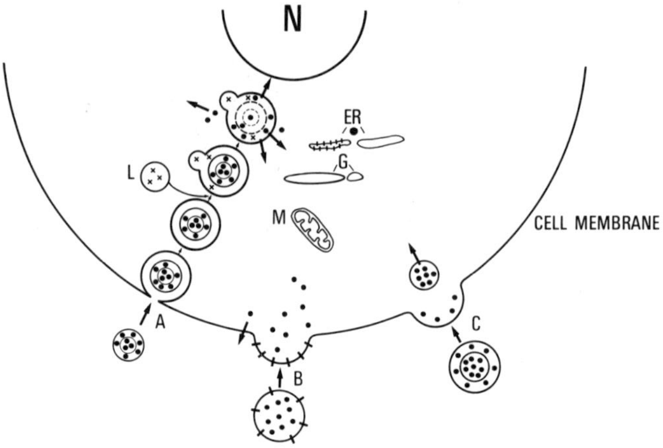

FIGURE 3. Possible mechanisms for cell–liposome interaction. (A) Endocytosis of the liposome is followed by fusion of the endocytic vacuole containing the liposome with a lysosome (L). Lysosomal phospholipases or other factors (crosses) disrupt the lipid bilayers of liposomes and free the entrapped agent (dots), which can then act either within the lysosome or, after its diffusion, in other cell compartments (e.g., N, nucleus; ER, rough and smooth endoplasmic reticulum; G, Golgi apparatus; M, mitochondrion). (B) after fusion of a monolamellar liposome with the cellular membrane, agents (rods) incorporated in the lipid bilayers of the liposome are transferred onto the membrane of the cell. Agents (dots) in the aqueous space of the liposome are expelled into the cell's cytoplasm and reach other organelles. (C) After fusion of a multilamellar liposome, agents (dots) in the outer aqueous space of the liposome together with the inner core are expelled into the cell's cytoplasm. Free and entrapped agent molecules can then reach other organelles. (From Gregoriadis, 1976a).

of low molecular weight penetrate the lipid bilayers to reach their respective enzyme, has been found feasible *in vivo* (Neerunjun and Gregoriadis, 1976).

5. *Selective Targeting of Liposomes*

The selective influence of metabolic events by the use of liposomes as carriers of the relevant molecular species is beset by a number of formidable difficulties. Whether it is treatment of disease or the study of a certain cell function, liposomes are expected to come into intimate contact with a target and carry out their task. On first inspection, there are three problems to be dealt with (Gregoriadis, 1976a). First, it is necessary to possess a preparation of liposomes containing the active agent and also bearing a recognition probe capable of mediating selective and efficient contact of the agent with the target site. Second, liposomes of the appropriate size must be used to penetrate capillary walls and other anatomical barriers. Third, with target sites other than the liver and spleen, one must ensure that these two tissues do not remove most of the injected dose destined for cells in which association with liposomes is not rapid.

5.1. *Homing of Liposomes*

Attempts have been made to improve association of liposomes with target cells. It was shown, for instance, that positively charged liposomes could bind avidly to mouse leukemia (Magee and Miller, 1972) and HeLa (Magee *et al.*, 1974) cells and that the presence of certain phospholipids in the liposomal bilayers could influence to some extent tissue distribution *in vivo* (Jonah *et al.*, 1975). In other experiments, advantage was taken of the property of the Fc portion of the isologous aggregated IgM immunoglobulins to bind to receptors on the surface of phagocytes. Such immunoglobulins adsorbed on the liposomal surface were shown to mediate introduction of entrapped horseradish peroxidase into the lysosomes of phagocytes (Weissman *et al.*, 1975).

We have developed a general method for the selective introduction of liposome-associated agents into target cells (Gregoriadis, 1974; Gregoriadis and Neerunjun, 1975). The technique is based on the incorporation of anti-target-site (e.g., cell) IgG immunoglobulin onto the liposomal surface by the use of coupling procedures or simply by hydrophobic bonding between the relevant moieties of the lipid bilayers and the proteins. Using this method it was possible to specifically direct liposomes containing bleomycin, a cytotoxic glycopeptide used in cancer chemotherapy, to a variety of normal and malignant cells. Thus, in experiments (Gregoriadis and Neerunjun, 1975) in which human skin fibroblasts, HeLa, and AKR-A mouse leukemia cells were incubated in the presence of [^{14}C]cholesterol-labeled liposomes containing ^{111}In-labeled bleomycin and carrying on their surface radioiodinated anti-cell IgG, uptake by cells of all three radioactivities, representing the carrier, the entrapped drug, and the homing probe (i.e., IgG), in that order, was greatest when liposomes were associated with the IgG corresponding to the cell studied (Figure 4).

Immunoglobulins are not the only macromolecules capable of mediating liposomal uptake by cells. Indeed, it seems that any molecule possessing specific affinity for the surface of target cells (c.g., trophic hormones, some lysosomal enzymes) could be a candidate for the role of a homing probe. This was shown to be true, for instance, in the case of desialylated fetuin, which together with a variety of other desialylated proteins binds specifically to the hepatic parenchymal cells and ends up by endocytosis into the cell's lysosomes (Gregoriadis, 1975). Bleomycin-containing liposomes bearing on their surface desialylated fetuin could, upon intravenous injection into rats, direct bleomycin to the parenchymal cells of the liver (Gregoriadis and Neerunjun, 1975). Since much of the drug radioactivity in these experiments as well as in those with cultured cells using IgG as the homing probe was recovered in the lysosome-rich fraction of the cells involved, it was suggested that homing molecules, upon binding onto the relevant receptors on the cell surface, mediate uptake and introduction of their liposomal moiety and its contents into the cell's interior.

5.2. *Penetration of Target Areas*

Attempts were also made to establish conditions for the passage of liposome-associated agents into areas of the body which liposomes, depending on the route of administration, can either reach only to a small extent or fail to enter altogether. Appropriate manipulations of the size, surface charge, and lipid composition of liposomes were shown to facilitate the penetration of liposome-associated agents into relevant areas in need of the agents (Dapergolas *et al.*, 1976). For instance, very small liposomes were capable of reaching tumor areas to a greater extent than larger liposomes (Table 2) apparently because of increased penetration of the vascular tumor barrier and also because of the decreased rate of elimination of such liposomes from the circulation (Juliano and Stamp, 1975), an event that

TABLE 2

Uptake of Radioactivity by Tumor Tissue of Mice Injected with Free and Liposome-Entrapped ^{111}*In-labeled Bleomycin*[a]

Strain and number of mice	Implanted tumor	Uptake by tumor (% ± S.D. of dose/g of tissue) Free bleomycin	Liposomal bleomycin (0.5-min sonication)	Liposomal bleomycin (20-min sonication)
C3H (8)	6C3HED	1.81 ± 0.47	3.92 ± 0.51	6.82 ± 0.94
Balb C (8)	Meth "A"	0.99 ± 0.39	2.38 ± 0.69	6.49 ± 1.39

[a]6C3HED cells were implanted in the intrascapular area and Meth "A" malignant tissue in the kidney of mice, which were later injected intravenously with ^{111}In-labeled bleomycin, free or entrapped in liposomes sonicated for 0.5 or 20 min. Measurements were carried out 48 hr after treatment. (From Dapergolas *et al.*, 1976).

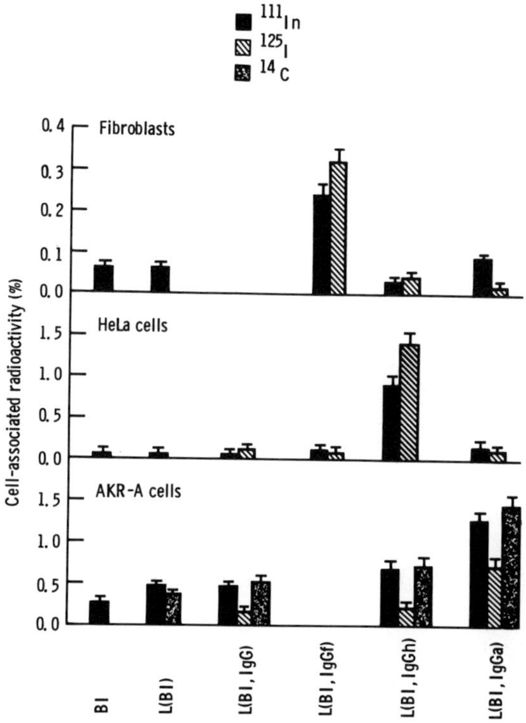

FIGURE 4. Cells were exposed to media containing a trace of ^{111}In-labeled bleomycin, free (Bl) or entrapped in liposomes (labeled with cholesteryl-^{14}C-palmitate when appropriate) also containing phosphate buffered saline, (L,Bl); ^{125}I-labeled nonspecific IgG, (L(Bl,IgG)); anti-human fibroblast IgG (L(Bl,IgGf)); anti-HeLa IgG (L(Bl,IgGh)); or anti-AKR-A IgG (L(Bl,IgGa)). Values are expressed as percent (mean ± standard error) of added radioactivity associated with cells. (From Gregoriadis and Neerunjun, 1975.)

would have allowed them longer contact with, and uptake by, other areas in the body (e.g., tumor tissue). In other experiments already discussed, changes in the lipid composition of liposomes given intragastrically allowed the introduction of glucose oxidase and of insulin into the blood circulation (Dapergolas *et al.*, 1976; Dapergolas and Gregoriadis, 1976).

5.3. *Control of Hepatic Uptake of Liposomes*

Administration of a large quantity of "empty" liposomes together with a small portion of liposomes carrying the active agent delays hepatic uptake of the latter liposomal population, presumably because of competition for endocytosis (Gregoriadis and Neerunjun, 1974). Delay in uptake has also been attained by the use of positively charged (Gregoriadis and Neerunjun, 1974) or small (Dapergolas *et al.*, 1976) liposomes, both of which are removed from the circulation at a decreased rate.

6. *Liposome-Entrapped Enzymes in the Treatment or Prevention of Disease*

A liposome-entrapped enzyme may act on its substrate either following the liberation of the enzyme by the disruption of the lipid bilayers or, in the case of low-molecular-weight substrates, after the inward penetration of the bilayers by the substrate itself. Both ways of enzyme action can find application in therapy (Gregoriadis, 1976a).

6.1. *Inherited Enzyme Deficiencies*

Enzyme replacement therapy is presently the recognized therapeutic approach for most of these disorders. However, the use of enzymes is related to a number of theoretical as well as practical problems (Gregoriadis, 1973b). As already mentioned, repeated administration of a foreign enzyme to the body can often elicit antibody response and dangerous allergic reactions. Further, there exists the possibility of enzyme action on substrates located in sites other than the target area. The action of sugar hydrolases, intended to act on lysosomal substrates, on glycoproteins of the blood is one example. In addition, enzymes can be inactivated prematurely in the blood or intracellularly by proteases.

Since liposomes are capable of transporting materials into the cell lysosomes (Figure 3A) treatment of lysosomal storage diseases, in which because of a lysosomal enzyme deficiency relevant substrates accumulate in the lysosomes of a variety of tissues, seems an attractive possibility. This is even more relevant to those disorders in which the afflicted tissues are the liver (parenchymal and Kupffer cells) and the spleen. In work with model lysosomal storage conditions, it has been shown that liposomes can indeed help in introducing hydrolases into the target area in need of the enzyme. For example, exposure of sucrose-loaded cells to liposome-entrapped invertase was followed by the localization of liposomes in the lysosomes (which contained the sucrose) and the release of the entrapped invertase, which could then hydrolyze sucrose to glucose and fructose, both diffusible through the lysosomal membranes. Because of the absence of free invertase activity in the media during the incubation period, it was postulated that the entrapped enzyme was introduced into the cell lysosomes via endocytosis of the liposome carrier (Gregoriadis and Buckland, 1973). Similar results were obtained with fibroblasts from Pompe's disease patients, the glycogen contents of which was diminished upon exposure to amyloglucosidase-containing liposomes (Scherphof *et al.,* 1975) and with dextran-loaded rats, in which treatment with liposome-associated dextranase led to considerable loss of dextran from the liver (Colley and Ryman, 1974).

Application of this principle in therapy is, however, certain to run into serious difficulties. For instance, it is unlikely that straightforwardly made liposomal enzyme preparations will have any beneficial effect in patients with neural tissue (e.g., Tay-Sach's disease) and cardiac or skeletal muscle (e.g., Pompe's disease) involvement. These tissues exhibit little endocytic acitvity, and uptake is deemed to be negligible unless modified liposomes capable of homing to specific cells are made. Even so, it would most probably be necessary to delay hepatic uptake of the bulk of the injected dose for as long as possible so as to allow such liposomes to penetrate endothelial and other membranes surrounding diseased cells. In addition to this, it is possible that even with actively endocytosing tissues, loaded lysosomes of storage disease cells might be unable to fuse properly with liposome-containing endocytic vacuoles and acquire the missing enzyme. Another obstacle in enzyme replacement therapy is the unavailability of some therapeutic enzymes. In Gaucher's disease, for instance, the missing glucocerebroside:β-glucosidase can only be replaced by the same enzyme extracted from normal human tissues. Purification (Brady *et al.,* 1965) of the enzyme from the tissue of choice, human placenta, is time consuming and cumbersome, but affinity chromatography techniques (Braidman and Gregoriadis, 1976) have shortened the time required and also yielded realistic quantities of the partially purified enzyme, of which more than 50% can be entrapped in liposomes composed of egg lecithin, cholesterol, and phosphatidic acid (Braidman and Gregoriadis, 1976). A further difficulty encountered with injected enzymes of foreign origin is the development of antienzyme antibodies which can subsequently interact with their antigen and have a dually detrimental effect on therapy, namely, inactivation of the enzyme in the circulation and, as already mentioned, development of anaphylactic reactions. As it happens, both these untoward effects can be circumvented with liposome-entrapped enzymes, thanks to the protective action of lipid bilayers, which prevent contact of the extraliposomal circulating immunoglobulins with the intraliposomal antigen (Gregoriadis and Allison, 1974; Neerunjun and Gregoriadis, 1976).

Assuming that the liposome-entrapped enzyme does eventually find its way into the target cell lysosomes, its task in degrading materials within the lysosomes could be jeopardized by yet another threat, namely, the lysosomal proteases, which will, sooner or later, degrade the enzyme. A solution to this problem might be to increase the survival of the administered enzyme by appropriate modification of its structure. Work in this laboratory has shown that extension of the intralysosomal life span of enzymes can be brought about by cross-linking liposome-entrapped enzymes with glutaraldehyde, which seems capable of penetrating the lipid bilayers of liposomes (Gregoriadis and Neerunjun, 1974).

According to Figure 3B, liposomes could, under appropriate conditions, fuse with cells *in vivo* and transport enzymes in the cell cytoplasm, an action that could perhaps contribute to the correction of enzyme deficiencies affecting other cell substructures, such as endoplasmic reticulum and Golgi apparatus. Finally, reverse action of liposomes (i.e., penetration of substrates of low molecular weight into the lipid bilayers containing respective enzymes) could prove important in the treatment of certian metabolic disorders, such as phenylketonuria and galactosemia. It is hoped that administration of liposome-entrapped enzymes capable of transforming unwanted materials (e.g., phenylalanine and galactose-1-phosphate) in the extracellular space to metabolically acceptable ones could be helpful to patients, assuming of course that penetration of the substrates into the liposomes containing the enzymes occurs rapidly. Furthermore, intragastric administration of liposomal enzymes, shown to be capable of reaching the blood circulation (Dapergolas *et al.,* 1976), should further facilitate chronic enzyme treatment (Figure 5).

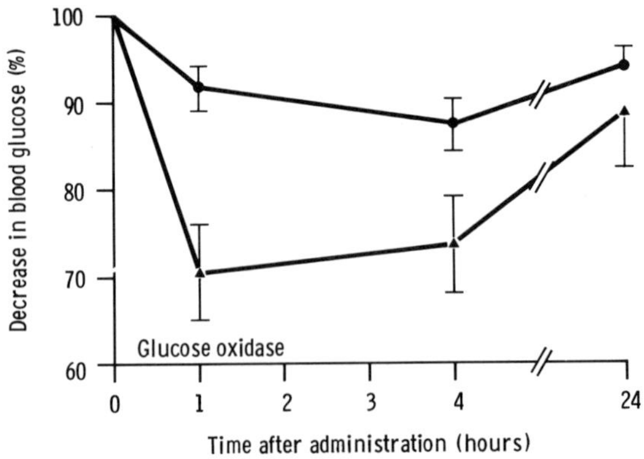

FIGURE 5. Rats were given intragastrically 25 mg of glucose oxidase, free (circles) or entrapped in phosphatidyl inositol liposomes. Blood glucose levels are expressed as percent (± standard error) of those measured before treatment. (From Dapergolas *et al.*, 1976.)

6.2. *Neoplasms*

The survival of certain malignant cells is dependent upon the exogenous supply of L-asparagine. In animal models, administration of asparaginase leads to low asparagine levels in the circulation, rapid tumor regression, and complete cure (Broome, 1968). The use of asparaginase in the treatment of acute lymphocytic leukemia in man has, however, met with only moderate success, and aggressive or long-term treatment can be hampered by anaphylactic and other toxic reactions and also by the development of resistance to the enzyme (Capizzi *et al.*, 1970). Possible advantages offered by liposome-entrapped *Erwinia* asparaginase have been investigated in this laboratory using C3H mice bearing 6C3HED asparagine-sensitive lymphoma cells. Results obtained (Neerunjun and Gregoriadis, 1976) suggest that both free and liposome-entrapped enzymes are active in effecting complete cures, which are preceded by decreased levels of asparagine in the blood and tumor tissue. This implies that asparagine can penetrate the lipid bilayers of circulating liposomes to reach the entrapped enzyme. However, penetration of liposomes into the tumor cells (e.g., see Table 2), and subsequent action of the enzyme on asparagine within the cells, cannot be excluded. In hypersensitivity experiments, whereas intravenous injection of the free enzyme to preimmunized mice led to anaphylactic shock and death, similar mice injected with entrapped asparaginase did not show any signs of illness (Table 3). Further, preimmunized tumor-bearing mice injected with a therapeutic dose of nonentrapped asparaginase also died but again, treatment of such animals with the entrapped asparaginase did not produce anaphylactic shock (Neerunjun and Gregoriadis, 1976). It is apparent, therefore, that surrounding of the enzyme by the lipid layers of liposomes prevents antiasparaginase antibodies in the blood from meeting their antigen (asparaginase), thus rendering the use of asparaginase in cancer treatment immunologically safer. It is not yet known what effect presentation of the enzyme in the entrapped form will have on the development of resistance to treatment.

TABLE 3
Serum Sickness in Preimmunized Mice Injected Intravenously with Erwinia Asparaginase

Number of mice	Treatment	Serum sickness	Dead
10	Free asparaginase, 1000 IU	3	7
10	Entrapped asparaginase, 1000 IU	0	0

6.3. *Diabetes Mellitus*

Efforts in this laboratory for the introduction of liposome-associated insulin into the blood circulation by the intragastric route have met with considerable success (Dapergolas and Gregoriadis, 1976). Of particular interest, perhaps, is the observation that intragastrically given insulin associated with dipalmitoyl lecithin liposomes is capable of lowering blood glucose levels during a 24-hr period to about 50% of those before treatment (Figure 6). In view of both the ramifications which introduction of insulin into the portal circulation could have on the progress of

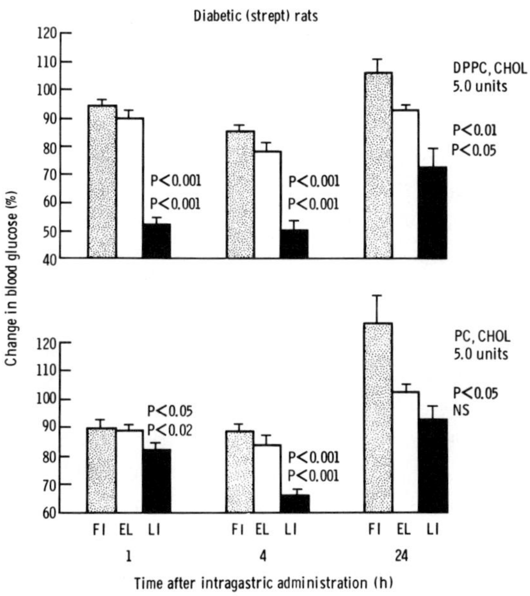

FIGURE 6. Diabetic rats were treated intragastrically with 5 units of insulin, free (shaded bars), or entrapped in liposomes (filled bars) composed of dipalmitoyl phosphatidylcholine (DPPC) and cholesterol or egg phosphatidylcholine (PC) and cholesterol, or with the respective liposomes without insulin (open bars.) Values are % ± S.E. of glucose levels measured before treatment in individual rats. P is probability of significant difference between groups treated with entrapped and free insulin (upper) and between groups treated with entrapped insulin and liposomes devoid of insulin (lower) (From Dapergolas and Gregoriadis, 1976).

diabetic patients, and the avoidance of the daily intramuscular injection, the importance of liposomes as potential carriers of the hormone and of other similar substances cannot be overestimated.

6.4. *Treatment of Local Disease*

Work on the fate of locally injected liposome-entrapped proteins has shown that while small (sonicated) liposomes are rapidly cleared from the site of injection (rat testicle) to enter the circulation or the relevant draining lymph nodes, large (unsonicated) liposomes persist in the injected area, where they eventually disintegrate and slowly release entrapped materials (Segal *et al.,* 1975). Locally injected liposomes could therefore prove useful in the treatment of local diseases. An example is gout, in which because of increased uric acid levels in the plasma and other pools of the body, leucocytes interact with crystals of monosodium urate monohydrate to give rise to inflammatory episodes. It would be of interest to see whether local application of uricase entrapped in large liposomes would have any beneficial effect. Furthermore, as the liver is the main site of uric acid synthesis, it would be worthwhile to investigate the effect of a locally injected mixed population of small uricase-containing liposomes, which are expected to diffuse rapidly and reach the liver, and of large uricase-containing liposomes, which would slowly release uricase for local action.

6.5. *Liposomes as Immunological Adjuvants*

a. Vaccines. Recent work suggests that the need for a safe and effective adjuvant for use in human and animal immunization programs could be satisfied by liposomes (Allison and Gregoriadis, 1974). Thus, antibody response in mice challenged intramuscularly, intravenously, and subcutaneously in the foot pad with diphtheria toxoid entrapped in negatively charged liposomes was several-fold greater than when the antigen was given in the free form (Table 4). In contrast, diphtheria toxoid entrapped in positively charged liposomes elicited less antibody than the same dose of free antigen. In this and other experiments, the adjuvant effect of liposomes with dicetyl phosphate in their structure was superior to that of phosphatidic acid liposomes. Although entrapment of antigens in liposomes facilitates the increase of antibody response at the same time and for the reasons discussed above, it offers protection from allergic reactions. For instance, mice immunized with diphtheria toxoid, when challenged by footpad injection of the same antigen, showed Arthus reaction at 4 hr, and there was no delayed hypersensitivity 24–72 hr later. Similarly, immune mice injected intravenously with free diphtheria toxoid developed serum sickness reaction and died of anaphylactic shock. Most significantly, no Arthus reaction or serum sickness were elicited in immune mice treated with liposome-entrapped diphtheria toxoid (Gregoriadis and Allison, 1974).

In other experiments, the effect of combined liposome-entrapped antigen and bacterial or chemical adjuvants on antibody response was studied (Allison and Gregoriadis, 1976). It was found that administration of liposomal diphtheria

TABLE 4

Serum Antibody Responses of Mice to Diphtheria Toxoid Administered Free or in Liposomes of Different Compositions[a]

Experimental group	Number of mice	Mode of administration	Route of administration	Primary Ab response	Secondary Ab response	Probability
a	15	Free	Intravenous	1.8	—	avb $p < 0.01$
b	15	−Liposomes (PA)	Intravenous	10.0	—	
c	6	Free	Subcutaneous	2.5	11.6	cvd $p < 0.01$
d	6	−Liposomes (PA)	Subcutaneous	6.7	13.3	dve $p < 0.01$
e	6	+Liposomes	Subcutaneous	0.0	11.3	cve $p < 0.10$
f	6	Free	Intramuscular	1.7	7.5	fvg $p < 0.10$
g	6	−Liposomes (PA)	Intramuscular	3.7	11.0	gvh $p < 0.05$
h	6	−Liposomes (DP)	Intramuscular	6.6	12.0	fvh $p < 0.01$

[a]Mice were of the TO strain, except in groups c–e, which were of the CBA strain. Positive (+) liposomes (0.5 mg of lipid) were composed of egg lecithin, cholesterol, and stearylamine in molar ratios 7:2:1; in negative (−) liposomes phosphatidic acid (PA), or dicetylphosphate (DP)-replaced stearylamine. Primary serum antibody responses were measured after 13 or 14 days; booster injections with 20 μg of antigen in the same form were given and secondary responses measured after an additional 10 days (groups c–e) or 14 days (groups f–h). Antibody responses were measured by indirect hemagglutination and expressed as the $\log_2$ IH titer (from Allison and Gregoriadis, 1974).

toxoid, together with heat-killed *Mycobacterum tuberculosis* or *Bordetella pertussis,* resulted in significantly increased antibody production. An antibody response similarly superior to that obtained with entrapped diphtheria toxoid alone was also obtained with entrapped diphtheria toxoid mixed with saponin. Further, and in contrast to conventional adjuvants, no granulomas were formed at the sites of injection of liposomes (Table 5).

In recent unpublished work (A. C. Allison and G. Gregoriadis) we have succeeded in obtaining enhanced immune response in guinea pigs immunized with liposome-entrapped cold virus hemagglutinin and neuraminidase, and antibody titers were further increased by mixing such liposomes with traces of saponin. Indeed, work by others (Almeida *et al.,* 1975) has shown that influenza virus subunits relocate on the liposomal surface in a fashion similar to that seen on the intact virus (Figure 7).

b. Immunotherapy of Tumors. An important application of the adjuvant effect of liposomes could be in cancer immunotherapy. Purification of tumor antigens and subsequent entrapment in liposomes could lead to increased specific immune

TABLE 5

Effect of Liposome Entrapment of Diphtheria Toxoid (DT) with or without Bordetella pertussis Organisms (BP), BCG, and Saponin on Primary and Secondary Immune Responses[a]

Mode of antigen administration	Primary response (IHA)	Probability	Secondary response (IHA)	Probability
Experiment 1				
a Free DT	8.0		12.0	
b Free DT + BP	10.2	avb $p < 0.01$	15.4	avb $p < 0.01$
c Liposomal DT	10.0	avc $p < 0.01$	15.6	
d Liposomal DT + BP	11.5	avd $p < 0.01$	19.7	avd $p < 0.01$
Experiment 2				
e Free DT	0.0		12.6	
f Free DT + BCG	0.0		14.4	
g Liposomal DT	1.4		12.8	
h Liposomal DT + BCG	7.2	gvh $p < 0.01$	17.4	gvh $p < 0.01$
Experiment 3				
j Free DT	0.0		5.4	
k Liposomal DT	0.0		7.5	
l Liposomal DT + 3 μg of saponin	0.0		6.5	
m Liposomal DT + 50 μg of saponin	0.0		12.6	$p < 0.01$

[a] In three separate experiments, groups of five VSBS/NIMR mice were inoculated with 20 μg of diphtheria toxoid (DT) free or entrapped in liposomes composed of egg lecithin, cholesterol, and dicetyl phosphate. Both preparations of diphtheria toxoid were also given with or without *Bordetella pertussis,* BCG, or saponin as shown. Mice were bled 13–14 days after the primary challenge, reinoculated, and bled again after an additional 10–13 days. Antibody responses were measured by indirect hemagglutination and expressed as the $\log_2$ IH titer (modified from Allison and Gregoriadis, 1976).

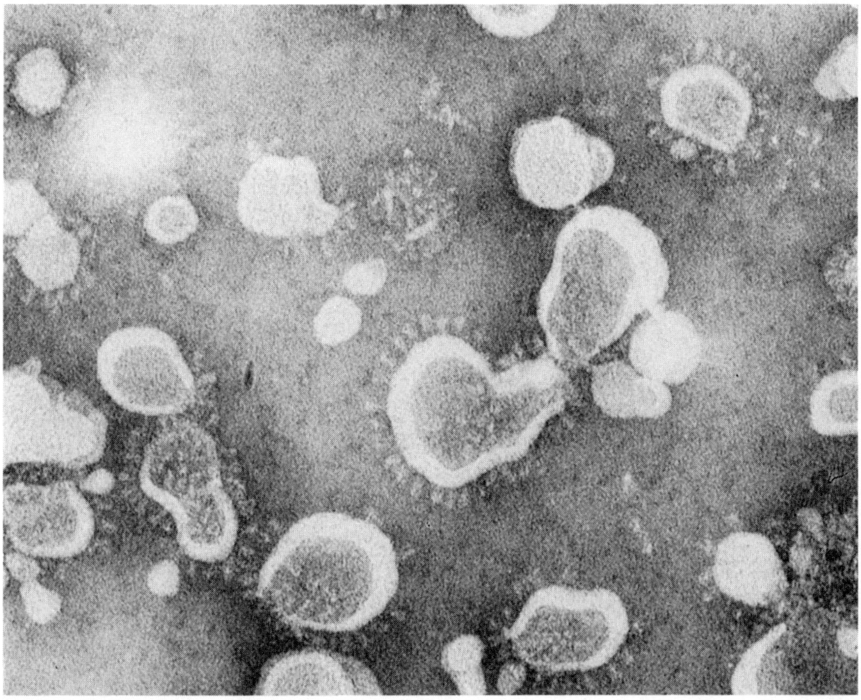

FIGURE 7. Virus subunits assayed on the surface of liposomes after sonication to form virosomes. x220,000. (From Almeida *et al.*, 1975.)

response. This could be accentuated by mixing liposomes with a nontoxic adjuvant, which would not only increase the specific immune response to the tumor antigens but would also enhance macrophage-mediated nonspecific immunity against tumors.

ACKNOWLEDGMENTS

I am grateful to Dr. A. C. Allison, C. D. V. Black, Dr. Isobel Braidman, G. Dapergolas, Diane E. Neerunjun, Susan Scott, Dorothy Seale, and Dr. E. Wisse for collaboration, and to Dr. June D. Almeida for allowing me to reproduce Figure 7.

7. *References*

Allen, C. S. S. J., 1974, Purification and properties of a phagocytosis promoting factor from serum, thesis, University of Illinois.

Allison, A. C., and Gregoriadis, G., 1974, Liposomes as immunological adjuvants, *Nature* **252**:252.

Allison, A. C., and Gregoriadis, G., 1976, Liposomes as immunological adjuvants, in: *Recent Results in Cancer Research* (G. Mathe and M. Sinnel, eds.), Springer-Verlag, Heidelberg.

Almeida, J. D., Brand, C. M., Edwards, D. C., and Heath, T. D., 1975, Formation of virosomes from influenza subunits and liposomes, *Lancet* **2**:899.

Bangham, A. D., Hill, M. W., and Miller, N. G. A., 1974, Preparation and use of liposomes as models of biological membranes, in: *Methods in Membrane Biology* (E. D. Korn, ed.), pp. 1–68, Plenum Press, New York.

Batzri, S., and Korn, E. D., 1975, Interaction of phospholipid vesicles with cells, *J. Cell Biol.* **66**:621.

Black, C. D. V., and Gregoriadis, G., 1976, Interaction of liposomes with blood plasma proteins, *Biochem. Soc. Trans.* **4**:253.

Brady, R. O., Kanfer, J. N., and Shapiro, D., 1965, The metabolism of glucocerebrosides; I. Purification and properties of a glucocerebroside-clearing enzyme from spleen tissue, *J. Biol. Chem.* **240**:39.

Braidman, I., and Gregoriadis, G., 1976, Preparation of glucocerebroside:β-glucosidase for entrapment in liposomes and treatment of patients with adult Gaucher's disease, *Biochem. Soc. Trans.* **4**:259.

Broome, J. D., 1968, Studies on the mechanism of tumour inhibition by L-asparaginase, *J. Exptl. Med.* **127**:1055.

Bruni, A., Leon, A., Toffano, G., and Boarato, E., 1976, Pharmacological effects of phosphatidyl serine liposomes, *Nature* **260**:331.

Capizzi, R. L., Bertino, J. R., and Handschumacher, R. E., 1970, L-Asparaginase, *Ann. Rev. Med.* **21**:433.

Colley, C. M., and Ryman, B. E., 1974, A model for a lysosomal storage disease and a possible method of therapy, *Biochem. Soc. Trans.* **2**:871.

Dapergolas, G., and Gregoriadis, G., 1976. Hypoglycaemic effect of liposome-entrapped insulin administered intragastrically into rats, *Lancet* **2**:824.

Dapergolas, G., Neerunjun, E. D., and Gregoriadis, G., 1976, Penetration of target areas in the rat by liposome-associated bleomycin, glucose oxidase and insulin, *FEBS letters* **63**:235.

de Barsy, T., Devos, P., and Van Hoof, F., 1975, The cellular distribution of liposomes in the liver of newborn rats, *Biochem. Soc. Trans.* **3**:159.

de Duve, C., de Barsy, T., Poole, B., Trouet, A., Tulkens, P., and Van Hoof, F., 1974, Lysosomotropic agents, *Biochem. Pharmacol.* **23**:2495.

Dunnick, J. K., McDougall, I. R., Aragon, S., Goris, M. L., and Kriss, J. P., 1975, Vesicle interactions with polyamino acids and antibody: *in vitro* and *in vivo* studies, *J. Nucl. Med.* **16**:483.

Eytan, G., Matheson, M. J., and Racker, E., 1975, Incorporation of biologically active proteins into liposomes, *FEBS Letters* **57**:121.

Fishman, Y., and Citri, N., 1975, L-Asparaginase entrapped in liposomes: preparation and properties, *FEBS Lett.* **60**:17.

Freer, J. H., Arbuthnott, J. P., and Bernheimer, A. W., 1968, Interaction of staphylococcal α-toxin with artificial and natural membranes, *J. Bacteriol.* **95**:1153.

Grant, C. W. M., and McConnell, H. M., 1973, Fusion of phospholipid vesicles with viable *Acholeplasma laidlawii, Proc. Natl. Acad. Sci. U.S.* **70**:1238.

Gregoriadis, G., 1973a, Drug entrapment in liposomes, *FEBS Letters* **36**:292.

Gregoriadis, G., 1973b, Molecular trojan horses, *New Scientist* **60**:890.

Gregoriadis, G., 1974, Structural requirements for the specific uptake of macromolecules and liposomes by target tissues, in: *Enzyme Replacement Therapy of Lysosomal Storage Diseases* (J. M. Tager, G. J. M. Hooghwinkel, and W. Th. Daems, eds.), pp. 131–148, North-Holland Publishing Co., Amsterdam.

Gregoriadis, G., 1975, Catabolism of glycoproteins, in: *Lysosomes in Biology and Pathology* (J. T. Dingle and R. T. Dean, eds.), pp. 265–294, North-Holland Publishing Co., Amsterdam.

Gregoriadis, G., 1976a, The carrier potential of liposomes in biology and medicine. *New Engl. J. Med.* **295**:704, 765.

Gregoriadis, G., 1976b, Enzyme entrapment in liposomes, in: *Methods in Enzymology* (K. Mosbach, ed.), Academic Press, Inc., New York.

Gregoriadis, G., and Allison, A. C., 1974, Entrapment of proteins in liposomes prevents allergic reactions in pre-immunised mice, *FEBS Letters* **45**:71.

Gregoriadis, G., and Buckland, R. A., 1973, Enzyme-containing liposomes alleviate a model for storage disease, *Nature* **244**:170.

Gregoriadis, G., and Neerunjun, E. D., 1974, Control of the rate of hepatic uptake and catabolism of liposome-entrapped proteins injected into rats: possible therapeutic applications, *Eur. J. Biochem.* **47**:179.

Gregoriadis, G., and Neerunjun, E. D., 1975, Homing of liposomes to target cells, *Biochem. Biophys. Res. Commun.* **65**:537.

Gregoriadis, G., and Ryman, B. E., 1972a, Fate of protein-containing liposomes injected into rats: an approach to the treatment of storage diseases, *Eur. J. Biochem.* **24**:485.

Gregoriadis, G., and Ryman, B. E., 1972b, Lysosomal localization of β-fructofuranosidase-containing liposomes injected into rats: some implications in the treatment of genetic disorders, *Biochem. J.* **129**:123.

Gregoriadis, G., Leathwood, P. D., and Ryman, B. E., 1971, Enzyme entrapment in liposomes, *FEBS Letters* **14**:95.

Gregoriadis, G., Putman, D., Louis, L., and Neerunjun, D., 1974a, Comparative fate and effect of non-entrapped and liposome-entrapped neuraminidase injected into rats, *Biochem. J.* **140**:323.

Gregoriadis, G., Swain, C. P., Wills, E. J., and Tavill, A. S., 1974b, Drug-carrier potential of liposomes in cancer chemotherapy, *Lancet* **1**:1313.

Gregoriadis, G., Dapergolas, G., and Neerunjun, E. D., 1976, Penetration of target areas in the rat by liposome-associated agents administered parenterally and intragastrically, *Biochem. Soc. Trans.* **4**:256.

Huang, L., and Pagano, R. E., 1975, Interaction of phospholipid vesicles with cultured mammalian cells: I. Characteristics of uptake, *J. Cell Biol.* **67**:38.

Jonah, M. M., Cerny, E. A., and Rahman, Y. E., 1975, Tissue distribution of EDTA encapsulated within liposomes of varying surface properties, *Biochim. Biophys. Acta* **401**:336.

Juliano, R. L., and Stamp, D., 1975, The effect of particle size and charge on the clearance rates of liposomes and liposome encapsulated drugs, *Biochem. Biophys. Res. Commun.* **63**:651.

Inbar, M., and Shinitzky, M., 1974, Increase of cholesterol level in the surface membrane of lymphoma cells and its inhibitory effect on ascites tumour development, *Proc. Natl. Acad. Sci. U.S.* **71**:2128.

Kawada, J., Kuwae, T., and Kurata, M., 1974, Interaction between thyroglobulin liposome and thyroidal lysosome, *Life Sci.* **13**:613.

Kimelberg, H. K., Mayhew, E., and Papahadjopoulos, D., 1975, Distribution of liposome-entrapped cations in tumor-bearing mice, *Life Sci.* **17**:715.

Leon, A., and Toffano, G., 1976, *Function and Metabolism of Phospholipids in Central Nervous System* (G. Porcellati, ed.), Plenum Press, New York.

Magee, W. E., and Miller, O. V., 1972, Liposomes containing antiviral antibody can protect cells from virus infection, *Nature* **235**:339.

Magee, W. E., Goff, C. W. Schoknecht, J., Smith, M. D., and Cherian, K., 1974, The interaction of cationic liposomes containing entrapped horseradish peroxidase with cells in culture, *J. Cell Biol.* **63**:492.

Mantovani, P., Pepeu, G., and Amaducci, L., 1976, *Function and Metabolism of Phospholipids in Central Nervous System* (G. Porcellati, ed.), Plenum Press, New York.

Neerunjun, E. D., and Gregoriadis, G., 1976, Tumour regression with liposome-entrapped asparaginase: some immunological advantages, *Biochem. Soc. Trans.* **4**:133.

Pagano, R. E., and Huang, L., 1974, Interaction of phospholipid vesicles with cultured mammalian cells, *Nature* **252**:166.

Pagano, R. E., and Huang, L., 1975, Interaction of phospholipid vesicles with cultured mammalian cells: II. Studies of mechanism, *J. Cell Biol.* **67**:49.

Papahadjopoulos, D., and Miller, N., 1967, Phospholipid model membranes: I. Structural characteristics of hydrated liquid crystals, *Biochim. Biophys. Acta* **135**:624.

Papahadjopoulos, D., Poste, G., and Mayhew, E., 1974a, Cellular uptake of cyclic AMP captured within phospholipid vesicles and effect on cell growth behaviour, *Biochim. Biophys. Acta* **363**:404.

Papahadjopoulos, D., Mayhew, E., Poste, G., 1974b, Incorporation of lipid vesicles by mammalian cells provides a potential method for modifying cell behaviour, *Nature* **252**:163.

Patel, H. M., and Ryman, B. E., 1974, α-Mannosidase in zinc-deficient rats: possibility of liposomal therapy in mannosidosis, *Biochem. Soc. Trans.* **2**:1014.

Patel, H. M., and Ryman, B. E., 1976, Oral administration of insulin by encapsulation within liposomes, *FEBS Letters* **62**:60.

Rogers, J. C., and Kornfeld, S., 1971, Hepatic uptake of proteins coupled to fetuin glycopeptide, *Biochem. Biophys. Res. Commun.* **45**:622.

Scherphof, G. L., Roerdink, F. H., and Zborowski, J., 1975, The use of liposomes to facilitate uptake of external substances by living cells: possible therapeutic applications, *Histochem. J.* **7**:508.

Segal, A. W., Wills, E. J., Richmond, J. E., Slavin, G., Black, C. D. V., and Gregoriadis, G., 1974, Morphological observations on the cellular and subcellular destination of intravenously administered liposomes, *Br. J. Exptl. Pathol.* **55**:320.

Segal, A. W. Gregoriadis, G., and Black, C. D. V., 1975, Liposomes as vehicles for the local release of drugs, *Clin. Sci. Mol. Med.* **49**:99.

Sessa, G., and Weissmann, G., 1970, Incorporation of lysozyme into liposomes, *J. Biol. Chem.* **245**:3295.

Surolia, A., Bachhawat, B. K., and Podder, S. K., 1975, Interaction between lactin from *Ricinus communis* and liposomes containing gangliosides, *Nature* **257**:802.

Weissmann, G., Brand, A., and Franklin, E. C., 1974, Interaction of immunoglobulins with liposomes, *J. Clin. Invest.* **53**:536.

Weissmann, G., Bloomgarden, D., Kaplan, R., Cohen, C., Hoffstein, S., Collins, T., Gotlieb, A., and Nagle, D., 1975, A general method for the introduction of enzymes, by means of immunoglobulin-coated liposomes into lysosomes of deficient cells, *Proc. Natl. Acad. Sci. U.S.* **72**:88.

Wisse, E., 1970, An electron microscopic study of the fenestrated endothelial lining of rat liver sinusoids, *J. Ultrastruct. Res.* **31**:125.

Wisse, E., and Gregoriadis, G., 1975, The uptake of liposomes by the rat liver RES., *J. Reticuloendothelial Soc.* **18**:10a.

Wisse, E., Gregoriadis, G., and Daems, W. T., 1976, The uptake of liposomes by the rat liver, in: *Proceedings of the 7th International Congress of the Reticuloendothelial Society* (M. R. Escobar, and S. Friedman, eds.), Plenum Press, New York.

15

Enzyme-Loaded Erythrocytes

Garret Ihler and Robert Glew

1. *Introduction*

Erythrocytes, especially human erythrocytes, have been very extensively studied because of their medical interest and because they are conveniently obtained for biochemical and physiological studies. Mammalian erythrocytes lack nuclei which are pinched off from the cells when they pass through narrow orifices while escaping from the bone marrow. Once matured, they lack ribosomes and have lost the capability of protein synthesis. Because of their inability to renew themselves, the lifetime of normal human erythrocytes is restricted to about 120 days. Cell energy metabolism is restricted to glycolysis since the cells do not contain mitochondria. Their basic function is to serve as a container for hemoglobin. In this article we discuss a procedure for utilizing erythrocytes as containers for foreign proteins and enzymes as well as hemoglobin.

2. *Enzyme-Loaded Ghosts*

If erythrocytes are suspended in a hypotonic solution, water enters more rapidly than small ions leave and hence the cell begins to swell. The erythrocyte first loses its biconcave shape and becomes a spherocyte and then, when its volume exceeds a critical level, holes or tears appear in the membrane. These holes are about 20–50 nm in diameter (Seeman, 1967) and are sufficiently large to allow proteins such as hemoglobin to escape. If salts are added back to the erythrocyte suspension to restore the isotonic state, the holes reseal and the erythrocyte becomes osmotically competent again.

Several investigators (Baker, 1967; Hoffman, 1958; Marsden and Ostling, 1959, Seeman, 1967) have shown that high-molecular-weight substances, including hemoglobin, ferritin, dextran, and colloidal gold, can enter the erythrocyte

Garret Ihler • Department of Biochemistry, University of Pittsburgh School of Medicine, Pittsburgh, Pa. Robert Glew • Department of Biochemistry, University of Pittsburgh School of Medicine, Pittsburgh, Pa.

through these holes during hemolysis. This suggested to us that enzymes might also enter the cell while the holes are open, providing a way of trapping foreign enzymes inside the cell. In the case of human diseases, such cells might be therapeutically useful in two types of situations. In the case of diseases involving the accumulation of intracellular substrates in phagocytic cells, the enzyme could be introduced into phagocytic cells during phagocytosis of the enzyme-containing erythrocyte. Alternatively, it should be possible to degrade circulating small molecules by circulaing enzyme-loaded cells, if the cell is permeable to the substrate.

We have shown that enzymes can be entrapped in red cells by hypotonic hemolysis and that the enzymes are active within the red cell (Ihler *et al.*, 1973). Erythrocytes are washed by sedimentation several times and suspended in 0.9% NaCl. The enzyme is dissolved in a low-ionic-strength buffer or alternatively in a dialyzed, membrane-free extract of red cells and the erythrocytes are diluted 1:3–1:5 into the enzyme-containing solution. Enzyme entry occurs in parallel with the escape of hemoglobin. Although maximum uptake is obtained 60 sec after dilution, in most experiments sodium chloride is added back at 30 sec to minimize damage to the erythrocytes. We usually monitor the hematocrit and the percent release of hemoglobin as convenient and simple parameters of cellular hemolysis.

If hemolysis is allowed to continue sufficiently long, the concentration of macromolecules is equal inside and outside the cell (Hoffman, 1958). Under our conditions of loading, the concentration of intracellular enzyme is about 30% of the theoretical equilibrium value after 30 sec. Failure to reach equilibrium does not present any serious difficulty, since the concentration of enzyme in the cell can be made arbitrarily high by merely increasing the extracellular concentration. Unincorporated enzyme can be recovered for reuse and repurified if desired. Under these nonequilibrium conditions, the holes in the membrane act as molecular sieves, and smaller proteins are preferentially incorporated. However, there is no difficulty in incorporating large proteins since *E. coli* β-galactosidase (540,000 daltons) is readily entrapped.

2.1. *Metabolic Properties of Ghosts*

There is an extensive literature on the properties of erythrocyte ghosts. In referring to this literature, it should be kept in mind that most investigators deliberately removed the contents of the erythrocyte as completely as possible. Schrier (1963) for example, showed that the level of all the glycolytic enzymes in ghosts is extremely low. One might erroneously anticipate that ghosts would be metabolically incompetent and hence incapable of survival in the circulation. However, Schrier's cells were deliberately washed extensively, and so we assumed that if the erythrocytes are diluted only 1:3, the level of intracellular enzymes should be at least one-third of normal, and probably higher.

Hemolyzed ghosts made by our initial procedure are capable of converting glucose to lactate at about 40% of the rate of unhemolyzed ghosts (Ihler *et al.*, 1975). This rate can be stimulated somewhat by supplementing the ghosts with hexokinase, adenine, and guanine. The ghosts are also capable of transporting sodium and potassium via the ouabain-sensitive ATP-requiring transport system using either glucose or inosine as energy sources at rates comparable to unhemolyzed erythrocytes. Thus, even without taking any special measures, ghosts prepared in this manner are nearly as competent metabolically as normal cells. However, their degree of normality can be further improved by using as the hypotonic medium a dialyzed extract of erythrocytes in which the concentration of the endogenous enzymes is equal to that of red cells (i.e., the extract is made by freezing and thawing packed erythrocytes and dialyzing the extract without dilution). Under these conditions, the intracellular concentration of erythrocyte enzymes should not change significantly during hypotonic hemolysis. It would, of course, be desirable to supplement the red cell extract with physiological concentrations of metabolites and cofactors.

2.2. *Transport of Substrates into Erythrocyte Ghosts*

One advantage of artifical cells, constructed for example of nylon, is that the pore size of the membrane can be varied (Chang, 1972). The pore size of the artificial cell can be made large enough to permit any molecule the size of glucose, sucrose, amino acids, or acetylsalicylic acid to rapidly enter by diffusion. In contrast, erythrocytes are generally relatively impermeable except to those substrates which have a carrier protein for facilitated diffusion. Among the substrates which enter red cells via facilitated diffusion are glucose, all of the amino acids, and such purine derivatives as hypozanthine and uric acid. Erythrocytes, on the other hand, are impermeable to sucrose and the rate of entry into red cells, even of permeable substances, is relatively slow. For example, uric acid penetrates nylon artificial cells with an equilibration half-time of 42.5 sec and penetrates red cells with a half-time of about 20 min.

The rate at which a substrate is acted upon by a particular enzyme will be dependent first on the rate at which the substrate enters the red cell, and second on the concentration of the enzyme within the red cell. Either of these factors may constitute the rate-limiting step, but in general it is possible to load the cell with a sufficient concentration of enzyme such that the rate-limiting step is entry rather than enzymatic reaction.

The rate of entry is, in turn, dependent on two factors, the extracellular concentration and the intrinsic rate constant for entry. The rate constant for entry of uric acid is 0.07 per min and the normal uric acid concentration in human plasma is 0.3 mM (5 mg%). When the intracellular substrate concentration is maintained near zero by an excess of enzyme, the rate of entry is maximal. For uric acid this would correspond to 21 μmole/min/liter red cells, which is an acceptable rate of entry. In contrast, the rate constant for asparagine uptake is 0.01 per min (Purpura *et al.*, unpublished results) and its concentration in blood is 4×10^{-6} M, so the rate of entry would be only 0.04 μmole/min/liter of red cells. This number is sufficiently low that red cells loaded with asparaginase would probably be useless for the treatment of asparaginase-sensitive leukemias. Direct injection of asparaginase, on the other hand, is routinely used in the treatment of leukemia and is successful because the enzyme can directly react with the substrate without waiting for it to be transported across a membrane.

2.3. *Degradation of Substrates in Ghosts*

At high concentrations of intracellular enzyme, the enzyme will be in excess, so the reaction rate will be limited by the rate of influx into the cell. At low concentrations, enzyme will be limiting, so the velocity of the reaction will be proportional to the enzyme concentration in the cell. The maximum effective enzyme concentration occurs when the enzyme is neither limiting nor in excess. This concentration of enzyme can be estimated using the formula.

$$\frac{dCi}{dt} = 0 = k(Co - Ci) - \frac{(\text{units enzyme})Ci}{K_m + Ci}$$

At equilibrium, the rate of influx of substrate will be matched by the rate of its enzymatic degradation.

We have experimentally verified this concept using uricase (Ihler *et al.*, 1975). When low levels of enzyme are employed, the rate is dependent on the enzyme concentration. When high levels are employed, the rate reaches a maximum which is identical with the maximal rate of entry of the substrate.

We have also verified that the reaction actually occurs within the red cell rather than being due to extracellular reaction by enzyme which might have leaked from the red cells. Entry of uric acid into erythrocytes is competitively inhibited by hypozanthine, and we have shown that 2 mM hypozanthine decreases the rate of reaction by enzyme-loaded erythrocytes in parallel with the decrease in rate of entry of uric acid, but does not have any effect on the reaction catalyzed by free enzyme.

It is possible that some enzymes might be inactive within erythrocytes, as a result of the presence of an inhibitor, absence of a required cofactor, insufficient regeneration of cofactor, instability of the enzyme due to an unfavorable internal environment, or product inhibition due to a slow rate of diffusion of the reaction product out of the erythrocyte. Most of these considerations are not especially worrisome, as enzyme isolated from any source probably can be used, since the antibody–antigen problem is in principle avoided by encapsulation. Generally, the properties of such enzymes will differ sufficiently that if enzyme from one source is unsatisfactory, enzyme from some other source will suffice. However, it is clearly desirable to use human enzymes, since they would be expected to be most compatible and since we have not yet experimentally ruled out the possibility of an antibody–antigen reaction.

Using uricase and uric acid as an example, we have calculated that about 150 ml of enzyme-loaded erythrocytes is capable of removing as much uric acid as the human kidneys. It should be emphasized again that the amount of red cells required to degrade some specified amount of substrate is not determined by the amount of enzyme per cell, but by the rate of entry of the substrate into the erythrocyte.

2.4. *Survival Characteristics of Ghosts*

Enzyme-loaded ghosts prepared as originally described (Ihler *et al.*, 1973) with hypotonic saline generally have poor survival characteristics. When injected into rabbits, rats, or chickens, the ghosts are rapidly cleared from the circulation, largely by the liver but to lesser extents by other organs as well (Ang *et al.*, unpublished results; Fiddler *et al.*., 1974). This is, of course, the desired result if the objective is to deliver the enzyme to reticuloendothelial cells (provided that the ghosts are sufficiently stable to avoid intravascular lysis, as has occurred occasionally). For other purposes it is essential that the cells have near-normal lifetimes.

Since *in vivo* survival studies are relatively difficult and time-consuming, we initially chose to utilize an *in vitro* assay involving active ion transport as a measure of cellular function. We have used both the entry of ^{45}RB and the exit of ^{23}Na introduced at the time of hemolysis) and have determined conditions that permit normal (oubaine-sensitive) active ion transport in human and chicken cells. Subsequent survival studies using catalase-loaded, ^{57}Cr-labeled chicken cells have resulted in survival times of 60–80% of unhemolysed control cells (Ang *et al.*, in preparation). Further experiments are in progress with various mammals to extend and confirm this result.

The loading conditions that we presently use are as follows. Two milliliters of washed erythrocytes (approximately 50% hematocrit) in 10 mM sodium phosphate buffer (pH 6.25) containing 0.9% NaCl is rapidly mixed with 7.2 ml of hypotonic medium at room temperature. The hypotonic medium consists of a dialyzed membrane-free extract of red cells prepared by sonication, to which is added the appropriate enzyme. After 30–45 sec, saline solution is added to a final concentration of 0.9%. Resealing is essentially instantaneous, as judged by the return of turbidity, but the suspension is allowed to incubate for an additional 5 min at room temperature. Inosine to 0.1 mM, guanosine to 0.1 mM, $MgCl_2$ to 0.2 mM, glucose to 10 mg/ml, and Tris buffer (pH 7.52) are then added. The cells are incubated at 37°C for 30 min with gentle shaking and then washed three times with 0.9% NaCl in 10 mM sodium phosphate buffer. We are currently attempting to further improve the survival characteristics by fortifying the hypotonic hemoglobin solution with glycolytic enzymes and small molecules such as glutathione and pyridine nucleotides.

For many purposes it is desirable to separate hemolysed cells from residual unlysed cells, and we accomplish this by equilibrium density centrifugation in ficol. Normal human cells have a calculated density of 1.2, which corresponds to about 40% ficol, whereas the calculated density for hemolyzed cells lacking most of the hemoglobin is about 1.04, which corresponds to about 10% ficol. These are also the values observed experimentally for their respective densities.

The entrapment of specific enzymes such as human liver β-glucocerebrosidase may require additional considerations. Solubilization of this enzyme requires detergent (deoxycholate), and so it is necessary to devise conditions compatible both with the enzyme and the red cell. We have found that relatively low concentrations (0.04%) of detergent will keep the enzyme solubilized without lysing the red cell, but we as yet do not know if the detergent has adversely affected their survival characteristics (DeLoach *et al.*, unpublished results).

2.5. *Phagocytosis of Erythrocytes*

Various heat treatments and exposure of the erythrocytes to mercurials and sulfhydryl reagents have been shown to preferentially direct erythrocytes either to the liver or the spleen, depending on the conditions (Rifkind, 1966). Thus, the possibility of tissue-specific or cell-type-specific uptake exists. We have shown that glutaraldehyde-treated enzyme-loaded erythrocytes are taken up almost exclusively by the liver. The reason for this presumably is that blood flow to the liver is very high relative to the spleen, so that grossly damaged cells are preferentially taken up by the liver. We were curious to see whether this result would apply to an "ectopic" liver provided by Morris hepatomas transplanted to the hind legs of rats. We found that the glutaraldehyde-treated erythrocytes were preferentially sequestered by the hepatomas relative to the host liver or spleen (Peters *et al.*, unpublished results), presumably owing in part to high blood flow through tumors. The implication of this observation to the possibility of using erythrocytes as vehicles for various chemotherapeutic agents is not clear, but this possibility certainly exists if the erythrocytes are phagocytized by the tumor cells. Further modifications of the enzyme-loaded erythrocytes by means such as neuraminidase, specific glycosidases acting on glycolipids and glycoproteins, and proteolytic enzymes may result in better control over the sites where the erythrocytes are sequestered.

Once the entrapped enzyme has been delivered to a lysosome, it presumably will be degraded (although lysosomal enzymes such as β-glucocerebrosidase should be much more stable than random enzymes). Various modifications of the enzyme by chemical pretreatment may stabilize the enzyme against degradation, thereby increasing its half-life. Such modifications would include blocking the N and C termini, cross-linking of the enzyme, and possibly delivering inhibitors of proteolytic enzymes along with the entrapped enzyme.

An example of a disease that is, in principle, treatable by replacement of the defective or missing enzyme is Gaucher's disease. Gaucher's disease is a typical glycosphingolipid lysosomal storage disease. In this and related disorders, a complex glycolipid accumulates in lysosomes of reticuloendothelial cells of affected individuals, usually in spleen, liver, and bone marrow. The complex lipid accumulates because hydrolases responsible for their degradation, most of which are glycoprotein in nature with relatively acidic pH optima, are markedly deficient. For example, glucocerebroside accumulates in the tissues of individuals with Gaucher's disease because they lack the enzyme β-glucocerebrosidase. This enzyme from human placenta has been purified and partially characterized (Pentchev *et al.*, 1973) and has been used in therapeutic trials in several patients with Gaucher's disease (Brady *et al.*, 1974).

3. *Evaluation of Results*

In addition to subjective and objective clinical signs, the evaluation of the efficacy of enzyme-loaded erythrocytes requires quantitative biochemical evidence. In the case of Gaucher's disease, currently the focus of much interest in terms of enzyme replacement, Brady and his collaborators (1974) found a large decrease in plasma and erythrocyte glucocerebroside concentration (presumably the effect of circulating enzyme) and a moderate (26%) decrease in the amount of

TABLE 1
Concentration of Various Glycosidases and β-Glucocerebroside in Normal Human Liver and Gaucher Spleen

	Normal liver			
Parameter	Range	Mean (μmoles/hr/mg)	S.D. (μmoles/hr/mg)	$\frac{2 \times \text{S.D.}}{\text{Mean}} \times 100$
β-Glucosidase	12.3–21.6	17.0	2.78	32.7
β-Arabinosidase	9.78–12.9	11.4	1.20	21.1
β-Glactosidase	36.3–48.1	43.8	4.11	18.7
β-Fucosidase	31.5–41.8	37.1	3.33	17.7
β-Mannosidase	59.5–98.5	80.4	12.0	29.8
β-*N*-Acetylgalactosaminidase	92.4–138	116	15.2	25.3
β-Glucocerebrosidase	23.0–41.8	25.6	8.41	65.6
	Gaucher spleen (mg/g wet wt.)			
β-Glucocerebroside	13.7–22.2	16.6	2.39	28.7

glycolipid stored in the liver. While these results are encouraging and support the idea that enzyme replacement can be effective in solubilizing abnormal storage material, additional studies are needed to verify this conclusion. Particularly troublesome and beyond the control of the clinical investigator is the sampling problem caused by tissue heterogeneity, as evidenced by the data contained in Table 1. To illustrate this point, recently one of us (R.G.) took eight random samples (2.0 mg) of normal human liver obtained at autopsy and nine samples from the spleen of a patient with Gaucher's disease obtained at surgery, and measured a number of tissue parameters. Although duplicate analyses of various enzymes and B-glucocerebroside in a single sample did not vary by more than 6%, there was considerable variation in the values obtained between specimens. Considering 2 S.D. as a reasonable confidence level (97%), it is noteworthy that for several of the enzymes assayed in the liver, such as β-glucosidase, β-mannosidase, and β-glucocerebrosidase, 2 S.D. represented 32.7, 29.8, and 65.6% of the mean value, respectively. Of particular interest was the finding that with respect to the glucocerebroside content in the Gaucher spleen, 2 S.D. represented 28.7% of the mean value. From these data it is clear that, owing to sampling problems and tissue heterogeneity, to establish a therapeutically significant clinical effect by any enzyme replacement attempt, one must study many patients or obtain relatively large changes (≥25–30%) in the quantity of storage material.

4. *Summary*

The use of human erythrocytes as carriers for enzyme-replacement therapy avoids some of the problems associated with carriers composed of substances foreign to the host. The erythrocyte can be used to deliver an enzyme to lysosomes for treatment of lysosomal storage diseases or to circulate an enzyme to degrade circulating small molecules. An additional possible therapeutic use is to deliver chemotherapeutic agents to erythrophagocytic tumors. The amount of enzyme that can be encapsulated is within the estimated requirements both for phagocytized and circulating erythrocytes. A variety of problems remain to be solved, however, before this procedure can be realistically considered to be therapeutically useful in medicine.

ACKNOWLEDGMENTS

This work was supported by National Institutes of Health Grants CA16003 and CA20813 to Garret Ihler and AM 17465 to Robert Glew.

5. *References*

Baker, R. F., 1967, *Nature* **215**:424–425.
Brady, R., Pentchev, P., Gal, A., Hibert, S., and Dekaban, A., 1974, *New Engl. J. Med.* **291**:989–993.
Chang, T. M. S., 1972, *Artificial Cells,* p. 52–66, Charles C. Thomas, Publisher, Springfield, Ill.
Fiddler, M., Thorpe, S., Krivit, W., and Desnick, R., 1974, Enzyme therapy III, *in vivo* fate of erythrocyte entrapped β-glucuronidase, in: *Enzyme Therapy in Lysosomal Storage Disease* (S. M. Tager, ed.), p. 182–185, North-Holland Publishing Co., Amsterdam.
Hoffman, J. F., 1958, *J. Gen. Physiol.* **42**:9–28.
Ihler, G., Glew, R. and Schnure, F., 1973, *Proc. Natl. Acad. Sci. V. S.* **70**:2663–2666.
Ihler, G., Lantzy, A., Purpura, J., and Glew, R., 1975, *J. Clin. Invest.* **56**:595–602.
Marsden, N. V. B., and Ostling, S. G., 1959, *Nature* **184**:723–724.
Pentchev, P., Brady, R., Hiffert, S., Gal, A., and Shapiro, D., 1973, *J. Biol. Chem.* **248**:5256–5261.
Rifkind, R. A., 1966, *Amer. J. Med.* **41**:711–722.
Schrier, S. L., 1963, *J. Clin. Invest.* **42**:756–766.
Seeman, P., 1967, *J. Cell. Biol.* **32**:55–70.

16

Enzyme Entrapment in Erythrocytes and Liposomes for the Treatment of Lysosomal Storage Diseases

R. J. DESNICK, M. B. FIDDLER, S. R. THORPE, AND L. D. STEGER

1. Introduction

Exploratory trials of enzyme replacement therapy have been stimulated by advances in the elucidation and characterization of the specific enzymatic defects in many inborn errors of metabolism, particularly the lysosomal storage diseases. The rationale for enzyme therapy for these disorders evolved from two fundamental observations: (1) the identification of lysosomes as the subcellular site of pathology, and (2) the elucidation of the basic role of the lysosome in cellular catabolism. Thus, it was reasoned that, after endocytosis, exogenous enzyme would be brought into contact with substrate by fusion of the various components of the lysosomal apparatus (de Duve, 1964).

Recent trials of enzyme replacement *in vitro* and *in vivo* support the feasibility of this rationale. The capacity of exogenous enzyme to gain access to and catabolize accumulated substrates was demonstrated *in vitro* when the appropriate purified enzyme was incubated in the media of cultured fibroblasts obtained from enzyme-deficient patients (Bach *et al.*, 1972; Hall *et al.*, 1973; Lagunoff *et al.*, 1973, O'Brien *et al.*, 1973; Porter *et al.*, 1971; von Figura and Kresse, 1974); and more importantly *in vivo* when highly purified enzymes from human sources were administered intravenously to patients with selected lysosomal storage diseases (Brady *et al.*, 1973, 1974; de Barsy *et al.*, 1973; Johnson *et al.*, 1973).

Critical review of the early and recent human trials has identified the major obstacles that must be overcome if enzyme replacement is to become an effective

---

R. J. DESNICK • Departments of Pediatrics, Genetics, and Cell Biology and The Dight Institute for Human Genetics, University of Minnesota, Minneapolis, Minn. MORRIS B. FIDDLER • Department of Genetics and Cell Biology and The Dight Institute for Human Genetics, University of Minnesota, Minneapolis, Minn. SUZANNE R. THORPE • Department of Pediatrics, University of Minnesota, Minneapolis, Minn. LYNN D. STEGER • Department of Genetics and Cell Biology and The Dight Institute for Human Genetics, University of Minnesota, Minneapolis, Minn.

therapeutic modality for selected inherited lysosomal storage diseases (Desnick *et al.*, 1975, 1976). The difficulties in these pilot trials included the short circulating and intracellular half-lives of the administered enzymatic activity, immunologic complications, and the inability to serially evaluate the physiologic and biochemical factors affecting the fate of the administered enzyme. Thus, effective enzyme therapy will require the development of enzyme and cellular strategies designed to satisfy the following requisites: (1) techniques for enzyme administration which afford maximal protection to the exogenous enzyme from potential physiologic and immunologic inactivation, as well as (2) delivery of the enzyme to the specific tissue and subcellular sites of pathologic substrate accumulation, and (3) the development of mammalian models to systematically assess the *in vivo* effectiveness of these strategies prior to human trials.

Recently, enzyme entrapment in biodegradable vesicles, such as autologous erythrocytes and liposomes, has been suggested as a strategy to protect the active enzyme from biodegradation and inactivation in the ciruclation and enhance its delivery to critical sites of substrate pathology (Chang, 1973; Fiddler *et al.*, 1974; Gregoriadis and Ryman, 1972; Ihler *et al.*, 1973). We describe here our efforts to assess the effectiveness of these enzyme entrapment strategies in a mammalian model system.

2. *In Vivo Fate of Unentrapped Enzyme*

2.1. *Mammalian Model System: β-Glucuronidase-Deficient Mice*

The inherent limitations of human experimentation mandate the development of mammalian model systems to assess the application of various enzyme entrapment techniques necessary to fulfill the requisites for effective enzyme therapy. Suitable models that permit the evaluation and manipulation of the factors that maximize enzyme stability, protection, target-tissue distribution, subcellular localization, and substrate metabolism have become a prime requisite for the systematic development of enzyme replacement techniques prior to human trials. In addition, model systems may also provide data regarding the potentialities for toxic or immunologic complications.

Thus, a murine model system was developed to serially evaluate and maximize entrapment methods for the delivery and protection of exogenous β-glucuronidase intravenously administered to C3H/HeJ Gus^h β-glucuronidase-deficient mice. A selective thermal inactivation assay allowed the sensitive and reliable discrimination of the bovine liver β-glucuronidase activity from the residual murine tissue activities; the assay was based on the differential thermal stabilities at 60°C of bovine (thermolabile) and murine (thermostable) activities (Thorpe *et al.*, 1974). This direct and convenient assay provided the ability to monitor the fate of the administered exogenous activity.

2.2 *In Vivo Fate of Unentrapped β-Glucuronidase*

Figure 1 shows the *in vivo* fate of unentrapped bovine β-glucuronidase administered intravenously into β-glucuronidase-deficient mice. A rapid clearance

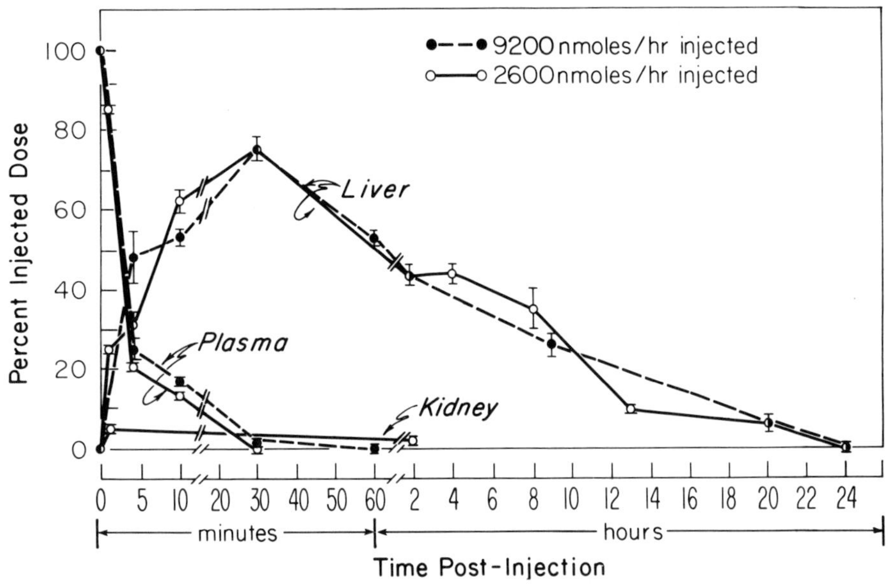

Figure 1. Fate of unentrapped bovine β-glucuronidase activity administered intravenously to β-glucuronidase-deficient mice. (From Thorpe *et al.*, 1974.)

from the circulation ($t_{1/2}$ ~ 3 min) was observed at either dosage administered. A reciprocal and rapid uptake of the injected enzyme was observed in the liver; approximately 25% of the bovine activity was recovered at 1 min and maximal hepatic recovery, 72% of dose, was detected at 30 min. A gradual and linear decrease in recovered hepatic activity to nondetectable levels was observed by 24 hr. Approximately 5% of dose was found in the kidneys at early time points; no exogenous bovine activity was detected in any other tissue, including spleen, lung, and brain.

Recent studies with unentrapped β-glucuronidase activities isolated from various bovine tissue sources have demonstrated the differential hepatic retention of these isozymes (Fiddler and Desnick, 1977). The hepatic retention of unentrapped splenic β-glucuronidase was significantly prolonged (approximately 2.5 times) compared to administered unentrapped renal and hepatic activities. These studies support the need to evaluate multiple tissues for the most stable and appropriate enzyme source for maximal tissue uptake and retention and efficient catabolism of accumulating substrates (Shapiro *et al.*, 1974; Brot *et al.*, 1974; Nicol *et al.*, 1975).

Table 1 shows the subcellular distribution of bovine and murine β-glucuronidase activities in liver obtained 2 hr after injection of the 2600-unit dose. Greater than 85% of the total exogenous activity recovered was in the mitochondrial-lysosomal fraction, indicating the *in vivo* localization of the administered enzyme in the lysosomally enriched fraction. In contrast, when an *in vitro* control mixture of bovine enzyme and murine liver homogenate was fractionated, 100% of the bovine activity was recovered in the soluble fraction, indicating the specific association of the intravenously administrered enzyme with the mitochondrial–lysosomal fraction.

TABLE 1
Hepatic Subcellular Distribution of Recovered Unentrapped Bovine β-Glucuronidase[a]

Time postinjection (hr)	Activity administered, units	Recovered enzymatic activity			% Subcellular fraction recovered of total fractions[b]		
		Source	Total homogenate, units	Total fractions, units	N	M + L	S
2	2600	Bovine	1030	840	12	88	0
		Murine	2680	2060			
In vitro control mixture[c]	7500	Bovine	7440	6950	0	0	100
		Murine	2130	2180			

[a] Bovine and murine sources designate thermolabile and thermostable activities recovered, respectively; thermolabile and thermostable activities determined as described in Thorpe *et al.* (1974). 1 unit (U) of enzymatic activity equals 1 nmole/hr of 4-methylumbelliferyl-β-D-glucuronide trihydrate hydrolyzed.
[b] Subcellular fractions obtained by differential centrifugation as described in Thorpe *et al.* (1975); N, nuclear (600*g* pellet), M + L, mitochondrial–lysosomal (14,000*g* pellet), and S, soluble fraction (14,000*g* supernatant).
[c] Bovine β-glucuronidase and a murine liver homogenate were mixed *in vitro* and incubated at 37°C for 10 min and then fractionated.

The rapid plasma clearance and hepatic uptake of unentrapped bovine β-glucuronidase in the murine system were similar to the findings in pilot trials using the appropriate highly purified enzymatic activity in patients with Tay-Sachs (Johnson *et al.*, 1973), Fabry's (Brady *et al.*, 1973), and Gaucher's (Brady *et al.*, 1974) diseases; however, subcellular localization of the administered human activities was not feasible. These observations identified the need to develop strategies for enzyme administration designed to protect the exogenous enzyme from bioinactivation processes as well as to optimize the delivery of the enzyme to target pathologic sites of substrate deposition.

3. *Administration of Erythrocyte-Entrapped Enzyme*

The demonstration that enzymes can be entrapped in human erythrocytes suggested the intriguing possibility that these biodegradable vesicles may be useful as *in vivo* carriers of exogenous enzyme (Ihler *et al.*, 1973). Therefore, we explored the feasibility of enzyme entrapment in autologous murine erythrocytes and evaluated the effectiveness of these enzyme-containing vesicles as a strategy for efficient enzyme delivery in our model system (Fiddler *et al.*, 1974; Thorpe *et al.*, 1975).

3.1. *Enzyme Entrapment in Erythrocytes*

Bovine β-glucuronidase was entrapped in autologous murine erythrocytes essentially by the hypotonic exchange method of Ihler *et al.*, (1973), as illustrated in Figure 2. In a typical loading experiment, bovine enzyme (62,000 U, 3.6 mg of

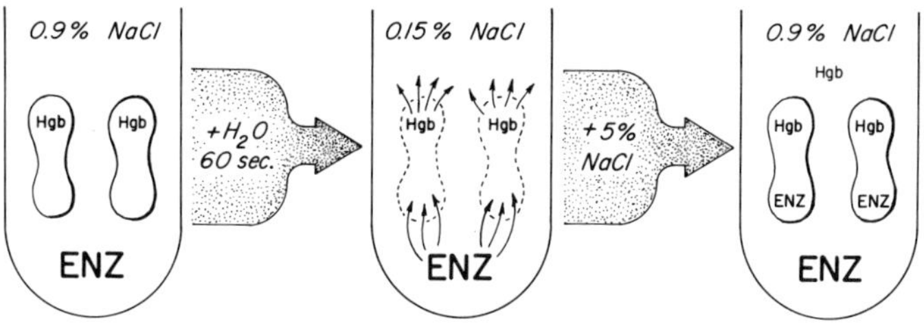

FIGURE 2. Entrapment of enzyme in erythrocytes, 3%–5% enzyme entrapped. Unentrapped enzyme recovered for reuse.

protein) in 0.6 ml of 0.02 M potassium phosphate buffer, pH 6.5, was mixed with 0.4 ml of packed, washed murine erythrocytes. Hypotonic exchange was induced by the rapid addition of 5.0 ml of distilled water and terminated after 60 sec by the addition of 1.3 ml of 0.83 M NaCl to restore isotonicity. The enzyme-containing erythrocytes were recovered by centrifugation at 600*g* for 20 min at 4°C; the cells were then washed repeatedly with 25 volumes of 0.15 M NaCl until no further activity was detected in the supernatant. The entrapped activity remained sedimentable after repeated washing of the erythrocytes with either 0.15 M or 0.3 M NaCl, but was readily released into solution following brief sonication of the cells.

Table 2 shows that approximately 4% of available bovine β-glucuronidase was routinely entrapped in murine erythrocytes under these conditions (Thorpe *et al.*, 1975). The enzyme-loaded erythrocytes retained normal discoid morphology and mean corpuscular volume. The percentage of enzyme entrapped and the morphologic characteristics compared favorably with similar findings reported for the entrapment of enzymes and other proteins in human erythrocytes (Ihler *et al.*, 1973; Whitaker *et al.*, 1974). The amount of protein entrapped has been shown to

TABLE 2
Recovery of β-Glucuronidase Activity in Hypotonic Exchange-Loaded Murine Erythrocytes[a]

Procedure	Bovine enzymatic activity added	Total enzymatic activity recovered	% Entrapped
	Packed erythrocytes (U/ml)		
Hypotonic exchange	155,000	6900	4.4
Hypotonic exchange	0	<0.01	0
Isotonic control	232,000	20	0
Packed erythrocytes (untreated)	0	16.5[b]	0

[a]All values represent the mean of triplicate determinations; 1 unit (U) equals 1 nmole/hr of 4-methylumbelliferyl-β-D-glucuronide trihydrate hydrolyzed.
[b]Mean endogenous β-glucuronidase activity in washed, packed murine erythrocytes (n = 12) determined as described in Thorpe *et al.* (1975).

be a function of the hypotonic exchange conditions, including degree and duration of hypotonicity, concentration of various ions, temperature, and the molecular weight or size of the molecule to be entrapped (Baker, 1967; Bodemann and Passow, 1972; Ihler *et al.,* 1973). Although the dilution factor required to induce hypotonic exchange establishes a maximal efficiency of entrapment, unentrapped enzyme was readily recovered and recycled for subsequent entrapment. Following the hypotonic exchange procedure, the erythrocytes were repeatedly washed with 0.15 M NaCl and the pooled supernatants were chromatographed over a Concanavalin A-Sepharose column at pH 6.0, which bound over 95% of the β-glucuronidase activity while excluding the major erythrocyte proteins. The bound enzyme was then easily batch-eluted at pH 7.0 with 1.0 M α-methyl-D-glucoside (Beutler *et al.,* 1975).

3.2. *In Vivo Fate of Erythrocyte-Entrapped β-Glucuronidase*

Figure 3 shows the time course for the blood clearance and tissue uptake after intravenous administration of erythrocyte-entrapped bovine β-glucuronidase into β-glucuronidase-deficient mice. Total recovery of administered activity from murine blood and tissues accounted for approximately 95% of dose at 1 and 2 hr post-injection. The activity was cleared from the circulation with a half-life of about 20 min and was no longer detectable at 2 hr. A concomitant uptake of the injected enzyme was observed in murine tissues, primarily the liver; approximately 30% of the bovine activity was recovered at 30 min and maximal hepatic uptake, more than 70% of dose, was detected at 2 hr. Notably, the recovered exogenous activity was maintained in hepatic tissue at this level from 2 to 13 hr.

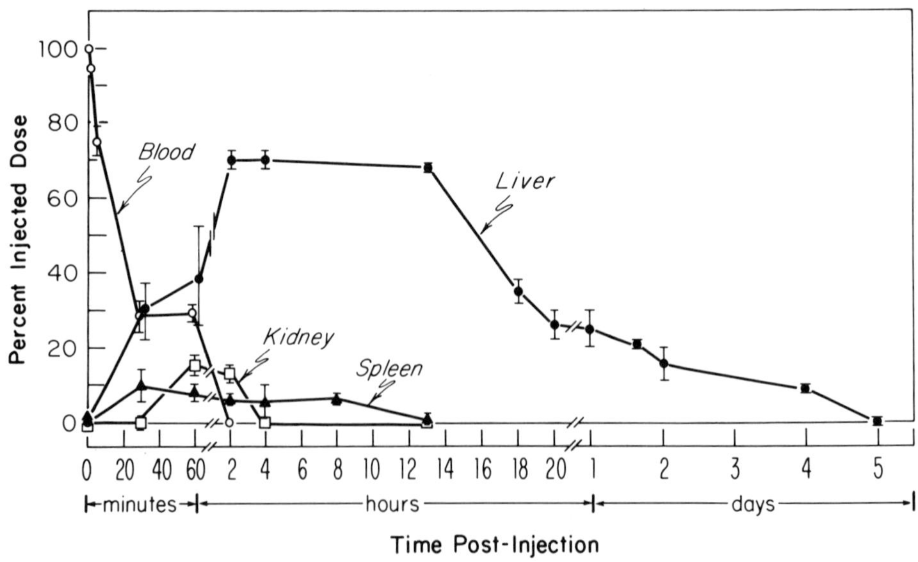

FIGURE 3. Fate of erythrocyte-entrapped bovine β-glucuronidase activity administered intravenously to β-glucuronidase-deficient mice. (From Thorpe *et al.,* 1975.)

TABLE 3
Hepatic Subcellular Distribution of Recovered Erythrocyte-Entrapped Bovine β-Glucuronidase[a]

Time postinjection (hr)	Activity administered, units	Source	Total homogenate, units	Total fraction, units	% Subcellular fraction recovered of total fractions: N	M + L	S
4	1000	Bovine	720	760	10	90	0
		Murine	3040	3240			
13	800	Bovine	617	550	0	87	13
		Murine	1720	1700			
24	834	Bovine	248	275	16	84	0
		Murine	2540	2730			
96	1330	Bovine	132	129	0	100	0
		Murine	2500	2390			
In vitro control mixture[b]	768	Bovine	704	705	71	0	29
		Murine	2620	2260			

(Column group heading: Recovered enzymatic activity)

[a]For details, see footnotes, Table 1.
[b]Erythrocyte-entrapped activity and a murine liver homogenate were mixed *in vitro*, incubated at 37°C for 10 min, and then fractionated.

The hepatic recovery of the bovine activity was then observed to decrease in a biphasic pattern characterized by an initial drop from the maximal level to 25% of dose at 20 hr followed by subsequent retention of 10–20% of dose for up to 5 days. This unique biphasic pattern was the result of erythrocyte entrapment and not a property of the bovine enzyme, since unentrapped activity was rapidly cleared from hepatic tissue (Figure 1). The biphasic pattern may reflect at least two subpopulations of enzyme-loaded erythrocytes (Bodemann and Passow, 1972; Thorpe *et al.*, 1975), one of which was more slowly processed, resulting in a prolonged intracellular retention of entrapped activity.

Subcellular fractionation and electron microscopic examination of hepatic tissue was carried out to determine the intracellular site of administered enzyme uptake. Table 3 shows the results of the subcellular fractionations of hepatic tissue obtained at various times after administration of erythrocyte-entrapped enzyme. At each time point, more than 80% of the recovered bovine activity was detected in the lysosomally enriched fraction. In addition, sonication of the lysosomally enriched fraction resulted in the release of all the bovine activity into the supernatant after centrifugation. Furthermore, when an *in vitro* control mixture of erythrocyte-entrapped enzyme and murine liver homogenate was fractionated, 70% of the bovine activity was found in the nuclear fraction. These results support the *in vivo* specificity of the lysosomal apparatus for erythrocyte-entrapped enzyme uptake. Ultrastructural examination of hepatic tissue at 2 hr postinjection revealed erythrocytes in single membrane-bound vacuoles in Kupffer cells, as shown in Figure 4; these morphologic studies indicated that the major sites of erythrocyte uptake were in cells of the reticuloendothelial system.

An important feature of the hepatic uptake of enzyme entrapped in erythrocytes is shown in Figure 5. A marked latency in the detection of administered

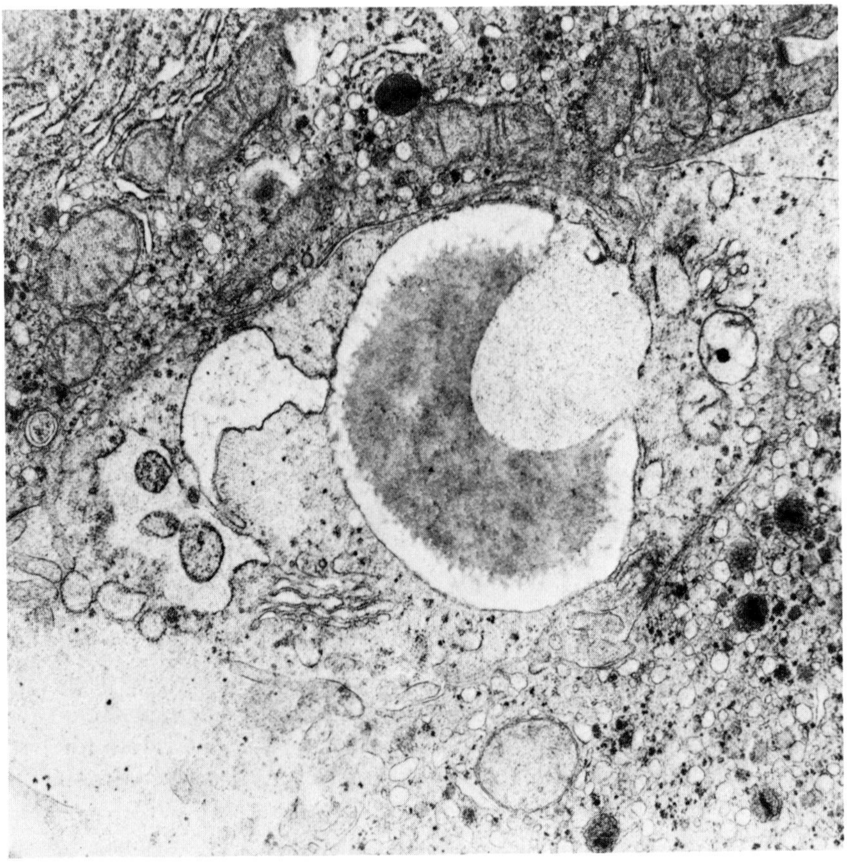

FIGURE 4. Electron photomicrograph of murine liver 2 hr after administration of erythrocyte-entrapped bovine β-glucuronidase. (Courtesy of Dr. H. L. Sharp.)

activity was observed when hepatic tissue homogenates obtained at sampling intervals up to 13 hr were not sonicated before enzyme assay. In fact, as much as threefold greater levels of exogenous activity were detected after sonication compared with unsonicated levels. In contrast, sonication did not increase the amount of exogenous activity recovered in hepatic homogenates obtained after 13 hr postinjection. This latency was unique since the standard assay was carried out under hypotonic conditions in the presence of detergent. Presumably, the latency for the detection of exogenous enzymatic activity reflected the intracellular release of enzyme from erythrocytes as they were being processed within lysosomes.

As shown in Figure 3, significant levels of bovine activity were recovered routinely in the kidneys, approximately 15% of dose up to 2 hr and low levels, less than 10% of dose, were found in the spleen. No exogenous activity was recovered in the lungs, bone marrow, or brain after administration.

The low levels of exogenous activity recovered in the spleen may reflect the fact that the murine splenic morphology differs from the sinusoid structure found in other mammals, including humans. In fact, we have found that the intravenous

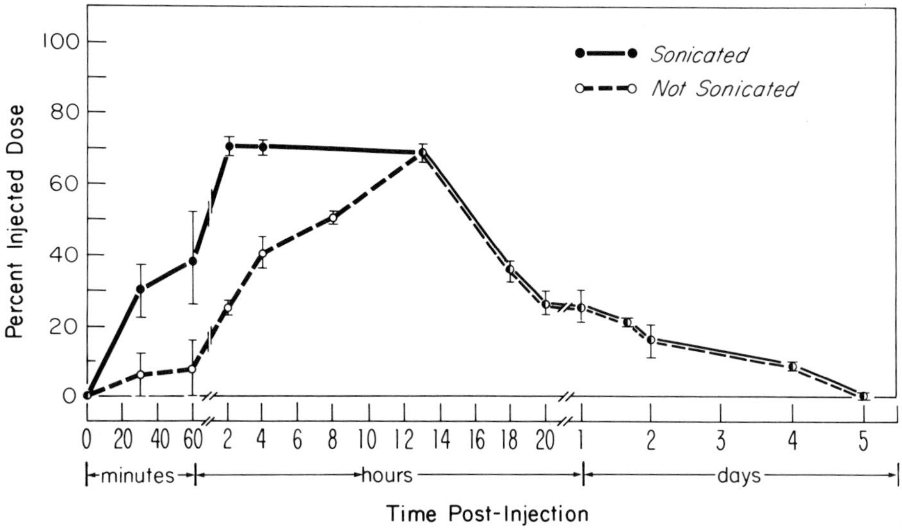

FIGURE 5. Effect of sonication on the hepatic recovery of erythrocyte-entrapped bovine β-glucuronidase. (From Thorpe *et al.*, 1975.)

administration of erythrocyte-entrapped [I^{131}]albumin into Swiss albino rats resulted in equal uptake by the liver and spleen. Thus, the splenic uptake data in the murine system may be only a minimal estimate of the uptake of autologous enzyme-loaded erythrocytes by the human spleen. In addition, it may be possible to specifically target erythrocytes to the spleen by modification of the erythrocyte surface. For example, Jacob and Jandl (1962) have shown that treatment of human and rat erythrocytes with sulfhydryl active compounds enhanced the *in*

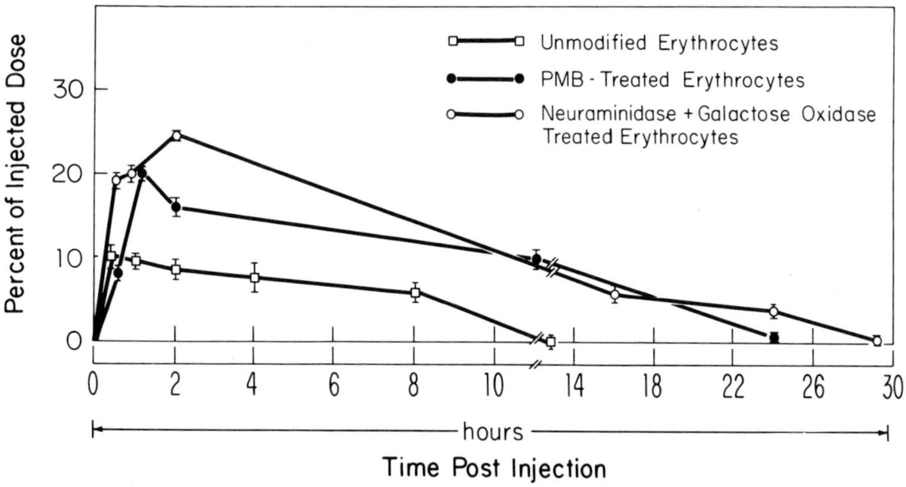

FIGURE 6. Splenic recovery of bovine β-glucuronidase activity entrapped in chemically and enzymatically modified murine erythrocytes following administration to β-glucuronidase-deficient mice. (From Fiddler and Desnick, 1977.)

vivo splenic sequestration of the treated erythrocytes. Figure 6 shows our results from experiments in which enzyme-loaded murine erythrocyte surfaces were either chemically (*p*-hydroxymercuribenzoate) or enzymatically (neuraminidase followed by galactose oxidase) modified prior to administration. Note that both the splenic uptake and the retention at 2 hr postinjection were increased about two- and threefold, respectively, following the administration of enzyme entrapped in these modified murine erythrocytes.

4. *Administration of Liposome-Entrapped Enzyme*

Liposomes, synthetic lipid spherules, have been considered attractive carriers for the delivery of enzymes entrapped in the aqueous compartments between their concentric lipid bilayers (Gregoriadis and Ryman, 1972; Sessa and Weissman, 1970). In addition, these biodegradable vesicles may provide protection of the entrapped enzyme (or other therapeutic agent) from bioinactivation in the circulation and/or immunologic surveillance. Furthermore, investigators have become intrigued with the possibility that the liposomal surface could be modified or coated with tissue-specific components which might promote the uptake of these vesicles by particular target cells and/or tissues (Gregoriadis, 1974; Weissman

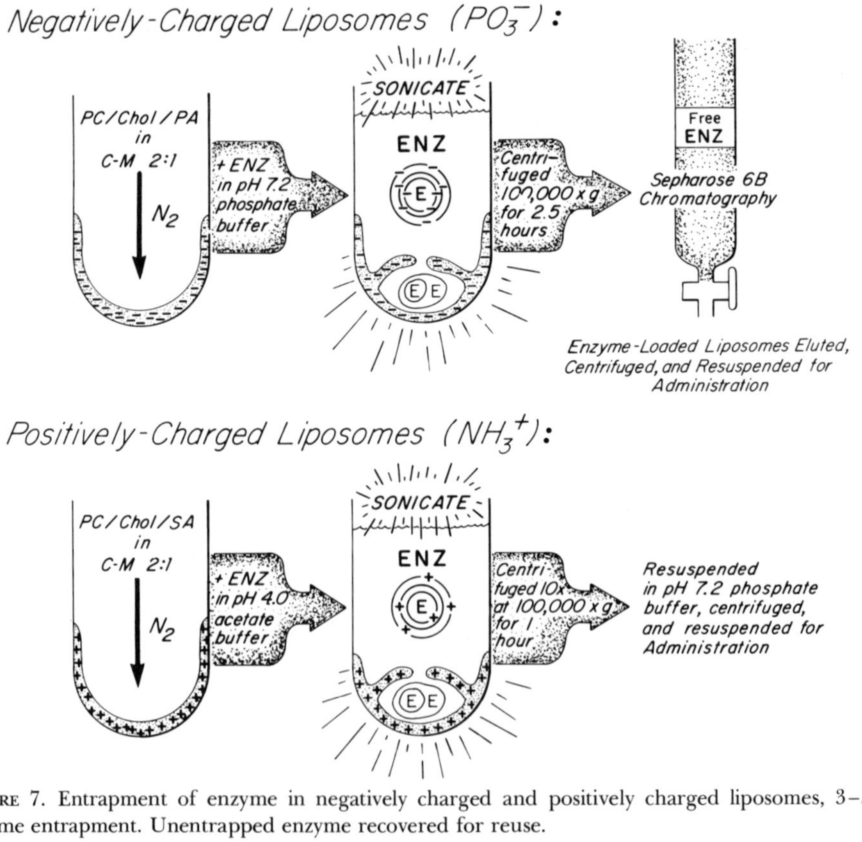

FIGURE 7. Entrapment of enzyme in negatively charged and positively charged liposomes, 3–5% enzyme entrapment. Unentrapped enzyme recovered for reuse.

et al., 1975). Encouraged by our observations using erythrocytes as enzyme carriers, we investigated the fate of β-glucuronidase entrapped in negatively and positively charged liposomes in our murine model system (Steger and Desnick, 1976).

4.1. *Enzyme Entrapment in Negatively and Positively Charged Liposomes*

The procedure for the entrapment of enzymes in negatively and positively charged liposomes is illustrated in Figure 7. Negatively charged liposomes were prepared essentially by the method of Gregoriadis and Leathwood (1971). Highly purified phosphatidyl choline, cholesterol, and negatively charged phosphatidic acid in a molar ratio of 7:2:1 (total lipid, 200 mg) were dissolved in chloroform/methanol (2:1 vol./vol.) and evaporated under nitrogen to form a lipid film. Bovine β-glucuronidase (500,000 U) in 15 ml of 3.3 mM potassium phosphate buffer, pH 7.2, was then added. When this mixture was sonicated, the enzyme became entrapped in the aqueous compartments between the lipid bilayers. To separate unentrapped enzyme, the mixture was centrifuged and the pelleted liposomes were resuspended in the buffer and applied to a Sepharose 6B column. The enzyme-loaded liposomes were eluted in the void volume, centrifuged, and resuspended in buffered saline; column fractions containing the unentrapped enzyme were pooled, concentrated, and dialyzed for subsequent liposome entrapment (Steger and Desnick,1976).

Positively charged liposomes were prepared similarly, except that stearylamine, a positively charged lipid, was used instead of phosphatidic acid. Enzyme entrapment in positively charged liposomes (phosphatidyl choline, cholesterol, and stearylamine in a molar ratio 7:2:1) was carried out at pH 4.0, below the isoelectric point of 5.1 for bovine β-glucuronidase, to avoid the formation of lipid–enzyme aggregates between the positively charged lipid film and the enzyme, which is negatively charged at pH 7.2. The positively charged liposomes were washed extensively by repeated resuspension in buffer and centrifugations (at 100,000 *g* for 1 hr) and the unentrapped enzyme present in the supernatant was recovered for reuse. Both positively and negatively charged liposomes were resuspended in buffered saline prior to intravenous injection (800–3500 U in a volume of 0.15–0.25 ml) in β-glucuronidase-deficient mice. Approximately 5% of the enzyme added to the lipid film was routinely entrapped in the negatively and positively charged liposomes (Steger and Desnick, 1976).

4.2. *In Vivo Fate of Negatively and Positively Charged Liposome-Entrapped β-Glucuronidase*

Figure 8a shows the *in vivo* fate of bovine β-glucuronidase entrapped in negatively charged liposomes (Steger and Desnick, 1976). Approximately 90–100% of the administered activity was recovered in murine blood and tissue for up to 48 hr postinjection. A rapid clearance of enzymatic activity from the circulation was observed with a concomitant uptake of activity in several tissues. The maximal hepatic recovery, approximately 75% of dose, was observed by 1 hr and was retained for 48 hr before gradually declining to nondetectable levels by 8 days.

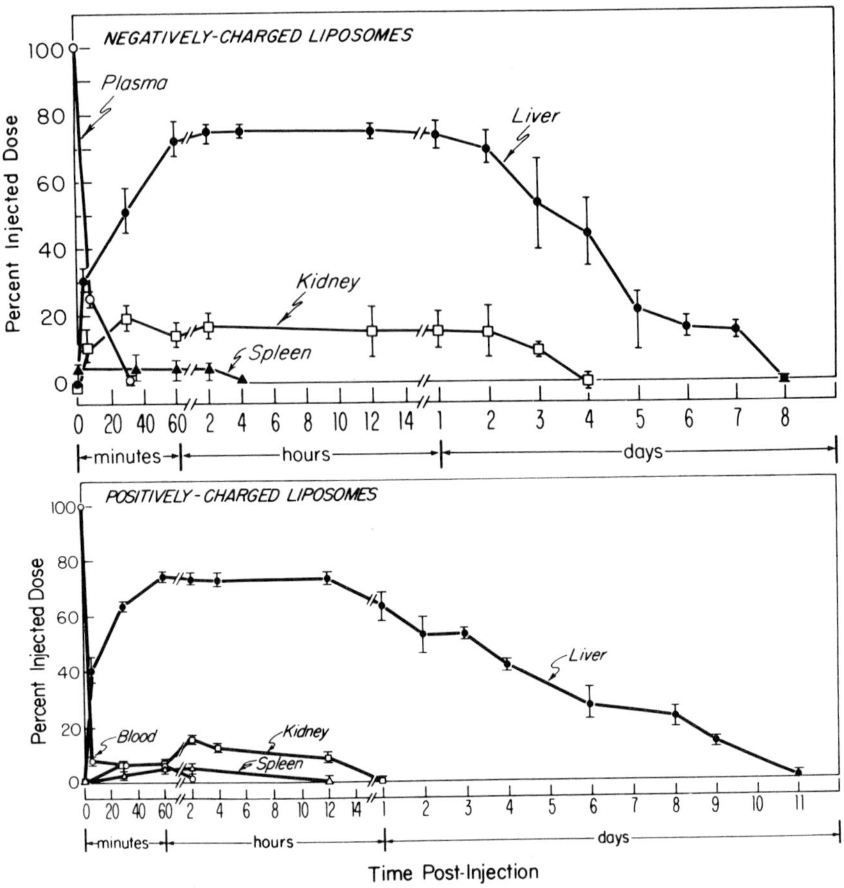

FIGURE 8. Fate of liposome-entrapped bovine β-glucuronidase activity administered intravenously to β-glucuronidase-deficient mice. *Upper graph:* negatively charged liposomes; *lower graph:* positively charged liposomes.

There was significant uptake and retention by the kidney; approximately 20% of dose was recovered for almost 4 days. Small amounts of β-glucuronidase were detected in splenic tissue at early time points, whereas no activity was observed in the lungs, bone marrow, or brain.

Figure 8b shows the tissue uptake and distribution of bovine β-glucuronidase entrapped in positively charged liposomes (Steger and Desnick, 1976). There was a rapid hepatic uptake of activity to a maximum level of 75% of dose by 1 hr, with a subsequent retention of activity for up to 11 days, 3 days longer than when entrapped in negatively charged liposomes. The renal uptake of enzyme administered in the positively charged vesicles was transient, in contrast to that observed with negatively charged liposomes. Less than 5% of dose was detected in splenic and pulmonary tissues up to 12 hr postinjection.

The hepatic subcellular distribution of recovered bovine activity entrapped in negatively and positively charged liposomes at several times postinjection is shown in Table 4 (Steger and Desnick, 1976). At each time point after injection of β-glucuronidase entrapped in negatively charged liposomes, approximately 70% of the recovered bovine activity was detected in the lysosomally enriched fraction. In contrast to the predominant lysosomal localization of bovine β-glucuronidase administered in negatively charged liposomes, hepatic subcellular fractionation revealed that the majority of exogenous activity entrapped in positively charged liposomes was recovered in the soluble, and to a lesser extent in the lysosomally enriched fractions up to 72 hr postinjection.

As shown in Table 4, subcellular fractionation of *in vitro* control mixtures of bovine activity entrapped in negatively or positively charged liposomes and murine hepatic homogenates demonstrated only 10% of the exogenous activity in the lysosomally enriched fractions. These data indicate that the *in vivo* distribution of exogenous activity was not attributable to a nonspecific assocation between the liposome-entrapped activity and a particular subcellular fractionation. Furthermore, these results suggest that the negatively charged liposomes are primarily taken up by the lysosomal apparatus, whereas the positively charged liposomes may, in part, deliver their contents to the cytoplasm.

5. *Biomedical Applications of Enzyme Entrapment: Enzyme Therapy*

Effective enzyme therapy for selected inborn errors of metabolism requires the delivery of sufficient quantities of the specific, active isozyme to the target tissue

TABLE 4
Hepatic Subcellular Distribution of Recovered Liposome-Entrapped Bovine β-Glucuronidase[a]

Enzyme entrapment	Time post injection (hr)	Activity administered, units	Recovered enzymatic activity: Source	Total homogenate, units	Total fraction, units	% Subcellular fraction recovered of total fractions: N	M + L	S
Negatively charged liposomes	4	600	Bovine	488	430	7	70	23
			Murine	2372	2418			
	24	800	Bovine	413	441	8	75	17
			Murine	3552	3339			
	96	800	Bovine	450	395	6	69	25
			Murine	2270	2250			
Positively charged liposomes	4	2000	Bovine	1500	1384	10	39	51
			Murine	5100	4839			
	24	2500	Bovine	1798	1622	1	33	66
			Murine	4247	4156			
	72	2600	Bovine	2002	1945	8	22	70
			Murine	6111	5940			
In vitro control mixtures[b]								
Negatively charged liposomes		1100	Bovine	1050	1051	2	10	88
			Murine	2140	1903			
Positively charged liposomes		3000	Bovine	3000	2961	28	10	62
			Murine	3410	2898			

[a] For details, see footnotes, Table 1.
[b] Liposome entrapped activity and a murine liver homogenate were mixed *in vitro*, incubated at 37°C for 10 min, and then fractionated.

and subcellular sites for efficient substrate metabolism. The availability of affinity chromatographic techniques for the isolation of specific enzymatic activities (Brot *et al.,* 1974; Radin *et al.,* 1974; Mapes and Sweeley, 1973; Norden and O'Brien, 1974; Beutler *et al.,* 1975) which can be chemically modified (Wold, 1973; Snyder *et al.,* 1974) to maximize *in vivo* stability has enhanced the feasibility of human enzyme therapy endeavors. The recent observations that intravenously injected unentrapped enzymes were rapidly cleared from the circulation and recovered primarily from the liver (Brady *et al.,* 1973; Brady *et al.,* 1974; de Barsy *et al.,* 1973; Johnson *et al.,* 1973) underscore the need for administration strategies that promote the delivery of enzyme to target tissue and subcellular sites while protecting it from bioinactivation processes.

Figure 9 shows the *in vivo* fate of intravenously administered unentrapped, erythrocyte-entrapped, and negatively and positively charged liposome-entrapped β-glucuronidase. Compared to unentrapped enzyme, erythrocyte-entrapped enzyme was markedly protected from rapid intracellular degradation. Although the maximal level of activity recovered from hepatic tissue was comparable to that observed for unentrapped enzyme, the activity administered in erythrocytes was maintained at that level up to 13 hr postinjection and retained in hepatic tissue as long as 5 days, 5 times longer than unentrapped enzyme. Even more striking was the protection afforded the enzyme by administration in negatively and positively charged liposomes; β-glucuronidase administered by these tech-

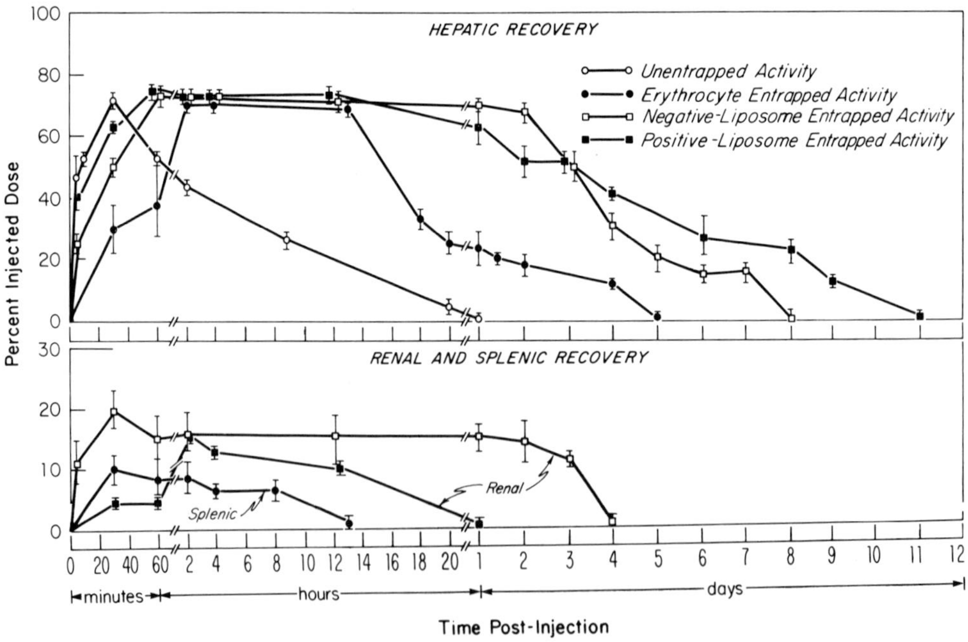

FIGURE 9. *Upper graph:* comparison of hepatic recoveries of unentrapped, erythrocyte-entrapped, and liposome-entrapped bovine β-glucuronidase intravenously administered to β-glucuronidase-deficient mice; *upper graph:* recovery of erythrocyte-entrapped and liposome-entrapped bovine β-glucuronidase from splenic and renal tissues, respectively.

niques was maintained at 70% of dose for almost 48 hr and retained in the liver up to 8 and 11 days postinjection, respectively.

Importantly, enzyme administered in either erythrocytes or liposomes provided a means of enzyme delivery to specific subcellular compartments. Hepatic subcellular fractionation of unentrapped, erythrocyte-entrapped, and negatively charged liposome-entrapped enzyme showed a primarily lysosomal localization of the recovered activity. Thus, these strategies should be useful for the treatment of patients with selected lysosomal storage diseases. In contrast, activity administered in positively charged liposomes was recovered predominantly in the soluble fraction, suggesting that this enzyme carrier may be effective for the replacement of defective cytoplasmic enzymes. Future studies using these entrapment vesicles should determine their metabolic effectiveness at these subcellular sites (e.g., lysosomal substrate catabolism).

In addition to the prolonged hepatic lifetime of the enzyme, these entrapment strategies provided enhanced enzyme delivery to sites other than the liver. As indicated in the lower graph of Figure 9, significant levels of enzymatic activity were recovered in splenic tissue after administration of enzyme in erythrocytes and in renal tissue follwing administration of enzyme in liposomes; neither of these tissues showed significant uptake of unentrapped enzyme. Furthermore, chemical and/or physical modifications of the membrane surfaces of erythrocyte and/or liposome carriers may direct them to target sites; the incorporation of specific carbohydrate, glycolipid, lipopolysaccharide, or glycopeptide moieties may provide the appropriate "address label" for specific tissue recognition and selective uptake (Ashwell and Morell, 1974). This approach is exemplified by the chemical and enzymatic modifications of the erythrocyte surface, which resulted in enhanced splenic uptake and retention of exogenous β-glucuronidase (Figure 6). Innovative approaches to "educate" liposomes by incorporating specific antibodies directed against selected cell or tissue membrane antigens into the liposomal (or erythrocyte) membrane may also "address" these enzyme carriers (Gregoriadis, 1974; Weissman *et al.*, 1975).

Evaluation of the potential immunologic consequences of β-glucuronidase administered unentrapped and entrapped in erythrocytes and liposomes has been undertaken in our laboratory (Steger *et al.*, 1977). Preliminary studies in mice sensitized to bovine β-glucuronidase by subcutaneous injections have demonstrated that the fate of β-glucuronidase entrapped in erythrocytes was not altered, whereas the fate of negatively charged liposome-entrapped enzyme was markedly altered. Following the injection of liposome-entrapped enzyme in sensitized mice, the hepatic recovery of exogenous activity was barely detectable (approximately 5% of dose) compared to 75% of dose observed in unsensitized mice at 24 hr postinjection. In contrast, the *in vivo* fate of erythrocyte-entrapped activity remained unchanged. These data suggest that liposomes have exposed antigenic sites of the enzyme which are able to complex with circulating antibodies. Further studies of liposomes as enzyme carriers are required to document their possible antigenicity (Gregoriadis and Allison, 1974; Allison and Gregoriadis, 1974; Uemura *et al.*, 1975); modification of the liposome surface by chemical (Weissmen *et al.*, 1975) or enzymatic (Gregoriadis, 1974) procedures may circumvent potential harmful immunologic responses. The use of a recipient's own erythrocytes to deliver active enzyme may avoid the potential immunologic and physiologic problems resulting from enzyme administration in synthetic carriers.

In summary, entrapment of enzymes in biodegradable carriers should have direct biomedical applications for the treatment of inherited enzyme deficiency diseases, as well as in the delivery of any therapeutic agent (e. g., antibiotics, cancer chemotherapy drugs, etc.) to the target site of pathology. Entrapment of enzymes or other agents in these vesicles provides the capability for delivery to a specific tissue or subcellular locale while affording protection from bioinactivation, toxic reactions, and/or immunolgic processes. Future application of entrapment techniqes should enhance the potential effectiveness of enzyme replacement endeavors in patients with selected inborn errors of metabolism resulting from defective lysosomal or cytoplasmic enzymatic activities.

ACKNOWLEDGMENTS

We wish to acknowledge Dr. Harvey L. Sharp for the electron microsopy studies, Dr. Steven D. Douglas for his helpful advice with the immunologic studies, and Ms. Carolyn S. Dullum for her excellent technical assistance. We also thank Ms. Ardys Ferman and Ms. Linda Walling for preparation of the manuscript.

This work was supported in part by grants from The National Foundation–March of Dimes (1-273), the American Heart Assocation (74-915), the National Institutes of Health (AM 15174), and the Clinical Research Centers Program of the Division of Research Resources, National Institutes of Health (RR-400). R. J. Desnick is a recipient of a National Institutes of Health Research Career Development Award (1 K04 AM 00042), M. B. Fiddler was a recipient of a National Science Foundation Pre-Doctoral Fellowship, S. R. Thorpe is a recipient of a National Institutes of Health Post-Doctoral Training Fellowship (1T22 GM 00024), and L. D. Steger is a recipient of a National Institutes of Health Pre-Doctoral Training Fellowship (2T01 MH 10679).

6. *References*

Allison, A. C., and Gregoriadis, G., 1974, Liposomes as immunological adjuvants, *Nature* **252**:252.

Ashwell, G., and Morell, A. G., 1974, The role of surface carbohydrates in the hepatic recognition and transport of circulating glycoproteins, *Adv. Enzymol.*. **41**:99.

Bach, G., Friedman, R., Weissman, B., and Neufeld, E. F., 1972, The defect in Hurler and Scheie syndrome: deficiency of α-iduronidase, *Proc. Natl. Acad. Sci. U.S.* **69**:2049.

Baker, R. F., 1967, Entry of ferritin into human red cells during hypotonic haemolysis, *Nature* **215**:424.

Beutler, E., Guinto, E., and Kuhl, W., 1975, Placental acid hydrolase purification on concanavalin-A-sepharose, *J. Lab. Clin. Med.* **85**:672.

Bodemann, H., and Passow, H., 1972, Factors controlling the resealing of the membrane of human erythrocyte ghosts after hypotonic hemolysis, *J. Membrane Biol.* **8**:1.

Boyer, S. H., Siggers, D. C., and Krueger, L. J., 1973, Caveat to protein replacement therapy for genetic disease: immunological implications of accurate molecular diagnosis, *Lancet* **ii**:654.

Brady, R. O., Tallman, J. F., Johnson, W. G., Gal, A. E., Leahy, W. R., Quirk, J. M., and Dekaban, A. S., 1973, Replacement therapy for inherited enzyme deficiency: use of purified ceramidetrihexosidase in Fabry's disease, *N. Engl. J. Med.* **289**:9.

Brady, R. O., Penchev, P. G., Gal, A. E., Hibbert, S. R., and Dekaban, A. S., 1974, Replacement therapy for inherited enzyme deficiency: use of purified glucocerebrosidase in Gaucher's disease, *N. Engl. J. Med.* **291**:989.

Brot, F. E., Glaser, J. H., Roozen, K. J., Sly, W. S., and Stahl, P. D., 1974, *In vitro* correction of deficient human fibroblasts by β-glucuronidase from different human sources, *Biochem. Biophys. Res. Commum.* **57**:1.

Chang, T. M. S., 1973, Immobilization of enzymes, absorbents or both within semipermeable microcapsules (artificial cells) for clinical and experimental treatment of metabolite related disorders, in: *Enzyme Therapy in Genetic Diseases, Birth Defects* (Original Article Series, Vol. 9) (R. J. Desnick, R. W. Bernlohr, and W. Krivit, eds.), pp. 66–75, The Williams & Wilkins Company, Baltimore.

de Barsy, T., Jacquemin, P., van Hoof, F., and Hers, H. G., 1973, Enzyme replacement in Pompe disease: an attempt with purified human acid α-glucosidase, in: *Enzyme Therapy in Genetic Diseases, Birth Defects* (Original Article Series, Vol. 9) (R. J. Desnick, R. W. Bernlohr, and W. Krivit, eds.), pp. 184–190, The Williams & Wilkins Company, Baltimore.

de Duve, C., 1964, From cytases to lysosomes, *Federation Proc.* **23**(2):1045.

Desnick, R. J., Krivit, W., and Fiddler, M. B., 1975, Enzyme therapy in genetic diseases: progress, principles and prospects, in: *The Prevention of Genetic Disease and Mental Retardation* (A. Milunsky, ed.), pp. 317–342, W. B. Saunders Company, Philadelphia.

Desnick, R. J., Thorpe, S. R., and Fiddler, M. B., 1976, Towards enzyme therapy for lysosomal storage diseases, *Physiol. Rev.* **56**:57.

Feinstein, R. N., 1973, Studies of acatalasemia, in: *Enzyme Therapy in Genetic Diseases, Birth Defects* (Original Article Series, Vol. 9) (R. J. Desnick, R. W. Bernlohr, and W. Krivit, eds.), pp. 55–61, The Williams & Wilkins Company, Baltimore.

Fiddler, M. B., and Desnick, R. J., 1977, Enzyme therapy VII: differential *in vivo* stabilities of bovine hepatic, renal and splenic β-glucuronidase activities, *Arch. Biochem. Biophys.*, in press.

Fiddler, M. B., Thorpe, S. R., Krivit, W., and Desnick, R. J., 1974, Enzyme therapy III: *in vivo* fate of erythrocyte entrapped β-glucuronidase, in: *Enzyme Therapy in Lysosmal Storage Diseases* (J. M. Tager, G. J. M. Hooghwinkel, and W. T. Daems, eds.), pp. 182–185, North-Holland Publishing Co., Amsterdam.

Gregoriadis, G., 1974, Structural requirements for the specific uptake of macromolecules and liposomes by target tissues, in: *Enzyme Therapy in Lysosmal Storage Diseases* (J. M. Tager, G. J. M. Hooghwinkel, and W. T. Daems, eds.), pp. 131–148, North-Holland Publishing Co., Amsterdam.

Gregoriadis, G., and Allison, A. C., 1974, Entrapment of proteins in liposomes prevents allergic reactions in pre-immunized mice, *FEBS Letters* **45**:71.

Gregoriadis, G., and Leathwood, R. D., 1971, Enzyme entrapment in liposomes, *FEBS Letters* **14**:95.

Gregoriadis, G., and Ryman, B. E., 1972, Fate of protein-containing liposomes injected into rats: an approach to the treatment of storage diseases, *Eur. J. Biochem.* **24**:485.

Hall, C., Cantz, M., Neufeld, E. F., 1973, A β-glucuronidase deficiency mucopolysaccharidosis: studies in cultured fibroblasts, *Arch. Biochem. Biophys.* **155**:32.

Ihler, G. M., Glew, R. H., and Schnure, F. W., 1973, Enzyme loading of erythrocytes, *Proc. Natl. Acad. Sci. U. S.* **70**:2663.

Jacob, H. S., and Jandl, J. H., 1962, Effects of sulfhydryl inhibition on red blood cells: II. Studies *in vivo*, *J. Clin. Invest.* **41**:1514.

Johnson, W. G., and Brady, R. O., 1972, Ceramide trihexosidase from human placenta, *Methods Enzymol.* **28**:849.

Johnson, W. G., Desnick, R. J., Long, D. M., Sharp, H. L., Krivit, W., Brady, B., and Brady, R. O., 1973, Intravenous injection of purified hexosaminidase A into a patient with Tay-Sachs disease, in: *Enzyme Therapy in Genetic Diseases, Birth Defects* (Original Article Series, Vol. 9) (R. J. Desnick, R. W. Bernlohr, and W. Krivit, eds.), pp. 120–124, The Williams & Wilkins Company, Baltimore.

Lagunoff, D., Nicol, D. M., and Pritzl, P., 1973, Uptake of β-glucuronidase by deficient human fibroblasts, *Lab. Invest.* **29**:449.

Mapes, C. A., and Sweeley, C. C., 1973, Preparation and properties of an affinity column absorbent for differentiation of multiple forms of α-galactosidase activity, *J. Biol. Chem.* **248**:2461.

Nicol, D. W. Lagunoff, D., and Pritzl, P., 1975, Differential uptake of human β-glucuronidase isoenzymes from spleen by deficient fibroblasts, *Biochem. Biophys. Res. Commun.* **59**:941.

Norden, A. G. W., and O'Brien, J. S., 1974, Binding of human liver β-galactosidases to plant lectins insolubilized on agarose, *Biochem. Biophys. Res. Commun.* **56**:193.

O'Brien, J. S., Miller, A. L., Loverde, A. W., and Veath, M. L., 1973, Sanfilippo disease type B: enzyme replacement and metabolic correction in cultured fibroblasts, *Science* **181**:753.

Porter, M. T., Fluharty, A. L., and Kihara, H., 1971, Correction of abnormal cerebroside sulfate metabolism in cultured metachromatic leucodystrophy fibroblasts, *Science* **172**:1263.

Radin, N. S., Hyun, J. C., and Misra, R. S., 1974, Cerebroside glucosidase: assay and purification, *Federation Proc.* **33**:1226.

Sessa, G., and Weissman, G., 1970, Incorporation of lysozyme into liposomes, *J. Biol. Chem.* **245**:3295.

Shapiro, L. J., Hickman, S. G., Hall, C. W., and Neufeld, E. F., 1974, Corrective and non-corrective forms of human α-L-iduronidase, *Amer. J. Human Genet.* **26**:79A.

Snyder, P. D., Jr., Wold, F., Bernlohr, R. W., Dullum, C., Desnick, R. J., Krivit, W., and Condie, R. M., 1974, Enzyme therapy II: purified human α-galactosidase A. Stabilization to heat and protease degradation by complexing with antibody and by chemical modification, *Biochim. Biophys. Acta* **350**:432.

Steger, L. D., and Desnick, R. J., 1976, Enzyme therapy VI: *in vivo* fate of bovine β-glucuronidase entrapped in negatively-and positively-charged liposomes, *Biochem. Biophys. Acta.*

Steger, L. D., Fiddler, M. B., Douglas, S. D., Thorpe, S. R., and Desnick, R. J., 1977, Immunologic evaluation of unentrapped, erythrocyte- and liposome-entrapped bovine β-glucuronidase administered to β-glucuronidase deficient mice, *J. Immunol.*, in press.

Thorpe, S. R., Fiddler, M. B., and Desnick, R. J., 1974, Enzyme therapy IV: a method for determining the *in vivo* fate of bovine β-glucuronidase in β-glucuronidase deficient mice, *Biochem. Biophys. Res. Commun.* **61**:1464.

Thorpe, S. R., Fiddler, M. B., and Desnick, R. J., 1975, Enzyme therapy V: *in vivo* fate of β-glucuronidase in β-glucuronidase deficient mice, *Pediat. Res.* **9**:918.

Uemura, K. I., Claflin, J. L., Davie, J. M., and Kinsky, S. C., 1975, Immune response to liposomal model membranes: restricted IgM and IgG antidinitrophenyl antibdoeis produced in guinea pigs, *J. Immunol.* **114**:958.

von Figura, K., and Kresse, H., 1974, Quantitative aspects of pinocytosis and the intracellular fate of *N*-acetyl-α-D-glucosaminidase in Sanfillipo B fibroblasts, *J. Clin. Invest.* **53**:85.

Weissman, G., Bloomgarden, D., Kaplan, R., Cohen, C., Hoffstein, S., Collins, T., Gotlieb, A., and Nagle, D., 1975, A general method for the introduction of enzymes, by means of immunoglobulin-coated liposomes, into lysosomes of deficient cells, *Proc. Natl. Acad. Sci. U. S.* **72**:88.

Whitaker, T. R., Sartiano, G. P., Hamelly, L. J., Jr., Scott, W. L., and Glew, R. H., 1974, Hypotonic exchange loading of human erythrocytes with ^{59}Fe-labeled rabbit hemoglobin, *J. Lab. Clin. Med.* **84**:879.

Wold, F., 1973, Chemical modification of enzymes, in: *Enzyme Therapy in Genetic Diseases, Birth Defects* (Original Article Series, Vol. 9) (R. J. Desnick, R. W. Bernlohr, and W. Krivit, eds.), pp. 46–54, The Williams & Wilkins Company, Baltimore.

17

Strategy for Enzyme Therapy: Immobilization in Hypoallergenic Gel Versus Entrapment in Red Blood Cell

STUART UPDIKE

1. Gel Immobilization System

The polyacrylamide gel immobilization system (Hicks and Updike, 1966; Updike and Hicks, 1967) provides a chemically inert and apparently nontoxic carrier of the enzyme. Polyacrylamide has a hydrocarbon backbone with amide groups as shown in Figure 1, and therefore is devoid of what normally are considered antigenic determinants (Kabat, 1966). To prevent excessive swelling of the gel, a cross-linking agent, *N,N'*-methylene-bisarcylamide is introduced, as shown in Figure 1. Purified enzyme can be entrapped in the gel during photopolymerization at room temperature. The polymerization reaction is exothermic, therefore, the reaction vial is placed in an ice bath as needed to dissipate heat. The gel is then mechanically fragmented and wet-sieved to provide the right particle size.

Pore size has been studied theoretically (Ornstein, 1964; Raymond and Nakamichi, 1962) and experimentally by permeability determinations (White, 1960). The studies established that the polymer strands of polyacrylamide form a random "brush-heap" structure, and that the average pore size varies inversely with the square of the polymer concentration. However, choosing a polyacrylamide gel recipe that gives an average pore size smaller than the diameter of the molecules of protein to be entrapped does not completely prevent washout of this protein from the gel. Theoretic and experimental studies of pore size provide only estimates of *average* pore size. Some pores will always be larger, and therefore, some leaking out of entrapped protein from the gel will always occur. We have studied washout of radioactively tagged protein as a function of percent polymer and percent cross-linking reagent in the polymerization recipe. Data from a typical study are presented in Figure 2. Most enzyme and antibody molecules have a

STUART UPDIKE • Department of Medicine, University of Wisconsin Center for Health Sciences, Madison, Wisconsin

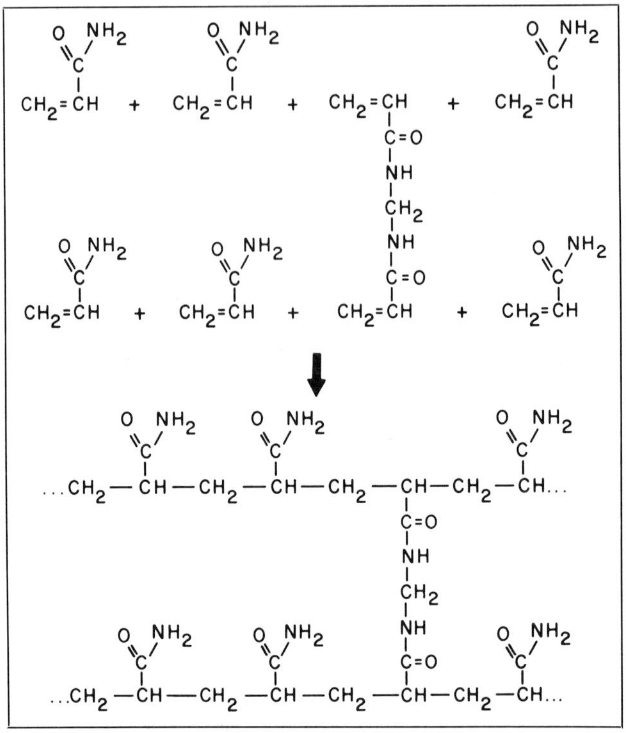

FIGURE 1. Chemical structure of polyacrylamide gel cross-linked with *N,N'*-methylenebisacrylamide.

molecular weight and size substantially larger than that of albumin. Accordingly, gel entrapment is substantially more effective than is shown for albumin in Figure 2.

In Figure 3, gel-entrapped enzyme is schematically presented. Note that the encaged enzyme is protected from high-molecular-weight interferences such as endogenously produced proteolytic enzyme or antibody, while substrate diffuses freely into and out of the gel. Immobilized enzyme has less activity compared to the same amount of free enzyme, presumably because some active sites are hindered. There is no covalent bonding between encaged enzyme and gel polymer matrix.

1.1. Asparaginase, Therapeutic Enzyme

The highly purified bacterial enzyme asparaginase has been shown to have antitumor action. It is based on destruction of an essential tumor tissue nutrient, the amino acid asparagine. For example, the cancer cell of childhood leukemia characteristically is unable to synthesize asparagine (Wriston and Yellin, 1973). Remission of this type of leukemia from asparaginase therapy has been demonstrated (Tan, 1972, Oettgen *et al.*, 1967). Unfortunately, the remission is usually short-lived or incomplete. Further therapy with this enzyme is complicated by life-threatening allergic reactions (Oettgen *et al.*, 1970; Land *et al.*, 1972).

1.2. *Stability of Immobilized Enzyme*

Immobilization of asparaginase does not change its stability with respect to temperature as studied at 55, 60, and 65°C (Updike, 1973). However, the gel entrapment process does protect this enzyme from proteolysis (pronase 4×10^{-3} mg/ml) (Updike, 1973). We have done similar studies to show that the gel immobilized form of this enzyme is protected from attack from bacteria and specific antibody.

1.3. *Intraperitoneal Injection of Gel-Protected Asparaginase*

The serum half-life of asparaginase injected into rodents is only about 2.6 hr (Broome, 1968). However, asparaginase entrapped in particles (wet-sieved 40–60

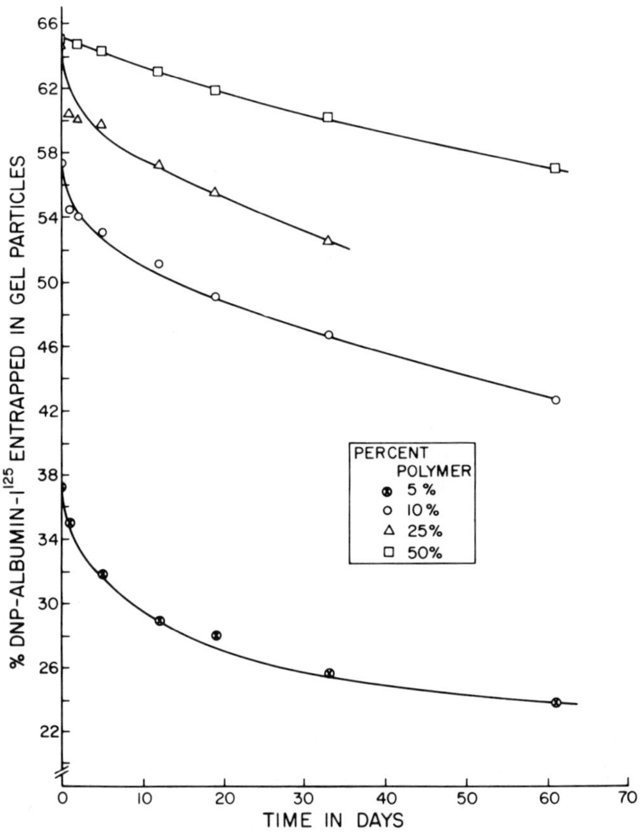

FIGURE 2. Washout of radioactively labeled bovine serum albumin (mol. wt. = 68,000) as a function of polymer concentration and degree of cross-linking. The albumin was first dinitrophenylated to prevent aggregation into higher-molecular-weight complexes. The gel particles were fragmented, washed, and sieved to 40–60 mesh size over a 1-hr interval before the washout study was begun. Thus, at time zero, most of the immediate rapid washout had already occurred. The percent polymer weight/volume is given in the key. The ratio of monomer to cross-linking agent was 4:1.

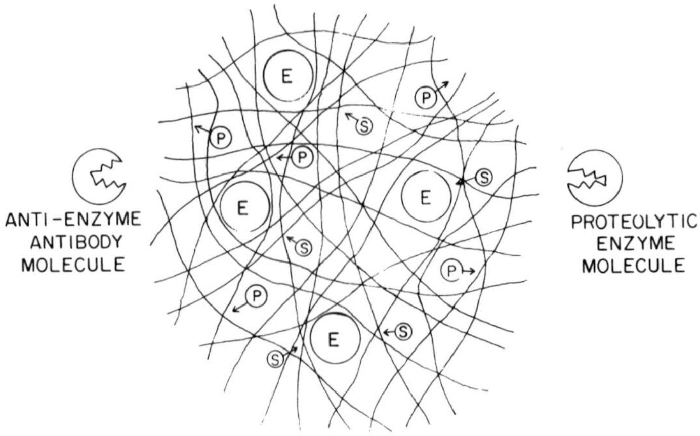

FIGURE 3. Schematic presentation of protection against high-molecular-weight interferences by gel entrapment of enzyme.

mesh) injected into the peritoneal cavity of rats showed an activity half-life of about 9 days. This estimate of *in vivo* half-life was made by carefully dissecting out the injected gel particles from the peritoneal cavity of sacrificed animals serially at various intervals. These particles were then placed back into the same miniature chromatographic column (Updike, 1973) used to initially estimate immobilized enzyme activity for repeat assay of enzyme activity. Measurable asparaginase activity was present in the particles isolated from the peritoneal cavity at 4 weeks.

Unfortunately, intraperitoneal injection of asparaginase active gel particles was associated with only transient and incomplete suppression of substrate, as determined by serial measurements of the asparagine level in plasma. Apparently, the intraperitoneal route of injection did not provide enough contact between the intraperitoneal enzyme and the substrate occupying the extracellular fluid compartment. Therefore, we elected to develop and evaluate an intravenous injection configuration of gel–enzyme treatment.

1.4. *Gel-Entrapped Enzyme Particles Suitable for Intravenous Injection*

Preparing gel particles suitable for intravenous injection was more difficult than initially anticipated. Gel particles greater than 5 μm in diameter injected intravenously will plug the microcirculation of the lung. This is not surprising considering that red blood cells, which are more compliant, are typically 7 μm in diameter. Gel particles 4 μm or smaller do not plug the microcirculation.

First attempts at obtaining uniformly small gel particles involved mechanically fragmenting gel slabs first in a blender and then ultrasonically. Then a system of differential centrifugation was used to remove particles greater than 5 μm in diameter. This was only partially successful. Typically a gel preparation would retain 0.3–1% particles greater than 5 μm. Upon IV injection in the rat, these gel-particle preparations would cause the animal to wheeze. Intravenous injection of greater than 0.8 ml of packed particle volume suspended in 1.6 ml of buffer solution would cause an immediate acute respiratory death owing to plugging of the microcirculation of the lung of the rat, as confirmed by study of postmortem microscopic sections of lung tissue. We concluded that we had not yet developed a satisfactory method to produce enzyme-active gel particles uniformly below 5 μm. This was finally achieved by ultrasonically nebulizing the gel enzyme recipe and then photocatalytically polymerizing the nebulized droplets, as shown in Figure 4. Spherical, enzyme-active gel particles resulted. These particles were uniformly less than 5 μm in diameter and could be injected without causing respiratory distress in the experimental animal.

1.5. *Intravenous Injection of Gel-Protected Asparaginase*

The amount of polyacrylamide gel particles intravenously injected into an experimental animal is most conveniently expressed as the percent packed particle volume using the same procedure and equipment (centrifuge) used to report packed red blood cell volume. Asparaginase-active polyacrylamide gel particles were prepared with 260 international enzyme units/ml of packed particle volume using the method described above for obtaining uniformly small (5 μm or less) particles. We have measured at intervals serum asparagine level in rats injected intravenously with the asparaginase-active gel particles. Each rat received 0.3 ml, which is approximately 78 units of immobilized asparaginase. A prompt fall in serum asparagine was noted which lasted for at least 24 hr. At 24 hr the experimental rats measured 6 nmoles/ml, as compared to 55 ± 13 nmoles/ml for

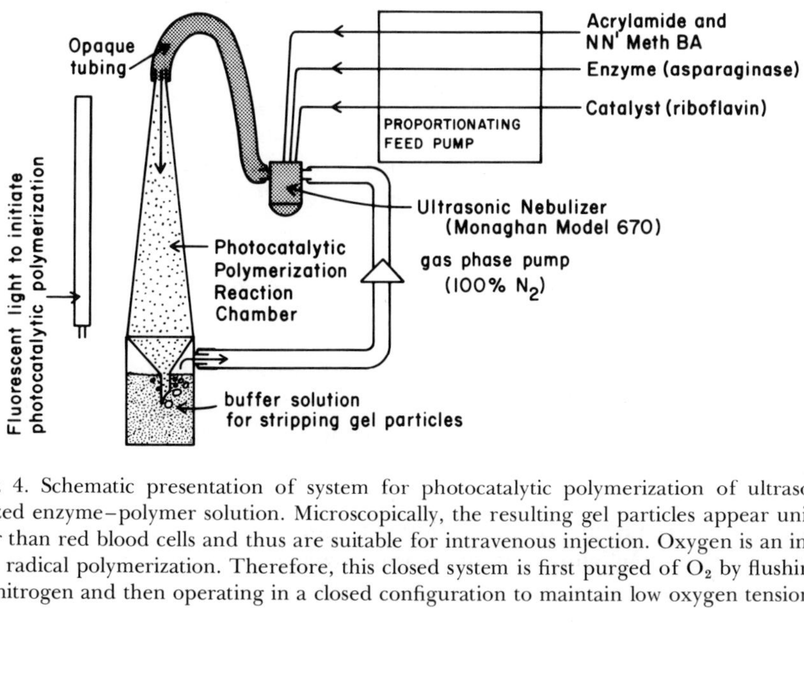

FIGURE 4. Schematic presentation of system for photocatalytic polymerization of ultrasonically nebulized enzyme–polymer solution. Microscopically, the resulting gel particles appear uniformly smaller than red blood cells and thus are suitable for intravenous injection. Oxygen is an inhibitor of free radical polymerization. Therefore, this closed system is first purged of O_2 by flushing with 100% nitrogen and then operating in a closed configuration to maintain low oxygen tension.

the control rats. Unfortunately, this profound drop in the serum asparagine level was sustained for 72 hr or longer in less than half the rats.

Control rats injected with 0.3–0.5 ml of packed particle volume of polyacrylamide appeared healthy, gained weight, and lived as long as noninjected rats. Only those rats injected with gel particles made from photocatalytic polymerization of nebulized polymer were uniformly small enough to allow injection without wheezing occurring due to plugging of the pulmonary microcirculation. On autopsy of these animals, a spleen twice normal size was found as early as 3 weeks. The liver was less obviously enlarged. Microscopic examination of the rat organs showed deposition of the polyacrylamide gel in the reticuloendothelial system, primarily spleen and liver. The amount of gel present could be semiquantitatively estimated by inspection of these tissues. The amount and appearance of the gel particles and tissues appeared remarkably the same in animals sacrificed at 2 and 16 months. Our impression is that the polyacrylamide gel is not broken down by any of the enzymes of the rat. This presents a serious problem. Without biodegradation of this polymer, continued use of this therapeutic modality would necessarily induce an iatrogenic gel storage disease.

We concluded that clearly a more biodegradable polymer carrier is needed. We have had some experience entrapping asparaginase in fibrin gel followed by fragmenting of the gel into particles small enough to be suitable for intravenous injection. Unfortunately, the plasminogen system is rapidly activated and the fibrin gel particles have a half-life of only a few minutes. Our search for an immunologically inert and moderately biodegradable synthetic polymer continues.

2. *Red Blood Cell Entrapment System*

Therapeutic enzyme can be loaded by diffusion into hypotonically lysed red blood cell ghosts and resealed as described by Ihler *et al.* (1973). Lysis is carried out hypothermically in hypotonic media. The resealment process occurs spontaneously when the enzyme-loaded, hemoglobin-depleted red blood cell ghosts are returned to a physiological normotonic media at 36°C.

2.1. *Red Blood Cell Asparaginase Entrapment Procedure*

Asparaginase (MSD E.C. 2 Lot L594269-0-51) was entrapped in red blood cells by resealing the ghosts of red blood cells following the methods of Bodemann and Passow (1972) and Ihler *et al.* (1973). Forty milliliters of fresh heparinized whole blood was centrifuged at 1000*g* for 10 min. The plasma and buffy coat were removed and the cells were washed three times in cold (6°C) Hank's basic salt solution (HBSS). The packed cells (Hct = 80–90%) were mixed rapidly in twice their volume of cold hemolyzing solution containing the enzyme. The hemolyzing solution was made by simply diluting the enzyme to be entrapped in distilled water. After a 5-min equilibration in the cold, sufficient concentrated cold HBSS was added to restore isotonicity. This suspension was warmed to 37°C and incubated at that temperature for 45 min. The resealed cells were collected by centrifugation at 1000*g* for 15 min and washed three times with isotonic HBSS to remove any untrapped enzyme (Updike *et al.*, 1976).

2.2. *In Vitro Evaluation of Enzyme-Loaded Red Blood Cells*

The cells loaded by this method were analyzed by a Type S Coulter Counter (Coulter Electronics, Inc., Hileah, Florida). In comparison with normal washed red blood cells, the cell volume of the loaded ghosts had decreased from 90 ± 4.0 S.E.M. μm^3 to 63 ± 4.0 S.E.M. μm^3. The corpuscular hemoglobin was reduced from 31 ± 0.5 S.E.M. pg/cell to 15 ± 0.5 S.E.M. pg/cell, and approximately 5% of the cells were lost during lysis, resealing, and washing.

Asparaginase was labeled with I^{125} by modifying the procedures by Greenwood *et al.* (1963) and Hunter and Greenwood (1962). When asparaginase was made 0.25 g/100 ml in the hemolyzing media, 20% of the enzyme protein was entrapped in the red blood cell. The concentration of intracellular I^{125}-labeled enzyme was 82.7% of the concentration expected theoretically if the enzyme had distributed homogeneously both intracellularly and extracellularly before resealing. Assayed enzyme activities (Meister, 1955) inside these asparaginase-loaded cells were 95–104% of activities expected from I^{125} determinations. Thus, the entrapment process caused no apparent loss of enzyme activity. However, it was shown that the substrate, asparagine, cannot cross the cell membrane to react with the entrapped asparaginase. No activity was seen unless the asparaginase-loaded cells were relysed. Asparagine probably fails to cross the RBC membrane because of its cationic charge.

2.3. *In Vivo Evaluation of Asparaginase-Loaded Red Blood Cells in Monkeys*

Monkeys were injected with asparaginase (1850 IU/Kg) in free solution (three control monkeys) and entrapped in RBC (three experimental monkeys). The I^{125} asparaginase-loaded cells prepared as above were labeled with ^{51}Cr (Mollison and Veal, 1955) and returned intravenously into each experimental monkey. Control monkeys were given injections of free asparaginase in HBSS and ^{51}Cr-labeled control cells. The control RBC's were treated identically to the loaded cells except they were not lysed and resealed. Over a 30-day period, serial determinations of serum and whole blood for I^{125}, ^{51}Cr, and enzyme activity (Cooney *et al.*, 1970) and substrate level (asparagine method of Cooney *et al.*, 1970) were made.

The half-life data determined from the survival of ^{51}Cr-labeled resealed cells and control washed red cells are shown in Figure 5. Note that a rapid loss of cells was followed by a slower, more gradual exponential loss. This gradual loss, which appears linear on the semilogarithmic graph, was used to estimate the half-life. A half-life of about 8 days was obtained for both enzyme-loaded and control RBC.

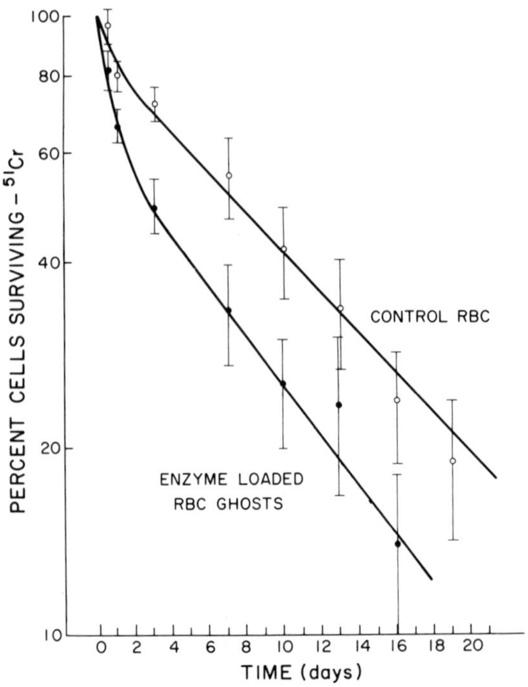

FIGURE 5. Survival in enzyme-loaded and control RBC. Open-circle data points represent control unlysed RBC washed and incubated identically as the resealed ghost RBC. Closed-circle data points represent enzyme-loaded RBC with an asparaginase concentration and activity of 3.9 mg and 780 IU/ml of RBC ghost volume. TBC were labeled with chromium 51. Each point represents blood samples drawn from three monkeys. Brackets show ±1 S.E.M.

The initial more rapid loss of the resealed cells as compared to the controls is probably due to the presence of some crenated and leaky, unsealed cells. However, it should be noted that some initial rapid disappearance of ^{51}Cr is normal for monkeys as well as for humans (Jones *et al.,* 1956; Huser, 1970). Since normal values in the literature for *rhesus* monkey red blood cell half-life is about 22 days (Huser, 1970). We believe that incubation and washing in our experiments have significantly shortened the half-life of both the enzyme-loaded and control RBC.

Our results show that after 10 days the circulating enzyme activity entrapped in red blood cells was two orders of magnitude higher than the circulating enzyme activity maintained when asparaginase was injected as a free solution as shown in Figure 6. Our results show further that markedly depressed or undetectable levels of circulating asparagine last twice as long after a single injection of RBC-entrapped asparaginase than after a single injection of the free enzyme, as shown in Figure 7. This figure compares the substrate concentrations of the six monkeys used in the experiment. Monkeys receiving the RBC-loaded enzyme had lowered asparagine for more than 19 days, whereas those receiving injection of the free enzyme showed asparagine substrate suppression for only 10 days.

We believe that *in vivo* lysis of the asparaginase loaded cells slowly released the entrapped enzyme and maintained a higher enzyme concentration for the longer interval.

3. Summary and Conclusions

In summary, resealed RBC ghosts show promise as a slow-release vehicle for enzyme therapy. Substantial protection against degradation of the therapeutic enzyme by endogenous proteolytic enzyme or by antibody complexing appears probable. For those therapeutic enzymes that theoretically should be targeted at the reticuloendothelial system, immobilization in RBC may offer the additional advantage of release directly into the reticuloendothelial system. The partial uptake of the radioactively tagged RBC was seen by gamma-counter scan over the spleen and liver in the monkeys used in these experiments.

We conclude the entrapment of enzyme in RBC offers an attractive new strategy for enzyme therapy. This approach appears particularly advantageous for treatment of the enzyme-deficiency-based storage diseases and the asparagine-dependent cancers of the reticuloendothelial system.

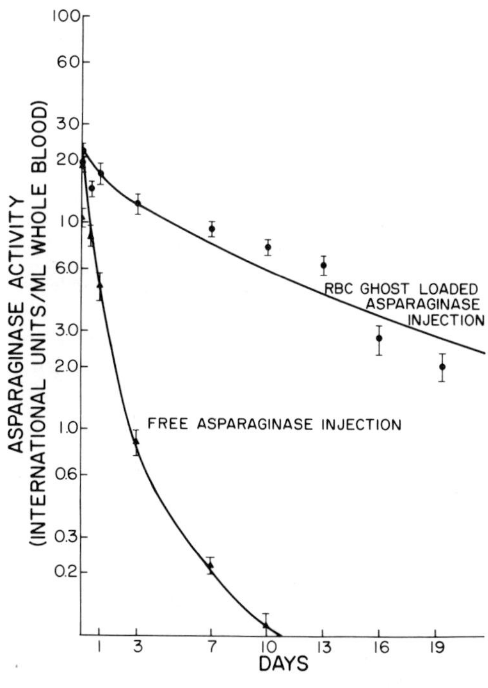

FIGURE 6. Asparaginase concentration. Circle data points represent asparaginase injected entrapped in red cells. Triangle data points represent asparaginase injected in Hank's buffered salt solution. Each point represents blood samples drawn from three monkeys. Activities of asparaginase were assayed by the method of Cooney *et al.* (1970). Brackets show ±1 S.E.M.

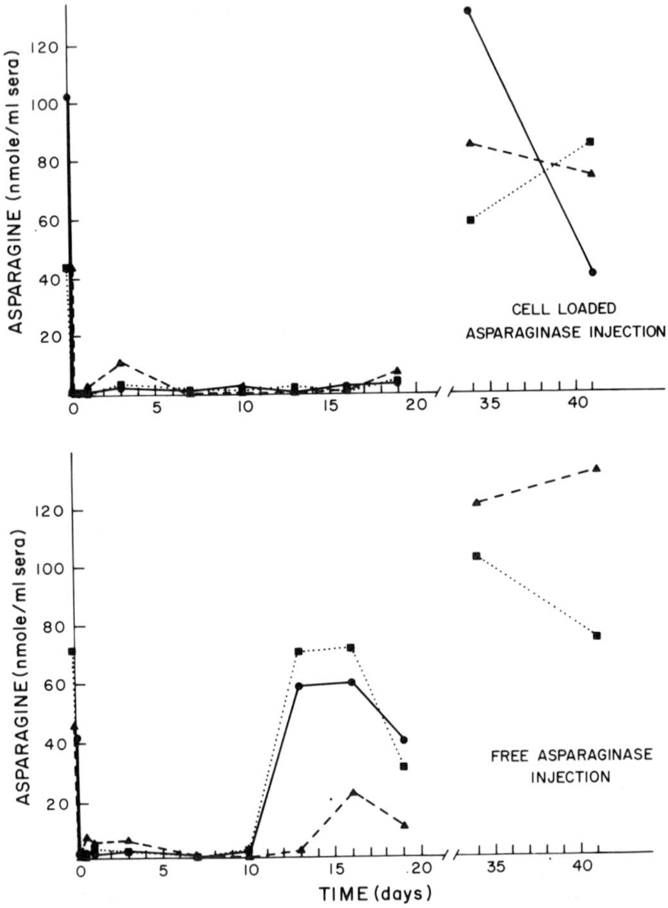

FIGURE 7. Asparagine levels in monkey sera. Figure shows asparagine levels in sera from six monkeys each receiving 1850 IU/Kg as a single intravenous injection. The top graph shows the levels in three monkeys receiving asparaginase entrapped in red cells. The lower graph shows asparagine levels in three monkeys receiving free asparaginase in solution. Assay for asparagine followed the method described by Cooney *et al.* (1970).

4. *References*

Bergsma, Daniel, 1973, *Enzyme Therapy in Genetic Diseases, Birth Defects* (Original Article Series, Vol. 9), pp. 1–233, The Williams & Wilkins Company, Publisher.

Bodemann, H., and Passow, H., 1972, *J. Membrane Biol.* **8**:1.

Broome, J. D., 1968, Factors which may influence the effectiveness of L-asparaginases as tumor inhibitors, *Brit. J. Cancer* **22**:595.

Cooney, D. A., Capizzi, R. L., and Handschumacher, R. E., 1970, Evaluation of L-asparagine metabolism in animals and man, *Cancer Res.* **30**:929.

Greenwood, F. C., Hunter, W. M., and Glover, J. S., 1963, *Biochem. J.* **89**:114.

Grossi, C. E., Cliffton, E. E., and Cannamela, D. A., 1954, *Blood* **9**:310.

Hersh, Evan M., 1971, *Transplantation* **12**:368.

Hicks, G. P., and Updike, S. J., 1966, The preparation and characterization of lypholized polyacrylamide enzyme gels for chemical analysis, *Anal. Chem.* **38**:726.
Hunter, W. M., and Greenwood, F. C., 1962, *Nature* **194**:495.
Huser, H., 1970 *Atlas of Comparative Primate Hematology,* pp. 78–85, Academic Press, Inc., New York.
Ihler, G. M., Glew, R. H., and Schnure, F. W., 1973, *Proc. Natl. Acad. Sci. U.S.* **70**:2663.
Innerfield, I., Angrist, A., and Benjamin, J. W., 1952 *Gastroenterology* **20**:630.
Jones, N. C. H., and Mollison, P. L., 1956, *Clin. Sci.* **15**:207.
Kabat, E. A., 1966, The nature of an antigenic determinant, *J. Immunol.* **97**:1.
Kryle, L., Arnoldi, C., and Kupperman, H. S., 1967, *Ann. N.Y. Acad. Sci.* **68**:178.
Land, V. J., Sutow, W. W., Fernbach, D. J., Lane, D. M., and Williams, T. E., 1972, *Cancer* **30**:339.
Meister, A., 1955, *Methods Enzymol.* **2**:380.
Mollison, P. L., and Veal, N., 1955, *Brit. J. Haematol.* **1**:62.
Oettgen, H. F., Old, L. J., Boyse, E. A., Campbell, H. A., Philips, F. S., Clarkson, B. D., Tallal, L., Leeper, R. D., Schwartz, M. K., and Kim, J. H., 1967, *Cancer Res.* **27**:2619.
Oettgen, H. F., Stephenson, P. A., Schwartz, M. K., Leeper, R. D., Tallal, L., Tan, C. C., Clarkson, B. D., Golbrey, R. B., Krakoff, I. H., Karnofsky, D. A., Murphy, M. L., and Burchenal, J. H., 1970, *Cancer* **25**:253.
Ornstein, L., 1964, Disc electrophoresis: I. Background and theory, *Ann. N. Y. Acad. Sci.* **121**:321.
Raymond, S., and Nakamichi, M., 1962, Electrophoresis in synthetic gels, *Ann. Biochem.* **3**:23.
Tan, Charlotte, 1972, *Hospital Practice* **1**:99.
Updike, S. J., 1973 in: *Enzyme Therapy in Genetic Diseases, Birth Defects* (Original Article Series, Vol. 9) (D. Bergsma, ed.) p. 79, The Williams & Wilkins Company, Publisher.
Updike, S. J., and Hicks, G. P., 1967, Reagentless substrate analysis with immobilized enzymes, *Science* **158**:270.
Updike, S. J., Wakamiya, R. T., and Lightfoot, E. N., 1976, *Science* **193**:681.
White, M. L., 1960, The permeability of an acrylamide polymer gel, *J. Phys. Chem.* **64**:1563.
Wriston, J. C., and Yellin, T. O., 1973, *Adv. Enzymol.* **39**:185.

18

Immobilized Enzymes for Therapeutic Applications and for Large-Scale Production of Biologically Active Compounds

ICHIRO CHIBATA, TETSUYA TOSA, AND TAKAO MORI

1. *Introduction*

Recently, the enzyme industry has been developing rapidly, through advances in the biochemistry, bioorganic chemistry, and techniques of enzyme production by microorganisms. Many kinds of enzymes have been widely used in the fields of medical, food, and synthetic chemical industries. However, these enzymes have been conventionally used in soluble form, which has some disadvantages. Immobilization of enzymes has been suggested as a potential method of solving these problems. In this chapter, studies on enzyme therapy using immobilized enzymes and the large-scale production of biologically active compounds are presented.

2. *Therapeutic Applications*

Although techniques for the production and purification of enzymes and studies on the reaction mechanisms of enzymes have been developed markedly in recent years, and the clinical use of enzymes in various diseases has been increasing, a number of problems limit the therapeutic use of proteins such as enzymes. For instance, repeated injections of enzymes induce the formation of antibodies in the recipient. As a result, anaphylactic shock and/or inactivation of the enzyme by immunological reactions may occur. Further, injected enzymes may be destroyed by proteolytic enzymes in blood.

ICHIRO CHIBATA • Research Laboratory of Applied Biochemistry, Tanabe Seiyaku Co. Ltd., Osaka, Japan TETSUYA TOSA • Research Laboratory of Applied Biochemistry, Tanabe Seiyaku Co. Ltd., Osaka, Japan TAKAO MORI • Research Laboratory of Applied Biochemistry, Tanabe Seiyaku Co. Ltd., Osaka, Japan

To solve these problems, the use of immobilized enzymes has been widely attempted (Mori *et al.*, 1976; Cooney *et al.*, 1975; Inada *et al.*, 1975; Chong and Chang, 1974; Nadler and Updike, 1974; Mori *et al.*, 1974; Horvath *et al.*, 1973; Mori *et al.*, 1973; Allison *et al.*, 1972; Mori *et al.*, 1972a; Boguslaski and Janik, 1971; Chang, 1971a; 1971b; Gregoriadis *et al.*, 1971; Weetall, 1970). Chang and coworkers have extensively studied the therapeutic use of enzymes microencapsulated by semipermeable polymer membranes. In this section we present our studies on the immobilization of asparaginase, which has been employed for the suppression of lymphosarcoma. In these studies we used the asparaginase from *Proteus vulgaris,* which differs immunologically from the enzyme from *Escherichia coli* (Tosa *et al.*, 1971b). The *P. vulgaris* asparaginase was purified in high yield from cell extract in our research laboratory by general purification techniques (Tosa *et al.*, 1972) or affinity chromatography (Tosa *et al.*, 1974b). The enzyme was found to have potent antitumor activity for 6C3HED-OG cells as well as *E. coli* enzyme (Tosa *et al.*, 1971b).

2.1. *Microencapsulated Asparaginase*

For the purpose of direct administration of enzyme, we carried out microencapsulation of asparaginase and investigated its enzymatic properties and possibility in its clinical use.

a. Screening of Microencapsulation Methods. We tested several microencapsulation methods, such as phase separation, interfacial polymerization, and drying in aqueous solution. The interfacial polymerization method using 1,6-hexanediamine and sebacoylchloride as a reactive monomer introduced by Chang *et al.*, (1966) was found to be most suitable for the microencapsulation of enzyme. Because the preparation procedure is easy, the yield of enzyme activity after microencapsulation is high and the particle size of the microcapsule can be easily controlled. Thus, to obtain active microencapsulated asparaginase suitable for therapeutic use, we studied in detail the conditions for polymerization reactions (Mori *et al.*, 1972a). Nylon microcapsules containing asparaginase were obtained in 37% yield under the optimal conditions. Besides nylon microcapsules, we succeeded in the preparation of polyurea microcapsules containing asparaginase by using 1,6-hexanediamine and 2,4-toluenediisocyanate instead of sebacoylchloride as a reactive monomer. This new microcapsule was prepared in 42% yield and was superior to the nylon microcapsule with respect to mechanical strength. To obtain the active microencapsulated asparaginase, the presence of proteins such as albumin, casein, or hemoglobin in the polymerization mixture is essential. The function of these proteins is not clear. However, from the fact that the protective effect varies with the protein species added in the polymerization medium and that the microcapsule is not formed in the absence of protein, it may be considered that these proteins are not only useful in the prevention of reactions between the enzyme–protein and sebacoylchloride, but also effectively stabilize the emulsion by giving a suitable viscosity to aqueous droplets. Further, the addition of L-aspartic acid, the product of asparaginase reaction, in the polymerization medium enhances the enzyme activity of the microcapsule as a result of the protective effect of the acid for the enzyme.

b. Characteristics and Possibility in the Clinical Use of Microencapsulated Asparaginase. To determine the applicability of microencapsulated asparaginase for clinical use, we studied the characteristics of nylon and polyurea microcapsules (Mori *et al.,* 1973). It was found that the pH optimum of both microencapsulated asparaginases for enzyme reaction is the same as that of native enzyme, but the apparent Michaelis constants of both microcapsules are about 100 times higher than that of the native one. Further, the stability of both microcapsules in storage and in repeated use was studied. Storage stability of the microcapsule was the same as that of the native one, but stability of the microcapsule after repeated use was high, and no loss of activity was observed after it had been used 5 times. To obtain further information on the stability of microencapsulated enzymes, the effect of proteases on the microencapsulated asparaginase was investigated and compared with that of the native one. These results are shown in Table 1. From the table, it is clear that by microencapsulation the resistance to attack by proteolytic enzymes is very much enhanced. These results on the stability mentioned above indicate that asparaginase is completely encapsulated in the semipermeable nylon or polyurea membrane and does not leak out from the microcapsule.

In practical use of microencapsulated asparaginase, it is necessary to clarify the acute toxicity and immunological reaction of the microcapsule membrane. A nylon or polyurea microcapsule was administered into the intraperitoneal cavity of a rat at appropriate intervals, and the acute toxicity was tested by measuring the growth rate of the animal. As a result, no difference was observed between the growth rate in the microcapsule and that in a physiological saline solution used as a control.

Further, antibodies were not produced in either microcapsule membrane. However, the microcapsule membrane consists of synthetic polymer, and decomposition of such microcapsules when administered into the living body is very slow.

TABLE 1

Effect of Proteases on the Enzyme Activity of Native and Microencapsulated Asparaginases

Proteases[a]	Remaining activity (%)[b]: Native asparaginase	Nylon microcapsules containing asparaginase	Polyurea microcapsules containing asparaginase
Chymotrypsin	4	111	96
Pronase-P	12	106	107
Trypsin	28	104	100

[a]For each 1 mg of asparaginase, 10 μg of chymotrypsin, 20 μg of Pronase-P, or 500 μg of trypsin was used. Treatment with these proteases was carried out at 37°C for 10 min.
[b]The activity of asparaginases before treatment was taken as 100%.

Microcapsules that remain in the body for a long period may become toxic, so a microcapsule membrane must be developed which is nontoxic and will decompose after a specified period.

2.2. *Polyacrylamide Gel-Entrapped Asparaginase*

For the clinical application of enzymes by an extracorporeal shunt system, we prepared immobilized asparaginase by entrapping it in polyacrylamide gel-lattice, which is physically more stable than the microcapsule membrane described above. In the case of this immobilized asparaginase, the same enzyme action may be expected as in the case of microencapsulated enzyme.

a. Preparation of Immobilized Asparaginase. To prepare immobilized asparaginase suitable for use in an extracorporeal shunt system, we investigated in detail conditions for the immobilization of the enzyme by polyacrylamide gel (Mori *et al.*, 1974). Immobilization procedures of asparaginase are briefly described as follows. To 4 ml of 0.1 M phosphate buffer (pH 7.0) containing asparaginase are added a protective substance, 750 mg of acrylamide monomer, 40 mg of *N,N'*-methylenebisacrylamide, 0.5 ml of 5% β-dimethylaminopropionitrile, and 0.5 ml of 1% potassium persulfate solution. The reaction mixture is stood at 23°C for 10–30 min. After completion of polymerization, the stiff gel is made in appropriate particle size, and then the gel particles are thoroughly washed with the buffer solution and water. For preparation of the active immobilized asparaginase, protection of the enzyme during immobilization is necessary. As shown in Table 2, in the polymerization reaction, the absence of one of the amino acids results in an immobilized asparaginase having no enzymatic activity. Active immobilized asparaginase can be obtained in good yield by the addition of one of the suitable amino acids to the polymerization reaction medium.

b. Properties of Immobilized Asparaginase. To study the possibility of the immobilized asparaginase in therapeutic use, the enzymatic properties of the immobilized enzyme were investigated and compared with those of the native enzyme. The immobilized asparaginase shows the highest activity at pH 7.0, whereas the native one is most active at pH 8.0. When the immobilized enzymes are prepared using polyacrylamide gel, the shift in optimum pH is also observed in several enzymes. The reason for the shift in optimum pH by immobilization is generally attributed to the polyelectrolytic nature of the insoluble enzyme carrier. However, amides are electrostatically neutral, and polyacrylamide gel does not exhibit a polycationic nature. Accordingly, the reason for the shift of optimum pH of the immobilized asparaginase is not clear. The apparent Michaelis constant of the immobilized asparaginase is about 200 times higher than that of the native enzyme.

The stability of the immobilized asparaginase packed in a jacketed column (10 ml in volume) was studied. The column was maintained at 37°C, and 0.1 M L-asparagine solution (pH 7.0) was passed through the column at a flow rate of 12.5

TABLE 2
Protective Effects of Various Substances on Immobilization of Asparaginase

Substances added	Immobilized asparaginase Activity (μmoles/min)	Yield[a] (%)
None	0	0
L-Asparagine (150)[b]	22.6	32
D-Asparagine (150)	20.4	29
L-Glutamine (500)	23.7	34
D-Glutamine (200)	21.7	30
L-Na-aspartate (500)	20.3	29
D-Na-aspartate (500)	21.3	30
L-Na-glutamate (1000)	17.8	25
D-Na-glutamate (500)	17.8	25
L-Isoleucine (200)	16.5	24
L-Phenylalanine (200)	7.4	11
L-Valine (200)	15.3	22
Casein (50)	0	0
Sodium sulfate (200)	0	0

[a]The total activity of asparaginase used was taken as 100%.
[b]Values in parentheses show the amount (mg) of added substances which gave the highest activity.

ml/hr. As a result, this column retained 40% of the initial activity after 15 days of continuous operation. To confirm that the asparaginase is firmly entrapped in the gel-lattice and is not leaking out, the column effluent was further incubated at 37°C for 1 hr, and it was determined whether there would be an increase in ammonia. No increase was observed.

To obtain further information regarding the stability of the immobilized asparaginase, the effect of proteases on the two forms of immobilized asparaginase was studied. It was found that native asparaginase and DEAE-cellulose–asparaginase complex were considerably inactivated by incubation with proteases, but the asparaginase entrapped in polyacrylamide gel was not affected. To prove the effectiveness of the immobilized asparaginase in blood, fresh rabbit blood was passed through the immobilized asparaginase column at 37°C at a space velocity of 0.6. L-Asparagine in blood was completely decomposed by the column, and formation of thrombus was not observed during the passage of heparinized blood through the column.

2.3. *Asparaginase Tube Prepared by Polyacrylamide Gel*

To prepare immobilized asparaginase suitable for therapeutic use by extracorporeal shunt system, we used an asparaginase tube, which is considered to be superior to the immobilized asparaginase column with respect to resistance in the bloodstream (Mori *et al.,* 1976).

a. Preparation of the Asparaginase Tube. To determine the optimal conditions for the preparation of the asparaginase tube, we investigated the effects of concentration of the acrylamide monomer, protective substances, and the tube shape (Mori *et al.,* 1976). A physically stable and active asparaginase tube was obtained under the following conditions. To 4 ml of 0.1 M phosphate buffer (pH 7.0) containing asparaginase are added 750 mg of acrylamide monomer, 60 mg of *N,N′*-methylenebisacrylamide, 300 mg of L-serine as a protective substance, 0.5 ml of 5% β-dimethylaminopropionitrile, and 0.5 ml of 1% potassium persulfate solution. One milliliter of this reaction mixture is poured into a glass tube in which a glass rod coated with Tween-20 is set vertically in the center. The glass tube is stood at 23°C for 5 min. After completion of polymerization, a glass rod is pulled out. The hollow asparaginase tube obtained is cooled by steeping in ice water and washed by passing 0.1 M borate buffer (pH 8.0) and water through the tube. The inside diameter, thickness of the gel layer, and length of the asparaginase tube are 3 mm, 0.8 mm, and 10.48 cm, respectively. A sectional diagram and photograph of the asparaginase tube are shown in Figures 1 and 2, respectively. In the case of the asparaginase tube, the activity of the tube depends on the inner surface area (i.e., the catalytic area) when a constant amount of asparaginase is used for the preparation of the tube. This fact indicates that enzyme reaction occurs on the surface region of gel, and the diffusion of substrate or product in the gel is a limiting step for the enzyme reaction rate. Accordingly, the asparaginase tube is inferior to the granular type of immobilized asparaginase with respect to enzyme activity, but the former is superior to the latter with respect to physical properties.

b. Characteristics and Application of the Asparaginase Tube. The enzymatic properties of the asparaginase tube were next compared with those of the native enzyme. The optimum pH of the asparaginase tube used for enzyme reaction is about 1 pH unit more acid than that of the native asparaginase. The effect of substrate concentration on enzyme activity was studied, and the apparent Michaelis constant of the asparaginase tube was about 200 times higher than that of the native asparaginase. As we observed for the microencapsulated asparaginase (Mori *et al.,* 1972a), Allison *et al.,* (1972) and Cooney *et al.,* (1975) observed the same phenom-

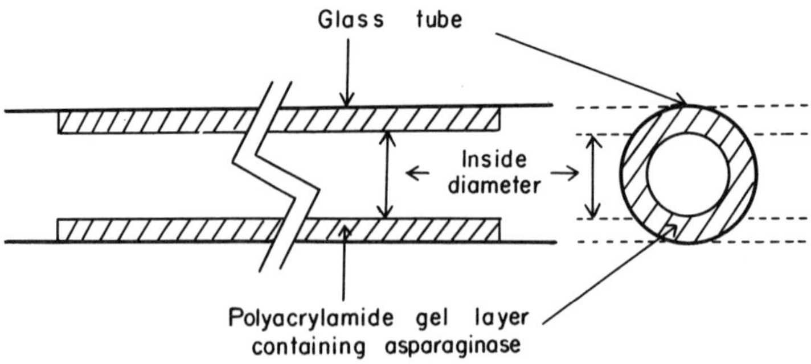

FIGURE 1. Sectional diagram of asparaginase tube.

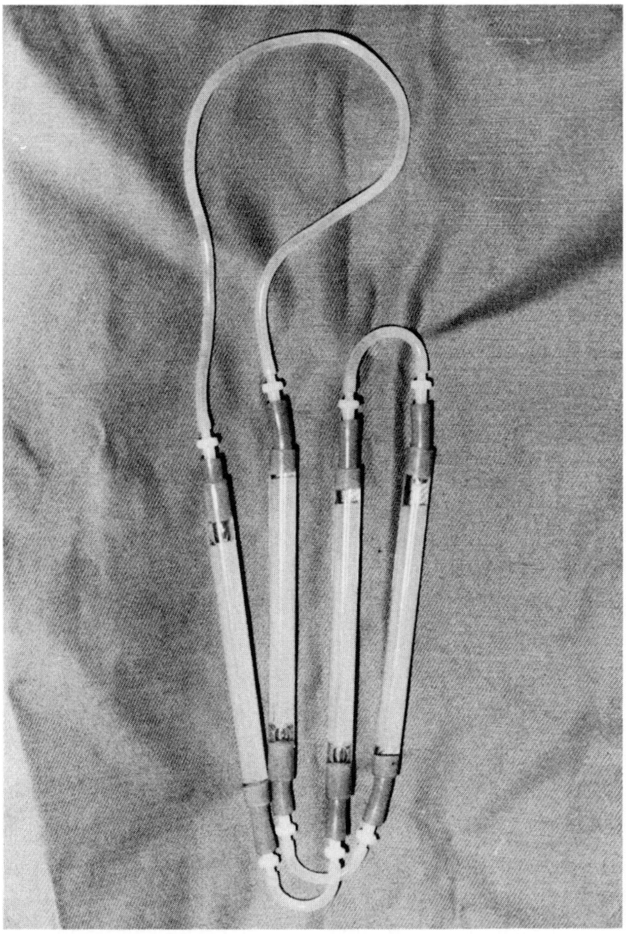

FIGURE 2. Asparaginase tube.

ena in the asparaginase covalently bound to nylon tube and the Dacron vascular prosthesis, respectively. The reason for the increase in K_m value is explained by the limitation in the permeation rate of the substrate and product through the polymer gel lattice or membrane.

Further, to clarify the effectiveness of the asparaginase tube, a model experiment of an extracorporeal shunt system was performed using the asparaginase tube. That is, 500 ml of 32 μM L-asparagine solution was circulated through the asparaginase tubes, having an activity of 600 μmoles/hr at 37°C at a flow rate of 2.4 liters/hr. Diagram for this circulation system is shown in Figure 3. After appropriate circulation, L-asparagine and L-aspartic acid in circulation solution were determined by the NAD-coupled method of Cooney *et al.*, (1970). L-Asparagine was completely decomposed after circulation for 3 hr. On the other hand, animal experiments on extracorporeal shunts using the asparaginase tube were per-

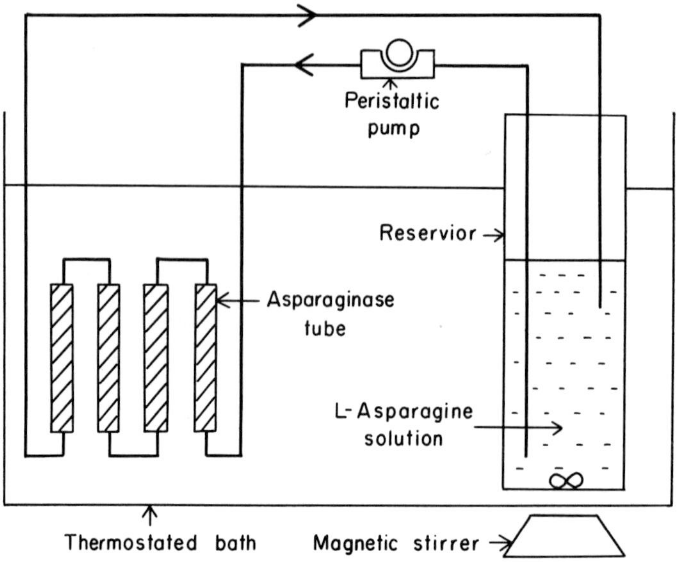

FIGURE 3. Model extracorporeal shunt system using asparaginase tube.

formed by Taguchi and Nakano at the Research Institute for Microbial Disease, Osaka University, as follows. A dog (beagle strain, 6 kg of body weight) was anesthetized with nembutal. The heparin and Ringer's solution were administered to the animal. The blood was circulated through the asparaginase tube having an activity of 800 μmoles/hr from artery to vein at the dog's thigh. The blood circulation was carried out at 37°C at a flow rate of natural bloodstream. After appropriate circulation, blood was drawn at the inlet of the asparaginase tube. The L-asparagine and L-aspartic acid in the plasma obtained were assayed by Cooney's method, described above. The concentration of L-asparagine in the blood plasma lowered to 25% of the initial concentration after circulation for 2 hr, but it reached an even plateau when circulation was continued further. These results may be due to the inductive formation of L-asparagine synthetase *in vivo*.

As described above, three kinds of immobilized asparaginase show peculiar properties, respectively. The characteristics of these immobilized asparaginases are summarized in Table 3. Among these preparations, asparaginase entrapped in a tube is resistant to attack by proteolytic enzymes and is not inactivated by its antibody. In the case of immobilized asparaginase attached to a nylon tube (Allison *et al.*, 1972) or a Dacron vascular prosthesis (Cooney *et al.*, 1975), the enzyme is in direct contact with proteolytic enzymes or antibody in the blood and may be inactivated. No formation of thrombus in the asparaginase tube or side effects during extracorporeal shunt were observed. These excellent properties indicate promise for the asparaginase tube for clinical use in lymphosarcoma therapy.

In this section we described the preparation and characteristics of three types of immobilized asparaginase and discussed their applicability to therapy of lymphosarcoma. As in the case of immobilized asparaginases, the extracorporeal

TABLE 3
Characteristics of Native Asparaginase and Various Immobilized Asparaginase

Characteristics	Native asparaginase	Immobilized asparaginase		
			Polyacrylamide gel entrapped	
		Microencapsulated[a]	Granule type	Tube type
Optimum pH	8.0	8.0	7.0	7.0
K_m (μM)	18	2000	3300	4000
Stability against:				
Proteases	Low	High	High	High
Mechanical shock	—	Low	High	High
Half-life in blood	Short	Long	Long	Long
Immunological troubles	Yes	No	No	No
Resistance to bloodstream	—	High	High	Low
Application system	Direct administration	Direct administration	Extracorporeal shunt	Extracorporeal shunt

[a] Polyurea membrane.

shunt system using immobilized enzymes generally seems to be superior to the direct administration method, for the following reasons: (1) a number of immobilized enzymes other than micoencapsulated enzymes are applicable for the extracorporeal shunt system; (2) as enzyme–protein does not enter the body, side effects may not occur; and (3) the time required for therapy can be arbitrarily controlled. Therefore, the extracorporeal shunt system using immobilized enzymes sees more promising than direct administration for therapeutic use in various diseases and medical emergencies.

3. *Production and Purification of Biologically Active Compounds*

3.1. *Production of L-Amino Acids by Immobilized Aminoacylase*

Recently, utilization of L-amino acids for medicines, food, and animal feed has been growing, and economical production of optically active amino acids is needed. At present, fermentative and chemical synthetic methods are employed for the industrial production of L-amino acids instead of the conventional method of isolation from protein hydrolysate. Amino acids produced by chemical synthetic methods are optically inactive racemic mixtures of L- and D-isomers. To obtain the L-amino acid from a mixture of the DL-form, optical resolution is necessary. Among many optical resolution methods, the enzymatic method using mold aminoacylase is one of the advantageous procedures for the industrial production of optically pure L-amino acids. A chemically synthesized acyl-DL-amino acid is asymmetrically hydrolyzed by the enzyme as follows:

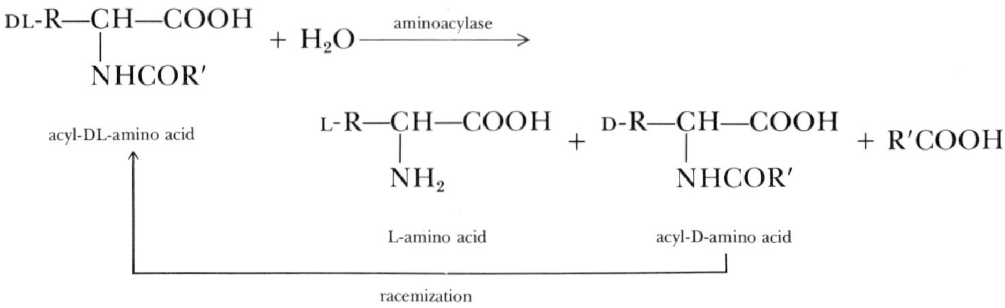

After the enzyme reaction, L- amino acid and unhydrolyzed acyl-D-amino acid can be easily separated by the difference in their solubilities. Acyl-D-amino acid is racemized by the usual method and reused for the resolution procedure. This mold aminoacylase has versatile substrate specificity, and many kinds of L-amino acids can be produced.

As this enzyme reaction had been carried out in batch process from 1954 to 1969 in Tanabe Seiyaku Co. Ltd., there were some disadvantages for industrial purpose. For instance, it was necessary to remove enzyme–protein by pH and/or heat treatments for isolation of L-amino acid from the reaction mixture. As the enzyme could not be reused, the method was uneconomical. Further, as a complicated purification procedure was necessary for removal of contaminating proteins and coloring materials, the yield of L-amino acids was lowered. Also, much labor was necessary for batch operation. To overcome these disadvantages, we studied the immobilization of aminoacylase and the continuous optical resolution of DL-amino acids using columns packed with immobilized enzymes. Since 1969 industrial production of L-amino acids by this immobilized enzyme reaction system has been carried out at Tanabe Seiyaku Co. Ltd.

a. Screening of Immobilized Aminoacylase Suitable for Industrial Purpose. To obtain the immobilized aminoacylase preparation suitable for industrial purpose, a number of immobilization methods were tested (Chibata *et al.*, 1972; Mori *et al.*, 1972b; Sata *et al.*, 1971; Tosa *et al.*, 1966a, 1966b, 1967). The results showed that relatively active and stable immobilized aminoacylases could be obtained by ionic binding to DEAE-cellulose (Tosa *et al.*, 1966a, 1966b) or DEAE-Sephadex (Tosa *et al.*, 1967), covalent binding to iodoacetyl cellulose (Sato *et al.*, 1971), and entrapping into polyacrylamide gel lattices (Mori *et al.*, 1972b). For the industrial application of immobilized aminoacylase, it is necessary to satisfy a number of conditions. Thus, to select the most suitable preparation, we compared the enzymatic properties, characteristics, and economy of these immobilized aminoacylases. The results are summarized in Table 4.

As the results, immobilized DEAE-Sephadex–aminoacylase was chosen as one of the most advantageous enzyme preparations for industrial use because (1) preparation is easy, (2) operational stability is high, (3) regeneration of deteriorated immobilized aminoacylase is possible, and (4) immobilization cost is low.

b. Continuous Production of L-Amino Acids. To discover the most suitable conditions for continuous production of L-amino acids, we investigated the chemical

TABLE 4
Comparison of Enzymatic Properties[a] and Characteristics of Typical Immobilized Aminoacylases

Enzymatic properties and characteristics	Native aminoacylase	Immobilized aminoacylases: Ionic binding to DEAE-Sephadex	Covalent binding to iodoacetylcellulose	Extrapment by polyacrylamide
Optimum pH	7.5–8.0	7.0	7.5–8.5	7.0
Optimum temperature (°C)	60	72	55	65
Optimum Co^{2+} (mM)	0.5	0.5	0.5	0.5
K_m (mM)	5.7	8.7	6.7	5.0
Operational stability[b] (half-life, days)	—	65 (50°C)	—	48 (37°C)
Preparation	—	Easy	Difficult	Medium
Binding force	—	Medium	Strong	Strong
Regeneration[c]	—	Possible	Impossible	Impossible
Cost of immobilization[d]	—	Low	High	Moderate

[a] Data are from the results with acetyl-DL-methionine as a substrate.
[b] The time required for 50% of the enzyme activity to be lost.
[c] Regeneration of deteriorated immobilized aminoacylase column after operation for a long period.
[d] Compared from the basis for unit production of L-amino acids.

engineering aspects of enzyme column system: that is, the pressure drop of the enzyme column, the relation between the flow rate of the substrate solution and reaction rate, and the influence of column dimension (Chibata *et al.*, 1972; Tosa *et al.*, 1967, 1969a, 1969b, 1971a). The most efficient and automatically controlled enzyme reaction system was found to be as shown in Figure 4. Industrial production of such L-amino acids as L-methionine, L-phenylalanine, and L-valine has been carried out in our plants since 1969. By employing this system, the average overall production cost of the amino acids was reduced to about 60% of that of the conventional batch process using the soluble enzyme, owing to marked reduction in the amounts of substrate and aminoacylase and in the cost of labor. This is the first industrial application of immobilized enzyme in the world.

3.2. *Production of Biologically Active Compounds by Immobilized Microbial Cells*

a. Production of L-Aspartic Acid by Immobilized Escherichia coli Cells. L-Aspartic acid is widely used in medicines and food additives, and has been industrially produced by fermentative and enzymatic methods from fumaric acid and ammonia by the action of aspartase:

$$\underset{\text{fumaric acid}}{HOOCCH{=}CHCOOH} + NH_3 \xrightarrow{\text{aspartase}} \underset{\text{L—aspartic acid}}{HOOCCH_2\underset{\displaystyle NH_2}{\underset{|}{C}}HCOOH}$$

As this reaction had been carried out as a batch process, there were some disadvantages for industrial purposes, just as in the case of native aminoacylase. So

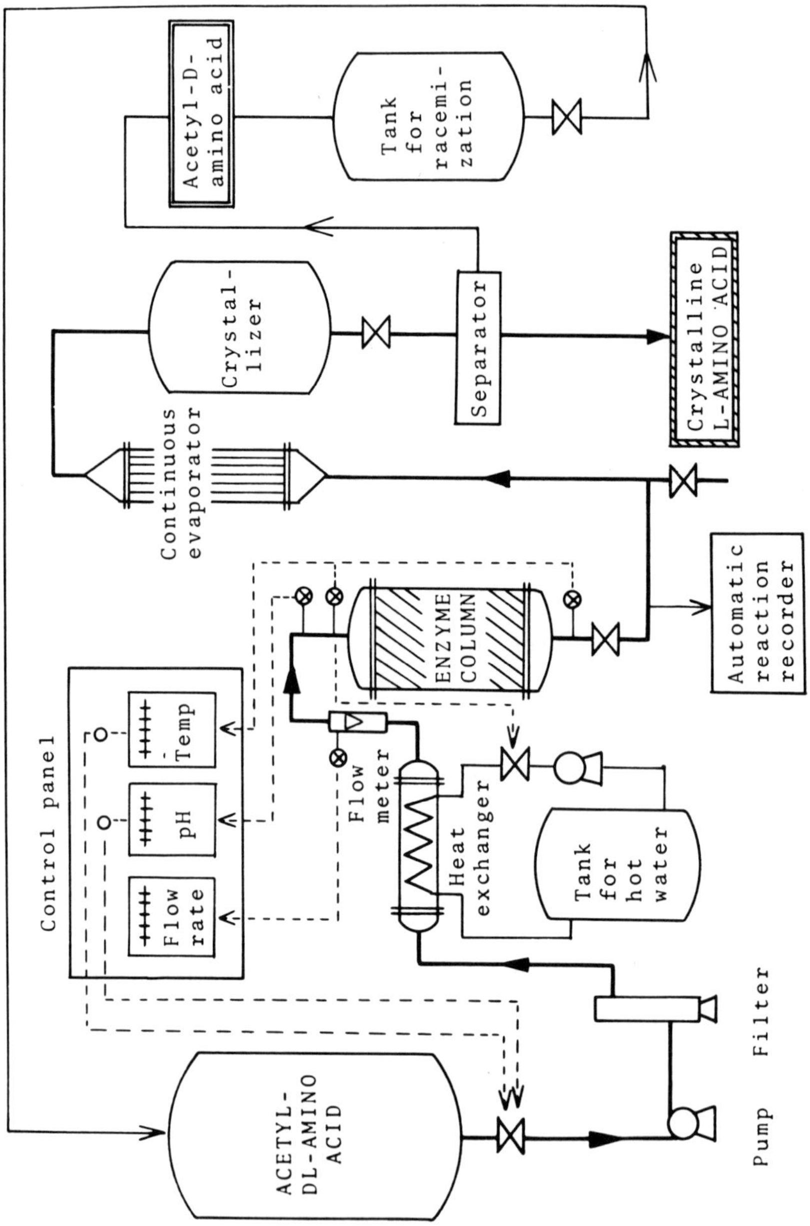

FIGURE 4. Flow diagram for continuous production of L-amino acid by immobilized aminoacylase.

we studied the continuous production of L-aspartic acid using immobilized aspartase (Tosa *et al.*, 1973). Active immobilized aspartase was obtained by entrapment into a polyacrylamide gel-lattice, but this immobilized enzyme was relatively unstable (i.e., its half-life was 27 days at 37°C) (Tosa *et al.*, 1973). Further, it was necessary to extract the enzyme from microbial cells before immobilization. Therefore, this immobilized aspartase was not considered to be satisfactory for industrial purposes. We felt that if the microbial cells having the enzyme activity could be immobilized, these disadvantages might be overcome. We studied the immobilization of whole microbial cells from this point of view.

Screening of Immobilization Methods. To prepare the immobilized microbial cells suitable for industrial production of L-aspartic acid, several immobilization methods for *E. coli* cells having high aspartase activity were tested (Chibata *et al.*, 1973a, 1974). The most active immobilized *E. coli* cells were obtained by entrapping the cells in polyacrylamide gel lattice.

Preparation of Immobilized E. coli Cells. In order to prepare the most efficient immobilized *E. coli* cells by the acrylamide gel method, we investigated the concentration of acrylamide monomer, bifunctional reagents, and the number of cells to be entrapped. As a result, the following optimum conditions were decided upon. Ten kilograms of *E. coli* cells are suspended in 40 liters of physiological saline. To this suspension are added 7.5 kg of acrylamide, 0.4 kg of *N,N'*-methylenebisacrylamide, 5 liters of 5% β-dimethylaminopropionitrile, and 5 liters of 2.5% potassium persulfate. The mixture is allowed to stand at below 40°C for 10–30 min and the resulting stiff gel is formed as 3-mm cubes. The aspartase activity of immobilized *E. coli* obtained under these conditions is 1310 μmoles/hr/g of wet cells.

An interesting phenomenon was observed when the immobilized *E. coli* cells were suspended at 37°C for 24–48 hr in 1 M ammonium fumarate solution, pH 8.5, containing 1 mM Mg^{2+}: the activity increased about 10 times. This activation is caused by an increase in membrane permeability for substrate and/or product due to autolysis of *E. coli* cells in the gel-lattice. Of course, even if lysis of the entrapped cells were to occur, the aspartase does not leak out from the gel-lattice, although the substrate and the product easily pass through the gel-lattice. The enzyme activities of native aspartase, immobilized aspartase, *E. coli* cells, and immobilized *E. coli* cells are summarized in Figure 5. It is clear that the immobilized *E. coli* cells are superior to the immobilized aspartase in respect of enzyme activity.

Continuous Production of L-Aspartic Acid. To determine the most suitable conditions for continuous production of L-aspartic acid from ammonium fumarate by a column packed with the immobilized *E. coli* cells, we investigated the enzymatic properties of the immobilized cells (Chibata *et al.*, 1973a, 1974; Tosa *et al.*, 1974a). The optimum pH of the immobilized cells for enzyme reaction is 8.5, whereas the intact cells are most active at pH 10.5.

The native aspartase is activated by Mn^{2+}, but intact and immobilized cells are not activated by this metal ion. On the other hand, bivalent metal ions such as Ca^{2+}, Mg^{2+}, and Mn^{2+} show protective effects against heat inactivation of aspartase activities of intact and immobilized cells. Furthermore, these bivalent metal ions have stabilizing effects on the aspartase activity of the immobilized cells during the continuous enzyme reaction.

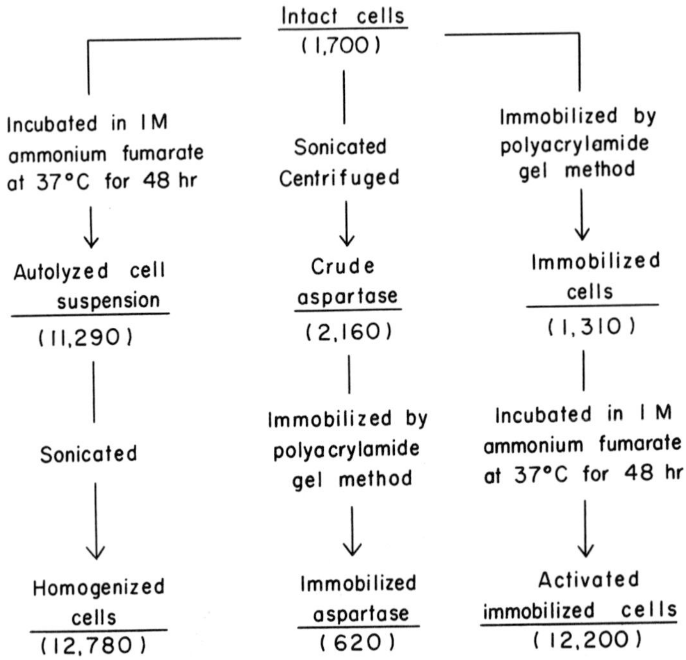

FIGURE 5. Schematic comparison of aspartase activity of various enzyme preparations per unit of intact cells. Numerical values in parentheses are aspartase activities (μmoles/hr) obtained from 1 g of intact cells.

The stability of the immobilized *E. coli* column was investigated by continuously passing a 1 M substrate solution for a long period at a specified temperature. The deterioration of the activity depends on the temperature. The half-life of the column is estimated to be 120 days at 37°C as shown in Table 5. The relationship between the flow rate of substrate and the formation of L-aspartic acid, the effect

TABLE 5
Operational Stability of Immobilized Microbial Cells

Microorganisms	Enzymes	Addition of metals	Operational stability (half-life,[a] days)
Escherichia coli	Aspartase	Ca^{2+}, Mg^{2+}, Mn^{2+}	120 (37°C)
Pseudomonas putida	L-Arginine deiminase	—	140 (37°C)
Achromobacter liquidum	L-Histidine ammonia-lyase	Ca^{2+}, Co^{2+}, Mg^{2+}, Zn^{2+}	180 (37°C)
Escherichia coli	Penicillin amidase	—	17 (40°C); 42 (30°C)
Brevibacterium ammoniagenes	Fumarase	—	55 (37°C)

[a] The time required for 50% of the enzyme activity to be lost.

of column dimensions on the reaction rate, and the formation of heat of reaction were also studied.

According to this information the most efficient reactor system using immobilized *E. coli* cells was designed. This new system has been in industrial operation at Tanabe Seiyaku Co. Ltd. since 1973. As a result, the overall production cost of this system is reduced to about 60% of the conventional batch process using intact cells. This is considered to be the first industrial application of immobilized microbial cells in the world.

b. Production of L-*Citrulline by Immobilized Pseudomonas putida Cells.* L-Citrulline is used for medicines and is industrially produced from L-arginine by the action of microbial L-arginine deiminase in a batch process:

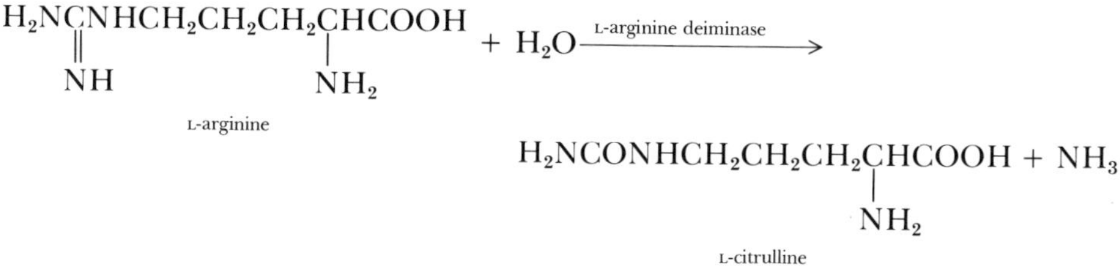

For the continuous production of L-citrulline, we (Yamamoto *et al.*, 1974a) studied the immobilization of *Pseudomonas putida* cells having high L-arginine deiminase activity, and successfully carried out immobilization by the polyacrylamide gel method as in the case of *E. coli* cells.

To find out the optimum conditions for the continuous production of L-citrulline from L-arginine by a column packed with immobilized cells, the relationship between flow rate of substrate solution and formation of L-citrulline, and the stability of the column were studied. It was found that the reaction was completed, when 0.5 M L-arginine hydrochloride solution (pH 6.0) was passed through the column at 37°C at a space velocity of 0.26. From the effluent of the column, L-citrulline was easily obtained in good yield. This column is very stable, and the half-life is estimated to be 140 days at 37°C as shown in Table 5. This technique is considered to be more advantageous for the mass production of L-citrulline than the batch method using microbial broth.

c. Production of Urocanic Acid by Immobilized Achromobacter liquidum Cells. Urocanic acid is used as a sun-screening agent in the pharmaceutical and cosmetic fields and is produced industrially from L-histidine by the action of L-histidine ammonia-lyase in a batch process using microbial broth or extracted enzyme:

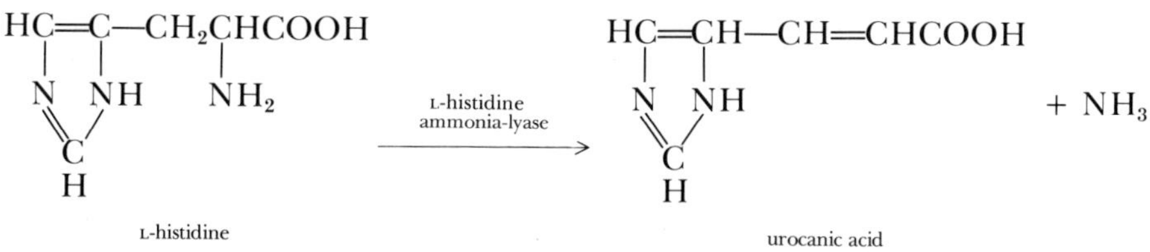

For the purpose of continuous production of urocanic acid, we tested immobilization of several microorganisms by the polyacrylamide gel method and compared the enzyme activities of the immobilized cells (Yamamoto *et al.,* 1974b). *Achromobacter liquidum* was found to show the highest activity after immobilization and to be most suitable for the industrial production of urocanic acid. Although this organism has urocanase activity converting urocanic acid to imidazolone propionic acid, the activity is easily stopped by heat treatment of the cells at 70°C for 30 min before immobilization. Further, in the case of intact cells, the rate of formation of urocanic acid was very low in the absence of cetyltrimethylammonium bromide (CTAB) owing to the low permeability of substrate or product through the cell wall. However, after the immobilization of cells, the permeability was enhanced in the absence of CTAB, as shown in Figure 6.

Optimum conditions for the continuous production of urocanic acid were studied by using a column packed with the immobilized cells. When 0.25 M L-histidine solution (pH 9.0) containing Mg was passed through the column at 37°C at a space velocity of 0.06, the reaction was completed. From the column effluent, urocanic acid was easily obtained in good yield by simple pH treatment. This column is very stable, and the half-life during the long period of operation is estimated to be 180 days at 37°C, as shown in Table 5. This process is more efficient for the industrial production of urocanic acid than the batch process using extracted enzyme or microbial broth.

d. Production of 6-Aminopenicillanic Acid by Immobilized Escherichia coli Cells. 6-Aminopenicillanic acid (6-APA) is used as a starting material for the synthetic penicillin and industrially produced from penicillin by the action of penicillin amidase in a batch process using microbial cells or extracted enzyme:

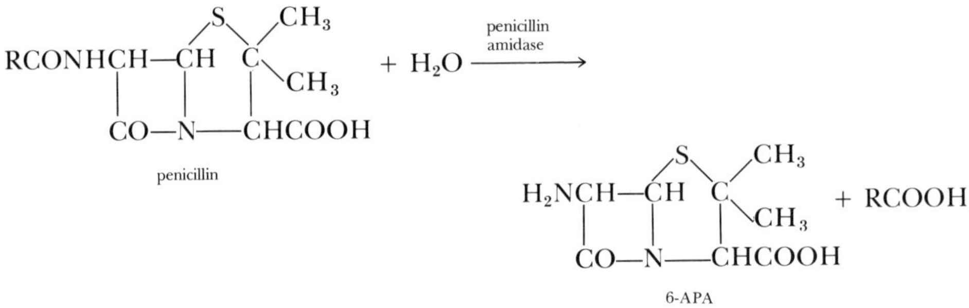

To improve the process, many studies have been done on the continuous production of 6-APA using a column packed with immobilized penicillin amidase. However, in this case, it is necessary to extract the enzyme from microbial cells before the immobilization process, and in general the extracted penicillin amidase is unstable.

To determine the more advantageous method, we investigated the immobilization of *E. coli* cells having high penicillin amidase activity (Sato *et al.,* 1976). These microbial cells contain penicillinase activity, which decomposes both penicillin and 6-APA. Specific inactivation of penicillinase activity is very difficult, but penicillinase activity is much lower than penicillin amidase activity. Therefore, optimum conditions for the continuous production of 6-APA without removing the penicillinase activity can be chosen as follows. When 0.05 M penicillin G

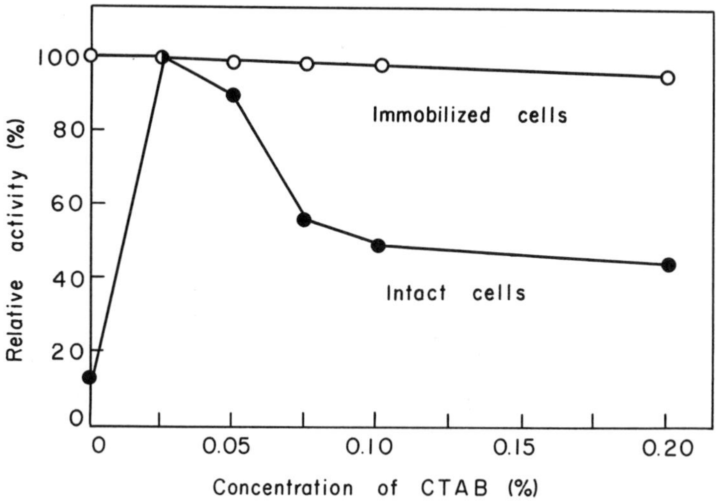

FIGURE 6. Effect of CTAB on rate of formation of urocanic acid. Enzyme reaction was carried out in the presence of specified concentrations of CTAB.

solution (pH 8.5) was passed through the immobilized *E. coli* cell column at 30°C at a space velocity of 0.12, 6-APA was efficiently produced. From the column effluent 6-APA was obtained in about 80% yield.

As this column is relatively stable, as shown in Table 5, the cost of enzyme–catalyst is reduced. Further, labor cost is reduced, as the production process of 6-APA can be automatically controlled. Therefore, this technique is considered to be more advantageous than the continuous method using immobilized penicillin amidase.

e. Production of L*-Malic Acid by Immobilized Brevibacterium ammoniagenes Cells.* L-Malic acid is used in the pharmaceutical field as an antidote for hyperammonemia and as a component of amino acid infusion. This acid is industrially produced from fumaric acid by the action of fumarase in a batch process using microbial broth:

$$\underset{\text{fumaric acid}}{HOOC{-}CH{=}CH{-}COOH} + H_2O \xrightarrow{\text{fumarase}} \underset{\text{L-malic acid}}{HOOC{-}\underset{\displaystyle OH}{\underset{|}{CH}}{-}CH_2{-}COOH}$$

For the continuous production of L-malic acid, we (Chibata *et al.*, 1975) studied by the polyacrylamide gel method the immobilization of several microorganisms having high fumarase activity and found that *Brevibacterium ammoniagenes* was most active before and after immobilization. However, these immobilized cells formed succinic acid as a by-product, and separation from L-malic acid is very difficult. Therefore, success for industrial production of pure L-malic acid rests in prevention of succinic acid formation during enzyme reaction. Various treatments of intact or immobilized cells were carried out, and treatment with such detergents as deoxycholic acid, bile acid, and bile extract was found to be very effective, as shown in Table 6. These detergents also strikingly enhanced the formation of L-malic acid by the immobilized cells. Readily available bile extract is considered to be most suitable for industrial prupose. The most effective conditions were studied and settled as follows. The immobilized cells are stood in 1 M sodium fumarate (pH 7.5) containing 0.3% bile extract at 37°C for 20 hr.

Optimum conditions for the continuous production of L-malic acid were as follows. When 1 M sodium fumarate solution (pH 7.0) was passed through a column packed with bile-extract-treated immobilized cells at 37°C at a space velocity of 0.23, the reaction reached equilibrium with about 80% conversion of fumaric acid to L-malic acid. From the effluent, fumaric acid was removed by acidifing, and L-malic acid was separated in about 70% yield from a raw material, fumaric acid. The stability of this column is shown in Table 6. This production system has been industrially operated at Tanabe Seiyaku Co. Ltd. since 1974. We are satisfied with it both as to economical efficiency and product quality.

From our experience industrially with one immobilized enzyme system and two immobilized microbial cell systems, we can conclude that the enzyme reactions by immobilized microbial cells are promising and advantageous in the following cases: (1) when enzymes are intracellular, (2) when enzymes extracted from the cells are unstable, (3) when enzymes are unstable during and after immobilization procedure, (4) when microorganism contains no interfering enzymes or when interfering enzymes are readily inactivated or removed, and (5) when substrates and products are not higher-molecular-weight compounds.

3.3. Purification of Coenzyme A by Immobilized Proteins

Reduced coenzyme A (CoA), which plays a very important role in the metabolism of carbohydrates and lipids, has been purified from microbial cells or fermentation broth by the method of precipitation with heavy metals or by ion-exchange chromatography. However, these methods are not advantageous for industrial production of reduced CoA because the procedure is very complicated and the yield is low. Recently, affinity chromatography has been employed as an advantageous method for the purification of biological substances. We tried the technique for the purification of reduced CoA (Matuo *et al.*, 1974; Chibata *et al.*, 1973b).

a. Screening of Immobilized Proteins Suitable for Purification of Reduced CoA. When CoA is immobilized as ligand for affinity chromatography, the immobilized CoA may be used for the purification of enzymes requiring CoA as a cofactor. Conversely, when enzymes requiring CoA are immobilized as ligands, the immobilized enzymes may be used for the purification of CoA. However, selection and purification of the enzymes requiring CoA are not so easy. If a specific ligand such as protein showing an affinity to CoA can be obtained by the use of immobilized CoA, it becomes possible to prepare the affinity adsorbent for CoA purification by immobilizing the above mentioned specific protein.

For the immobilization of CoA, CNBr-activated Sepharose 6B was prepared

TABLE 6
Effect of Detergent Treatments on Formations of L-Malic Acid and Succinic Acid by Immobilized B. ammoniagenes

Treatment[a]	Percent	L-Malic acid (mmoles/hr/g of cells)	Succinic acid (mole % to L-malic acid)[c]
Control[b]		0.49	—
No addition		0.99	2.5–5.0
Cetylpyridinium chloride	0.02	4.57	2.5–5.0
	0.16	3.07	1.0–2.5
Sodium lauryl sulfate	0.02	6.05	1.0–2.5
SL-10	0.02	1.22	2.5–5.0
Triton X-100	0.20	5.36	>5.0
Bile extract	0.20	7.48	<0.2
Bile acid	0.20	6.57	<0.2
Deoxycholic acid	0.20	7.38	<0.2

[a]Stood at 37°C for 20 hr in 1 M Na-fumarate (pH 7.5) containing detergent.
[b]Immediately after preparation of the gel.
[c]Estimated by paper chromatography after the reaction reached equilibrium.

essentially by the method of Axén *et al.*, (1967) and used as a carrier. The structure of immobilized CoA is shown in Figure 7.

To prepare proteins showing an affinity for CoA, bacteria accumulating CoA were selected as the ligand source and CoA-affinity proteins were isolated from the dialyzed extracts of various bacteria using the immobilized CoA column. Of the bacteria tested, *Sarcina lutea* was found to be most suitable as the source of ligand, because the dialyzed extract contained a large amount of CoA-affinity protein and could be used as a ligand without chromatographic separation by an immobilized CoA column. Therefore, an affinity adsorbent for CoA purification is prepared by immobilization of dialyzed extract from *S. lutea.*

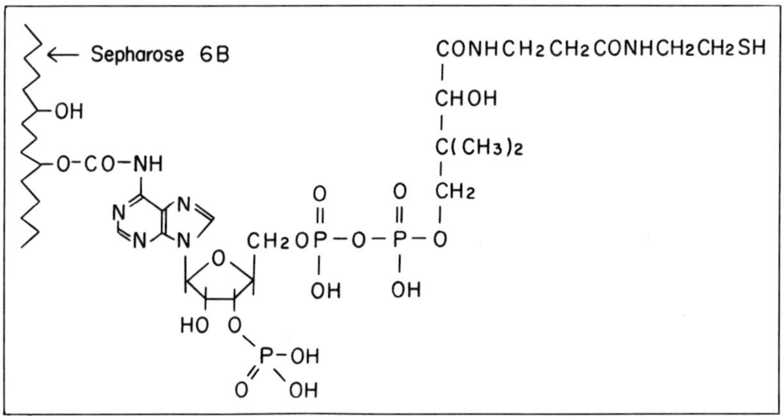

FIGURE 7. Suggested structure of immobilized CoA.

b. Preparation of Immobilized CoA-Affinity Protein. Optimum conditions for the preparation of immobilized affinity adsorbent are as follows. CNBr-activated Sepharose 6B (1 g of packed weight) is added to 20 ml of the dialyzed extract solution (20 mg of protein) from *S. lutea.* After gentle shaking at 5°C for 6 hr at pH 8.5, the pH is adjusted to 6.0 and the mixture is shaken at 5°C for 12 hr. To the mixture 1 g of CNBr-activated Sepharose 6B is added and incubated at 5°C for 12 hr with gentle shaking. The mixture is filtered and washed with 20 ml of 1 M NaC1. The resulting adsorbent contains 13.2 mg of protein per 2 g of Sepharose 6B.

c. Purification of Reduced CoA by an Immobilized CoA-Affinity Protein Column. To clarify the specificity of immobilized CoA-affinity protein column, CoA, dephospho-CoA, ATP, ADP, and AMP are separately charged on the column. The ionic strength (*I*) of the fraction showing the maximum content of these compounds is measured, and the resulting values are superimposed in Figure 8. All compounds except for AMP are adsorbed on the column of $I = 0.01$ and eluted by increasing ionic strength with NaC1. In this chromatography, two kinds of CoA-affinity proteins are observed; one shows the affinity to reduced CoA in low *I* ($I = 0.01$) and the other shows it in high *I* ($I = 0.04$). However, when the column is

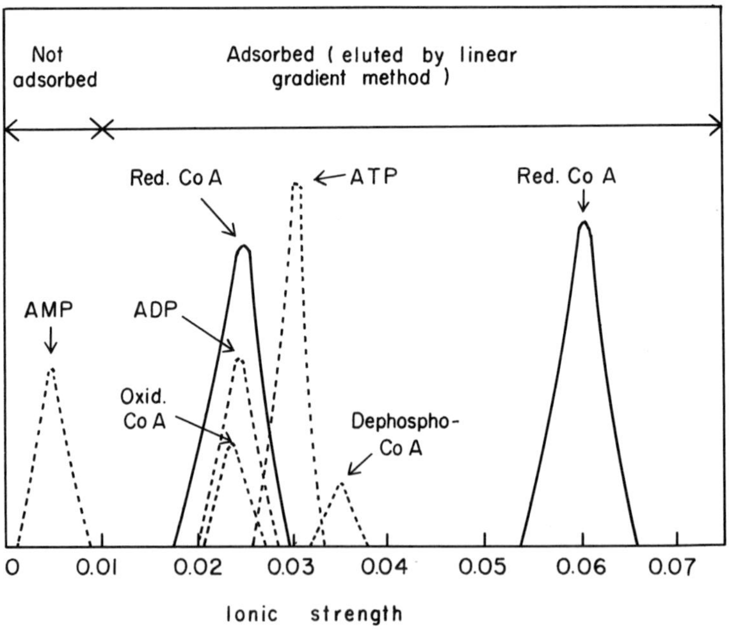

FIGURE 8. Adsorption specificity of immobilized CoA-affinity protein column. The column (1 cm × 5 cm) was equilibrated with sodium acetate buffer (pH 6.0, $I = 0.01$). Chromatographic patterns of the respective compounds were superimposed. Compounds in eluates were determined spectrophotometrically by measuring the absorbance at 260 nm. The horizontal axis indicates the *I* of the fractions eluted, and the vertical axis indicates the arbitrary of absorbance at 260 nm.

equlibrated with sodium acetate buffer (pH 6.0, $I = 0.04$), reduced CoA is selectively adsorbed. By this affinity technique, reduced CoA (92% purity) is obtained in 94% yield from crude CoA (5% purity).

The adsorbing capacity of the immobilized CoA-affinity proteins to reduced CoA is generally 85 μg/ml of the adsorbent under operating conditions of pH 6–7 and 10–25°C. The capacity of the adsorbent is not changed by storage at 5°C and pH 6.0 for 1 month.

This principle and techniques for the purification of CoA by immobilized CoA-affinity proteins are applicable to the separation of other biologically active substances, especially in cases where the selection and/or preparation of their ligands are difficult.

ACKNOWLEDGMENTS

We express our deep gratitude to T. Sato, R. Sano, K. Yamamoto, and Y. Iwasawa of our research laboratory, and to Y. Matuo of Osaka University, for their cooperation in the studies presented here and for their helpful advice and discussion.

4. *References*

Allison, J. P., Davidson, L., Hartman, A. C., and Kitto, G. B., 1972, Insolubilization of L-asparaginase by covalent attachment to nylon tubing, *Biochem. Biophys. Res. Commun.* **47**:66.

Axén, R., Porath, J., and Ernback, S., 1967, Chemical coupling of peptides and proteins to polysaccharides by means of cyanogen halides, *Nature* **214**:1302.

Boguslaski, R. C., and Janik, A. M., 1971, A kinetic study of microencapsulated bovine carbonic anhydrase, *Biochem. Biophys. Acta* **250**:266.

Chang, T. M. S., 1971a, The in vivo effects of semipermeable microcapsules containing L-asparaginase on 6C3HED lymphosarcoma, *Nature* **229**:117.

Chang, T. M. S., 1971b, Stabilization of enzymes by microencapsulation with a concentrated protein solution or by microencapsulation followed by cross-linking with glutaraldehyde, *Biochem. Biophys. Res. Commun.* **44**:1531.

Chang, T. M. S., MacIntosh, F. C., and Mason, S. G., 1966, Semipermeable aqueous microcapsules: I. Preparation and properties, *Can. J. Physiol. Pharmocol.* **44**:115.

Chibata, I., Tosa, T., Sato, T., Mori, T., and Matuo, Y., 1972, Preparation and industrial application of immobilized aminoacylases, in: *Proceedings of IV International Fermentation Symposium: Fermentation Technology Today,* p. 383, Society of Fermentation Technology, Osaka.

Chibata, I. Tosa, T., Sato, T., Mori, T., and Yamamoto, K., 1973a, Continuous enzyme reactions by immobilized microbial cells, in: *Enzyme Engineering,* Vol. 2 (E. K. Pye and L. B. Wingard, Jr., eds.), p. 303, Plenum Press, New York.

Chibata, I., Tosa, T., and Matuo, Y., 1973b, Immobilization of coenzyme A and its application, in: *Enzyme Engineering,* Vol. 2 (E. K. Pye and L. B. Wingard, Jr., eds.), p. 229, Plenum Press, New York.

Chibata, I., Tosa, T., and Sato, T., 1974, Immobilized aspartase-containing microbial cells: preparation and enzymatic properties, *Appl. Microbiol.* **27**:878.

Chibata, I., Tosa, T., and Yamamoto, K., 1975, Industrial production of L-malic acid by immobilized microbial cells, in: *Proceedings of 1975 Enzyme Engineering Conference at Portland,* Plenum Press, New York.

Chong, E. D. S, and Chang, T. M. S., 1974, *In vivo* effects of intraperitoneally injected L-asparaginase solution and L-asparaginase immobilized within semipermeable nylon microcapsules with emphasis on blood L-asparaginase, body L-asparaginase, and plasma L-asparagine levels, *Enzyme* **18**:218.

Cooney, D. A., Capizzi, R. L., and Handschumacher, R. E., 1970, Evaluation of L-asparagine metabolism in animals and man, *Cancer Res.* **30**:929.

Cooney, D. A., Weetall, H. H., and Long, E., 1975, Biochemical and pharmacologic properties of L-asparaginase bonded to dacron vascular prosthesis, *Biochem. Pharmacol.* **24**:503.

Gregoriadis, G., Leathwood, P. D., and Ryman, B. E., 1971, Enzyme entrapment in liposomes, *Federation Eur. Biochem. Soc. Letters* **14**:95.

Horvath, C., Sardi, A., and Woods, A. J., 1973, L-Asparaginase tubes: kinetic behavior and application in physiological studies, *J. Appl. Physiol.* **34**:181.

Inada, Y., Hirose, S., Okada, M., and Mihama, H., 1975, Immobilized L-asparaginase embedded in fibrin polymer, *Enzyme* **20**:188.

Matuo, Y., Tosa, T., and Chibata I., 1974, Purification of coenzyme A by affinity chromatography, *Biochem. Biophys. Acta* **338**:520.

Mori, T., Sato, T., Matuo, Y., Tosa, T., and Chibata, I., 1972a, Preparation and characteristics of microcapsules containing asparaginase, *Biotech. Bioeng.* **14**:663.

Mori, T., Sato, T., Tosa, T., and Chibata, I., 1972b, Studies on immobilized enzymes: X. Preparation and properties of aminoacylase entrapped into acrylamide gel-lattice, *Enzymologia* **43**:213.

Mori, T., Tosa, T., and Chibata, I., 1973, Enzymatic properties of microcapsules containing asparaginase, *Biochem Biophys. Acta* **321**:653.

Mori, T., Tosa, T., and Chibata, I., 1974, Preparation and properties of asparaginase entrapped in the lattice of polyacrylamide gel, *Cancer Res.* **34**:3066.

Mori, T., Sano, R., Iwasawa, Y., Tosa, T., and Chibata, I., 1976, Preparation, characteristics, and application of asparaginase tube, *J. Solid-Phase Biochem.* **1**:15.

Nadler, H. L., and Updike, S. J., 1974, Gel entrapment of enzyme: a strategy for enzyme correction, *Enzyme* **18**:150.

Sato, T., Mori, T., Tosa, T., and Chibata, I., 1971, Studies on immobilized enzymes: IX. Preparation and properties of aminoacylase covalently attached to halogenoacetylcellulose, *Arch. Biochem. Biophys.* **147**:788.

Sato, T., Tosa, T., and Chibata, I., 1976, Continuous production of 6-aminopenicillanic acid from penicillin by immobilized microbial cells, *Eur. J. Appl. Microbiol.* **2**:153.

Tosa, T., Mori, T., Fuse, N., and Chibata, I., 1966a, Studies on continuous enzyme reactions: I. Screening of water-insoluble aminoacylase, *Enzymologia* **31**:214.

Tosa, T., Mori, T., Fuse, N., and Chibata, I., 1966b, Studies on continuous enzyme reactions: II. Preparation of DEAE-cellulose–aminoacylase column and continuous optical resolution of acetyl-DL-methionine, *Enzymologia* **31**:225.

Tosa, T., Mori, T., Fuse, N., and Chibata, I., 1967, Studies on continuous enzyme reactions: IV. Preparation of a DEAE-sephadex–aminoacylase column and continuous optical resolution of acyl-DL-amino acids, *Biotech. Bioeng.* **9**:603.

Tosa, T., Mori, T., Fuse, N., and Chibata, I., 1969a, Studies on continuous enzyme reactions: V. Kinetics and industrial application of aminoacylase column for continuous optical resolution of acyl-DL-amino acids, *Agr. Bio. Chem. (Tokyo)* **33**:1047.

Tosa, T., Mori, T., and Chibata, I., 1969b, Studies on continuous enzyme reactions: VI. Enzymatic properties of the DEAE-sephadex-aminoacylase complex, *Agr. Bio. Chem. (Tokyo)* **33**;1053.

Tosa, T., Mori, T., and Chibata, I., 1971a, Studies on continuous enzyme reactions: VIII. Kinetics and pressure drop of aminoacylase column, *J. Fermen. Technol.* **49**:522.

Tosa, T., Sano, R., Yamamoto, K., Nakamura, M., Ando, K., and Chibata, I., 1971b, L-Asparaginase from *Proteus vulgaris, Appl. Microbiol.* **22**:387.

Tosa, T., Sano, R., Yamamoto, K., Nakamura, M., and Chibata, I., 1972, L-Asparaginase from *Proteus vulgaris:* Purification, crystallization, and enzymic properties, *Biochemistry* **11**:217.

Tosa, T., Sato, T., Mori, T., Matuo, Y., and Chibata, I., 1973, Continuous production of L-aspartic acid by immobilized aspartase, *Biotech. Bioeng.* **15**:69.

Tosa, T., Sato, T., Mori, T., and Chibata, I., 1974a, Basic studies for continuous production of L-aspartic acid by immobilized *Escherichia coli* cells, *Appl. Microbiol.* **27**:886.

Tosa, T., Sato, T., Sano, R., Yamamoto, K., Matuo, Y., and Chibata, I., 1974b, Characteristics and applications of *N*-(ω-aminohexyl)-L-aspartic acid-sepharose as an affinity adsorbent, *Biochem. Biophys. Acta* **334**:1.

Weetall, H. H., 1970, Insolubilized L-asparaginase implant: a preliminary report, *J. Biomed. Mater. Res.* **4**:597.

Yamamoto, K., Sato, T., Tosa, T., and Chibata, I., 1974a, Continuous production of L-citrulline by immobilized *Pseudomonas putida* cells, *Biotech. Bioeng.* **16**:1589.

Yamamoto, K., Sato, T., Tosa, T., and Chibata, I., 1974b, Continuous production of urocanic acid by immobilized *Achromobacter liquidum* cells, *Biotech. Bioeng.* **16**:1601.

19

Artificial Kidney, Artificial Liver, and Detoxifiers Based on Artificial Cells, Immobilized Proteins, and Immobilized Enzymes

T. M. S. Chang

1. Introduction

The standard artificial kidney for patients with chronic renal failure and detoxifiers for patients with drug intoxication being used at present are based on the principle of hemodialysis. All hemodialysis systems have in common the following basic principle. A semipermeable membrane separates two compartments. The patient's blood flows through one compartment. Here, permeant molecules from the patient's blood cross the semipermeable membrane by diffusion along electrochemical gradients or by ultrafiltration along hydrostatic gradients to the other compartment, where they are washed away by a very large volume of dialysate (200–300 liters). The rate of removal of waste metabolites depends on a number of factors; one is the total surface area of the membrane available for diffusion, another is the permeability of the membrane. Although hemodialysis has been successfully used for the treatment of patients with loss of kidney function or with drug intoxication, there are a number of problems. The general transport characteristics of the membrane are such that treatments of 6–12 hr 3 times a week is required. The present hemodialysis machine is the size of a washing machine, bulkly and extremely expensive. It is not efficient in the detoxification of many types of drug intoxication. Furthermore, it is not effective for the treatment of liver failure. As a result, extensive efforts have been concentrated on new approaches.

A new approach for the construction of an artificial kidney, artificial liver, and detoxifiers was proposed and demonstrated (Chang, 1966, 1972a) based on the principle of artificial cells (Chang, 1964). This chapter will discuss the studies in this area carried out in this laboratory.

T. M. S. Chang • Departments of Physiology and Medicine, McGill University, Montreal, Canada

2. *Basic Principles*

It is well known that the smaller a particle, the larger is the surface/volume relationship. If we calculate the total surface area in artificial cells of different diameters, the result obtained is quite striking. Thus, 10 ml of 20-μm-diameter microcapsules or 33 ml of 100-μm-diameter microcapsules have a total surface area (about 2.5 m^2) which is larger than that available in a conventional hemodialyzer machine (1 m^2). In addition, the membrane thickness of microcapsules (less than 500 Å) is much less than that used in standard artificial kidneys (250,000 Å). If everything else remains the same, the rate of net movement of permeant molecules across a membrane (ds/dt) along a given concentration gradient (ΔCs) is proportional to the surface area (A) and inversely proportional to the membrane thickness (dx). Let us now consider what this means if we pack and retain 33 ml of 100-μm-diameter microcapsules in a shunt and allow blood from the patient to recirculate and come in direct contact with the microcapsules. In this 33 ml of artificial cells, the total membrane area available for diffusion will be 2.5 times that of a hemodialyzer and the membrane thickness is 1/500 that of the hemodialyzer. This means that on this basis alone, metabolites from blood flowing past these microcapsules can cross the membrane into the microcapsules 1000 times faster than in standard hemodialyzers. Of course, these calculations are grossly oversimplified; nevertheless, they serve to give an approximate idea of the comparative transport efficiency. The compact volume is such that something would have to be placed inside the artificial cells to convert or trap the metabolites equilibrating rapidly across the membranes. If something is placed inside these microcapsules to trap or convert the metabolites which cross the membrane, we would have the basis for a small, compact, artificial organ for the removal of waste metabolites or toxic substances from the patients. Two approaches have been tested: (1) microencapsulation of enzymes to convert waste metabolites, and (2) albumin-coated microencapsulated adsorbents to remove waste metabolites and toxins. These will be discussed in the following two sections.

3. *Use of Microencapsulated Enzymes for the Conversion of Waste Metabolites and Toxins*

A feasibility test involved the use of 10 ml of microencapsulated urease, 90 μm in mean diameter, retained in a shunt chamber 15 cm in height and 2 cm internal diameter (Chang, 1966) (Figure 1). The microcapsules are retained by a nylon screen on either side. Heparinized blood from anesthetized dogs recirculates through the shunt to come in direct contact with the microcapsules. Urea from blood diffusing rapidly into the microcapsules is converted by urease into ammonia. Each shunt, when attached to dogs of 10 kg, was effective in removing blood urea from the systemic circulation so that after 90 min of hemoperfusion, the systemic blood urea level fell to 50% of the initial level (Figure 2). Blood urea is enzymatically converted to blood ammonia which can be removed by a combined system of microencapsulated urease and ammonia adsorbent (Chang, 1966, 1972a) (Figure 3). This demonstration of the feasibility of a model immobilized

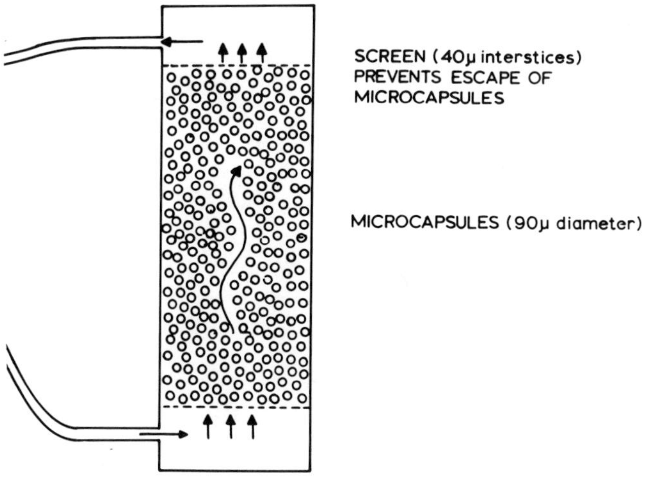

FIGURE 1. Artificial cells in an extracorporeal shunt chamber perfused by recirculating blood from animal. (From Chang, 1966, 1972a).

enzyme, urease, to remove a typical uremic metabolite, urea, has resulted in further analytical and experimental study by a number of other workers. Thus, the feasibility of using a column to convert blood urea in the human body was analyzed theoretically (Levine and LaCourse, 1967). This theoretical analysis supported our experimental results that for the removal of urea, it is sufficient to

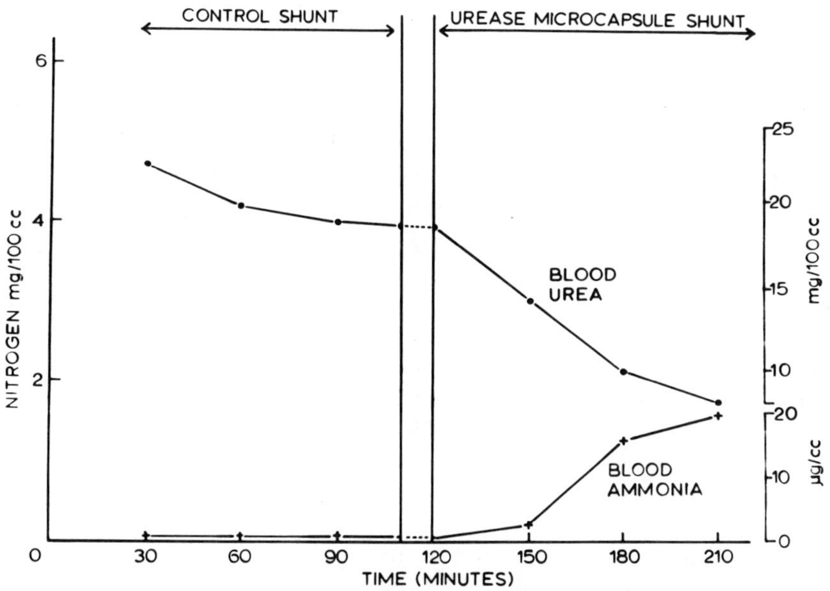

FIGURE 2. Extracorporeal shunt containing urease-loaded artificial cells. Effect on systematic arterial blood ammonia and blood urea level. (From Chang, 1966, 1972a.)

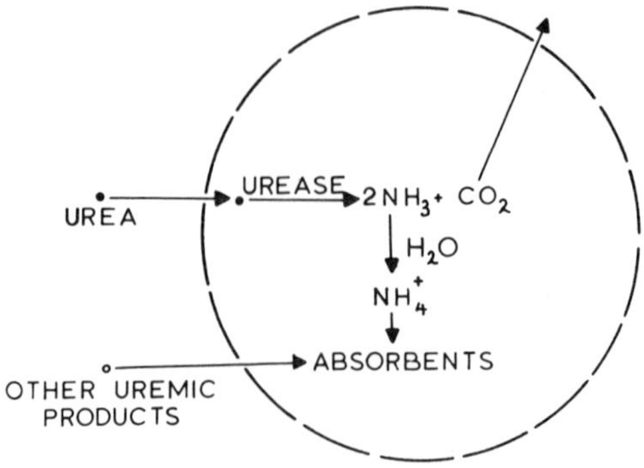

FIGURE 3. Microencapsulation of urease with adsorbent. Urea is converted enzymatically to ammonia, which is then removed by the adsorbent.

use microencapsulated urease in an extracorporeal shunt, 2 cm in diameter and 10 cm in length. The only delay in practical application of this approach will be the availability of a high-capacity ammonia adsorbent. The presently available adsorbents do not have the required capacity.

Since some waste metabolites can equilibrate rapidly from the body fluid into the intestinal tract, another approach was also studied. Microencapsulated urease was administered into the intestinal tract (Chang, 1972a, 1976a; Chang and Loa, 1970; Chang and Poznansky, 1968) (Figure 4). It was found in these studies that when microencapsulated urease and ammonia adsorbent was administered into the intestinal tracts of animals, the blood urea level can be lowered to 60.6 ± 5.0% of the control level (Chang, 1972a, 1976a; Chang and Loa, 1970). Another group has extended this approach by stabilizing urease by adsorption before microencapsulation (Gardner *et al.,* 1971). In this study, microencapsulated urease and ammonia adsorbent was fed to dogs by mixing with food. They were able to demonstrate that this resulted in a significant decrease of the blood urea levels in the animals. Further extensions along this line were carried out by other groups (May and Li, 1972; Asher *et al.,* 1975). Here, they used a liquid membrane system in which urease was microencapsulated within a liquid membrane which also contained buffer systems, to allow for a very high internal pH. This way, urea diffusing across the liquid membrane is converted by the enclosed urease to ammonia. Owing to the high pH of the liquid-membrane-microencapsulated system, most of the ammonia is present in an ionized form which cannot diffuse out of the liquid membrane. One advantage of the liquid membrane approach is that ammonia adsorbent is not required because the ammonia so formed is entrapped by the liquid membrane. However, this has to be tested in animal studies, especially to examine the stability of the liquid membrane emulsion in the presence of bile salts and other contents of the gastrointestinal tract. Oral administration of microencapsulated enzymes, though much more convenient than extracorporeal hemoperfusion, is limited by the fact that although urea can diffuse quickly across into the gastrointestinal tract, many important waste metabolites and toxins in the body fluid do not equilibrate with sufficient speed into the intestinal tract.

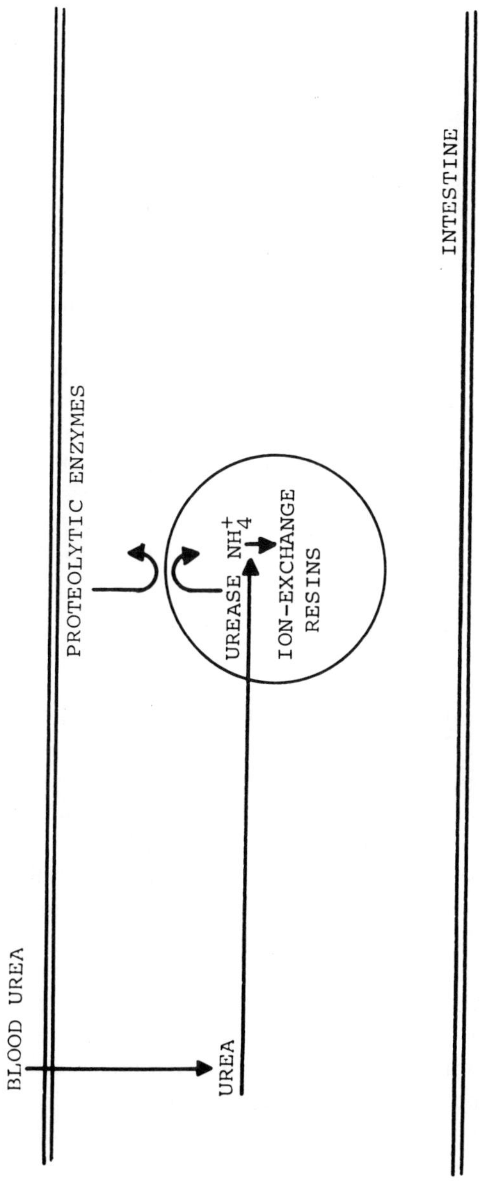

FIGURE 4. Schematic representation of the action of urease-loaded artificial cells in the gastrointestinal tract. (From Chang, 1972a.)

Another extension of the immobilized urease approach is the suggestion (Chang, 1966) that by placing the enzymes and adsorbents in the external dialyzing fluid of an artificial kidney, the enzymes and the ammonia adsorbent could also act on urea, diffusing across the standard hemodialysis membrane. This suggestion of using enzymes in the dialysate compartment of the artificial kidney has now been incorporated into one of the components of the dialysate regeneration system. In this system, immobilized urease in combination with ammonia adsorbent is used to remove urea diffusing from blood into the dialysate compartment of the standard hemodialysis machine, thereby cutting down the need for using large volumes of dialysate solution (Gordon *et al.*, 1969). However, the advantage of the microencapsulated system for direct blood hemoperfusion is the much larger surface area and ultrathin membrane that allows for a compact system.

In summary, the microencapsulation of urease as a model enzyme has demonstrated the basic effectiveness of the use of artificial cells containing enzymes for the construction of artificial organs for the removal of a typical waste metabolite, urea. For practical applications, further development is required to study the encapsulation of other enzyme systems. Work is already in progress. Thus, microencapsulated uricase is effective in converting uric acid (Chang, 1964). Actual *in vivo* tests have been successfully carried out using uricase immobilized in glass beads (Venter *et al.*, 1975). In this study, uricase immobilized in glass beads was tested in an extracorporeal shunt system in Dalmatian dogs. Here, 6 g of uricase–glass beads were effective in lowering the systemic blood uric acid level from 2.5 to 0.5 mg% in 4 hr. Microencapsulation of liver cells, liver extract, and multienzyme systems are other approaches to this end (Chang, 1965, 1972a).

4. *Immobilization of Albumin on Microencapsulated Adsorbents*

Experiments carried out here involve the microencapsulation of such adsorbents as ion-exchange resins and activated charcoal inside the artificial cells to remove waste metabolites and drugs from blood which equilibrate rapidly across the membrane (Chang, 1966, 1972a, 1975). Thus, artificial cells containing ion-exchange resins have been used experimentally in animals to remove blood ammonia (Chang, 1966, 1972a). Activated charcoal is very efficient in removing waste metabolites and toxins, however, when, in contact with blood, it releases embolizing particles and removes platelets, and thereby cannot be used clinically by itself (Yatzidis, 1964). Artificial cells containing activated charcoal have been developed in this laboratory and used to treat patients with chronic renal failure, acute intoxication, and liver failure (Chang, 1966, 1969, 1972a, 1975, 1976a, 1976b; Chang and Malave, 1970; Chang *et al.*, 1971, 1972, 1973a, 1973b, 1974, 1975, 1976). Activated charcoal inside these artificial cells is prevented from releasing particles, but waste metabolites and toxic materials can equilibrate across the membrane very rapidly to be removed by the activated charcoal within the artificial cells. The one that is best for clinical use involves the direct coating of activated charcoal with cellulose nitrate followed by a second coating with human

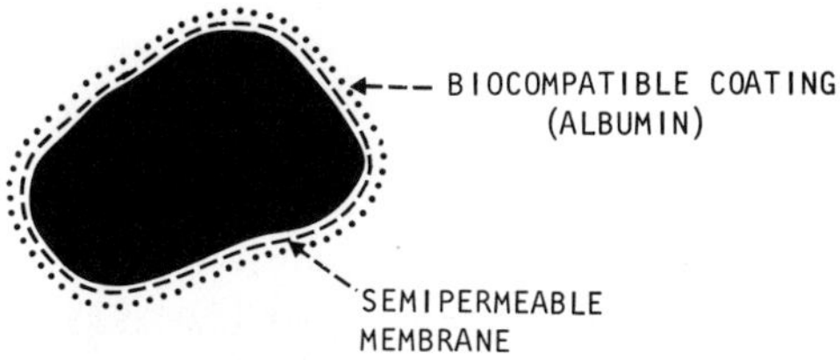

Figure 5. Charcoal granule microencapsulated by a 500-Å-thick coating of cellulose nitrate (semipermeable membrane) and a second ultrathin layer of albumin.

albumin (ACAC) (Chang, 1969, 1974) (Figure 5). Cellulose nitrate forms a semipermeable membrane to prevent charcoal embolism and allow for rapid equilibration of waste metabolites and toxins. Immobilization of albumin as a second coating makes the surface compatible to platelets and other blood elements (Chang, 1969, 1974). The control and the effects of hemoperfusion in experimental dogs over charcoal granules (AC), cellulose-nitrate-microencapsulated activated charcoal (CAC), or albumin-coated cellulose-nitrate-microencapsulated activated charcoal (ACAC) are shown in Figures 6–9. Albumin coating significantly decreased the fall of platelet level. Figures 10 and 11 showed that ACAC, but not CAC or AC, has no fibrin entrapment of formed elements of blood after hemoperfusion (Chang, 1974). Other centers are also using this approach of albumin-coated microencapsulated activated charcoal for use in patients (Blume *et al.*, 1976; Odaka *et al.*, 1976). Other centers have tried to coat the granules with various polymers and omit the second albumin coating. However, without the immobiliza-

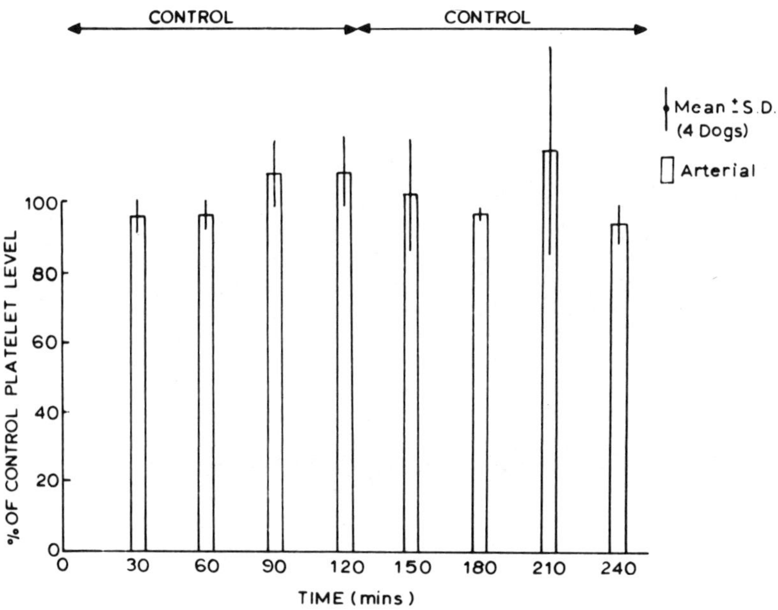

Figure 6. System platelet levels in four dogs expressed as mean and standard deviation. Two control periods of 120 min each are shown. (From Chang, 1974.)

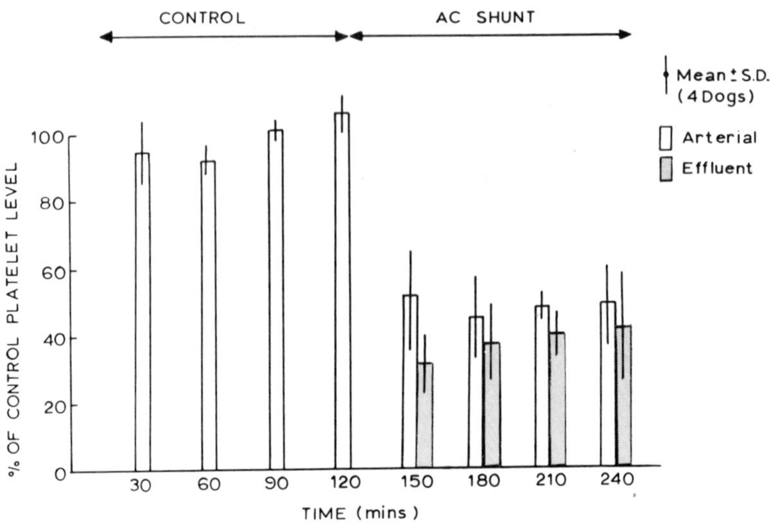

FIGURE 7. Systemic blood platelet levels in five dogs expressed as mean and standard deviation. Control period of 120 min followed by a 120-min test period of hemoperfusion through shunts containing 300 g of activated charcoal granules (AC). During the hemoperfusion, both the arterial systemic blood platelet levels and also the platelet levels of blood leaving the shunt chamber were followed. (From Chang, 1974.)

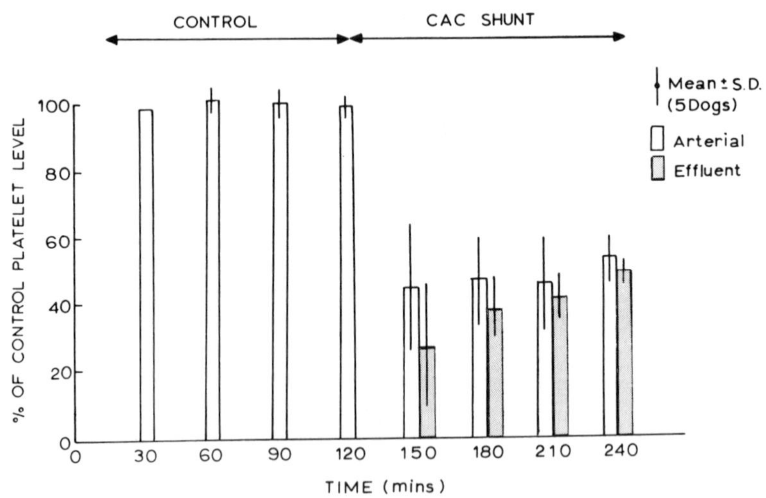

FIGURE 8. Systemic blood platelet levels in five dogs expressed as a mean and standard deviation. A control period of 120 min was followed by a 120-min test period of hemoperfusion through 300 g of cellulose-nitrate-coated activated charcoal (CAC). During the hemoperfusion, both the arterial systemic blood platelet levels and also the blood platelets levels of blood leaving the shunt were followed. (From Chang, 1974.)

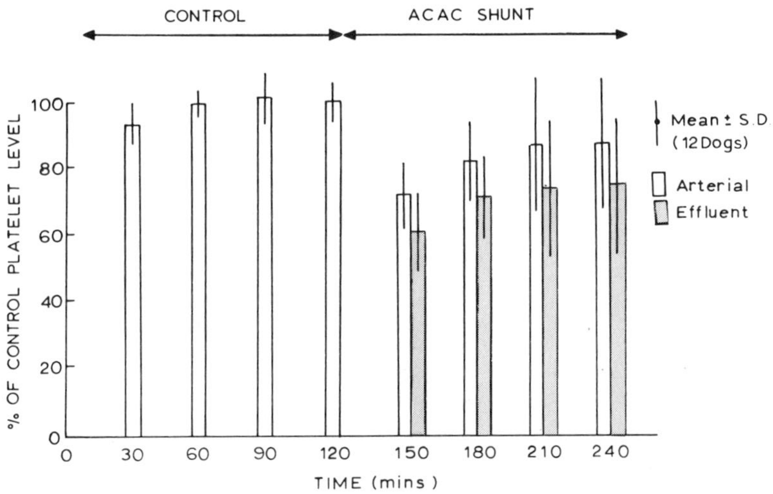

FIGURE 9. Systemic platelet levels in 12 dogs expressed as mean and standard deviation. A control period of 120 min was followed by a 120-min test period of hemoperfusion through the extracorporeal shunt chamber containing 300 g of albumin–cellulose-nitrate-coated activated charcoal (ACAC). During the hemoperfusion period both the arterial levels and the efferent levels of the blood leaving the shunt were followed. (From Chang, 1974.)

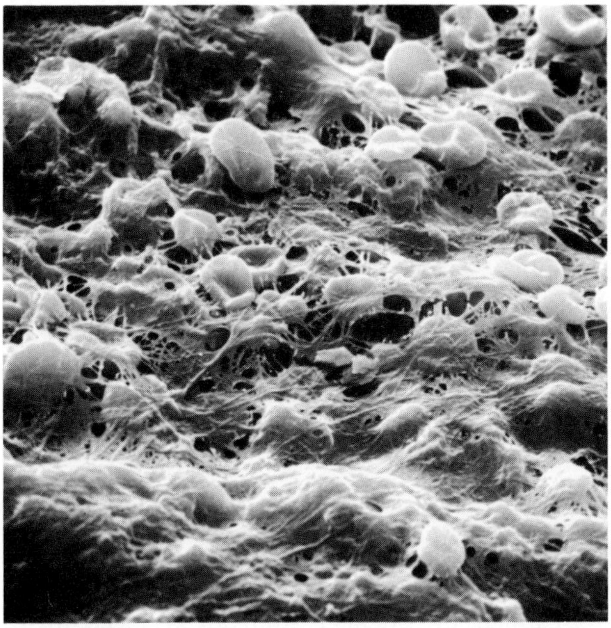

FIGURE 10. Scanning electron microscopic picture of cellulose-nitrate-coated activated-charcoal granules (CAC) after coming in contact with blood. Note the platelet adhesion, fibrin formation, and entrapment for formed elements of blood. (From Chang, 1974.)

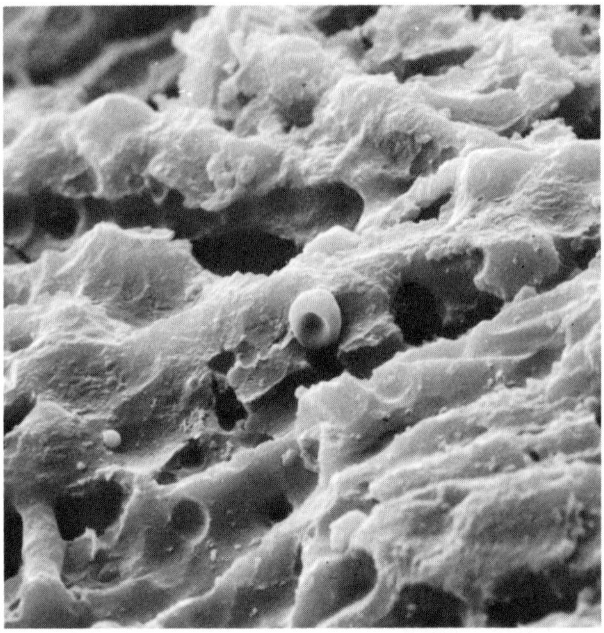

FIGURE 11. Scanning electron microscopic picture of albumin–cellulose-nitrate-coated activated-charcoal granules (ACAC) after coming in contact with blood. There is no platelet adhesion, fibrin formation, or entrapment of formed elements of blood. The cell at the center of the picture is a red blood cell for comparison of magnification (From Chang, 1974).

tion of albumin on the surface, there is a much greater loss of platelets from the systemic circulation after hemoperfusion. What is more important is that in the absence of albumin, aggregation of platelets may result in hypotension in some cases of severe hepatic failure. Preadsorption of albumin on capillary dialysis membrane can also decrease clotting tendencies.

At present, 300 g of ACAC of 1 to 5 mm in diameter are used in the extracorporeal shunt for each treatment. Figure 12 compares the effectiveness of this 300 g of ACAC to that of a "standard" hemodialysis machine being used at present in the hospital. Clearance is a terminology used to describe the effectiveness of a detoxifier in "clearing" the blood of a given substance and is based on the following equation:

$$\text{clearance} = \frac{C_A - C_E}{C_A} \cdot F$$

where C_A and C_E are the concentration of a given substance in the blood entering (C_A) and leaving (C_E) the detoxifier and F is the blood flow rate. Clearance results obtained in the treatment of patients is summarized in Figure 12. It is seen that with increasing molecular weight of the solutes, the clearance for the standard hemodialyzer decreases rapidly. In the case of the albumin-coated microencapsulated charcoal system (ACAC), much higher clearance is still obtained with large molecules of 5000 molecular weight. This is very important, because recently there has been strong evidence to support the theory that toxic metabolites in uremia, and perhaps liver failure, may be in the molecular weight range 300–5000. Protein-bound molecules in the blood are also removed much more effectively by the ACAC. This could be due either to the high affinity of activated charcoal for some of these molecules, or the presence of albumin coating, which binds the protein-bound molecule from blood and transfers it to the charcoal.

The ACAC system is very effective for the treatment of acute intoxication (Chang, 1969, 1975, 1976b; Chang *et al.*, 1973a, 1973b). Table 1 compares its effectiveness to standard hemodialyzers. In particular, many of the protein-bound drugs which are removed with great difficulty by standard hemodialyzer are effectively removed by the ACAC system.

In the treatment of liver-failure patients with grade IV hepatic coma first started here (Chang, 1972b), sufficient toxic materials are removed during hemoperfusion, to result in the temporary recovery of consciousness (Chang, 1972b, 1975, 1976b; Chang and Migchelsen, 1973). The use of this approach for the treatment of liver failure, although promising, is at present still not completely

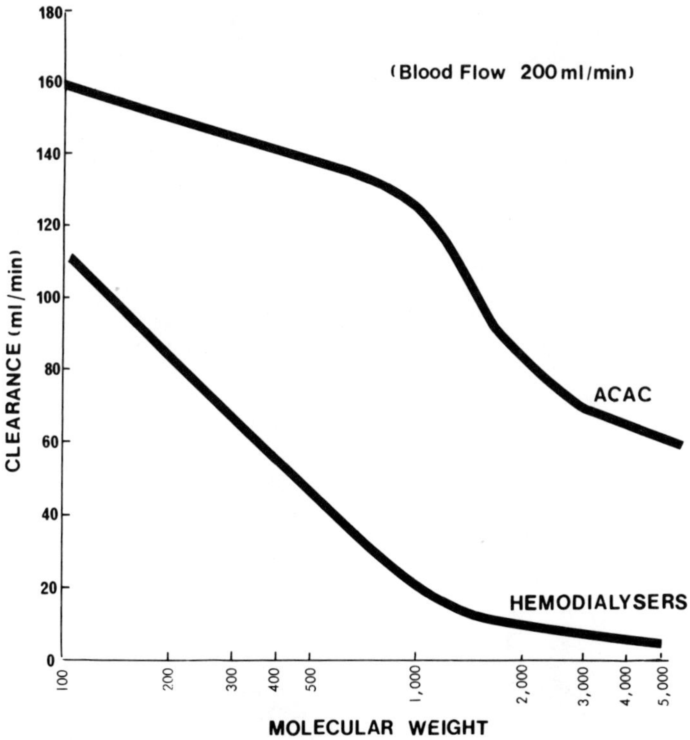

Figure 12. Schematic representation of clearance of different-molecular-weight substances in patient treatment. ACAC (albumin-coated microencapsulated charcoal) compared to a standard 1-m^2 hemodialyzer.

TABLE 1
Clearance of Drugs in Patients (ml/min)

Drug	ACAC	Hemodialyzer
Methyprylon	230	80
Methaqualone	230	29
Glutethimide	210	60
Phenobarbital	228	100
Salicylate	150	80

conclusive, and study using the ACAC system is being continued in this and other laboratories (Blume *et al.*, 1976; Odaka *et al.*, 1976; Ota *et al.*, 1976).

The ACAC system has been used for more than 6 years here for the treatment of patients with chronic renal failure. A total of 14 patients have been treated by hemoperfusion. Of the 14, 5 have been treated on a long-term basis of

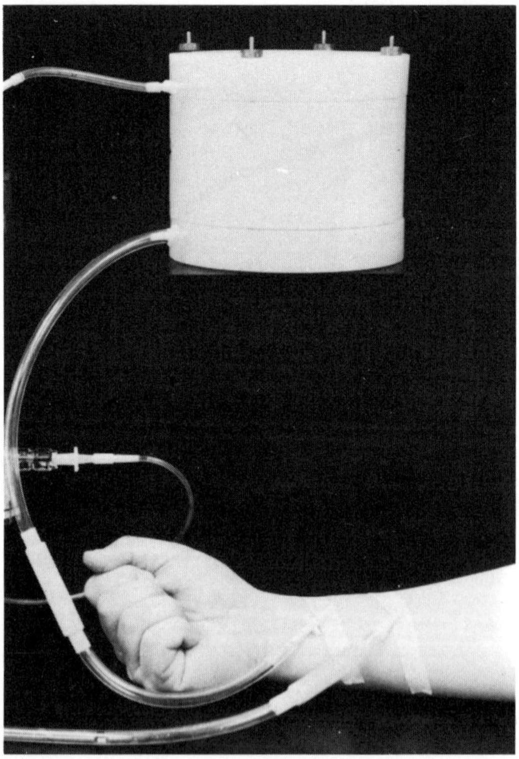

FIGURE 13. Extracorporeal chamber containing 300 g of albumin-coated microencapsulated charcoal being used for the treatment of patients with renal failure, liver failure, and acute intoxication.

between 4 and 8 months before returning to their standard hemodialysis regime. It has been demonstrated in the studies of chronic-renal-failure patients that hemoperfusion with the ACAC microcapsule artificial kidney for 2 hr is as effective as the standard hemodialysis treatment of 6–8 hr, especially for the removal of the larger uremic molecules and for keeping the patient symptom-free. Urea, phosphate, and potassium could be removed by oral adsorbents. Recently, a very small efficient ultrafiltrator became available for the removal of fluid and NaCl (Silverstein *et al.,* 1974). This ultrafiltrator does not require dialysate or negative pressure, but depends on the hydrostatic pressure of the arterial pressure. We have therefore combined the ACAC hemoperfusion in series with this small ultrafiltrator to form a portable, compact, dialysate-free artificial kidney (Chang, 1976b; Chang *et al.,* 1975, 1976) (Figure 13). The ACAC hemoperfusion, the ultrafiltrator, and the oral adsorbents will form a complete system for the treatment of chronic renal failure.

5. *Discussion*

Thus, the use of albumin-coated microencapsulated activated charcoal has already been demonstrated in human studies to be effective for the construction of artificial kidneys, artificial livers, and detoxifiers for the treatment of patients with chronic renal failure, acute intoxication, and liver failure. In this approach, the immobilization of albumin on the surface plays an important role. This has lent further support to the basic principle of the use of immobilized proteins and artificial cells for the construction of artificial organs. However, it would be expected that immobilized enzyme systems, when fully developed, would be much more specific and effective than adsorbent systems, since enzyme systems can much more selectively remove metabolites. Furthermore, enzymes are not capacity-limited. Unfortunately, the exact toxic metabolites in renal failure and liver failure are still not known. Thus, for the time being, the nonspecific albumin-coated microencapsulated adsorbent approach is more applicable. Although the more specific enzyme approach has great potential, it will have to be used in combination with adsorbents until more basic knowledge is available in regard to the toxic metabolites of renal failure and liver failure. A recent approach here is to microencapsulate a high concentration of albumin which can effectively bind metabolites and drugs. Cross-linked albumin spheres can also be prepared to remove protein-bound metabolites or drugs.

6. *References*

Asher, W. J., Bovee, K. C., Frankenfeld, J. W., Hamilton, R. W., Henderson, L. W., Holtzapple, P. G., and Li, N. N., 1975, Liquid membrane system directed toward chronic uremia, *Kidney Int.* **7**:S409.

Blume, V. Helmstaedt, D., Sybrecht, G., Baldamus, C., Submann, P., Heyer, V., Schmidt, E., and Schmidt, F. W., 1976, Hemoperfusions therapie des akuten leberversagens, *Lesser-Zuschriften* **14**:559.

Chang, T. M. S., 1964, Semipermeable microcapsules, *Science* **146**:524.

Chang, T. M. S., 1965, Semipermeable Aqueous Microcapsules, Ph.D. Thesis, McGill University, Montreal, P.Q., Canada.

Chang, T. M. S., 1966, Semipermeable aqueous microcapsules ("artificial cells"): with emphasis on experiments in an extracorporeal shunt system, *Trans. Amer. Soc. Artificial Internal Organs* **12**:13.

Chang, T. M. S., 1969, Removal of endogenous and exogenous toxins by a microencapsulated adsorbent, *Can. J. Physiol. Pharmacol.* **47**:1043.

Chang, T. M. S., 1972a, *Artificial Cells,* Charles C Thomas, Publisher, Springfield, Ill.

Chang, T. M. S., 1972b, Hemoperfusion over microencapsulated adsorbent in a patient with hepatic coma, *Lancet* **2**:1371.

Chang, T. M. S., 1974, Platelet interaction with albumin-coated surfaces, *Can. J. Physiol. Pharmacol.,* **52**:275.

Chang, T. M. S., 1975, Microencapsulated adsorbents for uremia, liver failure, and intoxication, *J. Kidney Int.* **7**:5378.

Chang, T. M. S., 1976a, Microcapsule artificial kidney including updated preparative procedures, *J. Kidney Int.* (in press).

Chang, T. M. S., 1976b, Hemoperfusion alone and in series with dialysis and ultrafiltration for patients with uremia, intoxication, and liver failure, *J. Kidney Int.* (in press).

Chang, T. M. S., and Loa, 1970, Urea removal by urease and ammonia adsorbents in the intestine, *The Physiologist* **13**:70.

Chang, T. M. S., and Malave, N., 1970, The development and first clinical use of semipermeable microcapsules (artificial cells) as a compact artificial kidney, *Trans. Amer. Soc. Artificial Internal Organs* **16**:141.

Chang, T. M. S., and Migchelsen, M., 1973, Characterization of possible "toxic" metabolites in uremia and hepatic coma based on the clearance spectrum for larger molecules by the ACAC microcapsule artificial kidney, *Trans. Amer. Soc. Artificial Internal Organs* **19**:314.

Chang, T. M. S., and Poznansky, 1968, Semipermeable microcapsules containing catalase for enzyme replacement in acatalasemic mice. *Nature* **218**:243.

Chang, T. M. S., Johnson, L. J., and Ransome, O., 1967, Semipermeable aqueous microcapsules: IV. Nonthrombogenic microcapsules with heparin-complexed membranes, *Can. J. Physiol. Pharmacol.* **45**:705.

Chang, T. M. S., Pont, A., Johnson, L. J., and Malave, N., 1968, Response to intermittent extracorporeal perfusion through shunts containing semipermeable microcapsules, *Trans. Amer. Soc. Artificial Internal Organs* **15**:163.

Chang, T. M. S., Gonda, A., Dirks, J. H., and Malave, N., 1971, Clinical evaluation of chronic, intermittent, and short term hemoperfusions in patients with chronic renal failure using semipermeable microcapsules (artificial cells) formed from membrane-coated activated charcoal, *Trans. Amer. Soc. Artificial Internal Organs* **17**:246.

Chang, T. M. S., Gonda, A., Dirks, J. H., Coffey, J. F., and Lee-Burns, T., 1972, ACAC microcapsule artificial kidney for the long term and short term management of eleven patients with chronic renal failure, *Trans. Amer. Soc. Artificial Internal Organs* **18**:465.

Chang, T. M. S., Coffey, J. F., Lister, C., Taroy, E., and Stark, A., 1973a, Methaqualone, methyprylon, and glutethimide clearance by the ACAC microcapsule artificial kidney: *in vitro* and in patients with acute intoxication, *Trans. Amer. Soc. Artificial Internal Organs* **19**:87.

Chang, T. M. S., Coffey, J. F., Barre, P., Gonda, A., Dirks, J. H., Levy, M., and Lister, C., 1973b, Microcapsule artificial kidney: treatment of patients with acute drug intoxication, *Can. Med. Assoc. J.* **108**:429.

Chang, T. M. S., Migchelsen, M., Coffey, J. F., and Stark, A., 1974, Serum middle molecule levels in uremia during long term intermittent hemoperfusions with the ACAC (coated charcoal) microcapsule artificial kidney, *Trans. Amer. Soc. Artificial Internal Organs* **20**:364.

Chang, T. M. S., Chirito, E., Barre, P., Cole, C., and Hewish, M., 1975, Clinical performance characteristics of a new combined system for simultaneous hemoperfusion-hemodialysis-ultrafiltration in series, *Trans. Amer. Soc. Artificial Internal Organs* **21**:502.

Chang, T. M. S., Chirito, E., Barre, P., Cole, C., Lister, C., and Resurreccion, E., 1977, Clinical clearance evaluation of a new combined system of ACAC hemoperfusion for blood purification and ultrafiltration for fluid removal, *J. Dialysis* (in press).

Gardner, D. L., Falb, R. D., Kim, B. C., and Emmerling, D. C., 1971, Possible uremic detoxification via oral-ingested microcapsules, *Trans. Amer. Soc. Artificial Internal Organs* **17**:239.

Gordon, A., Greenbaum, M. A., Marantz, L. B. McArthur, M. J., and Maxwell, M. H., 1969, Adsorbent based low volume recirculating dialysate, *Trans. Amer. Soc. Artificial Internal Organs* **15**:347.

Levine, S. N., and LaCourse, W. C., 1967, Materials and design consideration for a compact artificial kidney, *J. Biomed. Mater. Res.* **1**:275.

May, S. W., and Li, N. N., 1972, The immobilization of urease using liquid-surfactant membranes, *Biochem. Biophys. Res. Commun.* **47**:1179.

Odaka, M., Tobato, Y., Kobayashi, H., Soma, M., Nomura, Y., Hirassawa, H., Sato, H., Suenaga, E., and Nabela, K., 1976, Coated charcoal hemoperfusion, *J. Kidney Int.* **10**:197.

Ota, K., Ohta, T., Kobayashi, M., Yoshida, S., Kaneko, I., Agishi, T., and Sugchara, M., 1976, Coated charcoal hemoperfusion, *J. Kidney Int.* **10**:197.

Silverstein, M. D., Ford, C. A., Lysaght, M. J., and Henderson, C. W., 1974, Treatment of severe fluid overload by ultrafiltration, *New Engl. J. Med.* **291**:747.

Sparks, R. E., Mason, N. S., Meier, P. M., Litt, M. H., and Lindan, O., 1971, Removal of uremic waste metabolites from the intestinal tract by encapsulated carbon and oxidized starch, *Trans. Amer. Soc. Artificial Internal Organs,* **17**:229.

Venter, J. C., Venter, B. R., Dixon, J. E., and Kaplan, N. O., 1975, Extracorporeal immobilized enzyme reactors, *Biochem. Med.* **12**:79.

Yatzidis, D., 1964, A convenient hemoperfusion microapparatus for charcoal for the treatment of endogenous and exogenous intoxication, *Proc. Eur. Dialysis Transplant Assoc.* **1**:83.

20

Removal of Bilirubin from Blood by Affinity-Competition Chromatography over Albumin-Agarose Gel

PAUL D. BERK, PAUL H. PLOTZ, AND BRUCE F. SCHARSCHMIDT

1. *Introduction*

Largely as the result of advances in artificial kidney design, there have been dramatic improvements in the well-being and survival of patients with acute and chronic renal disease during the past several decades (Klebba *et al.,* 1974; Maher *et al.,* 1974). In contrast, despite the introduction of numerous new therapeutic approaches, there has been little improvement over the same period in the survival of patients with acute fulminant hepatic failure (Benhamou *et al.,* 1972). As a result of the continuing bleak prognosis for this condition, hepatologists are increasingly considering the feasibility of some sort of artificial device to perform temporarily some of the excretory functions of an acutely failing liver (Berk *et al.,* 1976a, 1976b). Both laboratory and clinical interest has focused on extracorporeal perfusion through columns of sorbents such as charcoal or synthetic resins, which remove a wide range of molecular species from blood with variable efficiency and relatively low specificity. In the setting of fulminant hepatic failure, the low specificity of such systems is, in a sense, desirable since the particular biochemical abnormalities requiring correction are not precisely known (Breen and Schenker, 1972; Schenker *et al.,* 1974; Fischer, 1974; Zieve, 1975). The most extensive clinical experience has been with charcoal hemoperfusion (Chang, 1972; Chang and Migchelson, 1973; Gazzard *et al.,* 1974a; Williams, 1976), but in view of the lack of suitable randomized controls, the effectiveness of this procedure remains unclear (Berk *et al.,* 1976a).

In contrast to the generalized hepatic dysfunction seen in fulminant hepatic failure, we have been faced on several occasions with gravely ill jaundiced patients in whom, except for an inability to excrete bilirubin, resulting from a hereditary deficiency of the conjugating enzyme bilirubin-UDP glucuronyl transferase (Berk

PAUL D. BERK AND BRUCE F. SCHARSCHMIDT • Section on Diseases of the Liver, Digestive Disease Branch, National Institutes of Health, Bethesda, Maryland PAUL H. PLOTZ • Arthritis and Rheumatism Branch, National Institutes of Health, Bethesda, Maryland

et al., 1974), liver function was entirely normal. This condition is known as type I congenital nonhemolytic jaundice, or Crigler-Najjar syndrome (Arias *et al.,* 1969). Although hyperbilirubinemia is widely believed to lead to the development of brain damage only in the neonatal period, there are increasing numbers of reports of bilirubin encephalopathy well beyond this period and even into adulthood in patients with Crigler-Najjar syndrome (Rosenthal *et al.,* 1956; Blumenschein *et al.,* 1968; Bloomer *et al.,* 1971; Blaschke *et al.,* 1974), and the sudden development of bilirubin encephalopathy in such individuals constitutes a grave medical emergency. Since a wide variety of measures aimed at reduction of plasma bilirubin concentration in such patients has proved unsuccessful (Blaschke *et al.,* 1974; Berk *et al.,* 1975), we were intrigued by the possibility that an extracorporeal device could be designed to remove bilirubin from whole blood or plasma of such individuals with moderate to high efficiency and, in contrast to previously employed sorbents, high specificity. Such a device would be of value not only in the management of patients with Crigler-Najjar syndrome, but also potentially as a new approach to the management of neonatal hyperbilirubinemia. There is also a possible role for such a device as one component of a more complex extracorporeal system for use in generalized hepatic failure.

We have, accordingly, studied the use of a modification of the technique of affinity chromatography for the selective removal of bilirubin from whole blood and plasma, both *in vitro* and *in vivo.* In the remainder of this chapter we shall describe both the theoretical and practical aspects of this approach. Data concerning its effectiveness in removing bilirubin from several animal models and selected aspects of the biocompatibility of such a system will be reviewed, and its potential clinical uses and remaining obstacles to clinical application will be considered.

2. *Rationale*

Unconjugated bilirubin, a lipid-soluble product of heme catabolism, is transported to the liver from its peripheral sites of formation as a very tight bilirubin–albumin complex. Although each molecule of albumin can bind three or more molecules of bilirubin (Jacobsen, 1969), under physiological conditions the molar ratio of bilirubin to albumin only occasionally exceeds 1, and virtually never exceeds 2. Hence, the first binding constant, which is the largest, is what ordinarily determines the concentration of unbound bilirubin in plasma. Because of the difficulties in preparing pure bilirubin entirely free of trace contaminants, reported values for the first dissociation constant of the bilirubin–albumin complex vary considerably with the method employed. The best available data suggest that it is of the order of 10M^{-1} (Jacobsen, 1969) indicating that the concentration of free bilirubin in plasma must, ordinarily, be very small. In addition, the very high binding constant implies that a single bilirubin molecule which has dissociated from albumin is statistically unlikely to remain free long enough to diffuse to and through a standard dialysis membrane before meeting and binding to another albumin molecule. Hence, the high-affinity constant of bilirubin for albumin explains why bilirubin cannot be removed from the body by either standard hemodialysis (Willson *et al.,* 1972) or by peritoneal dialysis, even when albumin has been added to the dialysis fluid (Grollman and Odell, 1962).

In contrast to the relatively long diffusion path required for dialysis across a membrane, if there were albumin molecules within free diffusion distance of the circulating plasma albumin, but immobilized so that they would not move with the circulating plasma stream, they should compete with the circulating albumin for any available free bilirubin and thereby remove bilirubin from the circulation. If the immobilized albumin were contained within a hydrogel structure, net transfer of bilirubin from circulating to immobilized protein would be expected to occur until a new steady state was achieved, with equilibration of bilirubin between free and immobilized protein.

Removal of bilirubin from a solution by perfusion through columns containing high-affinity binding sites (e.g., albumin molecules) immobilized on an inert matrix represents an application of the basic technique of affinity chromatography (Cuatrecasas and Anfinsen, 1971). When the bilirubin in solution is already complexed to circulating albumin, as it is in blood or plasma, the kinetics of bilirubin removal will be determined not merely by the affinity of the immobilized binding site, but also by competition between this immobilized site and similar sites on nonimmobilized circulating albumin. Under these circumstances, the term "competition chromatography" most aptly describes the physicochemical processes involved.

3. *Preparation of Albumin-Agarose Gels*

In order to test the effectiveness of affinity/competition chromatography for bilirubin removal, various approaches to the immobilization of albumin on inert matrixes were examined. After a series of experiments with various coupling methods and gels, we settled on a modification of the cyanogen bromide method of Axen *et al.,* (1967) to prepare conjugates of human serum albumin (HSA) with agarose (Plotz *et al.,* 1974). Forty grams of the selected agarose was weighed while wet, then washed on a scintered glass funnel with distilled water. The gel was added with an equal weight of water to a TPX beaker (Nalge Co., Nalgene Labware Div., Rochester, N.Y.). A stirring bar, a thermometer, and a pH electrode were added and the beaker was placed in a hood on a magnetic stirring table. Cyanogen bromide (Eastman, Rochester, N.Y.), 10 g, was added and the pH brought to about 10.5 with 2.5 N sodium hydroxide. For the next 10 min, 2.5 N NaOH and small amounts of ice were added to the reaction mixture to keep the pH at 10–11.5 and the temperature at 18–22°C. Then a large quantity of ice was added, the activated gel was rapidly filtered and washed with at least 25 volumes of ice-cold 0.1 M sodium bicarbonate, and then added to a beaker containing HSA at 50–100 mg/ml in 0.1 M sodium bicarbonate. The beaker was rotated gently at 4°C overnight, and on the following day the gel was washed with 0.14 M sodium chloride/0.1 M sodium phosphate, pH 7.2 (phosphate-buffered saline, PBS); 1.0 M NaCl/0.1 M sodium acetate, pH 5.0; 0.2 M borate saline, pH 8.4; PBS; and in recent preparations, 50% ethanol in water (vol./vol.) followed by PBS. The washed gel was then sieved to remove fine particles and stored in a TPX container in the icebox.

The amount of albumin bound was determined by Kjeldahl analysis of a sample of dried gel. Usually, the agarose bound between 30 and 50 mg of HSA/g of wet wt. This is equivalent to 30–50 mg of HSA/ml (since the density of the wet gel is about equal to water), which is the same as the concentration of HSA in normal plasma. Because of the critical temperature and pH requirements of the coupling reaction, it has been necessary to make the albumin-agarose gels (AAG) in small lots of not more than 100 g.

Coupling at other pH's from 4 to 11 did not improve the albumin binding to agarose over that obtained in 0.1 M sodium bicarbonate. The prior conjugation of phlorglucinol to the beads with epichlorhydrin did not increase the amount of albumin bound (Porath and Sundberg, 1972). Use of the carbodiimide reaction (Sheehan and Hess, 1955; Khorona, 1955) to couple a second albumin molecule to those already covalently bound to the agarose increased the amount of albumin bound but did not substantially improve bilirubin binding capacity of the gels (*vide infra*). This suggests that the addition of the second albumin molecule in some way blocked the bilirubin binding site on the first. Acrylamide beads conjugated with hydrazine hydrate and coupled to albumin via the acyl azide (Inman and Dintzis, 1969) bound less albumin than cyanogen bromide activated agarose.

During preliminary evaluation of a variety of agarose matrices, the albumin-binding capacities of Sepharose 6B, 65-325 mesh (Pharmacia Fine Chemicals, Inc., Piscataway, N.J.) and Bio-Gel A5M, 100–200 mesh in both cross-linked and non-cross-linked forms (Bio-Rad Laboratories, Richmond, Calif.) were found to be superior to those of Sepharose 2B and 4B, and Bio-Gel A1.5M and A50M. Subsequently, Bio-Gel A5M proved to have slightly greater bilirubin binding capacity than Sepharose 6B, and to permit better flow of whole blood, presumably because of its larger and more uniform bead size. Accordingly, it has been used in all the *in vivo* studies described below. Sepharose 6B and Bio-Gel A5M were used interchangeably, however, in initial *in vitro* studies.

4. *In Vitro Binding Studies*

4.1. *Bilirubin Binding*

When aliquots of either jaundiced plasma from patients with Crigler-Najjar syndrome or solutions of bilirubin in 5% human serum albumin are passed over a column of ablumin-agarose gel, a large proportion of the bilirubin is bound to the beads and is not eluted by prolonged washing with PBS (Figure 1). After the PBS washing, the column remains yellow. To establish that the coupled albumin, and not the agarose, was responsible for the bilirubin binding, we passed equal quantities of jaundiced plasma over plain agarose, agarose conjugated with human gamma globulin, and agarose-HSA. As Table 1 demonstrates, only agarose-HSA bound any significant amount of bilirubin (Plotz *et al.,* 1974).

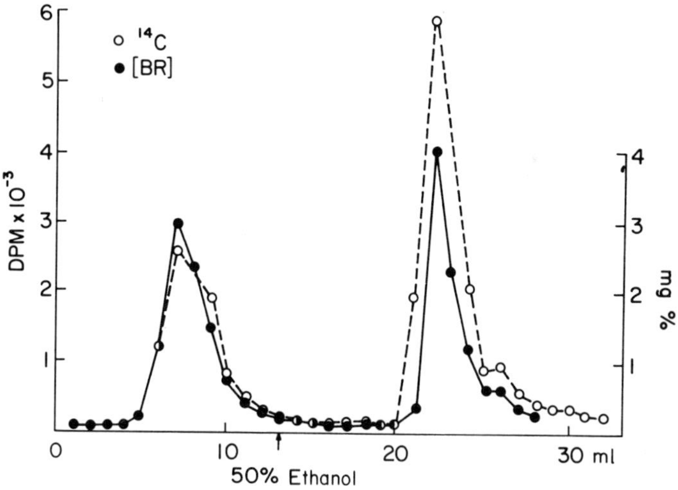

Figure 1. Binding of bilirubin to albumin-agarose gel, and subsequent elution by ethanol. One milliliter of plasma containing 180 μg of bilirubin and labeled with a tracer dose of [^{14}C]bilirubin was applied to a column containing 3 g of agarose-human serum albumin gel, and eluted with phosphate-buffered saline followed by 50% (vol./vol.) ethanol in water. Bilirubin was measured by diazotization and radioactivity. The lower chemical values in the ethanol eluate are due to partial degradation of bilirubin to diazo-negative products prior to measurement. (From Plotz *et al.*, 1974; reproduced by permission.)

The amount of bilirubin bound to agarose-HSA is proportional to the initial concentration in the plasma (Figure 2). The dependence of bilirubin on initial plasma concentration must be borne in mind when assaying lots of gel for binding capacity, either by flow-through or batch equilibration techniques. At any particular plasma bilirubin concentration, there is apparent saturation of gel when increasing volumes of plasma are presented to a column of fixed size (Figure 3). Assuming that one molecule of HSA has one strong binding site for bilirubin

Table 1
Requirement of Coupled Albumin for Binding of Bilirubin to Agarose[a]

Gel	Bilirubin bound (μg/g of gel)
Sepharose 6B	0.9
Sepharose 4B: human γ-globulin	0.3
Sepharose 6B: human serum albumin	71.8

[a] 1 ml of plasma containing 240 μg of bilirubin was passed over 3 g of the appropriate agarose. Amount bound represents the difference between the bilirubin added and the amount recovered in phosphate-buffered saline washes.

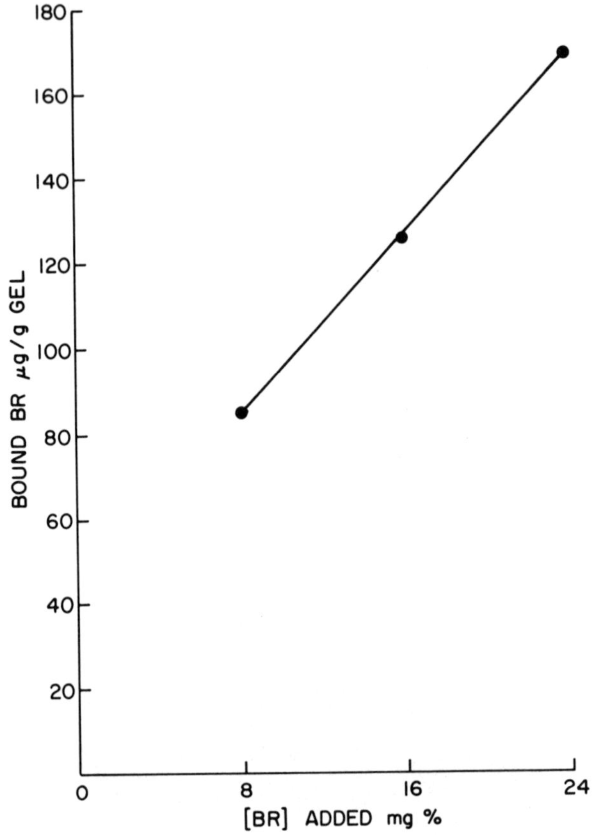

FIGURE 2. Relationship of plasma bilirubin concentration to bilirubin binding by albumin-agarose gel (AAG). Three-milliliter portions of plasma, each containing 240 μg of bilirubin/ml, were diluted with increasing volumes of normal plasma and then passed over identical columns of AAG. Each column was exposed to the same amount, but a different concentration, of bilirubin.

(Jacobsen, 1969), one can calculate that agarose-HSA with 40 mg of HSA/g of gel should strongly bind up to 300 μg of bilirubin/g. When exposed to plasma with an initial concentration of 24 mg of bilirubin/100 ml, some gels have bound more than 150 μg of bilirubin/g of gel, suggesting that the albumin lost less than half of its binding capacity as a result of coupling (Plotz *et al.*, 1974).

The bound bilirubin could be completely eluted by either normal plasma or a solution of HSA at 40 mg/ml, but since the need to regenerate columns with plasma components would decrease their utility, we sought simpler means to remove the bilirubin. Neither 1 M sodium chloride nor 1 M sodium chloride–0.1 M sodium acetate, pH 5.0, removed any of the bound counts or color. Ethanol in water, 50% (vol./vol.), however, removed all the bound material (Figure 1). In 55 experiments in which the amount bound (amount added minus amount in the PBS washes) and the amount eluted by 50% ethanol could be compared, the ratio

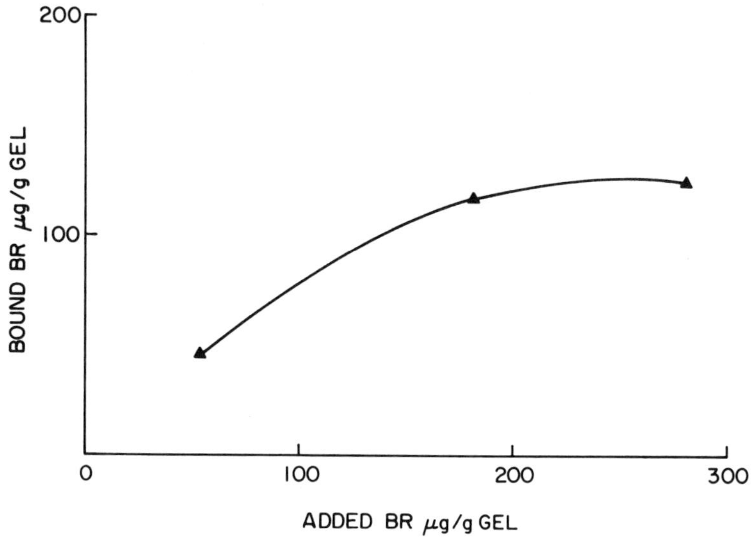

FIGURE 3. Apparent saturation of albumin-agarose gel. One-, 3-, and 5-ml aliquots of plasma (180 μg of bilirubin/ml) were passed over identical columns of 3 g of gel. (From Plotz *et al.,* 1974; reproduced by permission.)

of eluted/bound was 0.99 ± 0.093 (S.D.). The bilirubin binding capacity of the gel was found to be stable, despite repeated use and ethanol elution, and after more than 6 months of storage in a standard refrigerator (Table 2).

Recent studies have indicated that 76% (vol./vol.) ethanol, chosen because of its bacteriostatic properties (*vide infra*), was as effective as 50% (vol./vol.) ethanol for elution, and did not decrease subsequent bilirubin-binding capacity of the gel.

4.2. *Other Substances*

In addition to bilirubin, albumin binds a variety of other substances, of which hormones, bile acids, and drugs represent three major classes. In studies described in detail elsewhere (Plotz *et al.,* 1974), it was established that cortisol in plasma was retarded during passage through AAG but could be ultimately quantitatively eluted with prolonged PBS washing. In contrast, approximately 25% of a tracer dose of radiolabeled thyroxine in plasma bound to the column and could not be eluted either with PBS or 1 M NaCl–0.1 M sodium acetate, pH 5.0. This fraction was readily eluted with 50% ethanol.

Of the various drugs studied, salicylate behaved much like cortisol, in that it could be quantitatively eluted with prolonged PBS washing. Digoxin, which does not bind significantly to albumin, and digitoxin, which does (Smith, 1972), were studied at plasma levels in the ranges associated with clinical toxicity. No digoxin was retained by the beads. By contrast, virtually all of the digitoxin was retained and was eluted by 50% ethanol but not by 1 M sodium chloride–0.1 M sodium acetate, pH 5.0. Although digitoxin in plasma was relatively effectively removed *in*

TABLE 2
Effects of Prolonged Storage and Repeated Reuse of Albumin-Agarose Gel on Bilirubin Binding Capacity[a]

Age of gel (days)	Proportion of bilirubin bound (% of day 1 value)
1	100
7	110
15	114
36	116
91	122
192	143

[a]One milliliter of jaundiced plasma was passed over a single 10-g agarose column on six separate occasions from 1 day to more than 6 months after preparation of the gel. After elution with 50% (vol./vol.) ethanol on each occasion, column was stored in a refrigerator prior to subsequent reuse.

vitro, in vivo perfusion of digitoxin-intoxicated rats over columns of AAG (see below) removed only small quantities of the drug, presumably because of its wide distribution and tissue fixation. Sulfobromophthalein (SBP) and indocyanine green (ICG), two cholephelic anions whose transport in plasma and hepatic excretion are in many ways similar to bilirubin, were also tightly bound by AAG, and eluted only with 50% ethanol.

The binding by the column of the major human bile acids cholic acid (trihydroxy), chenodeoxycholic acid (dihydroxy), and lithocholic acid (monohydroxy) and their respective glycine and taurine conjugates have also been examined in detail (P. H. Plotz, unpublished). The degree of binding depended solely on the nature of the bile acid, not on the presence or absence of conjugating amino acids. The degree of removal by AAG columns varied inversely with the number of hydroxyl groups, as expected from published data on the relative affinities of various bile acids for albumin (Rudman and Kendall, 1957). Thus, cholic acid and its conjugates are retarded by the column but are completely elutable with PBS. In contrast, a significant fraction of lithocholic acid and its conjugates is bound and can only be eluted with ethanol.

5. *Removal of Bilirubin in Vivo*

5.1. *Studies of Jaundiced Rats*

The initial *in vivo* studies of bilirubin removal by AAG were performed in jaundiced rats, which were used because of their small size and ready availability, and because several columns of suitable size could be prepared from a single batch of gel. Homozygous jaundiced Gunn rats and Sprague-Dawley rats with surgically ligated bile ducts were chosen as models of unconjugated and conjugated hyperbilirubinemia, respectively (Scharschmidt *et al.,* 1974).

In the extracorporeal circuit used (Figure 4) blood leaves the animal through a carotid cannula, passes an inflow sampling site, filters down through a column containing the albumin-bound agarose beads, and passes through an outflow sampling site before returning through a jugular cannula. Flow in the circuit is controlled by a peristaltic pump. In the rat studies, each hemoperfusion was carried out for 60 min at a flow rate of 1 ml/min. All animals received systemic heparinization during the procedure.

Figure 5 illustrates the actual appearance of the translucent gel prior to hemoperfusion, as blood initially filters through the gel, during hemoperfusion, and at the end of hemoperfusion as the red cells are flushed back into the animal with saline, revealing the gel, which is now colored yellow by the bound bilirubin. The lower right panel in Figure 5 shows elution of the bound bilirubin in a bright yellow band with 50% ethanol, returning the gel to its previous translucent state.

In order to determine the relationship between quantities of bilirubin removed during *in vivo* hemoperfusion and endogenous bilirubin pools, hemoperfusion of Gunn rats was preceded by isotopic studies of bilirubin kinetics. Approximately 72 hr prior to hemoperfusion, each Gunn rat was given an intravenous injection of 6–7 μCi of tritiated bilirubin, and a plasma radiobilirubin disappearance curve was defined as illustrated in Figure 6, employing a specially designed small-animal blood sampling apparatus (Scharschmidt and Berk, 1973). Using the standard isotope dilution equation, extrapolation of the final exponential of the plasma curve to zero time allows calculation of the total miscible

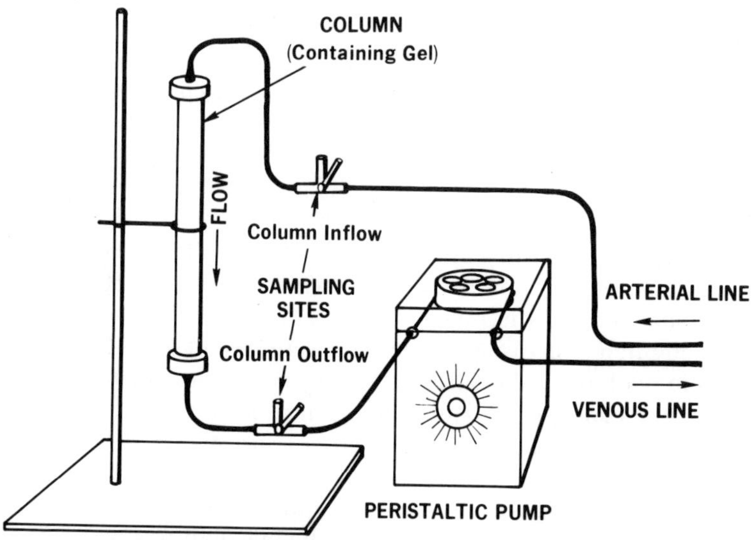

FIGURE 4. Extracorporeal hemoperfusion circuit employed in rat studies. (From Scharschmidt *et al.,* 1974; reproduced by permission.)

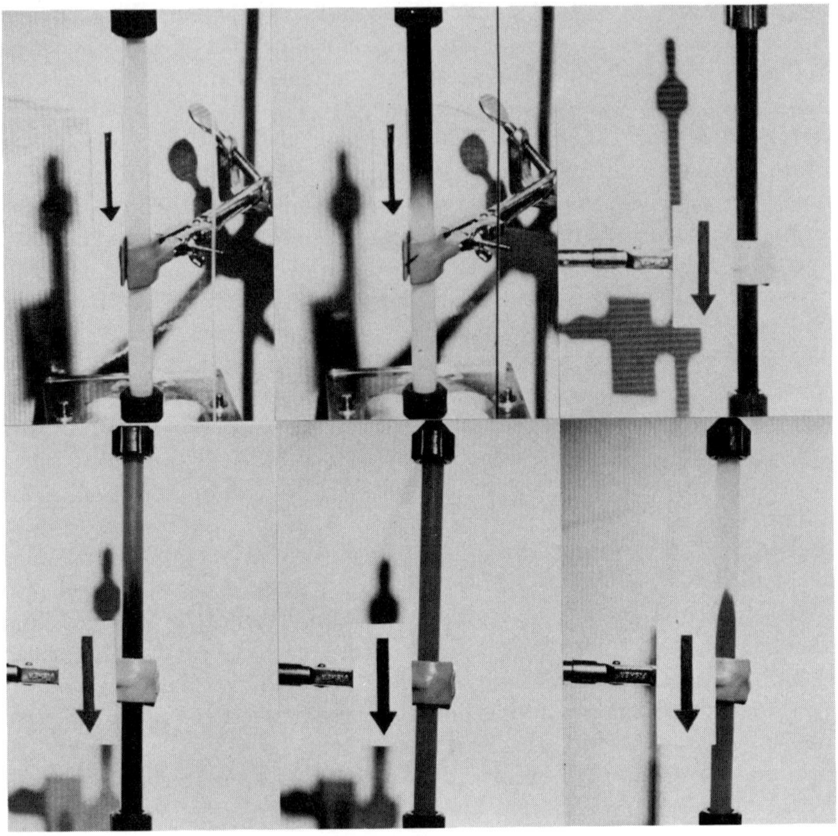

FIGURE 5. Appearance of an albumin-agarose column during hemoperfusion of a jaundiced Gunn rat, showing translucent gel before hemoperfusion (top left), as blood initially filters through the gel (top center), midway through hemoperfusion (top right), and at the end of hemoperfusion as blood is flushed back into the animal with saline (bottom left). After all blood is flushed from the column (bottom center), the yellow bilirubin can be quantitatively eluted from the column with 50% ethanol and water, returning the gel to its original translucent state (bottom right). (From Berk *et al.*, 1975; reproduced by permission.)

bilirubin pool, and extrapolation of the first 3–4 points taken during the 2 min following injection of the isotope allows calculation of the plasma pool. The amount of radioactivity and number of micrograms of bilirubin eluted from the column following hemoperfusion were then related to these body pool sizes. The plasma bilirubin pool size in the Sprague-Dawley rats was calculated as the product of plasma bilirubin concentration and the mean plasma volume, the latter having been determined by isotope dilution in a separate group of Sprague-Dawley animals of comparable size.

Results of five hemoperfusions in four Gunn rats are summarized in Table 3. The data indicate that (1) more than 97% of the bilirubin was removed from the first column volume of blood passing through the gel; (2) an average of slightly

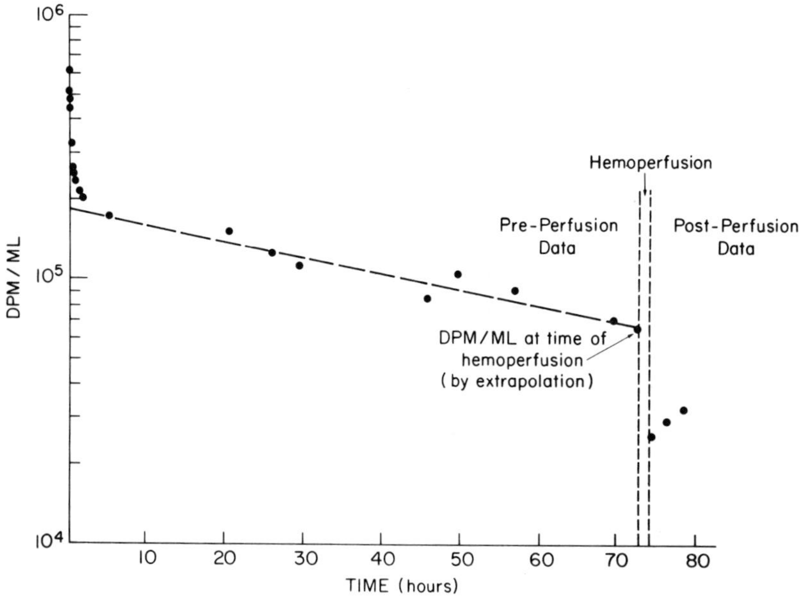

FIGURE 6. Plasma bilirubin disappearance curve in a jaundiced Gunn rat prior to, and immediately after, hemoperfusion. (From Scharschmidt *et al.*, 1974; reproduced by permission.)

TABLE 3
Removal of Bilirubin from Jaundiced Rats in Vivo by a 1-hr Hemoperfusion over Albumin-Agarose Gel[a]

Variables	Jaundiced Gunn rats (5 studies)	Rats with biliary obstruction (7 studies)
Initial bilirubin extraction (%)	97.5 ± 4.3	96.5 ± 4.1
Plasma bilirubin concentration in the rat (mg/100 ml)		
Before hemoperfusion[b]	8.2 ± 1.6	7.5 ± 4.1
After hemoperfusion	2.6 ± 0.5	4.9 ± 0.7
Bilirubin recovered from column (mg)	1.06 ± 0.26	1.02 ± 0.14
Body bilirubin pools removed (%)		
Plasma bilirubin pool[c]	96 ± 36	63 ± 15
Total miscible pool	23 ± 4	
Rat weight (g)	404 ± 24	635 ± 84

[a]All values are mean ± S.D.
[b]Plasma bilirubin values reported for Gunn rats are unconjugated, and for obstructed rats total bilirubin concentration. In the latter group, 92 ± 4% of the initial plasma bilirubin and 91 ± 4% of the bilirubin recovered from the columns was conjugated.
[c]Bilirubin pools in Gunn rats were determined by isotope dilution. Plasma bilirubin pool in the obstructed rats was calculated from the plasma bilirubin concentration, and measurements of plasma volume in similar animals.

more than 1 mg of bilirubin was recovered from the column after 1 hr of perfusion; and (3) the plasma bilirubin concentration in the intact animal fell more than 70%, from a mean of 8.2 to 2.6 mg/100 ml, during a 1-hr hemoperfusion. Recovery of isotope from the column eluate indicated that an average of 22% of the total miscible bilirubin pool and 96% of the plasma bilirubin pool initially present in the animals had been removed.

The results of seven hemoperfusions in five rats with biliary obstruction were entirely analagous (Table 3). Ninety-seven percent of the circulating conjugated bilirubin was removed from the first column volume of blood passing through the circuit, and more than 1 mg of bilirubin was recovered from the column following hemoperfusion. Plasma bilirubin concentration fell 35%, from 7.5 to 4.9 mg/100 ml. Because the total time of perfusion, flow rate, and bilirubin concentration before hemoperfusion were essentially the same for both the Gunn rat and the obstructed rats, the total bilirubin loads presented to the column in both groups were also similar. Given these similar loads of unconjugated and conjugated bilirubin, it is of interest that essentially the same amount of conjugated bilirubin (1 mg) was extracted by the column in the studies of rats with biliary obstruction as was unconjugated bilirubin from the Gunn rats. The lesser overall fall in bilirubin concentration in rats with biliary obstruction is probably explained by their greater body weight and correspondingly greater plasma and total body bilirubin pools. An estimated 63% of the plasma bilirubin pool was removed during the 1-hr hemoperfusion from the rats with biliary obstruction, and the gel was still removing 75% of bilirubin presented to it when the study was terminated.

A total of 34 animals have undergone 37 hemoperfusions. In those studies where the column volume was less than 50% of the estimated blood volume, the survival rate was 100% and the animals awoke shortly after the procedure without observable ill effects (Scharschmidt *et al.,* 1974).

5.2. *Studies in Neonatal Monkeys*

Because of the relatively limited binding capacity of the available gels, the studies described above were conducted with relatively large columns requiring that approximately 50% of the animal's blood volume be in the extracorporeal circuit. Although rats appear to tolerate such volume depletion, columns of comparable size could not be employed clinically without blood or plasma priming, which it was felt desirable to avoid. Accordingly, a new apparatus was designed, consisting of four small columns arranged in parallel. Circuitry was designed to permit perfusion of one column at a time, while the others are sequentially washed with saline, eluted with ethanol, and rewashed with saline. Sequential and repeated perfusion and elution of each column in the circuit provides a system of almost unlimited bilirubin binding capacity, while requiring less than 15% of the blood volume in the extracorporeal circuit at any time (Scharschmidt *et al.,* 1975a).

The new system was employed in several groups of neonatal *Rhesus* monkeys delivered prematurely by Caesarian section at an average gestational age of 150 days (normal gestation 155–165 days). *Rhesus* monkeys were selected for study because they develop an unconjugated neonatal hyperbilirubinemia similar to, although milder than, that seen in human neonates. Newborn animals were perfused with blood obtained from either the umbilical artery, femoral artery, or femoral vein and returned to the animal via the umbilical vein.

Compared with five control monkeys, which had a peak plasma bilirubin concentration averaging approximately 4 mg/100 ml, six hemoperfused neonates experienced a significant reduction in bilirubin levels, averaging 60%, after each of two perfusions at 20 and 26 hr of age (Scharschmidt *et al.*, 1975a). The effect of hemoperfusion was even more striking in groups of animals in whom endogenous hyperbilirubinemia was augmented by bilirubin infusion shortly after birth. In the group with the highest preperfusion bilirubin levels, the plasma unconjugated bilirubin concentration fell from a mean of 24.4 to 3.1 mg/100 ml as the result of the procedure. In the monkey studies, bilirubin recovered from the columns averaged 93% of the estimated preperfusion plasma bilirubin pool.

Following return of surviving perfused monkeys to the colony, subsequent growth and development was indistinguishable from that of controls and other young animals in the same population. In contrast to the excellent survival rate in the rat studies, however, 6 of 13 perfused monkeys died during the procedure. Although the actual cause of death varied, perfusion-induced thrombocytopenia was a contributing factor in each case *(vide infra).* Four of nine control animals also died, indicating the extreme fragility of this particular animal model.

6. *Specificity and Biocompatibility*

6.1. *Removal of Soluble Blood Components Other Than Bilirubin*

It was expected that only substances with a known affinity for albumin would be removed during hemoperfusion through albumin-agarose gel. As might have been predicted, small decreases in serum calcium and plasma cortisol and thyroxine levels were observed in various studies of either *in vitro* or *in vivo* perfusion when the columns were primed only with heparinized saline. In contrast, and again as predicted, a wide variety of blood components without known affinity for albumin were essentially unaltered by passage of whole blood through albumin-agarose gel (Table 4). In particular, immunoelectrophoresis of serum from whole blood which had been passed over AAG *in vitro* showed no detectable loss of any of the multiple protein constituents (Plotz *et al.,* 1974).

Detailed studies of the quantitative removal, during *in vivo* perfusion, of physiologic blood constituents such as bile acids and free fatty acids are currently being conducted. In view of the known affinity of these substances for albumin, some degree of removal is to be expected. Detailed identification of individual bile acids and free fatty acids by gas–liquid chromatography will be required to establish the effects of *in vivo* perfusion on these physiologically important blood constituents.

TABLE 4
Blood Components Essentially Unchanged by Hemoperfusion

Chemistries	Coagulation tests
Electrolytes	Prothrombin time
Calcium, phosphorus	Partial thromboplastin time
Blood urea nitrogen	Thrombin time
Creatinine	Factor V
Uric acid	Factor VII
Glucose	
Serum iron/iron binding capacity	Miscellaneous
Protein-bound iodine	Serum protein electrophoresis
T_3/T_4	Serum protein immunoelectrophoresis
Acid phosphatase	Haptoglobin
Alkaline phosphatase	Ceruloplasmin
SGOT/SGPT	Lipids

6.2. *Effect of AAG Hemoperfusion on Formed Blood Elements*

Our original *in vivo* studies of biocompatibility were performed in heparinized Sprague-Dawley and Gunn rats. These indicated that during hemoperfusion: (1) Hemoglobin concentration and hematocrit measured in the animal either remained unchanged or decreased to a small degree compatible with the amount of blood sampling or hemodilution. Red cell morphology and osmotic fragility were unaltered. (2) Total leukocyte counts in the animal either remained unchanged or increased, with an increase in the percentage of young granulocytes. In contrast, measurements in the column effluent showed a selective loss of granulocytes. (3) Platelet counts in the animal were either unchanged or modestly decreased, whereas counts in the column effluent varied from unchanged to markedly decreased. Inspection of the gel and column components suggested that much of the loss of granulocytes and platelets occurring during rat studies occurred not in the gel itself but in the various nylon or stainless steel meshes used to retain the gel within the columns (Scharschmidt *et al.*, 1974).

As a result of these studies, we anticipated that albumin-agarose gel would be highly "biocompatible" and that its use would not be associated with the striking losses of platelets and significant decrease in circulating granulocytes seen during hemoperfusion over uncoated charcoal or various resins (Cohn *et al.*, 1960; Pallotta and Koppanyi, 1960; Dunea and Kolff, 1965; Barakat and MacPhee, 1970; Andrade *et al.*, 1971; Juggi, 1971; Gazzard *et al.*, 1974b; Willson *et al.*, 1974). Subsequent studies have proved that these hopes were too optimistic. There appears to be considerable species variation in the ability of formed blood elements to withstand extracorporeal circulation, with rat blood cells being particularly resistant to entrapment and destruction on columns, while those of man and the *Rhesus* monkey are much more sensitive. During the perfusion studies in neonatal monkeys cited above (Scharschmidt *et al.*, 1975a), platelet counts fell to an average of only 10% of baseline during a series of two perfusions. Furthermore, significant thrombocytopenia, accompanied by oozing from cutdown sites, persisted for up to 2 days. Total white cell counts also fell to an average of 40% of baseline, principally due to a selective loss of neutrophils. Unlike the platelet counts, white cell counts rebounded within 2 hr to values in excess of baseline, owing to a specific increase in neutrophils. Red blood cell losses could again be accounted for by external losses and hemodilution.

In evaluating possible means of reducing formed element losses, the observation was made during *in vitro* perfusion of human blood that such losses were appreciably less in blood anticoagulated with various calcium chelating agents than in heparinized samples. Subsequently, detailed *in vitro* studies demonstrated that addition of citrate, EDTA, or oxalate to normal, heparinized human blood essentially eliminates the loss of platelets and granulocytes during perfusion not only through albumin-agarose gel (AAG), but also through uncoated activated charcoal (AC), the neutral resin XAD-2, and the cation-exchange resin Dowex AG^R50W-X8 (Scharschmidt *et al.,* 1975b). Citrate reduced the percentage of platelets lost during perfusion of heparinized blood as follows: AC (56% → 4%); XAD-2 (87% → 5%); AG^R50W-X8 (90% → 6%); and AAG (83% → 6%). Prostaglandin E_1 (PGE_1) at a dose of 5μg/ml also markedly reduced platelet losses. Lower concentrations (down to 0.001 μg/ml) were equally as effective for some adsorbents. Citrate, the most promising of these agents, was used "regionally" (Figure 7) during hemoperfusion studies through AAG, uncoated AC, and XAD-2 in three *Rhesus* monkeys. Each perfusion was continued for 160 min. During the first 80 min only, citrate and calcium were infused into the column inflow and outflow lines, respectively. Results for all three perfusions were similar. During the first 80 min, platelet counts in the monkeys and column effluents averaged 115% and 95%, respectively, of the preperfusion value in the monkey, but fell promptly to 13% and 7%, respectively, after citrate was stopped. Each of the monkeys tolerated the procedure without apparent ill effect (Scharschmidt *et al.,* 1975b). "Regional" citrate was equally effective during trials of AAG hemoperfusion in several additional jaundiced *Rhesus* neonates and did not influence the capacity of the gel to bind bilirubin (Scharschmidt *et al.,* 1975a). Additional animal studies suggest that very high plasma citrate concentration are well tolerated provided

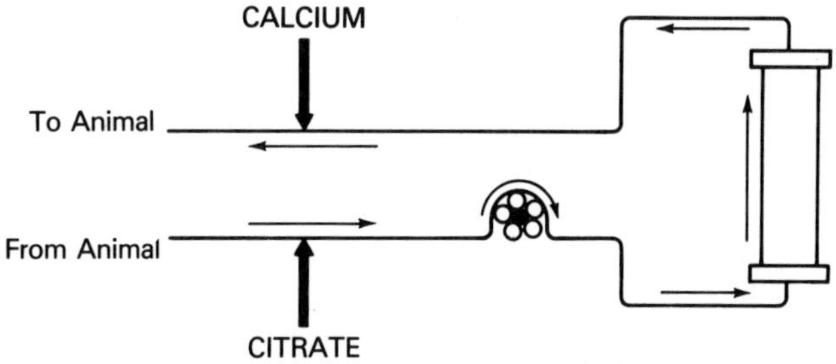

FIGURE 7. Technique for abolishing platelet and granulocyte losses during extracorporeal hemoperfusion. Addition of citrate on the "arterial" side of the column at a rate sufficient to produce a concentration of 3.6 mg of citrate anion/ml of perfused blood results in virtually complete sparing of formed elements in Rhesus monkeys perfused over albumin-agarose gel, uncoated charcoal, or the neutral resin XAD-2. The resulting citrate levels are well tolerated provided the ionized calcium is maintained within normal limits by calcium infusion on the "venous" side of the column.

that the calcium infusion rate into the column outflow line is adequate to maintain a normal ionized (not total) calcium concentration in the animal.

The general applicability of regional citrate anticoagulation may be of considerable importance, as platelet losses have considerably inhibited clinical studies of hemoperfusion over a variety of sorbents, particularly neutral and cation-exchange resins. This approach may be particularly suitable for use in systems combining sorbent perfusion with hemodialysis (Chang *et al.*, 1975). Under these circumstances, location of the hemodialyzer "downstream" from the sorbent column could facilitate removal of the infused citrate and maintenance of adequate ionized calcium levels.

7. *Unresolved Problems*

Although the use of albumin-agarose gel hemoperfusion has considerable appeal in certain clinical situations, a number of obvious problems require further investigation. These include:

7.1. *Sterilization of the Gels*

Optimal sterilization of an AAG extracorporeal system would involve autoclaving. To date, autoclaving, ultraviolet irradiation, formaldehyde, glutaraldehyde, phenol, and iodine have all been found to reduce appreciably or destroy completely the bilirubin binding capacity of such gels. Preliminary studies indicate that 76% (vol./vol.) ethanol effectively sterilizes AAG against many organisms, even after deliberate contamination with non-spore-forming pathogens such as *Pseudomonas*. However, 76% ethanol is relatively ineffective in sterilizing gels contaminated with spore-forming bacteria (e.g., *B. subtilis*) and would not be expected to be effective against viruses and fungi. A more generally effective sterilizing agent would therefore be desirable before AAG hemoperfusion is widely used. Aseptic preparation of AAG from sterile components may be feasible on a commercial scale.

7.2. *Embolization and Immunogenicity*

Particle emboli have been found in only one of the many animals perfused over AAG. In this animal, a *Rhesus* monkey whose blood was returned via the umbilical vein, large extracellular spheres, presumably of AAG, were found in some portal vein branches at autopsy. Careful examination of other tissues from that animal, as well as liver, lung, heart, and brain tissue from other animals similarly perfused, revealed no evidence of particle emboli. In preliminary studies, saline was washed through the fully assembled apparatus, and the effluent was passed through a 400-mesh nylon screen. Microscopic examination of the screen after perfusion showed them to be free of trapped AAG or particulate material. It is therefore likely that a defective screen or some other leak in the system caused microembolization in this one experiment. More effective means of detecting such leaks are clearly needed.

Although human albumin is not normally antigenic when infused into patients, the possibility that coupling to AAG might unmask new antigenic markers is a potential cause of concern if such altered albumin were to leach from the column. Loss of human albumin from albumin-agarose gel into perfused neonatal monkeys would be expected to result in formation of antibodies to human albumin. Three of the perfused monkeys were tested at 5–7 months of age for the presence of such anti-HSA antibodies. One of these three monkeys was retested 3 months after a second AAG hemoperfusion at 7 months of age. In no instance were anti-HSA antibodies detected. Further studies of this nature are required.

7.3. *Problems of Scale*

Studies described above have been conducted on small columns suited to the perfusion of animals with body weight up to 1200 g. Considerably larger columns would be required for effective bilirubin removal in man. Unpublished studies in several laboratories, including our own, suggest that merely scaling up to larger columns of the same geometry as is currently employed (long, thin cylinders) results in unacceptable pressure drops across the column and poor blood flow. It is likely that alterations in column geometry may alleviate this problem. Alternatively, other types of matrixes or surfaces for immobilizing albumin may be more suitable for large-scale, high-flow systems, even if their bilirubin binding capacity per unit volume or surface area is inferior to that of AAG.

8. *Future Applications*

We foresee a number of future uses for competition/affinity chromatography employing albumin-agarose gels. These include, on the one hand, purely laboratory procedures in which such gels are used in the selective purification of albumin-bound ligands from complex biological fluids. An example of such an application is the purification of conjugated bilirubin from bile by a simple three-step procedure in which AAG chromatography is employed as the first and critical step (Wolkoff *et al.,* 1976).

More dramatic potential applications of AAG would be in extracorporeal perfusion of patients in whom dangerously excessive quantities of albumin-bound metabolites have accumulated in the blood. The most obvious such application would be in the Crigler-Najjar syndrome, the condition that first stimulated our interest in the field. If the system employed in rats and monkeys can be successfully scaled up, we have calculated that three hemoperfusions per week would remove the entire week's bilirubin production in such a patient (Scharschmidt *et al.,* 1974). Therefore, chronic hemoperfusion, on a schedule not very different from that employed in chronic hemodialysis, might maintain circulating unconjugated bilirubin concentrations in such patients well below the critical levels associated with brain damage.

Extracorporeal AAG hemoperfusion via the umbilical vessels remains a potentially useful approach to management of neonatal hyperbilirubinemia. Although subject to the same catheter complications as exchange transfusion, it avoids the additional risks associated with administration of exogenous blood products. The extent of its future use in this situation may depend on the outcome of studies currently being conducted to assess the long-term effectiveness of phototherapy, a simpler procedure whose widespread but uncontrolled use has raised some concern with regard to possible undesirable side effects.

Extracorporeal perfusion may also have a role in the management of certain anomalies of the biliary tract, such as biliary atresia. It has been postulated by some that an accumulation of lithocholic acid in this condition is one of the factors contributing to the development of cirrhosis. In view of the relatively effective binding of this secondary bile acid and its conjugates by AAG, hemoperfusion in this condition—in which hyperbilirubinemia per se has only cosmetic implications—might be directed at a reduction of potentially hepatotoxic bile salt concentrations.

Extracorporeal perfusion over a variety of sorbents has been proposed and/or used in the treatment of fulminant hepatic failure (Chang, 1972; Chang and Migchelson, 1973; Juggi, 1973; Gazzard *et al.*, 1974a,b) with uncertain effectiveness (Berk *et al.*, 1976a). Since the site of "toxin" accumulation in this situation is not established, and the nature of the critical "toxins" requiring removal is also unknown, such approaches are largely empirical. It is possible that albumin-bound metabolites are among the toxins requiring removal in fulminant hepatic failure, and under these circumstances AAG may play a role as one component of an extracorporeal system which might include other sorbents, hemodialysis, and hemodiafiltration.

AAG perfusion of jaundiced animals is providing some data of use in the design of other systems for use in fulminant hepatic failure. The time sequence of arousal following charcoal perfusion in fulminant hepatic failure suggests that awakening depends on the removal of the unknown toxin from a peripheral compartment (possible cerebrospinal fluid) which exchanges very slowly with the plasma. Under these circumstances the timing and duration of hemoperfusions is a critical factor in determining the rate of toxin removed from the slowly exchanging pool. Since neither the toxin nor its internal exchange rates are known, hemoperfusion regimes currently employed are entirely arbitrary. In contrast, the internal kinetics of bilirubin are well described in both man and experimental animals (Berk *et al.*, 1969; Berk *et al.*, 1970; Bloomer *et al.*, 1971; Scharschmidt *et al.*, 1973; Blaschke *et al.*, 1974). Accordingly, the effect of various hemoperfusion schedules on the bilirubin content of various extravascular bilirubin pools can be determined both experimentally and by computer simulation. Such studies may indicate particular hemoperfusion schedules especially effective in the depletion of slowly exchanging extravascular compartments. Application of such schedules in clinical trials of charcoal perfusion might improve the effectiveness of the latter in fulminant hepatic failure.

9. *Summary*

Perfusion of whole blood, plasma, and bile over columns of albumin-agarose gel selectively removes albumin-bound materials with an efficiency roughly proportional to the ligand–albumin affinity constant. *In vivo* studies in several animal model systems have demonstrated highly effective bilirubin removal, a result having several potential clinical and laboratory applications. Biocompatibility of such gels with regard to blood-formed elements can be dramatically improved by "regional" citrate anticoagulation, which also reduces platelet and granulocyte losses over charcoal and several resins. Since numerous other proteins, with various binding specificities, can be substituted for albumin, the technique of affinity/competition chromatography offers a powerful tool for isolation of numerous compounds from blood and other complex biologic fluids, both *in vivo* and *in vitro.*

10. References

Andrade, J. D., Kunitomo, K., Van Wagenen, R., Kastigir, B., Gough, D., and Kolff, J., 1971, Coated adsorbents for direct blood perfusion: hema/activated carbon, *Trans. Amer. Soc. Artificial Internal Organs* **17**:222–238.

Arias, I. M., Gartner, L. M., Cohen, M., Ben-Ezzer, J., and Levi, A. J., 1969, Chronic nonhemolytic unconjugated hyperbilirubinemia with glucuronyl transferase deficiency: clinical, biochemical, pharmacologic and genetic evidence for heterogeneity, *Amer. J. Med.* **47**:395–409.

Axen, R., Porath, J., and Ernback, S., 1967, Chemical coupling of peptides and proteins to polysaccharides by means of cyanogen halides, *Nature* **214**:1302–1304.

Barakat, T., and MacPhee, I. W., 1970, Experiments with an extracorporeal carbon column as a simplified artificial kidney, *Brit. J. Surg.* **57**:580–583.

Benhamou, J.-P., Rueff, B., and Sicot, C., 1972, Severe hepatic failure: a critical study of current therapy, in: *Liver and Drugs* (F. Orlandi and A. M. Jezequel, eds.), Academic Press, Inc., New York.

Berk, P. D., Howe, R. B., Bloomer, J. R., and Berlin, N. I., 1969, Studies of bilirubin kinetics in normal adults, *J. Clin. Invest.* **48**:2176–2190.

Berk, P. D., Bloomer, J. R., Howe, R. D., and Berlin, N. I., 1970, Constitutional hepatic dysfunction (Gilbert's syndrome): a new definition based on kinetic studies with unconjugated radio-bilirubin, *Amer. J. Med.* **49**:296–305.

Berk, P. D., Berlin, N. I., and Howe, R. B., 1974, Disorders of bilirubin metabolism, in: *Duncan's Diseases of Metabolism* (P. K. Bondy and L. Rosenberg, eds.), 7th ed., W. B. Saunders Company, Philadelphia.

Berk, P. D., Martin, J. F., Blaschke, T. F., Scharschmidt, B. F., and Plotz, P. H., 1975, Unconjugated hyperbilirubinemia: physiologic evaluation and experimental approaches to therapy, *Ann. Internal Med.* **82**:552–570.

Berk, P. D., Scharschmidt, B. F., Martin, J. F., and Plotz, P. H., 1976a, Artificial support systems for liver failure, *Kidney Inst.* **10**(7):S233.

Berk, P. D., Martin, J. F., Scharschmidt, B. F., and Plotz, P. H., 1976b, Current status of artificial hepatic support systems, in: *Progress in Liver Disease* (H. Popper and F. Schaffner, eds.), 5th ed., Grune & Stratton, Inc., New York.

Blaschke, T. F., Berk, P. D., Scharschmidt, B. F., Guyther, J. R., Vergalla, J., and Waggoner, J. G., 1974, Crigler-Najjar syndrome: an unusual course with development of neurologic damage at age 18, *Pediat. Res.* **8**:573–590.

Bloomer, J. R., Berk, P. D., Howe, R. B., and Berlin, N. I., 1971, Bilirubin metabolism in congenital nonhemolytic jaundice, *Pediat. Res.* **5**:256–274.

Blumenschein, S. D., Kallan, R. J., Storey, B., Natzschka, C., Odell, G. B., and Childs, B., 1968, Familial nonhemolytic jaundice with late onset of neurologic damage, *Pediatrics* **42**:786–792.

Breen, K. J., and Schenker, S., 1972, Hepatic coma: present concepts of pathogenesis and therapy, in: *Progress in Liver Disease* (H. Popper and F. Schaffner, eds.), Vol. 4, Grune & Stratton, Inc., New York.

Chang, T. M. S., 1972, Hemoperfusion over microencapsulated absorbent in a patient with hepatic coma, *Lancet* **2**:1371–1372.

Chang, T. M. S., and Migchelson, M., 1973, Characterization of possible "toxic" metabolites in uremia and hepatic coma based on the clearance spectrum for larger molecules by the ACAC microcapsule artificial kidney, *Trans. Amer. Soc. Artificial Internal Organs* **17**:246–252.

Chang, T. M. S., Chisito, E., Barre, P., Cole, C., and Hawish, M., 1975, Clinical performance characteristics of a new combined system for simultaneous hemoperfusion-hemodialysis-ultrafiltration in series, *Trans. Amer. Soc. Artificial Internal Organs* **21**:502–509.

Cohn, H. E., Weiss, A. J., Clark, J. E., and Nealon, T. F., 1960, An ion exchange resin artificial kidney, *Surg. Forum* **11**:15–17.

Cuatrecasas, P., and Anfinsen, C. B., 1971, Affinity chromatography, *Ann. Rev. Biochem.* **40**:259–278.

Dunea, G., and Kolff, W. J., 1965, Clinical experiences with the Yatzidis charcoal kidney, *Trans. Amer. Soc. Artificial Internal Organs* **11**:178–182.

Fischer, J. E., 1974, Hepatic coma in cirrhosis, portal hypertension, and following portocaval shunt, *Arch. Surg.* **108**:325–336.

Gazzard, B. G., Weston, M. J., Murray-Lyon, I. M., Flax, H., Record, C. O., Portman, B., Langley, P. G., Dunlop, E. H., Mellon, P. J., Ward, M. B., and Williams, R., 1974a, Charcoal hemoperfusion in the treatment of fulminant hepatic failure, *Lancet* **1**:1301–1307.

Gazzard, B. G., Buxton, B. H., Winch, J., Machado, A. L., Flax, H., and Williams, R., 1974b, Effects of haemoperfusion through charcoal or XAD-2 resin on an animal model of fulminant liver failure, *Gut* **15**:482–486.

Grollman, A. P., and Odell, G. B., 1962, Removal of bilirubin by albumin binding during intermittant peritoneal dialysis, *New Engl. J. Med.* **267**:279–282.

Inman, J. K., and Dintzis, H. M., 1969, The derivatization of cross-linked polyacrylamide beads: controlled introduction of functional groups for the preparation of special purpose, biochemical adsorbents, *Biochemistry* **8**:4074–4082.

Jacobsen, J., 1969, Binding of bilirubin to human serum albumin—determination of the dissociation constants, *FEBS Letters* **5**:112–114.

Juggi, J. S., 1971, In vivo studies with a cation exchange resin mixture in the removal of excessive ammonium from the extracorporeal circulation system, *Australian New Zealand J. Surg.* **41**:194–200.

Juggi, J. S., 1973, Extracorporeal cation exchange circuits in the treatment of hyperammonemia of liver failure, *Med. J. Australia* **1**:926–930.

Khorona, H. G., 1955, The use of dicyclohexylcarbodiimide in the synthesis of peptides, *Chem. Ind. (London)* August:1087–1088.

Klebba, A. J., Maurer, J. D., and Glass, E. J., 1974, Mortality trends for leading causes of death: United States, 1950–69 (National Center for Health Statistics: data from the National Vital Statistics System, ser. 20, no. 16), *U.S. Department of Health, Education and Welfare Publication No. (HRA) 74-1853,* Government Printing Office, Washington.

Maher, J. F., Nolph, K. D., and Bryan, C. W., 1974, Prognosis of advanced chronic renal failure, *Ann. Internal Med.* **81**:43–47.

Pallotta, A. J., and Koppanyi, T., 1960, The use of ion exchange resins in the treatment of phenobarbital intoxication in dogs, *J. Pharmacol. Exptl. Therapy* **128**:318–327.

Plotz, P. H., Berk, P. D., Scharschmidt, B. F., Gordon, J. K., and Vergalla, J., 1974, Removing substances from blood by affinity chromatography: 1. Removal of bilirubin and other albumin-bound substances from plasma and blood with albumin-conjugated agarose beads, *J. Clin. Invest.* **53**:778–785.

Porath, J., and Sundberg, L., 1972, High capacity chemisorbents for protein immobilization, *Nature New Biol.* **238**:261–262.

Rosenthal, I. M., Zimmerman, H. J., and Hardy, M., 1956, Congenital nonhemolytic jaundice with disease of the central nervous system, *Pediatrics* **18**:378–386.

Rudman, D., and Kendall, F. E., 1957, Bile acid content of human serum: II. The binding of cholanic acids by human plasma proteins, *J. Clin. Invest.* **36**:538–542.

Scharschmidt, B. F., and Berk, P. D., 1973, A simple device to facilitate rapid blood sampling in small animals, *Proc. Soc. Exptl. Biol. Med.* **143**:364–366.

Scharschmidt, B. F., Berk, P. D., and Waggoner, J. G., 1973, Bilirubin kinetics in the normal rat, *Clin. Res.* **21**:523 (abstract).

Scharschmidt, B. F., Plotz, P. H., Berk. P. D., Waggoner, J. G., and Vergalla, J., 1974, Removing substances from blood by affinity chromatography: II. Removing bilirubin from the blood of jaundiced rats by hemoperfusion over albumin-conjugated agarose beads. *J. Clin. Invest.* **53**:786–795.

Scharschmidt, B. F., Martin, J. F., Shapiro, L. J., Plotz, P. H., and Berk, P. D., 1975a, Hemoperfusion through albumin-agarose gel (AAG) for the treatment of neonatal jaundice in premature Rhesus monkeys, *Gastroenterology* **69**:864 (abstract).

Scharschmidt, B. F., Martin, J. F., Shapiro, L. J., Plotz, P. H., and Berk, P. D., 1975b, Use of calcium chelating agents and prostaglandin E_1 (PGE_1) to eliminate platelet losses during hemoperfusion through various adsorbents, *Gastroenterology* **69**:864 (abstract).

Schenker, S., Breen, K. J., and Hoyumpa, A. M., Jr., 1974, Hepatic encephalopathy: current status, *Gastroenterology* **66**:121–154.

Sheehan, J. C., and Hess, G. P., 1955, A new method for forming peptide bonds, *J. Amer. Chem. Soc.* **77**:1067–1068.

Smith, T. W., 1972, Contribution of quantitative assay technics to the understanding of the clinical pharmacology of digitalis, *Circulation* **46**:188–199.

Williams, R., 1976, Hepatic failure and the development of an artificial liver support system, in: *Progress in Liver Disease* (H. Popper and F. Schaffner, eds.), Vol. 5, Grune & Stratton, Inc., New York.

Willson, R. A., Webster, K. H., Hoffman, A. F., and Summerskill, W. H. J., 1972, Towards an artificial liver: in vitro removal of unbound and protein bound plasma compounds related to hepatic failure, *Gastroenterology* **62**:1191–1199.

Willson, R. A., Hofmann, A. F., and Kuster, G. G. R., 1974, Toward an artificial liver: II. Removal of cholephilic anions from dogs with biliary obstruction by hemoperfusion through charcoal and uncharged resins, *Gastroenterology* **66**:95–107.

Wolkoff, A. W., Scharschmidt, B. F., Plotz, P. H., and Berk, P. D., 1976, Purification of conjugated bilirubin: a new approach utilizing albumin-agarose gel affinity chromatography, *Proc. Soc. Extl. Biol. Med.* **152**:20.

Zieve, L., 1975, Metabolic abnormalities in hepatic coma and potential toxins to be removed, in: *Artificial Hepatic Support* (R. Williams and I. M. Murray-Lyon, eds.), pp. 11–26, Sir Isaac Pitman & Sons Ltd., London.

21

Membrane-Immobilized Liver Microsome Drug Detoxifier

William Cohen, William H. Baricos, Peter R. Kastl, and Robert P. Chambers

1. *Introduction*

The common occurrence of poisoning due to drug overdose has led to an increasing need for a safe, simple, effective, and inexpensive means of removing such drugs from the blood in the emergency treatment of drug-overdose victims. Various devices and techniques based on extracorporeal blood processing have been investigated. Physical techniques include hemodialysis (Kennedy *et al.*, 1969), charcoal adsorption (Widdop *et al.*, 1975; Chang *et al.*, 1973), and ion-exchange adsorption (Rosenbaum *et al.*, 1971; Medd *et al.*, 1974). Biological techniques include liver perfusion (McDermott and Norman, 1971), liver slice perfusion (Koshino *et al.*, 1975), perfusion over hepatic cell suspensions in direct contact with the blood (Eisman and Soyer, 1971; Soyer *et al.*, 1973), and perfusion through hollow fibers containing cultured hepatic cells on the outer surface of the fibers (Wolf and Munkelt, 1975). It is well accepted that the biological system responsible for detoxification of a variety of drugs is located in the smooth endoplasmic reticulum of hepatic cells which can be isolated as microsomes by cell fractionation. The microsomes contain a complex mixture of enzymes, including cytochrome P-450, which requires the coenzyme NADPH and molecular oxygen for drug detoxification. In general these detoxification enzymes modify toxins by increasing their polarity and consequently their aqueous solubility, thus decreasing their permeation into tissues (Mandel, 1971) and promoting excretion. The chemical modifications catalyzed by the microsomal drug detoxification enzymes include, among others, hydroxylation, demethylation, and conjugation reactions. Several of the microsomal detoxification enzymes have been purified (Lu and Levin, 1974; Brunner, 1975; Ziegler and Mitchell, 1972) and at least two have been immobilized by covalent bonding to insoluble particles (Brunner, 1975; Parikh *et al.*, 1976; Sofer *et al.*, 1975). One approach for utilization of these

William Cohen, William H. Baricos and Peter R. Kastl • Department of Biochemistry, Tulane University, New Orleans, Louisiana 70112 Robert P. Chambers • Department of Chemical Engineering, Auburn University, Auburn, Alabama 36830

enzymes in an extracorporeal drug detoxification system is to purify the individual enzymes and then reconstitute the multienzyme complex by binding to insoluble particles. An alternative approach, the one we have chosen, is to utilize the isolated microsomes themselves. This method has the advantage of simplicity and lower cost. More importantly, in contrast to the former approach, it assures that all the microsomal enzymes are present in proportions and molecular arrangement closely resembling the *in vivo* state. We have used an extracorporeal hollow fiber-based enzyme reactor in an attempt to develop a therapeutic technique suitable for treatment of drug and poison detoxification. Since all the protein components of the system are impermeable to the hollow fiber membranes, adverse immunological reactions will be avoided. In addition, the safety of blood processing using hollow fiber devices has been well established in recent years by extensive use in hemodialysis.

The extracorporeal shunt drug detoxifier system being developed is shown in Figure 1. This system differs from conventional hollow fiber artificial kidney

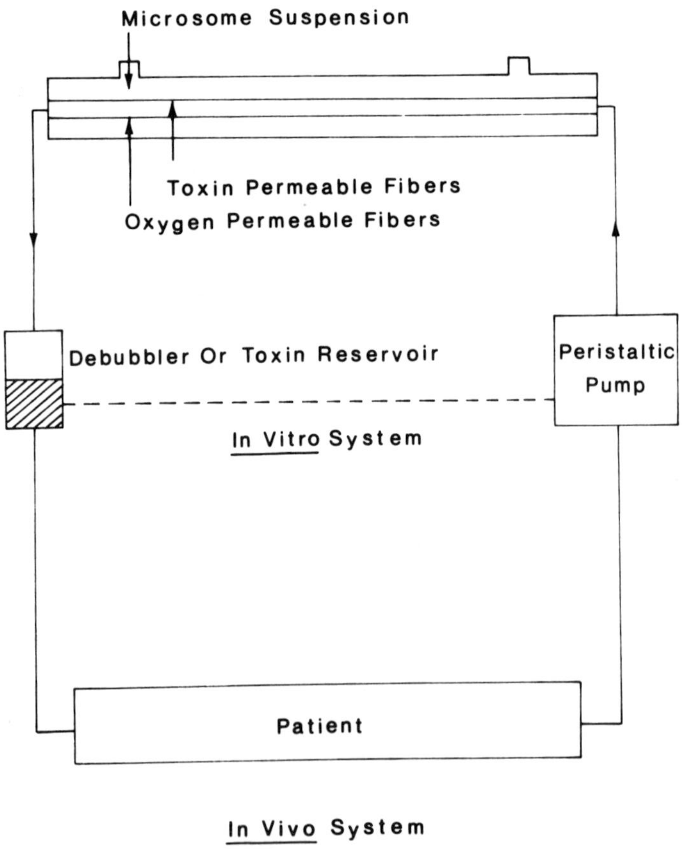

FIGURE 1. Proposed drug and poison detoxification system.

hemodialysis in several respects. First, it contains the multienzyme detoxification system in the sealed shell side of the hollow fiber dialyzer, eliminating the need for large volumes of dialyzing fluid. Second, the hollow fiber device (Bio-Fiber 80/5 cell culture tube, 50 cm^2 fiber surface area) contains two types of fibers: cellulose fibers, permeable to drugs and poisons, and silicone fibers, permeable to oxygen. The toxic substance and oxygen can thus permeate from the blood into the microsomal suspension. Following transformation by the microsomal enzymes, the modified toxin can reenter the bloodstream, where it will be removed by the kidney and excreted. An important advantage of using intact animal liver microsomes for detoxification is that this approach will also be applicable to drugs developed in the future since those drugs will also have to be detoxified by the microsomal enzyme system and/or excreted by the body. We have chosen to use two compounds, *p*-nitroanisole and hexobarbital, as model drugs in the development of the extracorporeal detoxification system; *p*-nitroanisole, because its detoxification (*O*-demethylation) can be continuously monitored spectrophotometrically, and hexobarbital, because poisoning by this drug is representative of the most common drug class encountered in drug overdose.

2. *Preparation and Characterization of Microsomal Drug Detoxification Enzymes*

The development of a practical microsome-based drug detoxifier system requires a highly active microsomal enzyme preparation which can be conveniently stored until needed. Since inactivated microsomes can be easily replaced by fresh microsomes, long-lasting functional stability is not an absolute requirement for successful detoxification, although the advantages of a system in which the microsomes are active for several hours are evident. While it has been known for some time that frozen microsomes retain much of their detoxification activity, routine storage of such preparations presents a variety of problems. We therefore carried out a study to investigate alternative storage techniques. We found that lyophilized microsomes, prepared and stored by the following method, were quite satisfactory.

Sprague-Dawley male rats averaging 250 g in body weight were pretreated by intraperitoneal injection of 75 mg/kg of fresh aqueous sodium phenobarbital on the first day and 100 mg/kg on the second and third days. The animals were fasted for 24 hr following the last dose, sacrificed by decapitation, and exsanguinated. The livers were immediately removed, washed, and placed in ice-cold 0.25 M sucrose solution, then minced and homogenized in 5 volumes of cold sucrose solution with a Potter-Elvehjem homogenizer. The homogenate was centrifuged at 10,000*g* for 30 min at 4°C. The supernatant was decanted and centrifuged at 105,000*g* for 1 hr. The resulting microsomal pellet was washed by resuspension and homogenization in cold sucrose and recentrifuged at 105,000*g* for 1 hr. The washed microsomes were resuspended in 2.5 volumes of cold sucrose and lyophilized directly in screw-cap vials using a Virtis Vac-Vial system. This equipment permits freeze-drying and *in situ* sealing under vacuum. Microsomes prepared by this procedure retained 85% of the activity of the fresh nonlyophilized microsomes when assayed with hexobarbital and 55% of the original activity toward *p*-nitroanisole. The specific activity was 9.0 munits/mg protein with hexobarbital and 0.9 munit/mg protein with *p*-nitroanisole.

Both *O*-demethylase and hydroxylase activity were found to be retained by the lyophilized microsomes after 6 months of storage at both 0°C and room temperature. The hexobarbital hydroxylase stability was estimated by comparing the activity of freshly prepared microsomes to those stored for 6 months. The *p*-nitroanisole *O*-demethylase activity was monitored at regular intervals (see Figure 2). The activity expressed as μM/min is equivalent to nmoles/ml/min. The best technique for the storage of microsomes previously reported (Borton *et al.,* 1974) required quick freezing in liquid nitrogen and storage at −85°C. For routine use, screw-cap bottles containing lyophilized microsomes have been stored in an evacuated desiccator with silica gel desiccant at room temperature. No significant loss of activity has been observed in batches typically stored for 2–3 months.

Microsomes contain at least three enzymes which hinder drug detoxification reactions by decreasing the effective concentration of the coenzyme NADPH, a necessary component of the microsomal detoxification system. Effects of these deleterious reactions may be decreased by use of extremely high concentrations of NADPH; however, specific inhibitors of these destructive enzymes have been shown to be more effective (Gillette, 1963; Gillette *et al.,* 1963). The pyrophosphatase and NADPase can be inhibited by sodium pyrohhosphate and nicotinamide, respectively, while an NADPH regeneration system such as glucose-6-phosphate dehydrogenase (D-glucose-6-phosphate: NADP oxidoreductase EC 1.1.1.49)/glucose-6-phosphate can reverse the effect of NADPH oxidase. Figure 3 shows the

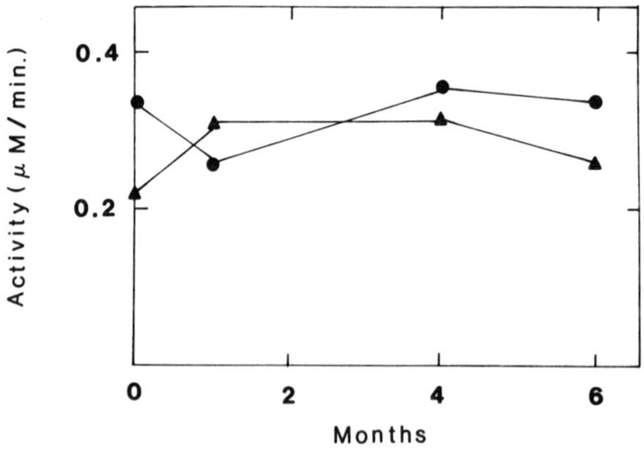

FIGURE 2. Microsomal *O*-demethylase activity as a function of storage time. ●, 0°C; ▲, room temperature. The *O*-demethylase activity was continuously monitored at 420 nm in a Cary 16K spectrophotometer in an assay solution containing 0.42 mg/ml of microsomal protein, 1.2 mM *p*-nitroanisole, 2.4 mM nicotinamide, 24 mM sodium pyrophosphate, 6 mM magnesium chloride, 6 mM glucose-6-phosphate, 4 units per ml of glucose-6-phosphate dehydrogenase, and 0.12 mM NADP in 100 mM sodium phosphate buffer, pH 7.8 at 25°C.

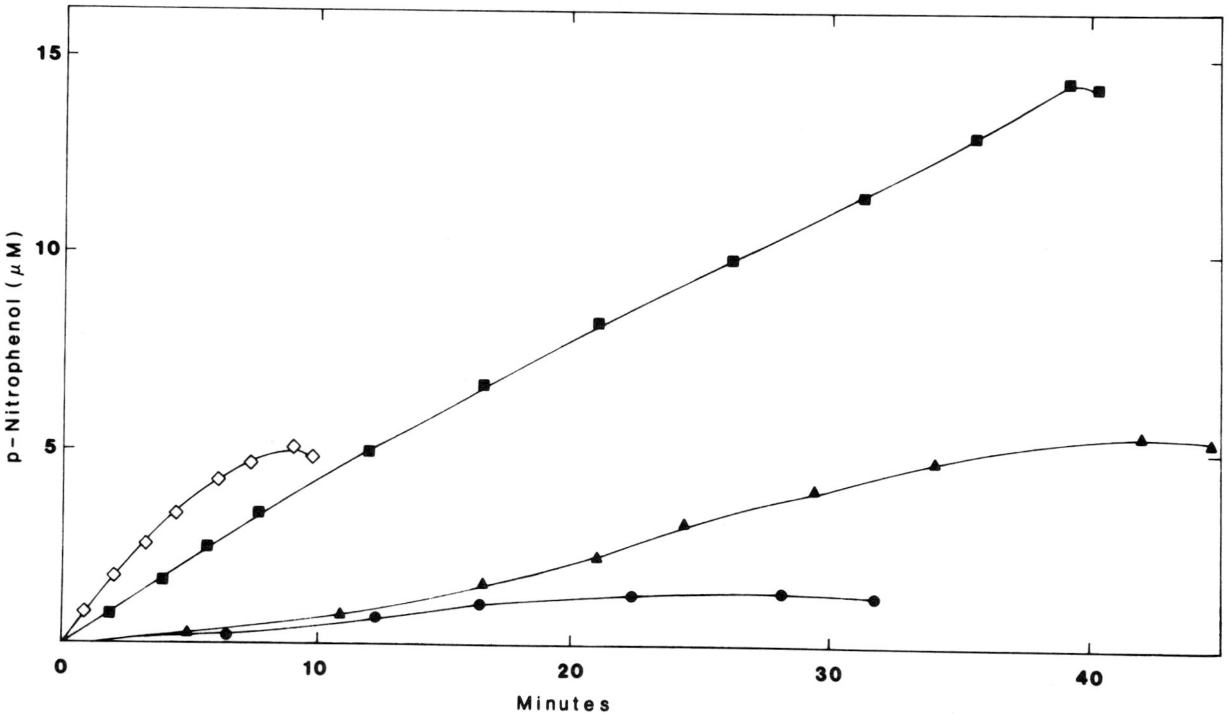

FIGURE 3. Effect of cofactor regeneration, pyrophosphate and nicotinamide on microsomal *O*-demethylation (0.73 mg of microsomal protein/ml). ■, Cofactor regeneration system, pyrophosphate (50mM), nicotinamide (2.4 mM); ◇, cofactor regeneration system replaced by 0.12 mM NADPH; ▲, pyrophosphate deleted; ●, pyrophosphate and nicotinamide deleted.

effect of these three reagents on microsomal *O*-demethylation. When these reagents are employed simultaneously, the *O*-demethylation is rapid and linear for 40 min, until the oxygen of the system is depleted. Deletion of any one of them adversely affects detoxification. Each of the reagents was optimized individually prior to this experiment. Significantly, it was observed that when these reagents were employed simultaneously and the oxygen supply maintained by slow stirring in an open beaker, the microsomes remained functional for up to 12 hr (not shown).

Pharmacologists have traditionally used glucose-6-phosphate dehydrogenase/glucose-6-phosphate as an NADPH regeneration system. Clearly this system of NADPH regeneration is not feasible for use in an extracorporeal drug detoxification system since glucose-6-phosphate and Mg^{2+} would have to be supplied at physiologically unacceptable concentrations. The use of glucose dehydrogenase (β-D-glucose:NAD(P) oxidoreductase, EC 1.1.1.47)/glucose appears ideally suited as an NADPH regeneration system, since ambient blood glucose can be used as the substrate. We have tested a commercially available crude glucose dehydrogenase (Sigma, 15 units/g) isolated from calf liver. Figure 4 shows the effectiveness of this regeneration system compared to that of glucose-6-phosphate dehydrogenase/glucose-6-phosphate. The slightly lower rate of *p*-nitroanisole *O*-demethylation with glucose-6-phosphate dehydrogenase is the result of partial inhibition by pyrophosphate.

In order to establish that oxygen can be supplied by hollow fibers to a microsomal detoxification system and that the resulting amount of drug metabolized will be increased, hexobarbital metabolism was investigated in the presence of an external source of oxygen (Biofiber 5 beaker containing 500 cm^2 of silicone/polycarbonate hollow fibers which were pressurized with 5-psi oxygen) and in the

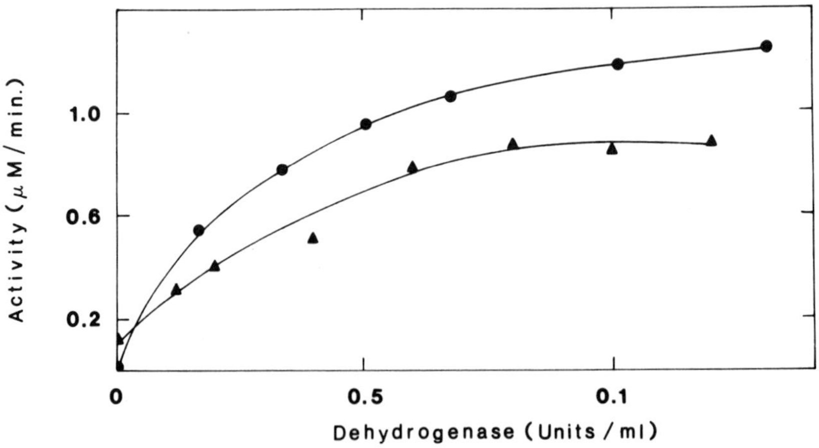

FIGURE 4. Microsomal *O*-demethylation as a function of NADPH regeneration enzyme concentration (0.77 mg of microsomal protein/ml). ●, Glucose dehydrogenase/5% glucose; ▲, glucose-6-phosphate dehydrogenase/glucose-6-phosphate (6mM), $MgCl_2$ (6mM).

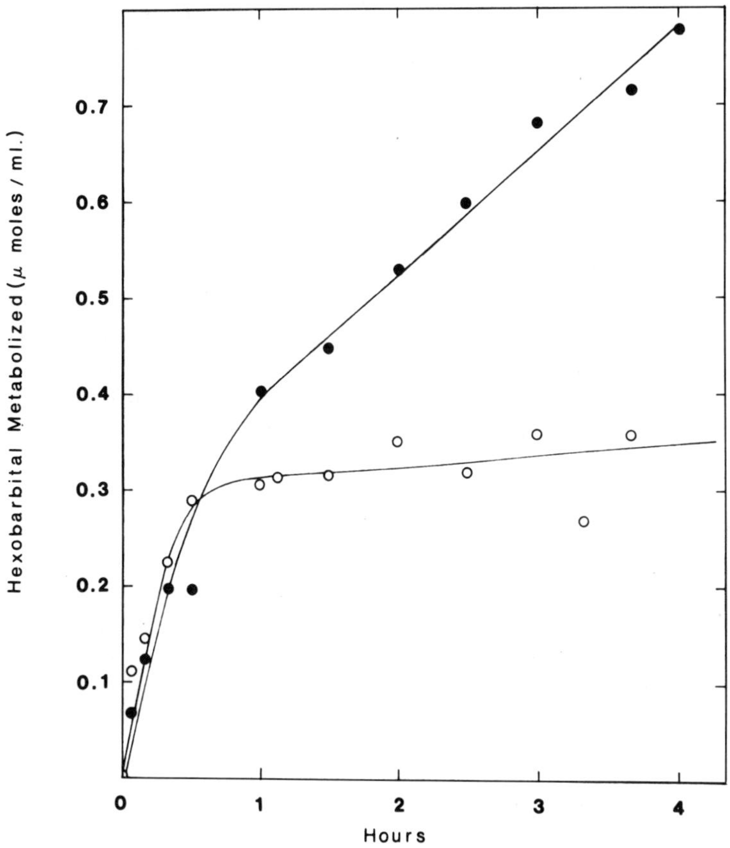

FIGURE 5. Effect of infused oxygen on hexobarbital detoxification in a Bio-Fiber 5 beaker (0.57 mg of microsomal protein/ml). ●, 5-psi Oxygen; ○, no added oxygen. Hexobarbital hydroxylase activity was determined by the method of Kupfer and Rosenfeld (1973) with tritium-labeled hexobarbital. The assay solution is the same as for *p*-nitroanisole *O*-demethylase assay except that 0.77 mM hexobarbital and 1.2×10^6 dpm of tritium-labeled hexobarbital were substituted for *p*-nitroanisole.

absence of an external oxygen source. The results plotted in Figure 5 dramatically show that without added oxygen the reaction rapidly exhausts the dissolved oxygen and stops, while the sample supplied with an external oxygen source continues to metabolize hexobarbital at a rapid rate.

3. *Experimental Simulation of in Vivo Drug Detoxification*

To simulate drug detoxification in a rat we used the system shown schematically in Figure 1. A 200-g rat has approximately 10 ml of of blood plus 20 ml of interstitial body fluids or a total of 30 ml of fluid in which drugs are rapidly equilibrated (for simplicity we neglected the equilibration of drugs with the adipose tissue). Thus, 30 ml of buffered barbiturate solution was circulated at a rate of 12 ml/min through the hollow fibers of a Bio-Fiber 80/5 cell culture tube containing equal numbers of cellulose acetate and silicone/polycarbonate fibers. The rate of hexobarbital metabolism was linear for 5 hr (Figure 6), at which time approximately 25% of the total hexobarbital (5.5 μmoles) was metabolized. This experiment demonstrates that efficient drug metabolism can be accomplished with this system and that oxygen as well as substrate from the circulating solution can effectively permeate the fibers. Initially, the 0.5 ml of microsome solution contains only 0.65 μmole of oxygen, yet 5.5 μmoles of hexobarbital is metabolized. When the oxygen supply was enriched by bubbling oxygen through the circulating solution, an equal amount of hexobarbital was metabolized in only 3 hr, indicating that oxygen diffusion is a rate-limiting parameter. In the presence of added oxygen, the rates of metabolism appear limited by diffusion of substrate and product, since the initial rates in the hollow fiber systems are much lower than those in control experiments carried out in the absence of membranes (see Figure 5). This experiment also indicates that NADPH regeneration is not limiting. Although we plan to use glucose dehydrogenase in future studies, glucose-6-phosphate dehydrogenase/glucose-6-phosphate was used to regenerate NADPH in this experiment because the relatively large quantities of the impure glucose dehydrogenase needed rapidly clog the membranes. In addition, we anticipate the use of a soluble, macromolecular, nonpermeable $NADP^+$, similar to that described by Lowe and Mosbach (1974), which will be retained on the shell side of the detoxifier.

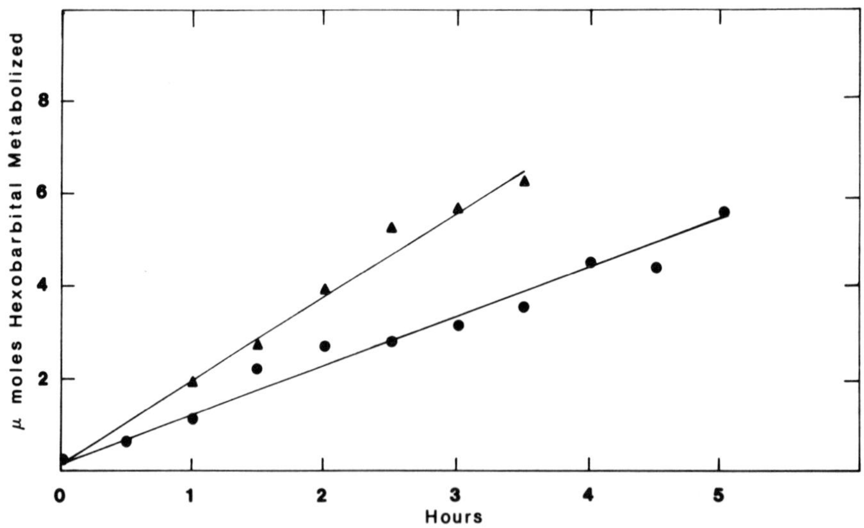

FIGURE 6. Hexobarbital metabolism in a Bio-Fiber 80/5 cell culture tube system (17 mg of microsomal protein). ▲, Oxygen bubbled through circulating solution; ●, no oxygen bubbled through circulating solution.

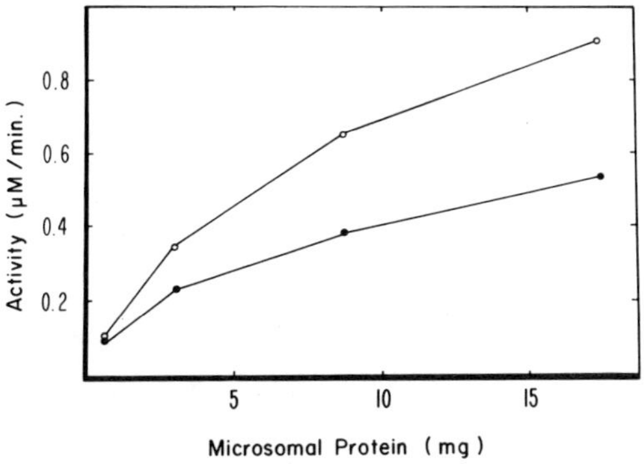

FIGURE 7. Hexobarbital metabolism as a function of microsome concentration in a Bio-Fiber 80/5 cell culture tube system. ○, Oxygen bubbled through circulating solution; •, no oxygen bubbled through circulating solution.

In Figure 7 the rate of hexobarbital hydroxylation is plotted as a function of microsome concentration with and without bubbling oxygen through the circulating solution. The nonlinear increase in reaction rate has not yet been analyzed, but it appears to be due to diffusional limitations possibly produced by clogging of the membranes at high protein concentration.

With no added oxygen, as much as 1.2 μmole of hexobarbital are metabolized per hour, equivalent to a clearance of 1.4 ml/hr. This clearance is increased more than 50% with oxygen addition. The clearance of 1.4 ml/hr is more than 10 times higher than that reported by detoxification with cultured cells on semipermeable hollow fibers (0.1 ml/hr clearance of bilirubin) in a system with similar surface area (Wolf and Munkelt, 1975). A direct extrapolation of the results from our experiments without added oxygen to those which may be achieved with the larger Bio-Fiber 80/ cell culture miniplant (a size quite similar to the conventional hollow fiber hemodialyzer, 10,000 cm^2 fiber surface area) would increase the clearance to 4.8 ml/min. This rate is within an order of magnitude of present clinical hemodialysis clearance of amylobarbitone (30 ml/min), butobarbitone (30 ml/min), and phenobarbitone (60 ml/min) (Widdop *et al.,* 1975). Clearly these preliminary results presented above are not yet equivalent to charcoal adsorption (Widdop *et al.,* 1975) or clinical hemodialysis. However, the concept of the microsomal detoxifier shows substantive promise. In addition, this system, unlike many of the other drug detoxification approaches, has the potential of being employed in artificial liver support.

ACKNOWLEDGMENTS

This investigation was supported by grants from the Edward G. Schlieder Educational Foundation, USPHS Postdoctoral Fellowship GM 00968 to P.R.K., and Avondale Shipyards–Louisiana Heart Association.

4. *References*

Borton, P., Carson, R., and Reed, D. J., 1974, Stability of suspensions of rat liver microsomes quick-frozen with liquid nitrogen and stored at −85°, *Biochem. Pharmacol.* **23**:2322.

Brunner, G., 1977, Refixation of solubilized and purified microsomal enzymes: a new approach towards an extracorporeal detoxification in man, in: *Third Int. Conf. Enzyme Eng.* (Reed College, Portland, Oregon, Aug. 3–8, 1975) (E. K. Pye and L. B. Wingard, eds.), Plenum Press, New York. (in press).

Chang, T. M. S., Coffey, J. F., Lister, E., Taroy, E., and Stark, A., 1973, Methaqualone, methyprylon, and gluthethimide clearance by ACAC microcapsule artificial kidney: *in vitro* and in patients with acute intoxication, *Trans. Amer. Soc. Artificial Internal Organs* **19**:87.

Eisman, B., and Soyer, T., 1971, Prosthetics in hepatic assistance, *Transplant Proc.* **3**:1519.

Gillette, J. R., 1963, Metabolism of drugs and other foreign compounds by enzymatic mechanisms, *Progr. Drug Res.* **6**:13.

Gillette, J. R., Grieb, W., and Seasame, H., 1963, Mechanisms of apparent inhibition of microsomal drug enzymes, *Federation Proc.* **22**:366.

Kennedy, A. C., Brigg, J. D., Young, N., Lindsey, R. M., Luke, R. G., and Campbell, D., 1969, Successful treatment of three cases of very severe barbiturate poisoning, *Lancet* **1**:995.

Koshino, I., Castino, F., Yoshida, K., Carse, C., Kambic, H., Scheucher, K., Kretz, A. P., Malchesky, P. S., and Nose, Y., 1975, A biological extracorporeal metabolic device for hepatic support, *Trans. Amer. Soc. Artificial Internal Organs* **21**:492.

Kupfer, D., and Rosenfeld, J., 1973, A sensitive radioactive assay for hexobarbitual hydroxylase in hepatic microsomes, *Drug Metabolism and Disposition* **1**:760.

Lowe, C. R., and Mosbach, K., 1974, The synthesis of adenine substituted derivatives of $NADP^+$ and their potential as active coenzymes and affinity adsorbents, *Eur. J. Biochem.* **49**:511.

Lu, A. Y. H., and Levin, W., 1974, The resolution and reconstitution of the liver microsomal hydroxylation system, *Biochim. Biophys. Acta* **344**:205.

Mandel, H. G., 1971, Pathways of drug transformation: biochemical conjugations, in: *Fundamentals of Drug Metabolism and Drug Disposition* (B. N. La Du, H. G. Mandel, and E. L. Way, eds.), pp. 149–205, The Williams & Wilkins Company, Baltimore.

McDermott, W. V., and Norman, J. C., 1971, Metabolic considerations in treatment of liver failure with particular reference to use of *ex vivo* perfused pig liver, *Transplant Proc.* **3**:1509.

Medd, R. K., Widdop, B., Braithwaite, R. A., Rees, A. J., and Goulding, R., 1974, Configuration of haemoperfusion and haemodialysis in the therapy of barbituate intoxication in dogs, *Arch. Toxicol.* **31**:163.

Parikh, I., MacGlashan, D. W., and Fenselay, C., 1976, Immobilized glucuronyltransferase for synthesis of conjugates, *J. Med. Chem.* **19**:296.

Rosenbaum, J. L., Kramer, M. S., Raja, R., and Boreyko, C., 1971, Resin hemoperfusion: a new treatment for acute drug intoxication, *New Engl. J. Med.* **284**:874.

Sofer, S. S., Ziegler, D. M., and Popovich, R. P., 1975, Application of an insolubilized hepatic microsomal drug oxidase to continuous production of *N*-oxides, *Biotechnol. Bioeng.* **17**:107.

Soyer, T., Lempinen, M., and Eiseman, B., 1973, *In vitro* extracorporeal liver slices and cell suspensions for temporary hepatic support, *Ann. Surg.* **177**:393.

Widdop, B., Medd, R. K., Braithwaite, R. A., Rees, A. J., and Goulding, R., 1975, Experimental drug intoxication: treatment with charcoal haemoperfusion, *Arch. Toxicol.* **34**:27.

Wolf, C. F. W., and Munkelt, B. E., 1975, Bilirubin conjugation by an artificial liver composed of cultured cells and synthetic capillaries, *Trans. Amer. Soc. Artificial Internal Organs* **21**:16.

Ziegler, D. M., and Mitchell, C. H., 1972, Microsomal oxidase IV: properties of a mixed-function amine oxidase isolated from pig liver microsomes, *Arch. Biochem. Biophys.* **150**:116.

22

Some *in Vivo* and *in Vitro* Studies of Biologically Active Molecules on Organic Matrixes for Potential Therapeutic Applications

L. S. Hersh and H. H. Weetall

1. *Introduction*

The medical application of an immobilized enzyme or antibody is defined as the use of bound biologically active materials (BBAM) to treat some disease or clinically defined state such as poisoning. The immobilized enzyme or antibody may be used outside the body, *ex vivo* or extracorporeally, or as an implant in the body, *in vivo*.

The use of an immobilized antibody or enzyme either *in vivo* or *ex vivo* requires intimate contact with body tissues. This immediately presents a difficult medical problem: how to avoid a severe inflammation or rejection reaction by the body. The approach used in hemodialysis can be used for BBAM. The BBAM can be coupled to sheets, plates, or tubing and the patient's blood is passed through this system. The main medical risks of this approach are damage to the blood components, releasing blood clots into the body, and an immune reaction. In addition, the blood must remain noncoagulated in the region of the BBAM.

An additional complication in the *ex vivo* approach to the use of BBAM is that access to the blood supply is required. In the case of hemodialysis, this is accomplished by periodic access to arteriovenous shunts or fistuli. This process is inherently associated with a number of inconveniences and problems. A bodily implant containing the BBAM could theoretically avoid this problem. However, the possibility of inflammation and rejection complicates this approach.

In view of the medical complications of the use of BBAM, why bother with this approach? Why not administer the free enzyme or antibody by intravenous

---

L. S. Hersh • Research and Development Laboratories, Corning Glass Works, Corning, New York
H. H. Weetall • Research Works, Corning, New York

(i.v.) administration. Of course, the reason is very obvious. Most of the available enzymes and the useful antibodies are heterologous proteins to humans. Therefore, injection of a high concentration of foreign proteins would most likely cause an immune reaction. The immune reaction could lead to an active anaphylaxis, a life-threatening state.

This chapter has a very limited scope; it is a review of our laboratory's effort in the medical use of BBAM. The organization of the chapter is based on the various elements of the overall process: the reactor design; the coupling methods; sterilization techniques; and *in vitro,* animal, and clinical results. Some consideration is given to general approaches to each topic, however, this work is mainly limited to our own results.

2. *Reactor Design*

2.1. *Physiological Considerations*

Two main considerations govern the design of an *ex vivo* reactor. (1) Blood in the reactor should be maintained in a liquid state. Thus, thrombogenesis and very low flow rates should be avoided. (2) Blood components should not be damaged by high shear rates, pressure fluctuations, and impact on solid surfaces. This section will consider both factors in a general manner and show how they help influence the choice of reactor type. In addition to considerations of blood compatibility, factors related to the efficiency of the reactor design must be evaluated. The kinetic characteristics of each of the reactors discussed in this section will be presented in a separate discussion.

Thrombogenesis is the formation of nonreversible masses consisting of formed elements, either red cells or platelets, intertwined in a fibrin network. The mechanism of thrombogenesis consists of cascading, enzymatically controlled reactions that may be triggered by a foreign surface. The rate and nature of surface-activated thrombogenesis depends on *both* the chemical nature of the foreign surface and the fluid motion of the blood in the vicinity of that surface (Leonard and Friedman, 1970).

Past studies have shown that fluid motion and thrombogenesis are interconnected. The results of these studies may be summarized as follows. (1) Turbulence and/or stasis and disruption of established flow patterns increase the probability of blood clotting. (2) Slow rates of blood flow enhances clotting, presumably by allowing local buildup of activated clotting elements. According to those facts, the ideal *ex vivo* (or *in vivo*) reactor should provide rigid flow channels, producing laminar flow, and the maximum velocity that provides laminar flow should be employed to aid in the prevention of thrombogenesis.

The chemical nature of the surface in contact with blood is also an important factor in activation of thrombi. This extremely complicated subject has been reviewed extensively (Mason, 1973) and will not be discussed here. Our approach was pragmatic; materials were chosen on the basis of either clinical usage or research studies indicating the relative nonthrombogenicity of the material.

There are actually two reasons for maintaining a maximum flow rate in an *ex vivo* reactor. One was mentioned above as a factor in preventing surface-activated thrombogenesis. The other reason is based on the rheological nature of blood. Steady-state viscosity measurements and cinephotomicrographic observations of transient behavior of blood flowing in small tubes demonstrate that a definite *structure* forms in blood at low and zero flow rates (Merrill, 1965). These studies show that as the pressure gradient is decreased to a critical value (less than the yield stress), blood flow will abruptly halt. If blood were a Newtonian liquid (viscosity independent of flow velocity), it would have continued to flow at a slower rate. Obviously, mass transfer in stalled blood would be minimal.

In summary of the previous discussion, maximum fluid-velocity-yielding laminar flow is recommended to help prevent loss of liquidity due to either thrombogenesis or thixotropic effects. Now, we will consider a factor that places an upper limit on the flow velocity in an *ex vivo* reactor: damage to either blood elements or the dissolved components in blood such as proteins and lipids. Most of the information about damage to blood components by artificial surfaces in *ex vivo* or *in vitro* flow systems is confined to red blood cells (Blackshear, 1972). These studies have indicated that cell-wall encounters are the dominant factor in red cell lysis below blood velocities of 2000 cm/sec. The rate of cell-wall-induced lysis depends on two factors: (1) frequency of cell-wall contacts, and (2) severity of cell-wall encounter. The latter may depend on the chemical and physical character of the wall alone or may include a flow contribution such as velocity gradient.

Blackshear (1972) has proposed a mechanism of wall-induced lysis based on observations of tethering of red cells to spots on a wall. When the tether breaks, the cell appears altered. Local wall velocity gradients, wall roughness, and wall stickiness determine the frequency of cells tethered and released. The once-tethered cell is presumably more susceptible to lysis.

The twin goals of maintaining blood liquidity and avoiding damage to particulate matter in blood suggests the following characteristics of an "ideal" reactor:

1. Laminar flow.
2. Velocity gradient should exceed 350 per sec.
3. Material in contact with blood should be relatively nonthrombogenic and "nonsticky" to blood particulates.
4. Smooth surfaces should be maintained (roughness size $<<$ cell size).
5. Minimum flow channel diameter of about 100 μm.
6. Avoidance of crushing or grinding action of support material.

This reactor design "checklist" can be used to select a reactor type for *ex vivo* or *in vivo* use.

2.2. *Choice of Reactor Type*

In general, there are four basic types of reactors that may be used in a continuous *ex vivo* shunt. These are displayed in Figure 1. These four types may be classified according to their rheological characteristics: (1) Variable relative motion between solid support material and the flowing solution: stirred tank

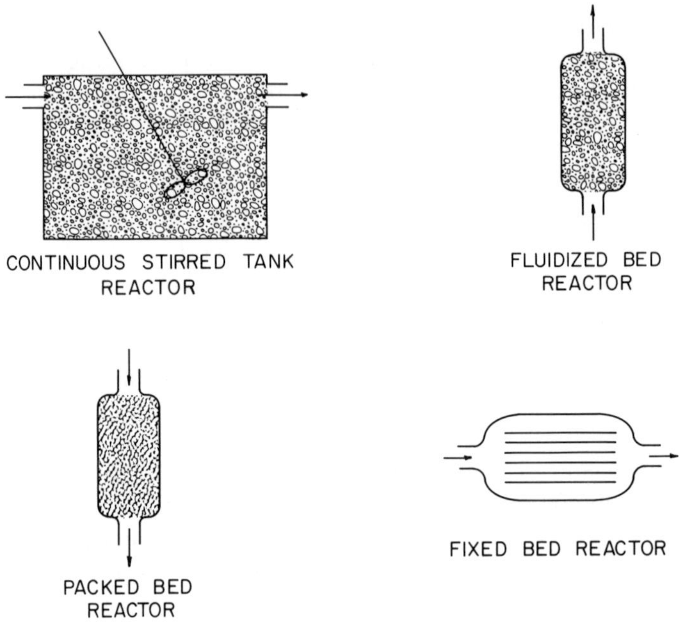

Figure 1. Reactor types.

(STR) and fluidized bed reactors (FLBR); (2) fixed relative motion between solid support and reacting solution: fixed bed (FBR) and packed bed reactors (PBR). The PBR support material will undergo compression and expansion in circuits involving pulsatile flow. Thus, the FBR is the only reactor type allowing complete control of the relative fluid motion between the solid carrier and the flowing solution. According to the reactor design checklist, rheological control is necessary to avoid both loss of blood liquidity and trauma to blood components.

A simplified analysis of some of blood's characteristics quickly leads to the choice of the FBR as the best approach to an ideal *ex vivo* or *in vivo* reactor design. The specific design of a FBR may vary from tubular to a random porous network. The choice of design will depend on whether *ex vivo* or *in vivo* use is planned, the maximum allowable pressure drop, and the minimum effective internal surface area that is required for the BBAM.

The initial *ex vivo* reactor design (Hyden, 1971) used in our studies is shown schematically in Figure 2. The fixed bed is a stacked array of 10 oval-shaped plates. The outside dimensions of the reactor are 13 × 23 × 3 cm. The plates measured 15 × 9 × 0.15 cm and had six 0.3-cm-diameter spacers, 0.05 cm high. Studies of streamlines using an injected dye showed points of separation at the inlet. There was no indication of areas of stagnation within the reactor chamber. All the surfaces in contact with blood were poly(methyl methacrylate) (PMMA). A recent report of relative thrombogenicity of various surfaces indicated that PMMA was rated highly compatible with blood. Mason *et al.* (1969) found that only 1 of 58 test materials had better blood compatibility than PMMA.

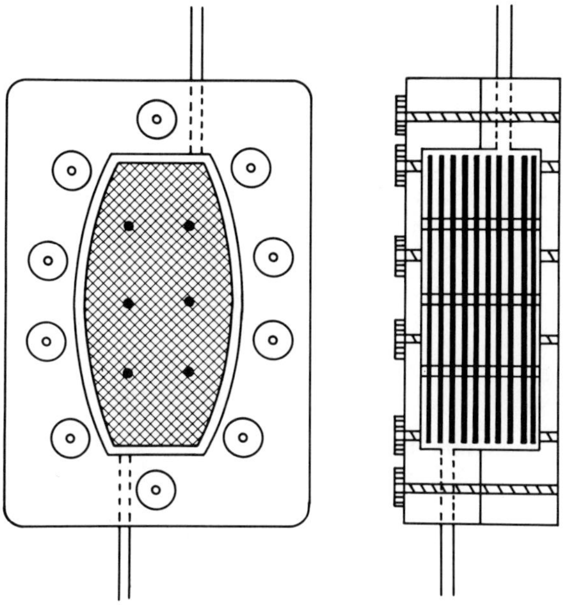

FIGURE 2. Fixed-bed poly (methyl methacrylate) reactor shown with 10 PMMA plates.

The next reactor scheme we evaluated represented an attempt to achieve greater efficiency at the cost of lowered blood compatibility. The reactor shown in Figure 3 has a diameter of 15 cm and is 5 cm deep. The carrier material is Dacron wool (3.0 denier, 1.5 in. long, type 54). About 1.5 g of wool is packed between two supporting screens. The 17-μm-diameter fiber has a surface area of about 1700 cm^2/g. The use of Dacron wool is preferred over nylon wool because of Dacron's greater resistance to compaction.

Some problems of increased rate of blood clotting were observed for the Dacron reactor compared to the PMMA plate reactor in several dog perfusion experiments. This was overcome by administering increased levels of the anticoagulant heparin. The filtering of blood particulate matter by the Dacron wool was easily eliminated by proper packing density.

2.3. *Coupling of Biologically Active Materials*

The main purpose of this section is to discuss the methods used to couple proteins to Dacron and PMMA, the two carrier materials used in our studies. The proteins coupled were the enzymes, L-asparaginase, α-galactosidase, urease, and digoxin antiserum. Some mention will be made of coupling mechanisms, but for the most part these reaction schemes are speculative.

The approach taken to couple proteins to PET and PMMA surfaces was very similar to bonding proteins to glass (Weetall, 1971). A bifunctional aminosilane was used as a coupling agent. The silane chosen, [$H_2N—CH_2CH_2CH_2Si(OC_2H_5)_3$]

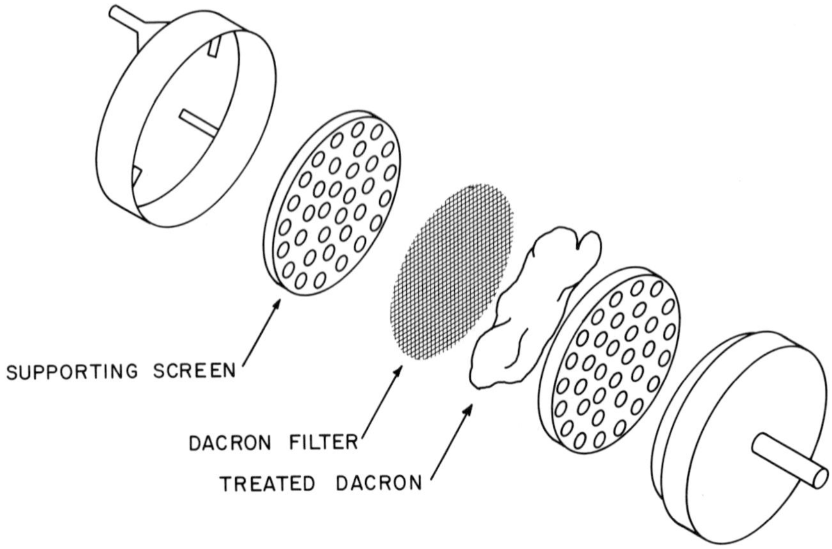

Figure 3. Packed-bed dacron wool reactor.

γ-aminopropyltriethoxysilane (γ-APTES) is only one of several amine-based silanes that are commercially available.

Ordinarily, the silanes are used to couple organic molecules to inorganic surfaces. The alkoxy or silanol end of the silane presumably reacts with the surface hydroxyl groups on inorganic surfaces (Plueddemann, 1970). For example, on silica surfaces the reactions are

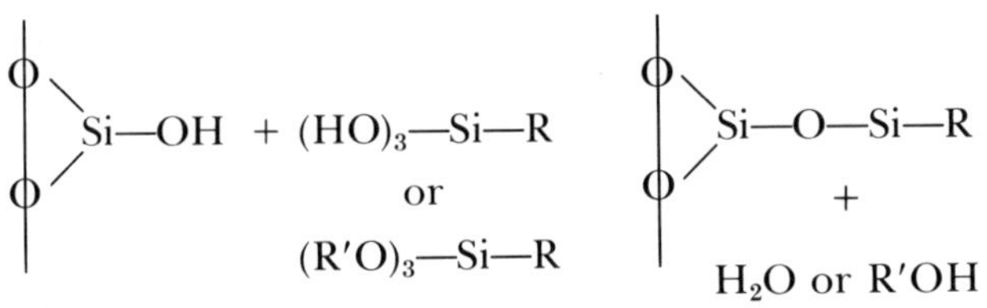

The organic end of the silane R is selected according to the nature of the coupled organic molecule (Plueddemann, 1970).

It is also known that in the case of γ-APTES, besides monomers, there are also dimers and trimers present in both polar or nonpolar solutions (Plueddemann, 1970). These dimers or trimers are linked by siloxane bridges, leaving free primary amine groups at each end. The diamino or triamino complexes can act as a coupling agent for organic to organic linkages.

The reaction of γ-APTES to PMMA was preformed by contacting the PMMA surfaces with aqueous solutions containing 1% (by volume) γ-APTES with pH adjusted to 3, using acetic acid. The surfaces were drained and then dried at 50°C in vacuum for 24 hr. After rinsing the modified PMMA with water, the amine groups were activated with a 5% (vol./vol.) solution of glutaraldehyde containing

0.15 M phosphate buffer (PB), pH 6.7 for 1.5 hr. Then, after thoroughly rinsing the excess glutaraldehyde, the PMMA surfaces were exposed to a solution containing 5–40 μ of L-asparaginase/ml of PB at 4°C for 24 hr.

The proposed reaction mechanism is

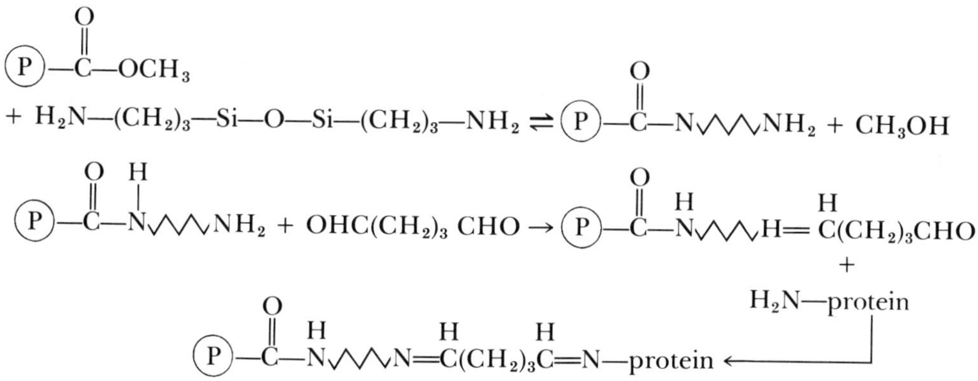

The imine group may be reduced with sodium borohydride. Our results indicated a 25% loss of L-asnase activity upon borohydride reduction with little improvement of the stability of bound L-asnase.

Some model coupling studies were performed using ^{14}C-L-tyrosine and PMMA discs. Three different coupling schemes were tried: (1) γ-APTES and glutaraldehyde, (2) same process plus reduction by sodium borohydride, and (3) an aryl amine silane diazo method (Weetall, 1971). The bound tyrosine was eluted in rabbit serum at 37°C with shaking for up to 140 hr. The results indicated that two distinct fractions of bound tyrosine were obtained by all three coupling methods. The more easily removed fraction had a half-life of 100 hr, while the second type of bound tyrosine had a half-life of about 300 hr. The more tightly bound tyrosine had an initial surface concentration of about 100 Å^2 per molecule.

The coupling of proteins to Dacron fibers was accomplished using a procedure similar to that used for PMMA. The main difference was the use of a nonpolar solvent toluene for γ-APTES and refluxing the γ-APTES/toluene with the Dacron for 18 hr. Toluene was used to obtain good wetting between solvent and the polyester fiber, while the higher temperature obtained in refluxing aided the penetration of the γ-APTES into the Dacron fiber. Upon penetration, the γ-APTES can react with the TiO_2 filler present in type 54 Dacron (about 0.25% by weight). The reaction between γAPTES and TiO_2 would be identifical to the reaction between γ-APTES and silica surfaces (Merker *et al.*, 1969). Merker *et al,* (1969) used this approach to react γ-APTES with the silica filler in silicone rubber.

2.4. *Sterilization of Bound Biologically Active Materials*

The final goal of using BBAM for *in vivo* experimentation requires that these surfaces be sterile. Sterility is the complete destruction of living organisms. This condition is operationally defined as the lack of growth of the original viable organisms under the optimum conditions of growth for that organism. The condition of safe sterilization is the *minimum* time–temperature–concentration relationship needed to ensure destruction of the *most* resistant forms of microbial life.

There are two main considerations associated with sterilizing BBAM: (1) The sterilization procedure must not degrade the biological activity of the coupled material below an acceptable minimum value, and (2) the same procedure must not cause an excessive physical deterioration of the carrier material. These restrictions eliminate some of the more common sterilization procedures, such as high-temperature steam and dry heat for PMMA. The approaches we attempted were the use of electron beam radiation (EBR), low-temperature ethylene oxide, and glutaraldehyde. The effects these methods had on test organisms and enzyme activity will be described in this section.

Low-energy electron beams can be used for surface sterilization of all commonly used plastics (Nablo and Hipple, 1972). The advantages of electron beam sterilization are (1) practically instantaneous sterilization, less than 1 μsec required for 2.5Mrad dose; (2) no significant changes in tensile strength, elongation, and notch impact strength of all conventional packaging materials (includes materials that are damaged by gamma-radiation sterilization), and (3) temperature elevations of material irradiated of about 15°C.

The main disadvantage of electron beam sterilization is rapid energy dissipation with penetration depth. The practical limit of 100-keV electrons is a penetration depth of about 100 μm. This is sufficient for the surface of coupled biologically active material, but the shallow penetration depth precludes the use of EBR for an active surface positioned in an enclosure.

L-Asparaginase (*E. coli* II) was coupled to PMMA plates using the method described in the section describing coupling procedures. Several plates were then inoculated with *B. subtilis* at a concentration of 5×10^8 microorganisms per plate. The plate and organisms were kept moist with 10^{-2} M Tris, pH 7.5. Results indicated that sterility was obtained at a total dose of 4.7 Mrad. The irradiations were carried out using 2.2-MeV electrons delivered at 10^{14} rad/sec. The sterile L-asparaginase-PMMA plate yielded 55% of the original activity.

The results obtained with EBR were excellent. A large fraction of enzyme activity was retained, and there was no sign of adverse effects on the PMMA plates. However, owing to the shallow penetration depth of the electron beams, the plates could not be sterilized *in situ*. Assembly of the active plates would have to occur under sterile conditions. It was this consideration that led us to try chemical sterilization by ethylene oxide and glutaraldehyde.

Ethylene oxide gas (EO) is a powerful alkylating agent and is active against all microbes, including viruses and bacterial spores (Ernst and Doyle, 1968). It is common practice to use ethylene oxide mixed with an inert gas at about 60°C. In order to avoid thermal deactivation of coupled enzymes, we used a 1% solution of EO in 10^{-2} M Tris at pH values of 5.0 and 7.5. At room temperature, it required 17 and 24 hr to sterilize 7.7×10^5 *B. subtilis* var. *niger* at those pH's, respectively. The immobilized *E. coli* L-asparaginase (L-asnase) on PMMA had 86 and 97% of the original enzyme activity at the former and latter conditions.

The aqeuous EO sterilization method was used in two human trials of the

PMMA-bound L-asnase (Sampson *et al.*, 1974). The procedure used was to ship the reactors containing the PMMA-L-asnase plates and 1% EO 10^{-2} M Tris, pH 7.5 solution. Then upon arrival at the hospital (about 24 hr later), the ethylene oxide solution was flushed out with at least 3 liters of sterile saline solution.

An alternative sterilization method, particularly for BBAM systems that use glutaraldehyde in the coupling process, is to utilize the sterilization properties of glutaraldehyde solutions. A 2% by weight solution of glutaraldehyde at pH 8 is 100% lethal to *B. subtilis* spores ATCC 6051 after 10 hr at 25°C (Boucher, 1972). These same conditions may be used for the coupling of glutaraldehyde to either the aminosilane-coated Dacron wool or PMMA plates. There was very little difference in the final amount of coupled protein using either the sterilizing conditions described above or the conditions described in the coupling section for glutaraldehyde activation (5% glutaraldehyde, pH 6.5, and 1.5 hr at 25°C).

The use of a second exposure of bound L-asnase to sterilizing conditions for glutaraldehyde resulted in a very large decrease in L-asnase activity. Approximately 25% of the original activity was left. Thus, in the case of immobilized L-asnase, glutaraldehyde sterilization can only be applied during the glutaraldehyde-activation step. From that point, all solutions entering the reactor would have to be filtered through a 0.25-μm filter to avoid contamination.

2.5 *In Vitro and in Vivo Results*

a. Theory. The section on reactor design discussed the medical compatibility of different reactor designs. Once a medically safe reactor type has been chosen, the design details are determined by kinetic considerations. The ideal goal is to have a reactor geometry that has minimum detrimental effect on the kinetic behavior of the enzyme.

One result of immobilizing an enzyme is to introduce bulk diffusion of either or both substrates and products. In other words, the observed overall initial rate of substrate depletion or product formation includes rate constants due to the enzymic reaction and the bulk diffusion rates of the substrate or substrates. In the case of microporous surfaces or gel-entrapped enzymes, the rates of diffusion in the pores or gel must also be considered.

The introduction of bulk diffusion rates into the enzymic process causes difficulties in the kinetic analysis of immobilized enzyme reactors. Only two extreme examples of reactor kinetics may be readily treated (Horvath *et al.*, 1973). The first case involves a reaction that is bulk-diffusion-controlled. The rate of reaction is completely determined by the rate of diffusion of substrate to the bound enzyme. The second case is completely determined by the same rate constants that describe the free enzyme. Here, the diffusion rates of substrate and product are much faster than the enzymic rate constants.

Most reported immobilized enzyme reactor kinetics involve cases somewhere in between the two limiting cases mentioned above. Even though these systems display a dependence on bulk diffusion and kinetic rates, it is usually possible to use a phenomenological approach. The reaction kinetics, in certain concentration ranges, may be described by pseudo-Michaelis–Menten (M-M) kinetics. The formal relationships are the same as for the free enzyme. The difference is that in the case of the immobilized enzyme, the constants K_m and k_3, the apparent Michaelis constant and the rate of decomposition of the enzyme–substrate complex, respectively, are generally functions of the flow rate in the reactor.

A useful form of M-M kinetics is the integrated expression

$$S_0 - S_t = k_3[\mathrm{E}]t + \overline{K}_m \ln\left(\frac{S_t}{S_0}\right) \tag{1}$$

where S_0 is the initial concentration of substrate, S_t is the substrate concentration after time t, and $\overline{K}_m$, k_3, and [E] are the apparent Michaelis constant, the rate constant for the decomposition of the enzyme–substrate complex, and the concentration of total enzyme, respectively.

The application of Eq. (1) to an immobilized-enzyme-catalyzed reaction requires that all elements of fluid move through the reactor at equal velocities. Under these conditions, usually referred to as "piston flow," Eq. (1) applies to each infinitesimal cross-section volume of the reactor and therefore expresses the total reaction taking place in each such volume during its passage through the reactor. Then t is the residence time of each fluid element in the reactor and S_0 and S_t are the concentrations of substrate entering and leaving the reactor, respectively. In addition, the residence time, t, is related to the void volume, V_v, and the flow rate through the reactor, Q, by

$$t = \frac{V_v}{Q} \tag{2}$$

The immobilized enzyme is considered to exert its catalytic action through the void volume, V_v, of the reactor. Thus, its "concentration" is equal to the total moles of bound enzyme, E, divided by V_v, or

$$\mathrm{E} = \frac{\mathrm{E}}{V_v} \tag{3}$$

Substituting the identities expressed in Eqs. (2) and (3) into Eq. (1), we have

$$(S_0 - S_t) = \frac{k_3\mathrm{E}}{Q} + \overline{K}_m \ln\left(\frac{S_t}{S_0}\right) \tag{4}$$

Further, if we let

$$P = \frac{S_0 - S_t}{S_0}$$

then

$$PS_0 = \overline{K}_m \ln(1 - P) + k_3\left(\frac{\mathrm{E}}{Q}\right) \tag{5}$$

The next sections will discuss the application of Eq. (5) to some *in vitro* and *in vivo* results. In particular, the dependence of P on flow rate will be demonstrated for the case where $\overline{K}_m$ is a function of reactor flow rate.

b. In Vitro Results. Figure 4 shows the reduction of $\bar{K}_m$ with increasing flow rate through the Dacron wool reactor containing *E. coli* II L-asparaginase. The $\bar{K}_m$'s at different flow rates were determined from Lineweaver–Burk plots at L-asparagine concentrations from 10^{-3} to 10^{-4} M in Tris buffer, pH 7.5, at 37°C. The determinations were made using a gradientless circulation method (Ko and Hersh, 1976). The total activity of L-asparaginase in the reactor was 54 IU (37 IU/g of Dacron) at 37°C.

The values of $\bar{K}_m$ plotted in Figure 4 range from a value of 8.5×10^{-4} M at a flow rate of 50 ml/min, to 6×10^{-5} M at a flow rate of 320 ml/min. The K_m for *E. coli* L-asparaginase in solution has been reported to be 1×10^{-5} M at 37°C and pH 7.5. Thus, at higher flow rates, the influence of substrate bulk diffusion rates on $\bar{K}_m$ is decreased, and $\bar{K}_m$ approaches the K_m of the free enzyme.

The $\bar{K}_m$ for the PMMA reactor containing *E. coli* L-asparaginase was obtained at only one flow rate, 150 ml/min (Hersh, 1974). The value obtained, 3×10^{-4} M, in nearly identical to the $\bar{K}_m$ measured for the Dacron reactor, 3.5×10^{-4} M at the same flow rate.

In the case of the PMMA reactor, the value of $\bar{K}_m$ was nearly the same for L-asparagine solutions in either Tris buffer at pH 7.5 or in canine blood. The Dacron wool reactor's $\bar{K}_m$, however, decreased by an order of magnitude in canine blood. The different kinetic behavior of the Dacron wool reactor is probably due to a thicker film of protein or fibrin coating the Dacron fibers than was observed on the PMMA surfaces.

Examination of Eq. (5) shows that the degree of reaction, P, depends on the flow rate, Q, in two ways: (1) varying Q will cause the residence time, t, to vary according to Eq. (2); and (2) $\bar{K}_m$ strongly depends on the value of Q, as shown in Fig. 4. The relationship between P and Q is sufficiently complex to warrant additional discussion.

A plot of P versus flow rate through the Dacron wool reactor is shown in Figure 5. P was calculated using Eq. (5) and the $\bar{K}_m$ versus flow rate data plotted in

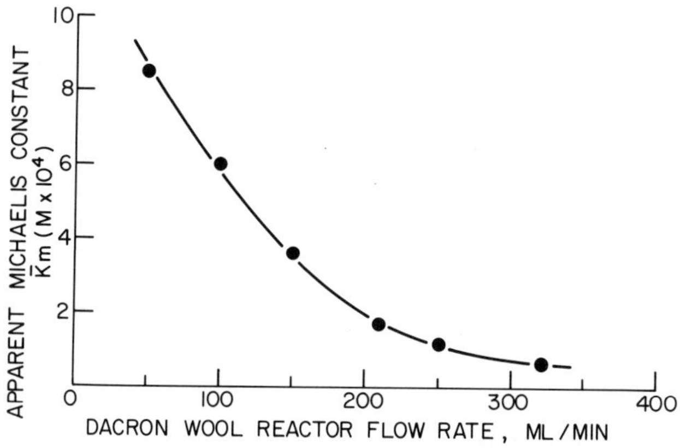

FIGURE 4. Plot of apparent Michaelis "constant" versus flow rate through Dacron wool reactor. $\bar{K}_m$ was obtained from Lineweaver–Burk plots using 10^{-2} M Tris/HCl, pH 7.5 at 37°C.

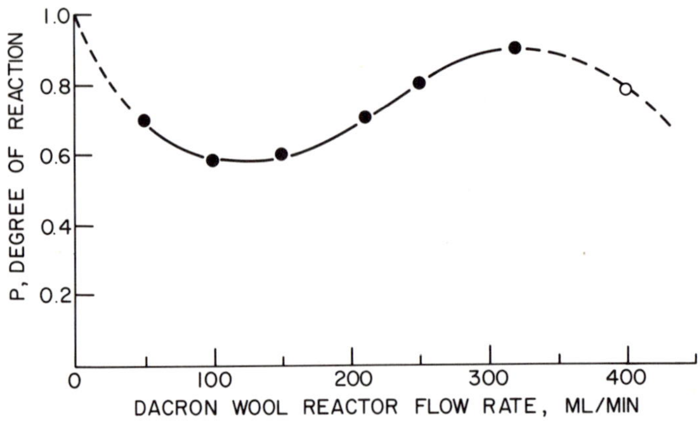

FIGURE 5. Calculated degree of reaction [$P = (S_o–S_t)/S_o$] plotted against flow rate through a Dacron wool reactor.

Figure 4. The value of k_3E was obtained from Lineweaver–Burk plots (Ko and Hersh, 1976). A single value for the maximum velocity was observed for all the flow rates studied. The initial L-asparagine concentration, S_0, was chosen at 5×10^{-5} M, the concentration present in normal serum.

The relationship between P and Q displayed in Figure 5 has some interesting characteristics. The dotted portion at low flow rates is an extrapolation to $P = 1$ at $Q = 0$. This must obviously be the case, since at zero flow rate the residence time is infinite. The dotted line at high flow rates is due to a calculated value of P at $Q =$ 500 ml/min, assuming that $\bar{K}_m$ has the same value at 400 ml/min obtained at 320 ml/min. At any rate, P must approach zero at infinite values of Q, since t, in that case, approaches zero.

The most interesting aspect of the dependence of P on Q shown in Figure 5 is that there are two regions that yield high values of P: a low-flow-rate region of about 25 ml/min and a higher-flow-rate region centered around 325–250 ml/min. The higher flow rates should be accompanied with improved blood compatibility according to our earlier discussion. The relationship between P and Q displayed in Figure 5 applies in a qualitative sense to any reactor that has a similar relationship between $\bar{K}_m$ and Q shown in Figure 4.

The use of a maximum value for P in a medical application of a BBAM reactor is important, since it means a minimum time of treatment for the depletion of a certain amount of substance.

c. *In Vivo Results.* The *in vivo* experiments with the PMMA and Dacron wool reactors involved a flow system schematically shown in Figure 6. Special precautions were taken to introduce blood into a saline solution-filled reactor. Blackshear (1972) has given evidence for minimized red blood cell surface encounter if a particle-free fluid precedes the introduction of blood into a conduit. The dogs or baboons were anesthesized using Nembutal (25 mg/kg) and heparin was administered at a dose level of about 500 IU/kg 30 min before perfusion was started.

An example of the *in vivo* results obtained with the PMMA reactor is given in Table 1. L-Asparagine levels were measured in samples taken at various times at the inlet and outlet of the reactor. Thus, the decrease in L-asparagine concentration must be due to action of the immobilized enzyme. The asparagine concentration was measured using a radiometric method (Cooney *et al.*, 1975).

Table 1 also tabulates calculated values of $\bar{K}^m$. These values of $\bar{K}_m$ were calculated using Eq. (5) for each measurement of S_0 and S_t. A separate measurement of the maximum velocity was performed at saturating levels of L-asparagine (Hersh, 1974). The average value for $\bar{K}_m$, 3.2×10^{-4} M, is similar to the $\bar{K}_m$ value obtained in Tris buffer, 6.3×10^{-4} M.

The PMMA reactor containing immobilized *Erwinia carotovora* L-asparaginase was also used in two human trials (Sampson *et al.*, 1974). The ethylene oxide method was used to sterilize the reactor. Regional heparinization was used to avoid blood clots. The levels of L-asparagine and L-aspartic acid, and L-asparaginase were monitored in samples taken from the arterial or inlet side of the reactor. The results are shown in Figure 7. The increase in aspartic acid levels do not exactly match the decrease in asparagine. However, a steady-state concentration of L-asparagine at 3×10^{-6} M was reached in about 1 hr of perfusion. This represents 6% of the original level of L-asparagine concentration. The serum asparagine level returned to normal values in about 3 hr after the perfusion ended.

The Dacron wool reactor was used in several canine studies. The results of one study are shown in Figure 8. The perfusion lasted for 2 hr. Arterial and venous samples were taken for L-asparagine and L-aspartic acid analysis (Cooney *et al.*, 1975). The arterial levels shown in Figure 8 are similar to those obtained

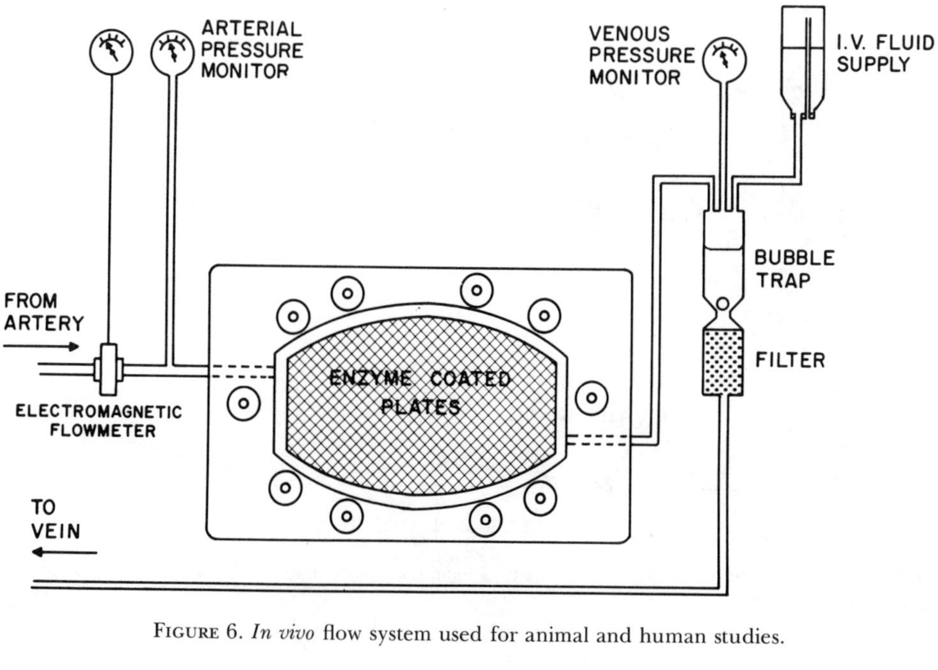

FIGURE 6. *In vivo* flow system used for animal and human studies.

TABLE 1
Concentration of L-Asparagine in a Baboon (P-1, C-6) Serum at Inlet and Outlet of PMMA Reactor During ex Vivo Experiment[a]

Time perfusion (min)	L-Asparagine (nM/ml) Inlet	Outlet	$\overline{K}_m \times 10^4$ M
0	19.8	—	—
5	22.7	11.9	4.9
20	10.0	5.2	5.0
25	14.7	0.3	0.8
30	29.0	1.4	1.0
35	9.7	4.2	4.0
40	2.9	1.2	3.7
45	7.3	1.7	2.2
50	8.9	1.1	1.5
60	6.2	1.8	2.6
120	6.1	3.6	6.2
			< 3.2 + 1.6 >

[a] $Vm = 7.2 \times 10^{-4}$ m/liter-min, Q = 175 Ml/min, V_v = 80 ml, 37°C.

with the PMMA reactor. The L-asparagine level decreased to about 10% of the initial value after 30 min of perfusion, and remained at this level for 2 hr. The asparagine level quickly started to increase after the perfusion was ended. Normal values of aspartic acid and asparagine were obtained about 12 hr after the perfusion ended.

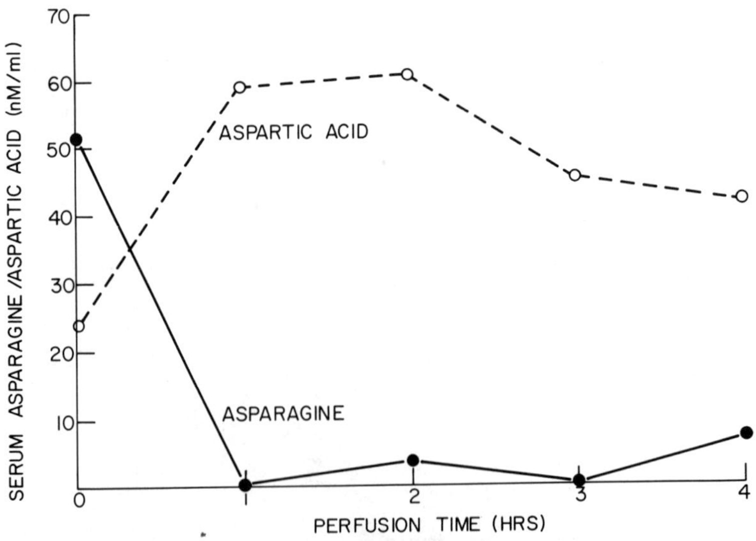

FIGURE 7. *In vivo* (human) results obtained using *Erwinia carotovora* L-asparaginase coupled to PMMA plates.

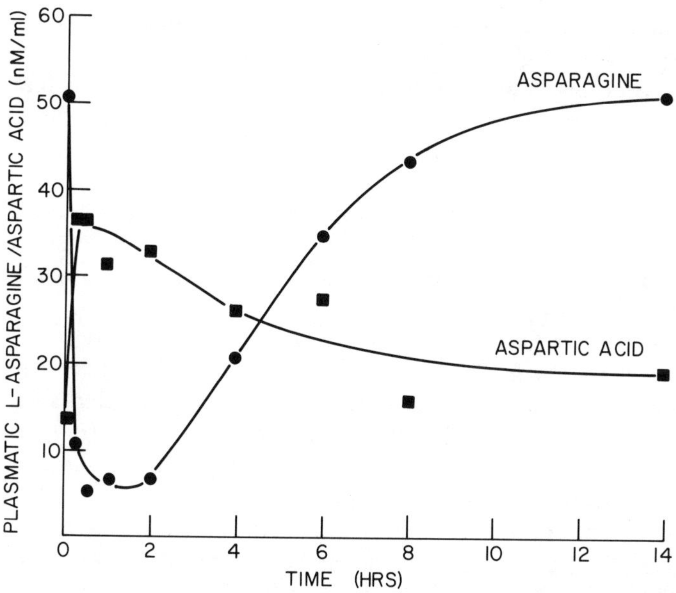

FIGURE 8. *In vivo* (canine) results obtained with *E. coli* L-asparaginase coupled to Dacron wool. Perfusion time was from 0 to 2 hr.

$\bar{K}_m$ values calculated for the perfusion shown in Figure 8 yielded an average value of 2.5×10^{-4} M. This was obtained at an average flow rate of 450 ml/min. This value is about 4 times the $\bar{K}_m$ obtained for the same reactor using Tris solutions at 320 ml/min. This increase in $\bar{K}_m$ is probably due to protein deposition on the Dacron fibers.

d. Antidigoxin Experiments. The Dacron wool reactor was also used in animal experiments as a digoxin trap. Antidigoxin antiserum was attached to the Dacron wool by the same method used for L-asparaginase coupling. *In vitro* testing of the Dacron–antidigoxin material indicated a total capacity of about 600 ng of digoxin/g of Dacron fiber. Binding studies with human γ-globulin bound to Dacron indicated that about 25% of the digoxin was nonspecifically adsorbed. Cross-reactivity tests with [^{125}I]angiotensin and [^{125}I]TSH with the antidigoxin–Dacron resulted in about 10% nonspecific adsorption.

The *in vivo* experiments were performed using pigs weighing 40 lb as subjects. The pigs were given digoxin doses for 5 days prior to the experiment. The initial blood digoxin level was about 2.5–3 ng/ml on the day of perfusion, Nembutal was used as an anesthetic. Heparin was applied intravenously 30 min prior to the start of shunt operation. Blood samples were taken at inlet and outlet positions 30 and 15 min before the perfusion and every 15 min thereafter. A total of three pigs were treated with the shunt. The third pig received 4.9×10^6 cpm of [^{125}I]digoxin 65 min before the treatment began.

The flow system was similar to that shown in Figure 6. To increase the total capacity for binding digoxin, four reactors were connected in series, yielding a total of 6 g of digoxin–Dacron conjugate. Thus, a total theoretical binding capacity of about 4 μg of digoxin was available. The reactors were also coated with heparin using a procedure involving an amino silane and ionic binding of heparin (Hersh *et al.,* 1971).

The blood samples were assayed for digoxin using a solid-phase radioimmunoassay procedure. Rabbit antidigoxin coupled to 1-μm-diameter porous glass and [^{125}I]digoxin was used in the assay. In all cases, the standard curves were linear from 0.5 to 10 ng of digoxin/ml of solution.

Figure 9 shows the levels of digoxin in the pigs' blood before and after passing through the reactor during the *in vivo* experiment. The perfusion lasted for 3.75 hr. Integration beneath the arterial and venous values yielded a total pickup of about 4 μg of digoxin by the 6 g of Dacron conjugate, or 670 ng of digoxin/g of Dacron. This is almost identical to the *in vitro* adsorption at saturation.

The concentration of digoxin in blood samples obtained at a distant site from the shunt connections is shown in Figure 10. The digoxin levels surprisingly increased from 3.5 ng/ml of plasma at the beginning of perfusion to 6.5 ng/ml of plasma at the end of the circulation. Immediately after the reactor was disconnected from the pig, the digoxin level slowly returned to the initial steady-state concentration. The data in Figure 10 indicate a return to the original digoxin level in about 4 hr postperfusion.

The results of the injection of 4.9×10^6 cpm of [^{125}I]digoxin are shown in

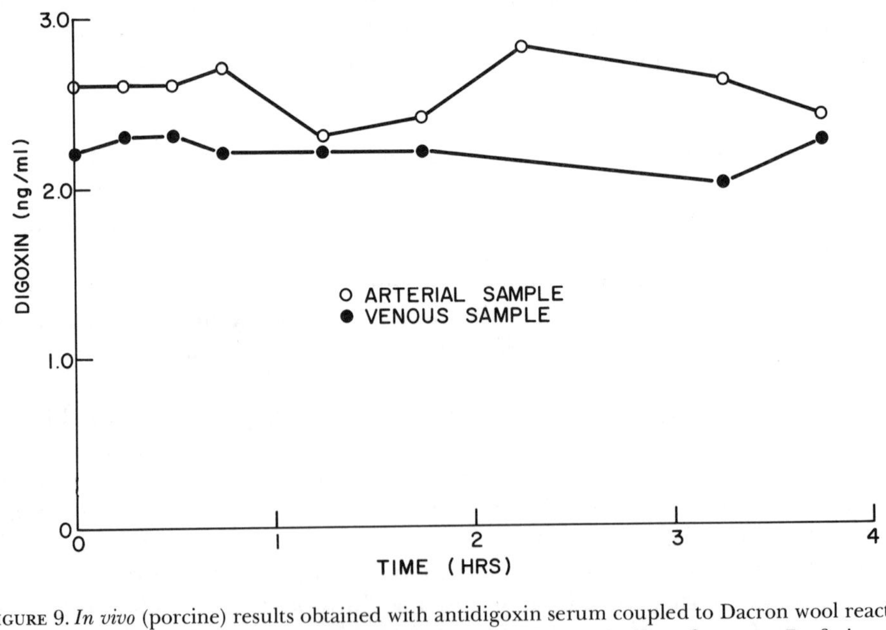

FIGURE 9. *In vivo* (porcine) results obtained with antidigoxin serum coupled to Dacron wool reactor. Serum samples were assayed at inlet (arterial) and outlet (venous) sides of reactor. Perfusion ran from 0 to 3.75 hr.

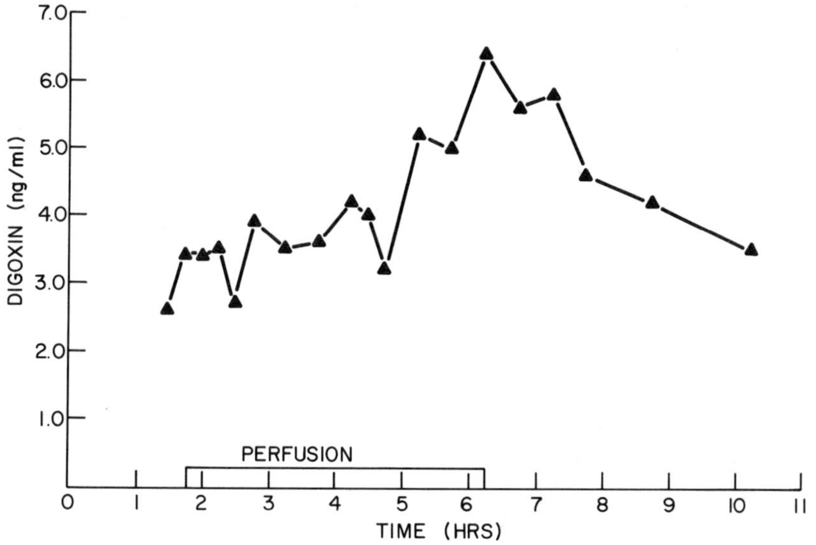

FIGURE 10. Serum levels of digoxin obtained at a distant site from shunt connection during porcine perfusion.

Figure 11. This was done 65 min prior to the connection of the shunt. The semilog plot shows the multiple exponential character of the decrease of labeled digoxin in the pig's plasma. The depletion of [^{125}I]digoxin is seen to have a rapid initial rate: 90% was removed during the first 10 min, and a 97% reduction of iodine-labeled digoxin in about 1 hr after the injection.

The same pig was killed after 2 hr of perfusion. Various tissue slices were monitored for ^{125}I radioactivity. The results are shown in Table 2. The cpm gram of tissue have been converted to ng total of digoxin/g of tissue in the next-to-last column. The percent distribution of digoxin indicates that digoxin is dispersed throughout the various organs and tissues of the pig. Unfortunately, the heart is a relatively poor second place for digoxin accummulation. The kidney has almost 4 times the digoxin contained in the heart.

TABLE 2
Distribution of [^{125}*I*] *Digoxin in Pig No. 3's Organs*

Organ	Weight	cpm	cpm/g	ng of Digoxin/g of tissue	%
Muscle	0.72	164	228	0,016	2
Lungs	0.12	120	1000	0.069	9
Kidney	0.71	4374	6161	0.428	57
Liver	0.37	451	1219	0.085	11
Spleen	0.53	323	609	0.042	5
Fat	0.75	42	56	0.004	1
Heart	0.63	1070	1698	0.118	15
Skin	0.56	137	245	0.017	2

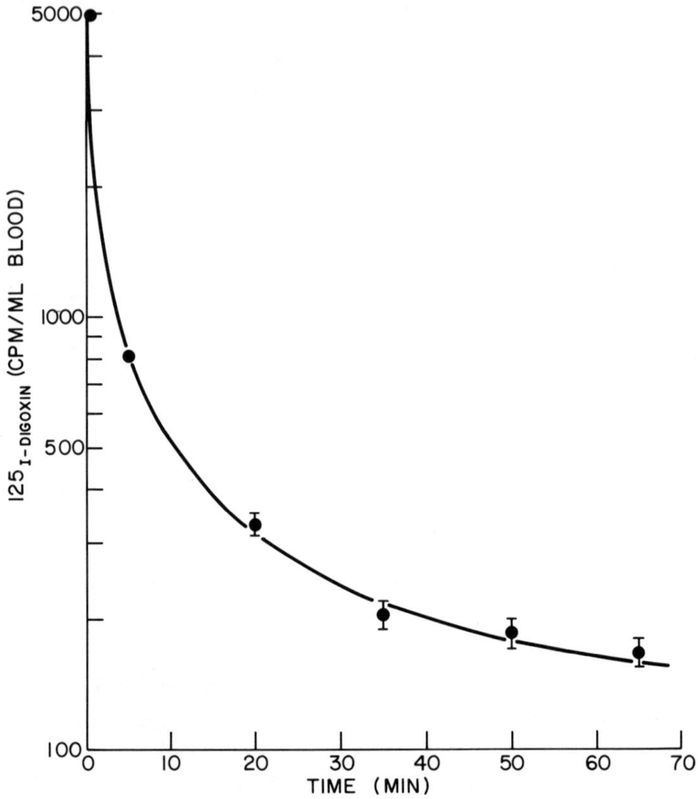

Figure 11. Serum levels of [^{125}I]-digoxin in porcine plotted against time after injection of 4.9 × 10^6 cpm.

The conclusion reached in this preliminary study was that because of the high ratio of total volume to the plasma volume distribution of digoxin, there was not sufficient adsorbent capacity in the three perfusions performed. To reduce the digoxin level in the pig from 3 to 0.5 ng/ml of plasma, with a total volume/plasma volume distribution ratio of 35 : 1, a total of about 100 g of Dacron fiber was required. This amount of Dacron wool, using the same reactor design, required a void volume of about 1 liter, almost the total blood volume of the pig. Obviously, a higher-specific-adsorbent-capacity material is required for digoxin removal. The same reasoning applies, of course, to human digoxin detoxification.

e. Studies with L*-Asparaginase Covalently Coupled to Vascular Prosthetics.* The L-asparaginase used in these studies (Cooney *et al.*, 1975) was donated by Merck, Sharp and Dohme to the NCI. The coupling technique using a silane coupling was described in the previous section. Initial *in vitro* studies were carried out to define the characteristics of the immobilized L-asparaginase. The velocity of the enzyme-bearing graft was observed to be a direct function of flow rate of the substrate through the graft when pure solutions of L-asparagine were used

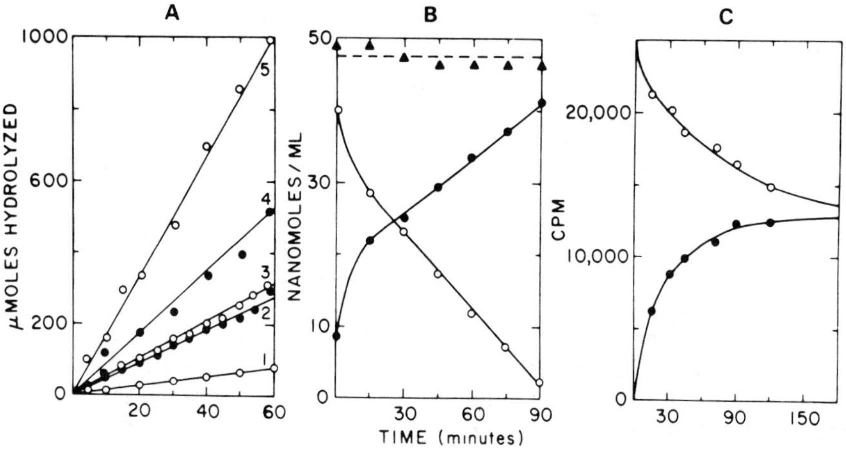

Figure 12. (A) Influence of substrate concentration and flow-rate on the hydrolysis of L-asparagine by L-asparaginase covalently coupled to a Dacron prosthesis. Line 1: 3 liters of L-asparagine 1.5×10^{-4} M in 0.05 M Tris HCl, pH 8.3, 37°C, was pumped through a 7-cm length of Dacron tubing bearing 20 IU of L-asparaginase at a flow rate of 0.2 liter/min. The substrate recirculated for the duration of the experiment. Line 2: conditions same as in line 1 except that flow rate was 0.6 liter/min. Line 3: conditions same as in line 1 except that flow rate was 0.7 liter/min. Line 4: conditions same as in line 1 except that concentration of L-asparagine was 1.5×10^{-3} M and flow rate 0.5 liter/min. Line 5: conditions same as in line 1 except that concentration of L-asparagine was 1.5×10^{-3} M and flow rate 0.7 liter/min. (B) Time course of hydrolysis of the L-asparagine in human plasma by a Dacron prosthesis to which L-asparaginase was covalently bound. One liter of human plasma at 37°C at a flow rate of 2 liters/min was pumped through a Dacron prosthesis bearing 10 IU of L-asparaginase. At 15-min intervals, 2-ml samples were removed, boiled for 20 min, and centrifuged at 104,000*g* for 30 min. The concentrations of L-aspartic acid and L-asparagine were measured spectrophotometrically in the supernatants. O, L-asparagine; ●, L-aspartic acid; and ▲, L-asparagine and L-aspartic acid. (C) Hydrolysis of L-asparagine in whole human blood by L-asparaginase covalently coupled to Dacron tubing. L-Asparaginase (10 IU) was covalently coupled to a Dacron prosthesis as described in the text. Human blood (200 ml) was anticoagulated with 1000 units of heparin; then 10 μCi 4-[^{14}c]L-asparagine was added. The blood was warmed at 37°C and pumped through the tubing at a flow rate of 100 ml/min. Samples of whole blood were removed at the times indicated in the figure and plasma was separated at 4°C. The plasmatic L-asparagine, L-aspartic acid, and 4-[^{14}C]L-aspartic acid were measured as described in the text. The concentration of L-aspartic acid in the zero-time sample was 5 nmoles/ml, and the concentration of L-asparagine at the same time was 35 nmoles/ml. At 90 min, the concentration of L-aspartic acid had risen to 40 nmoles/ml and the concentration of L-asparagine had fallen to 15 nmoles/ml. Since some hemolysis transpired during the course of the perfusion, the sum of the dicarboxylic amino acids, L-aspartic acid, and L-asparagine deviated somewhat from theory in the final sample. The open circles (○) represent 4-[^{14}C]L-asparagine; the closed circles (●) represent 4-[^{14}C]L-aspartic acid. During the course of the perfusion, a portion of the intimal surface of the graft near the outflow became coated with fibrinous material.

(Figure 12a). If plasma was substituted for buffer, the hydrolysis was no longer linear but showed a second slower rate after the initial observed rate (Figure 12b). With blood as the perfusate, an even more pronounced change in hydrolysis rate was observed (Figure 12c). It was later found that a film of protein or fibrin coated the graft in the last two examples, impairing diffusion of substrate and product from the catalytic site.

A series of experiments were carried out to determine the apparent K_m of the immobilized enzyme. For these studies, a series of substrates were tested, including L-asparagine, L-glutamine, D-asparagine, β-cyano-L-alanine, and others. In all cases the $\overline{K}_m$(app) of the enzyme-bearing graft was less than the soluble enzyme. In the case of L-asparagine, the difference was approximately two orders of magnitude. It was also noted that with extreme agitation the $\overline{K}_m$(app) of the prosthetic could be decreased by one order of magnitude, from 3×10^{-2} M to 3×10^{-3} M. It is probable that the decrease is, at best, due in part to decrease of the static or Nernst layer. Therefore, it is reasonable to assume that bulk diffusion control plays a major role in the observed kinetic differences.

The stereospecificity of native L-asparaginase from *E. coli* is not absolute. The D isomer will hydrolyze at a rate 5–7% of that observed with the L isomer. In the case of the immobilized preparation, it appears that additional stereospecificity was lost (Table 3). It was also noted that as the temperature was raised from 2°C to 56°C, the relative ratio of the velocity of the native enzyme for the L isomer to the D isomer increased from 1.75 to 5.98. In the case of the bound enzyme, the opposite was observed in that the ratio dropped from 28 to 9.3 as the temperature was increased. This would indicate that the bound enzyme cannot respond in a normal fashion to thermal perturbations. As expected, differences were observed in regard to pH optimum, ionic strength effects, and thermostability.

In vivo studies were carried out as follows. Enzyme-bearing grafts were installed in a common iliac artery of a series of dogs. These grafts remained patent for at least 10 days. The grafts failed to lower the level of plasma L-asparaginase. When the grafts were removed 3 weeks after placement, it was noted that all grafts had retained enzymatic activity but were completely covered with fibrinous deposits. Installation of an enzyme-bearing graft which had been heparinized produced only a transitory decrease in L-asparagine levels (Figure 12). A series of studies were initiated with BDF mice. Half the mice received an implant of enzyme-conjugated Dacron intraperitoneally while the remainder received grafts minus the enzyme, Both groups were inoculated intraperitoneally with 1×10^5 cells of leukemia 5178Y, sensitive to L-asparaginase. No difference in median survival times were observed. The grafts when removed were covered with tissue. When the tissue was stripped away, all enzyme-bearing grafts were found to have retained activity.

TABLE 3
Stereospecificity of Native and Immobilized L-Asparaginase Covalently Coupled to a Dacron Vascular Prothesis

Substrate	Relative rates of hydrolysis: Native enzyme	Relative rates of hydrolysis: Coupled enzyme
L-Asparagine	100	100
D-Asparagine	4.8	10.7
L-Glutamine	2.6	9.0
D-Glutamine	0.1	0.3

It is apparent from these studies that the implantation of this type of immobilized enzyme for long-term therapy is going to be difficult and may require a new and different approach. At this time, the most successful approach appears to be the extracorporeal shunt.

3. References

Blackshear, P. L., 1972, Hemolysis at prosthetic surfaces, in: *The Chemistry of Biosurfaces* (M. L. Hair, ed.), pp. 523–562, Marcel Dekker, Inc., New York.

Boucher, R. M. G., 1972, Advances in sterilization techniques; state of the art and recent breakthroughs, *Amer. J. Hospital Pharmacy* **29**:660–672.

Cooney, D. A., Weetall, H. H., and Long, E., 1975, Biochemical and pharmacologic properties of L-asparaginase bonded to Dacron vascular prosthesis, *Biochem. Pharmacol.* **24**:503–515.

Enst, R. R., and Doyle, J. E., 1968, Sterilization with gaseous ethylene oxide: a review of chemical and physical factors, *Biotechnol. Bioeng.* **10**:1–31.

Hersh, L. S,, 1974, L-Asparaginase from *E. coli* II and *Erwin. carot.* bound to poly(methyl methacryolate), *J. Polymer Sci., Symp.* **47**:55–67.

Hersh, L. S., Weetall, H. H., and Brown, I. W., 1971, Heparinized polyester fibers, *J. Biomed. Mater. Res. Symp.* **1**:99–104.

Horvath, C., Sardi, A., and Woods, J. S., 1973, L-Asparaginase tubes; kinetic behavior and application in physiological studies, *J. Appl. Physiol.* **34**:181–187.

Hyden, H., 1971, An extra-corporeal shunt apparatus for blood detoxification, *Arzneimittel-Forschung* **21**:1671–1675.

Ko, R. Y. C., and Hersh, L. S. 1976, A Dacron wool packed-bed extracorporeal reactor: a kinetic study of immobilized *E. coli* II L-asparaginase, *J. Biomed. Mater. Res.* **10**:249.

Leonard, E. F., and Friedman, L. I., 1970, Thrombogenesis on artificial surfaces: a flow reactor problem, *Chem. Eng. Progr. Symp. Ser.* **66**:59–71.

Mason, R. G., 1973, Blood compatibility of biomaterials: evaluation of a simple screening test, *Biomater. Med. Devices and Artificial Organs* **1**:131–139.

Mason, R. G., Scarborough, D. E., Sabina, R. S., Brinkhous, K. M., Ikenberry, L. D., Kearney, J. J., and Clark, H. G., 1969, Thrombogenicity of some biomedical materials: platelet–interface reactions, *J. Biomed. Mater. Res.* **3**:615–644.

Merker, R. L., Eylash, L. J., Mayhew, S. H., and Wang, J. Y. C., 1969, The heparinization of silicone rubber using amino-organo silane coupling agents, *Proc. Artificial Heart Conf.,* 29–39.

Merrill, E. W., 1965, Rheology of human blood and some speculations on its role in vascular homeostasis, in: *Biophysical Mechanisms in Vascular Homeostasis and Intravascular Thrombosis* (P. N. Sawyer, ed.), pp. 121–137, Appleton-Century-Crofts, New York.

Nablo, S. V., and Hipple, J. E., 1972, Low energy electron beams for surface sterilization and asceptic practice, *Soc. Plastics Eng. Conf.,* **TR-14**:1–19.

Plueddemann, E. P., 1970, Adhesion through silane coupling agents, *Soc. Plastics Ind., 25th Ann. Tech. Conf.,* **13-D**:1–10.

Sampson, D., Han, T., Hersh, L. S., and Murphey, G. P., 1974, Extracorporeal chemotheraphy with L-asparaginase in man, *J. Surg. Oncol.* **6**:39–48.

Weetall H. H., 1971, Enzymes immobilized on inorganic carriers by covalent attachment, *Res. Develop.* **22**:18–22.

23

Therapeutic Perspectives of Enzyme Reactors

FRED R. BERNATH, LAWRENCE S. OLANOFF, AND WOLF R. VIETH

1. *Introduction*

The interest in employing enzymes in various types of practical applications has existed for many years. It is only within the last decade, however, that biochemists, chemical and biochemical engineers, food scientists, microbiologists, organic chemists, medical researchers, and others have cooperated in a significant interdisciplinary effort to solve the complex problems associated with the actual utilization of these superior catalysts. This effort has resulted in the development of techniques for binding enzymes to insoluble supports, methods for the design, construction, and optimization of various types of immobilized enzyme reactors and other advances that have provided a basis for a new field called enzyme engineering. Although enzyme engineering has been primarily aimed at establishing industrial applications, the technology that has been developed in this field has much to offer in the area of therapeutic medicine.

In this paper we shall discuss the potential contributions of immobilized enzyme reactors to the enhancement of enzyme therapy. The major part of our discussion will concern our own experience with reactors containing *E. coli* L-asparaginase (L-asparagine amidohydrolase, EC 3.5.1.1), an enzyme that has been widely investigated in its soluble form as a chemotherapeutic agent for acute lymphocytic leukemia in humans (Wriston and Yellin, 1973). In addition to providing a potentially more effective alternative to soluble asparaginase therapy for leukemia, the immobilized enzyme reactors that we propose serve as models for the administration of any therapeutic enzyme that acts on substrates in the blood.

We begin by providing an historical perspective with a brief description of the evolution of enzyme applications in both industry and medicine. Although the problems that have limited the utilization of therapeutic enzymes are quite differ-

FRED R. BERNATH • Department of Chemical and Biochemical Engineering, Rutgers University, New Brunswick, N.J. LAWRENCE S. OLANOFF • Department of Chemical and Biochemical Engineering, Rutgers University, New Brunswick, N.J. WOLF R. VIETH • Department of Chemical and Biochemical Engineering, Rutgers University, New Brunswick, N.J.

ent than those inhibiting industrial applications, many of the proposed solutions have general value in both areas. As we shall see, this is especially true for immobilized enzyme reactors which have been developed primarily for industrial use.

2. *Development of Enzyme Technology—Promise of Things to Come*

2.1. *Advances in the Field of Enzymology*

Enzyme technology is still in the early stages of its development, despite the fact that our knowledge and study of these biocatalysts dates back to the 1700s, when scientists began investigating digestion. This is especially obvious when we compare the number and quality of existing uses of enzymes with the vast range of potential industrial and medical applications that should be possible with such highly active and specific catalysts. We have long recognized and appreciated the valuable function that enzymes provide for living organisms by mediating the chemical reaction of all life processes. Over the years, however, we have been hampered by a number of technological barriers in our attempts to mimic nature's extensive and efficient use of these unique catalysts.

One of the first major difficulties that early enzyme researchers faced was in obtaining preparations of enzymes that were sufficiently pure for basic experiments. The enzyme lysozyme, which is perhaps the most studied and best characterized of all enzymes, serves as a good model for this point. Although Alexander Fleming "discovered" lysozyme in 1922, it was not until the mid-1940s that adequate separation and purification techniques were developed to provide significant amounts of purified enzyme for research. After this barrier was overcome for lysozyme and other enzymes, and after a refinement of chemical and analytical techniques in the 1950s, there was a vast acceleration in enzyme research, resulting in a tremendous amount of basic knowledge. Primary amino acid sequences were determined, three-dimensional structures were observed, reaction mechanisms were elucidated, and enzymes were chemically synthesized. For the most part, these and other major advances have occurred only during the last two decades.

2.2. *Utilization of Enzymes*

Although the bulk of our knowledge concerning enzyme structure and function has been obtained in the relatively recent past, the use of enzyme extracts in practical applications in both medicine and industry can be traced back to the beginning of this century. In spite of the many years of practice, however, most enzyme applications, both past and present, can be characterized as unsophisticated processes utilizing inexpensive preparations of the simplest enzymes. In industry, for example, most uses are limited to crude preparations of microbial hydrolases in food, beverage, paper, textile, and leather processing (Wolnak, 1972). In medicine, several articles on trypsin appearing between 1908 and 1911 indicate an early interest in therapeutic enzymes, but true enzyme therapy probably dates from about 1950, if we discount the digestive aids (Abderhalden, 1961). As in industry, most therapeutic applications of enzymes are limited to simple hydrolases administered in treatment requiring a rather low level of specificity. Examples of current established therapeutic uses of enzymes are provided in excellent reviews by Innerfield (1960), Abderhalden (1961), and Wolf and Ransberger (1972).

The ultimate contribution of enzyme technology to humankind will occur when we are capable of utilizing the entire family of enzymes—oxidoreductases, transferases, hydrolases, lyases, isomerases, and ligases—in applications that take full advantage of their unique properties (i.e., their specificity, activity, and controllability). In industry this means utilizing enzymes to act with respect to specific bonds of key raw materials in the synthesis of a wide range of valuable products. In medicine it means the administration of specific enzymes to modify substrates that have been identified as the primary causes or effects of certain diseases. Despite the great advances that have occurred in enzymology, we have not yet reached the point where enzymes are widely utilized in these types of specific applications. The results of research conducted over the last few years, however, gives us cause to believe that we will soon overcome the remaining barriers that have inhibited such a development.

3. *Current Research and Future Vistas—Realization of the Potential*

3.1. *Industrial Applications*

Process economics has been a major factor in limiting the large-scale industrial use of enzymes. Enzymes for specific reactions are often very expensive to isolate and purify. In addition, in the soluble state they are essentially lost in the reaction mixture and, as such, are catalysts that cannot be regenerated or reused. The result is an expensive catalyst requirement that often makes the enzyme process prohibitively expensive. Something on the order of 1000 papers during the last 5 years describing various methods for binding many different enzymes to several different types of insoluble carriers have demonstrated, in theory at least, that the reuse problem is solved. The application of fundamental reaction engineering principles to the design and optimization of immobilized enzyme reactors for specific reactions will solve the problem in practice. The development of efficient enzyme reactors, plus the improvement of purification techniques and the design of better methods for cofactor utilization, should make the broad use of enzymes a reality in all phases of industry.

3.2. *Therapeutic Applications*

a. Limiting Factors. In medicine the limiting factors have been much more basic. Before a therapeutic enzyme can be used to act on a specific substrate, one must know what that substrate is and the role that it plays in a particular disease. Then it is necessary to obtain significant amounts of active, pure enzyme and to develop safe and efficient methods to deliver it to its site of action. Beyond that, according to Wolf and Ransberger (1972), the practical application of enzymes in the treatment of many different diseases depends on the experience of its application, the exact diagnosis, on continued monitoring, and on the interpretation of the results obtained. All these factors have been and continue to be inhibitors to the widespread application of specifically aimed therapeutic enzymes. They also represent key areas of current research.

b. Research into the Biochemical Basis of Enzyme Therapy. Extensive research conducted in the recent past to determine the biochemical bases of a number of diseases appears to be providing very encouraging results, especially for hereditary enzyme defects and certain types of cancer. With regard to the former, Brady *et al.* (1975) state that during the past 2 years there has been a significant shift in the emphasis of investigations in hereditary diseases from the elucidation of underlying enzymatic abnormalities to the therapy of these by enzyme replacement. In the cancer area a recent report from the Sloan-Kettering Institute (Hadden, 1974) describes current investigations of the neoplastic activities of asparaginase, glutaminase, phenylalinine ammonia lyase, carboxypeptidase G1, and other enzymes to degrade arginine, citrulline, thereonine, and isoleucine. Each of these enzymes acts by depriving a certain type of neoplastic cell of a significant amino acid nutrient. Regarding other abnormalities. Wolf and Ransberger (1972) state that in recent years a close relationship has also been established between enzymes and inflammation, fibrinolysis, thrombolysis, and metastasis. It is obvious from these and other reports that the biochemical basis of enzyme therapy has been and is being demonstrated for a number of diseases. For these diseases the limiting factors then become availability of the purified enzyme and the development of safe and effective delivery methods. These problems appear to be the major factors currently limiting the development of enzyme therapy.

c. Availability of Therapeutic Enzymes. The problems of availability and delivery have generally not been approached for a specific treatment until the need has arisen (i.e., only after the biochemical basis for enzyme therapy has been determined). As such, significant investigations in these areas have begun only recently. However, because of the recent successes in relating specific enzymes to specific diseases, they represent very active areas of research.

The difficulties involved in obtaining sufficient amounts of active, pure enzyme and the methods to overcome these difficulties are relatively straightforward. The methods are not so simple, however, and often require years to develop. If an enzyme is obtained from a bacterial source, such as *E. coli* L-asparaginase, the preparation obviously must be subjected to rigorous purification techniques to remove foreign proteins, pyrogens, and other toxic agents. These techniques could be quite tedious, and the resulting cost of the enzyme, for those who are not in a position to do their own purification, could be extremely high. For asparaginase, which has been studied intensely since the early 1960s, this problem is no longer a major one; the production and purification techniques have been developed and optimized, and there is currently an adequate supply of a reasonably priced enzyme. Researchers working with newer experimental enzymes, however, often face the same problems that were associated with investigations of asparaginase in earlier years—clinical trials are often limited by the amount and expense of available enzyme.

For enzymes that are derived from human or animal sources the immunological and toxicological problems are not as severe as with microbially derived enzymes; the major problem with these enzymes, however, is their scarcity. Often a very small amount of enzyme is present in a particular organ or body fluid, and techniques must be developed to recover most of it if sufficient amounts are to be provided. The experience of Roscoe Brady's group in its investigations involving the enzyme therapy of Gaucher's disease, a hereditary lipid storage disorder, provides an excellent example of a case in point. The yield of this group's complex procedure to purify human placental glucocerebrosidase was about 5% (Pentchev *et al.,* 1973), and, as a result, several months of tedious work were required to obtain sufficient quantities of the enzyme for single doses for two patients (Schmeck, 1974). Although the results were encouraging, continued therapy was prevented by limited supplies of the enzyme. The group is currently developing an affinity column for the enzyme since they feel that prospects of enzyme replacement studies will require more practical and efficient methods of isolation. Other researchers agree, and affinity chromatography appears to be the most promising new method for the preparation of therapeutic enzymes. As specific techniques are developed for each enzyme, the problem of availability will be improved somewhat. In all probability, however, other, novel methods will have to be applied to circumvent this problem satisfactorily.

d. Delivery of a Therapeutic Enzyme to Its Site of Action. In a soluble state therapeutic enzymes have been administered in external topical applications, in aerosol form, orally, and parenterally. The method of choice obviously depends strongly on on the type of disorder treated, but parenteral administration appears to be the most popular route of delivery, particularly via intravenous injection. This is especially true for the various types of cancer and the hereditary diseases which have shown so much promise lately as candidates for enzyme therapy.

A number of serious difficulties have been associated with the intravenous injection of soluble enzymes. Among the limiting problems are low active circulating times due to immunoinactivation, immunoprecipitation, and urinary elimination; toxic and antigenic side effects; and high dosage levels to compensate for low circulating times. In conjunction with limited availability and a high purity requirement for the enzyme, the last factor tends to make enzyme therapy extremely expensive. It also adds further complications in cases where toxic and immunological effects are observed. Every therapeutic enzyme administered in soluble form is probably severely limited by at least some of these problems; the key factor, however, may vary for each specific enzyme. For example, the major problems limiting soluble asparaginase therapy are rapid clearance rates from the systemic circulation (Wriston and Yellin, 1973) and toxic and antigenic side effects (Capizzi *et al.,* 1970; Oettgen, 1970). In the treatment of Gaucher's disease by the intravenous injection of glucocerebrosidase, Brady *et al.* (1975) feel that the probability of an anaphylactic shock is quite low. They postulate that this is so because most patients with lipid storage diseases contain some residual enzymatic activity and, furthermore, because intravenous injection of proteins is a poor route for the induction of antibodies (Heller and Siskind, 1973). Unfortunately, in this case, however, low circulating times again require large doses of very scarce enzyme.

To overcome the difficulties listed above, medical researchers have attempted to modify the form of injected enzymes in a number of different ways. Some methods involve an attempt to improve the resistance of the therapeutic enzyme to *in vivo* denaturation and degradation by increasing its molecular weight in the presence of stabilizing agents. Paillot *et al.* (1974) found that polymeric urate oxidase that had been cross-linked in the presence of albumin demonstrated a longer *in vivo* activity when injected into hens than did the monomer. Abuchowski *et al.* (1974) found that the covalent attachment of polyethlene glycols to available but nonessential groups renders enzymes nonimmunogenic and extends blood circulating life. Shier and Trotter (1975) showed that conjugates of conconavalin A and L-asparaginase were retained in the footpads of mice 95 times longer than the unmodified enzyme; unfortunately, however, the conjugates were completely ineffective in the treatment of asparaginase-sensitive lymphosarcoma. Other attempts have been directed at increasing *in vivo* lifetime of the injected enzyme by entrapping it within protective shells such as liposomes (Gregoriadis, 1975), erythrocyte ghosts (Ihler *et al.,* 1973) and synthetic microcapsules (Chang, 1974). The objective of the first two systems is to protect the enzyme until it reaches its site of action in liver, spleen, or other tissues. The last system allows the entrance of low-molecular-weight substrates, but not antibodies or other proteins, while it retains the therapeutic enzyme in the peritoneal cavity or the interstitial space.

All these methods are essentially still in experimental stages. Some of them are extremely promising, but it is too early to tell which may play a major role in establishing enzyme therapy on a much broader basis. Nevertheless, it is important that research involving these and other novel methods of delivery continue, so that the enzyme therapist may have a number of different vehicles in hand to meet the specific requirements of a particular disease.

The methods discussed in this section have one common disadvantage (i.e., they require repeated injections of large doses of the enzyme). Although they improve the situation, they may not satisfactorily solve the problem of availability. Also, at this time it is still not clear whether these methods can prevent the harmful effects that may be associated with the large amounts of enzyme that will be injected during the course of long-term therapy. Recent work shows that immobilized enzyme reactors may be able to solve these problems, and thereby assume an important position among potential delivery methods for therapeutic enzymes.

e. Immobilized Enzyme Reactors as Vehicles for Administering Therapeutic Enzymes. For the purpose of this discussion we shall define an immobilized enzyme reactor as any modular unit that contains within it an enzyme that is stably bound to an insoluble carrier. The carrier may exist in a number of different forms, such as a spiral-wound membrane, solid particles, tubes, and plates. The reactor may contain a protective shell around the carrier, or it may be constructed of the carrier material itself, as in the case of enzymes immobilized on the inside surfaces of tubes.

In its various forms an immobilized enzyme reactor can be utilized to administer therapeutic enzymes in at least two different ways: (1) by perfusing a patient's blood through an extracorporeal circuit that contains the reactor, and (2) by surgically implanting the reactor as a vascular prosthesis. In either case this method of delivery could offer a number of significant benefits to enzyme therapy. Immunological reactions to the foreign enzyme protein would be reduced by preventing the free circulation of enzyme molecules and, possibly, by masking antibody binding sites. Toxic effects would be lowered by the ability to remove toxic agents from the carrier while retaining the enzyme, or by masking the activity of toxic agents. Also, it might be possible to concentrate the effects of the therapeutic enzyme by implanting it in an optimum location. An additional benefit of an implantable device would be the ability to administer continuous, long-term treatment. Finally, the cost of treatment would be decreased by the ability to use a small amount of enzyme continuously or intermittently for long periods of time. This would be especially valuable for studies involving new experimental enzymes which have limited availability or high cost.

During the last few years a number of researchers have realized that immobilized enzyme reactors could greatly reduce the complications associated with soluble enzyme treatment, and have investigated many carrier materials in several different forms (Broun, 1974). Much of this work has been conducted with L-asparaginase which has been bound to nylon tubes (Allison *et al.*, 1972), poly(methyl methacrylate) sheets (Sampson *et al.*, 1972), polycarboxylic gels (Horvath *et al.*, 1973), Dacron (Cooney *et al.*, 1975), and other materials. These methods have generally suffered from low specific enzyme activity, relatively poor long-term enzyme stability, and poor biocompatibility. The three factors of activity, stability, and biocompatibility are of paramount importance in the success of immobilized enzyme therapy and depend strongly on the carrier material that is used. To date, no system has been successful enough to warrant an extensive clinical investigation.

In this paper we propose reconstituted collagen as a carrier material for therapeutic enzymes and present preliminary clinical results with L-asparaginase to support our proposal. In discussing the development of this material, the fabrication of the enzyme complex, the design of the reactor, and the application of the system, we shall outline the important factors that must be considered in the general development of any immobilized enzyme reactor for therapeutic applications.

4. *Collagen–Enzyme Complexes as Vehicles for Therapeutic Enzymes*

4.1. *Collagen as a Carrier for Enzymes*

During the last 5 years we have conducted a major study in our laboratory to investigate the potential of reconstituted collagen as a matrix material for the binding of various enzymes. In our work we have successfully immobilized over 20 different enzymes and have demonstrated that collagen has a number of advantages as an enzyme carrier (Bernath and Vieth, 1974). It is readily available from a number of sources, especially cattle and fish, and is inexpensive. The immobilization technique is simple, requiring only collagen, the enzyme, buffers, and in some cases a bifunctional cross-linking agent (Vieth and Venkatasubramanian, 1976). In addition, immobilization takes place at room temperature and in a mild, aqueous environment. Since collagen is a protein, it has a large number of polar and nonpolar amino acid residues which provide sites for strong cooperative noncovalent interactions between the enzyme and carrier. The hydrophilic, proteinaceous nature of the carrier also tends to have a stabilizing effect on the bound enzyme (Balaji, 1975). Collagen has a high swelling capacity, which provides an aqueous environment for the immobilized enzyme and reduces internal diffusional resistances. Finally, its fibrous nature provides a strong material that can be cast into a number of forms, and its chemical nature enables a wide variety of characteristics via controlled chemical modification.

We first bound *E. coli* L-asparaginase to reconstituted bovine collagen in 1973 (Venkatasubramanian *et al.,* 1974). The asparaginase–collagen preparation exhibited a high stable limit of activity and good reusability characteristics under *in vitro* testing conditions (retention of full activity over 4 months of storage at 4°C with intermittent use at 37°C with 4 mM L-asparagine at pH 8.5 as the substrate). At one point during the 4-month testing period the collagen–asparaginase membrane was subjected to a continuous reaction study. Figure 1 shows that after 5 days of operation the immobilized enzyme preparation had retained essentially all its initial activity. These activity and stability characteristics are superior to those of other immobilized forms of the enzyme which have been reported in the literature.

4.2. Collagen as a Biomaterial

Collagen has been extensively investigated as a biomaterial, and in its various forms it has found numerous clinical applications (Chvapil *et al.,* 1973). Some of its

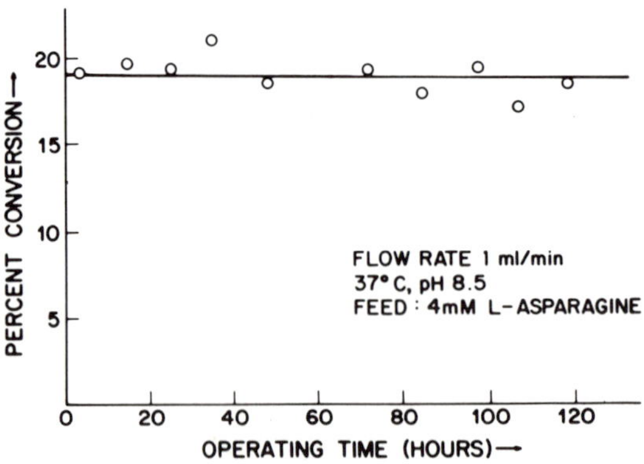

FIGURE 1. Stability of an asparaginase–collagen reactor during long-term continuous reaction studies.

uses include employment as a hemodialysis membrane, a suture material, a wound dressing, vascular prosthesis, and tubing for reconstruction surgery of hollow organs. Mechanically, collagen has demonstrated high tensile strength and low extensibility. It is easily fabricated into a number of medical products and shows excellent handling and reproducibility qualities. Furthermore, its versatile chemical nature allows controlled modification of semipermeability, solubility, swelling, and resorption characteristics.

Despite its biological origin, collagen is a weak antigen. Since its antigenic determinants probably lie within the randomly coiled telopeptide regions of the molecule, protease treatment of collagen has been successful in diminishing its antigenicity even further (Nishihara *et al.*, 1967). Bifunctional cross-linking agents also appear to be effective in reducing collagen's immune response (Miyata *et al.*, 1971). According to Stenzel *et al.* (1974) collagen antibodies have never been detected in patients' sera following implantation of cross-linked collagen biomaterials.

One of collagen's most valuable characteristics as a biomaterial is its extremely versatile thrombogenic nature. In the form of a flour or sponge it is an excellent hemostatic agent (Chvapil *et al.*, 1973). This is due primarily to its highly selective affinity for blood platelets which adhere to regularly oriented surface polar groups, to its ability to activate Hageman factor XII (probably via free carboxyl groups) and to its accessible hydroxylysine-bound galactose or glucosyl-galactose residues, which interact with the membrane-bound glucosyl transferase of human platelets. On the other hand, collagen products which come in direct contact with the blood may be fabricated in a relatively nonthrombogenic form by complexing with heparin (Rubin *et al.*, 1968) or by tanning with dialdehyde cross-linking agents (Stenzel *et al.*, 1974). Heparinized collagen does not activate factor XII but it does retain its affinity for platelets. Accoring to Salzman (1971), however, the platelet layer that forms on the surface of vascular implants soon disappears, and the surface that remains resists thrombus formation very effectively, even in low shear areas. Cross-linking with dialdehyde agents is probably effective because it blocks free amino groups (Wilner *et al.*, 1968) and limits the accessibility of galactose or glucosyl-galactose residues.

Collagen's excellent mechanical, nonantigenic, and nonthrombogenic properties are perhaps best demonstrated by the successful use of bovine arterial heterografts for the creation of arteriovenous fistulas in hemodialysis patients. These grafts, which are essentially dialdehyde cross-linked collagen tubes derived from carotid bovine arteries, were developed by Rosenberg *et al.* (1955). In 1971 Chinitz *et al.* (1972) were the first to utilize bovine collagen arterial heterografts in obtaining vascular access in hemodialysis patients. Since that time several other groups have successfully used these grafts for the same purpose, and their utilization has become the method of choice for vascular access in aged, diabetic, and chronically ill dialysis patients with severe peripheral vascular disease (Katzman *et al.*, 1975). The heterografts, which are approximately 6–9 mm in diameter and 18–46 cm in length, are generally implanted in the thigh or forearm and often accommodate a flow rate of 500 ml/min or more.

5. *Clinical Studies of Collagen-Immobilized* L*-Asparaginase Reactors: Reduction of Canine Serum Asparagine Levels*

5.1. *Rationale and Objectives*

It has been demonstrated independently that reconstituted bovine collagen is an excellent carrier material for immobilized enzymes and also an excellent biomaterial. It is our primary objective to take advantage of this dual capability by binding therapeutic enzymes to collagen and incorporating the resulting complexes into reactors that can be effectively utilized in the treatment of enzyme-sensitive diseases.

Our preliminary studies, which we describe here, involve the treatment of healthy nonleukemic dogs by extracorporeal perfusion through a collagen–L-asparaginase reactor. We chose to work with the enzyme L-asparaginase for a number of reasons. First, the enzyme has been extensively studied in its soluble form and a large number of data are available for direct comparison with our system. Second, intravenous injection has proved clinically ineffective, owing to the enzyme's toxicity and rapid clearance from the systemic circulation. Administration of collagen-immobilized asparaginase could bypass these limitations and provide an effective alternative for the treatment of acute lymphocytic leukemia in humans. The enzyme has been bound to a number of other carriers; while some of these systems are effective, others have low specific activity and relatively poor stability and biocompatibility. Data from these studies are also available in the literature for comparison with our results. Finally, L-asparaginase acts on a substrate that is contained in the systemic circulation as well as within internal tissue and bond marrow cells. As such, it provides an excellent general model for evaluating the clinical effectiveness of immobilized enzyme reactors in the administration of enzyme therapy.

The ultimate test of our system, of course, must go beyond an evaluation of its ability to deplete its substrate in the bloodstream; it must also be able to induce remission of the disease. The second phase of our studies, therefore, will involve the treatment of leukemic animals preliminary to the use of the collagen–asparaginase system in human therapy. Additional goals include the development of a collagen–enzyme vascular prosthesis as a means for continuous long-term therapy and the immobilization of other therapeutic enzymes.

5.2. *Materials and Methods*

a. Immobilization of L*-Asparaginase on Collagen Membranes. Escherichia coli* L-asparaginase was a generous gift of Merck, Sharp and Dohme, West Point, Pa. (Lyovac) and E. R. Squibb & Sons, Inc., New Brunswick, N.J. (Aspinase). Comminuted collagen from cowhide was obtained from the U.S. Department of Agriculture, Eastern Regional Research Center, Philadelphia. The details of the macromolecular complexation technique for immobilization have been presented elsewhere (Vieth and Venkatasubramanian, 1975). In brief, approximately 2000 IU of the soluble enzyme was added to 15.0 g of reconstituted bovine hide collagen dispersion, preswollen at pH 4.2 by the addition of lactic acid. The asparaginase–collagen mixture was homogenized in a Waring blender, and the resulting dispersion was deaerated for 2 hr under a vacuum of at least 25 in. of mercury. This dispersion was then cast on a Mylar sheet, and its wet thickness was uniformly adjusted to 0.25 mm by means of a mechanically operated Gardner knife. After drying, the membrane, which had an average thickness of approximately 0.1 mm, was manually removed from the Mylar support and immersed in a 1.0% solution of glutaraldehyde (pH adjusted to 7.0 by addition of $NaHCO_3$) for 30 sec. Any excess glutaraldehyde was removed by immediately washing the membrane for a period of 2 hr under cold running tap water.

b. Construction of the Biocatalytic Reactor Module and Extracorporeal Flow Circuit. The collagen–asparaginase membrane was placed over a meshed Vexar material and the two layers were tightly wound around a ⅛-inch-diameter Teflon rod (Figure 2). The Vexar mesh served as a spacer element and prevented the overlapping of contiguous layers of the membrane. The wound membrane was placed in a cylindrical tube constructed of Lexan, a polycarbonate plastic of proved biocompatibility. The reactor shell was 1 in. inner diameter, 6 in. long, and contained approximately 11 g (dry weight) of the collagen–enzyme complex. Caps, fitted with neoprene rubber "O" rings and measuring ¾ in. in length and 2 in. in diameter, were also constructed of Lexan. A hole was bored in each cap to gain entrance to and exit from the cylindrical reactor module. These holes were bored in a radial direction and tapered the width of the cap in an effort to

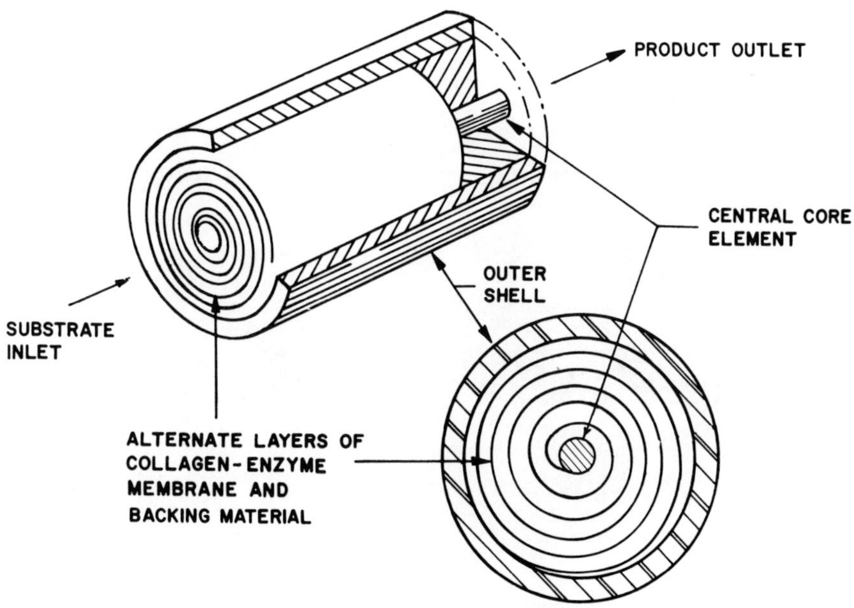

FIGURE 2. Spiral-wound membrane element housed in a reactor module. The central core is a Teflon rod, the backing material is meshed Vexar plastic.

minimize the possibility of turbulent eddy regions and thereby inhibit thrombus formation at these sites. Quarter-inch plastic connectors with lure-lock fittings were inserted into the holes and fixed with an acrylic cement. Silastic tubing was fitted over these connectors to place the reactor module in series with the rest of the extracorporeal flow circuit. Sampling valves were attached to the lure-lock fittings of the tubing connectors at both ends of the module. A picture of the reactor is shown in Figure 3.

The entrance of the reactor was attached by ¼-in. Tygon tubing to a peristaltic pump and a ¼-inch Biotronex extracorporeal electromagnetic flow transducer. The flow transducer was wired to a Biotronex BL-610-A, double-channel, pulsed-logic flowmeter. The exit line of the reactor was attached to a debubbler to remove any air bubbles from the blood before returning to the animal. The ¼-in. tubing at both ends was tapered through connectors to IV tubing, which was fitted into 18-gauge Abbocath-T catheters to complete the extracorporeal circuit, as shown in Figure 4.

c. In Vitro Activity Assay. The *in vitro* activity of the immobilized enzyme reactor was assayed by a Nessler's reagent technique, which measured production of ammonia, a product of the hydrolysis of asparagine. The module was operated

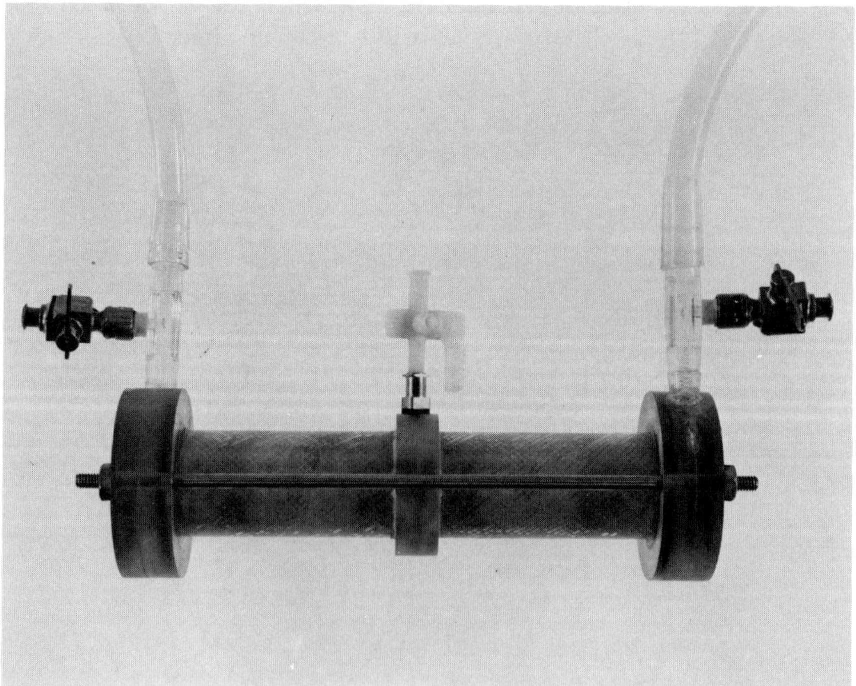

FIGURE 3. Reactor module (1 in. in diameter and 6 in. long) containing 11 g of collagen–asparaginase membrane complex.

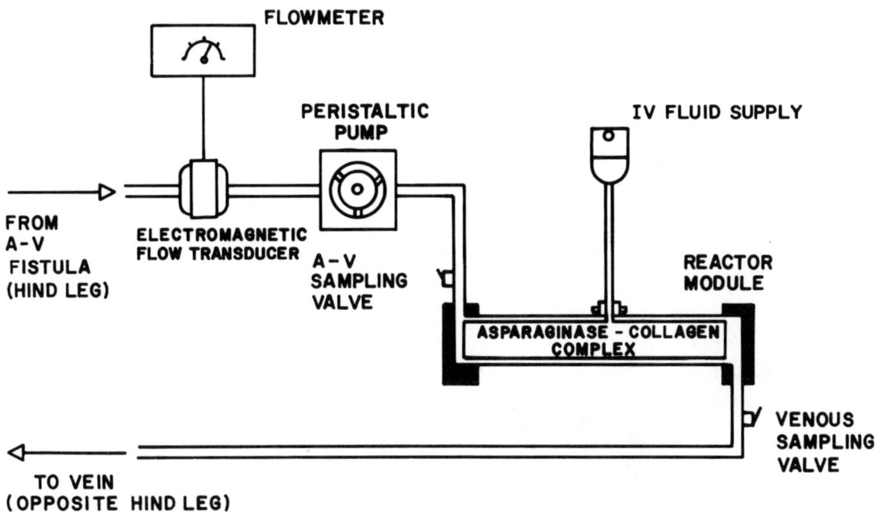

FIGURE 4. Schematic of the extracorporeal flow circuit. Intravenous fluid supply is utilized to prime the device and is disconnected during perfusion.

as a recycle reactor utilizing a substrate of 0.02 M asparagine (Sigma Chemical Co.) in 0.05 M Tris buffer at pH 8.5 and a temperature of 37°C. A peristaltic pump was used to maintain a circulating flow rate of approximately 50 ml/min. The substrate reservoir contained 500 ml of substrate, which was continuously mixed by a magnetic stirring device. The void volume of the reactor was 28 ml.

In a recycle reactor, the substrate spends only a fraction of the experimental time in the reactor. Therefore, the observed initial reaction rates, based on the concentration history of the reservoir, should be corrected to true catalytic activities by applying a correction factor which accounts for the residence time distribution between the reservoir and the reactor. For a perfectly mixed system the true reaction rate (volume basis) is given by (Constantinides *et al.*, 1973)

$$r_t = r_{obs} \frac{\tau_{Res} + \tau_R}{\tau_R}$$

where

r_t = true reaction rate

r_{obs} = observed reaction rate (initial slope of ammonia concentration versus time, μmoles/ml min)

τ_{Res} = residence time in the reservoir (volume of reservoir divided by the flow rate)

τ_R = residence time in the reactor (fluid volume of the reactor divided by the flow rate)

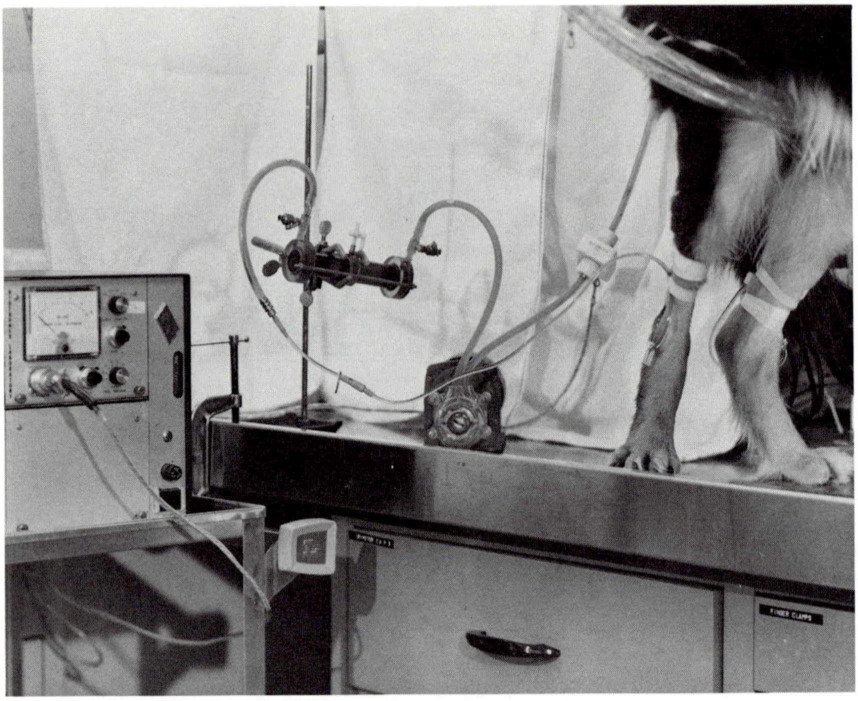

FIGURE 5. Perfusion apparatus in operation. The flowmeter is to the left.

d. Perfusion Studies. Eight healthy mongrel dogs weighing between 12 and 20 kg were prepared for perfusion by establishing in each animal an ateriovenous fistula between the saphenous vein and artery of a hind leg. The conscious dogs were secured in a modified Pavlov apparatus and heparinized with 7000–10,000 IU of sodium–heparin (Sigma Chemical Co.). One-half of the heparin was injected intravenously before the perfusion and the remainder was infused into the reactor with a lactate Ringer's solution which was used to prime the circuit. A catheter fitted with a sampling valve was inserted in the arterialized vein of the hind leg of the animal above the site of the arteriovenous fistula and an initial blood sample was withdrawn. This catheter was connected to the inlet line of the reactor and the outlet line was connected to a similar catheter inserted in a normal vein in the opposite hind leg. A picture of the experimental setup is shown in Figure 5. Blood was pumped through the circuit at a rate of 35–80 ml/min. Perfusions were run for 1–2 hr, during which time periodic samples of blood were withdrawn from the catheter sampling valve. These samples were used to measure serum asparagine and aspartic acid and to assess blood damage. After the perfusions, blood samples were obtained from the systemic circulation to determine the rate of recovery of asparagine and blood characteristics. After use, the module was cleaned by washing under cold tap water, infused with a saline solution of a broad-spectrum penicillin (Procaine Penicillin, G-Rugby Lab, New York, or Bicillin-C-R, Wyeth Labs, Philadelphia) and stored at 4°C. To examine any potential problems due to the device, eight control perfusions were performed with a blank collagen membrane before initiation of the immobilized enzyme studies. Also, three perfusions were run with no collagen membrane in the reactor to assess the blood damage characteristics of the tubing, reactor shell, and spacer material.

e. Blood Analysis. Serum asparagine and aspartic acid levels were assayed by the use of a Beckman Model 120 Amino Acid Analyzer according to a modified technique of Benson *et al.* (1967). Blood damage due to treatment with the extracorporeal reactor circuit was assessed by performing RBC, WBC, and platelet counts; checking hematocrit values; and measuring hemoglobin and fibrinogen content. Initially, analyses were performed manually at the laboratories of the Department of Surgery, Rutgers Medical School, CMDNJ. Later, blood samples were automatically processed by the Department of Hematology, Middlesex General Hospital, New Brunswick, N.J. Osmotic fragility and filterability of red blood cells were determined at the Department of Hematology, Rutgers Medical School, CMDNJ.

5.3. Results

a. Activity of Collagen–Asparaginase Complexes

In Vitro. The collagen–asparaginase membrane exhibited pseudo- zero-order kinetic behavior (i.e., a plot of ammonia concentration versus time was essentially linear for a major part of the reaction). Typical results for a new membrane are shown in Figure 6. After initial washing with buffer and a few batch contacts with

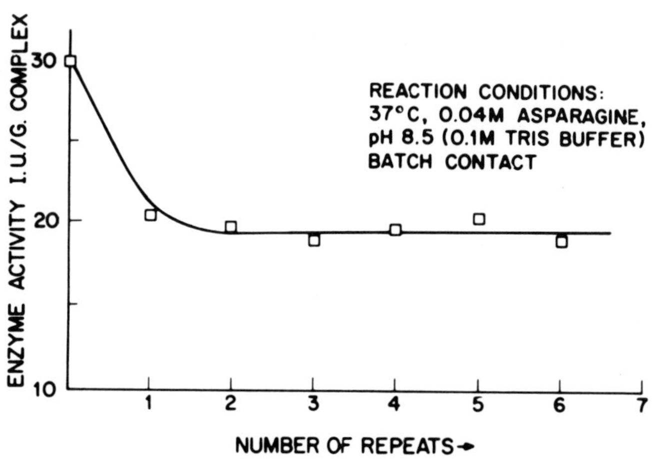

FIGURE 6. Enzymatic activity of a collagen–asparaginase complex when subjected to repeated batch runs.

fresh substrate, the membrane reaches a stable limit of activity which is maintained for a large number of repeated contacts. The level of the stable activity is strongly dependent on the concentration and time of glutaraldehyde treatment. The data in Figure 6 correspond to cross-linking conditions of 10% glutaraldehyde for 30 sec. For 1% glutaraldehyde and 30-sec treatment, initial activities are as high as 40.0–60.0 IU/g of collagen, with corresponding improvements in the level of stable activity. Other important variables are the ionic strength of the immobilization and cross-linking baths, and the level of drying after casting the membrane. High ionic strength and incomplete drying each result in decreased membrane activities.

The apparent K_m of the collagen–asparaginase complex (measured at pH 8.5 and 37°C) is 8.33×10^{-4}M. The pH-activity profile of the immobilized enzyme is very similar to that of soluble asparaginase, the pH optimum occurring between 8 and 8.5 for both cases (Venkatasubramanian *et al.,* 1974).

In Vivo—Reduction of Canine Serum Asparagine. Over a period of 4 months 13 perfusions were run with four different animals. The collagen–asparaginase membrane used for these experiments demonstrated a stable activity of 25 IU/g of collagen when subjected to *in vitro* assays as described above. The extracorporeal reactor, which contained 11 g of the collagen–enzyme complex, therefore contained approximately 275 IU of asparaginase activity.

The reduction of serum asparagine levels for a typical perfusion run are shown in Figure 7. Initial levels of asparagine varied between 20 and 80 nmoles/

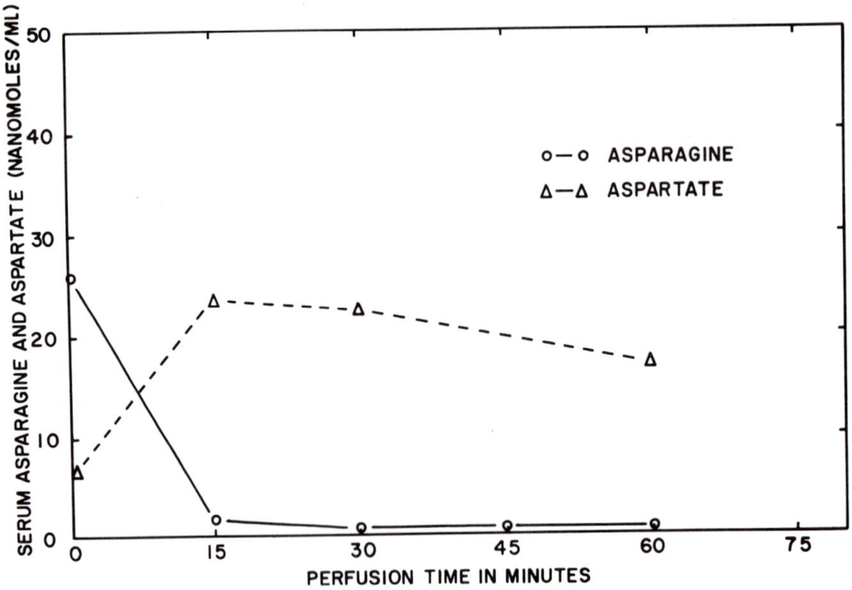

FIGURE 7. Serum asparagine and aspartate levels as measured in blood obtained from an arteriovenous sampling valve of a circuit inlet line (13th perfusion, dog J-6). Typical results.

ml. These were regularly reduced to trace levels (less than 0.5 nmole/ml) in 15–30 min. A corresponding increase in aspartic acid was observed also. There appeared to be some variability in the rates of rebound of serum asparagine following the perfusion, even for the same dog. For example, when 24-hr recovery samples were measured in dog J-6 for three different perfusions in which initial asparagine levels of 20, 34, and 80 nmoles/ml were reduced to trace levels in 30 min, serum asparagine levels were approximately 80%, 0%, and 0%, respectively, of the initial values. One other 24-hr recovery sample that was run using another dog showed a rebound to about 17% of initial serum asparagine levels.

b. Stability of Collagen–Asparaginase Complexes

In Vitro. Figures 1 and 6 represent data for the same collagen– asparaginase membrane. In addition to these experiments, this membrane was assayed periodically while being stored at 4°C over a 4-month period, and showed no significant loss in activity during that time.

In order to determine whether elution of asparaginase from the collagen–enzyme membrane is occurring, *in vitro* assays were halted during the course of the reaction by stopping the circulating pump and removing the enzyme reactor from the circuit. When this was done there was no significant increase in ammonia produced in the substrate reservoir during the next 60 min. In order to test for the possibility of autohydrolysis of the substrate, asparagine was incubated under the reaction conditions and samples were analyzed for ammonia by the Nessler's technique. Essentially no ammonia was produced during the 2-hr period of analysis.

In Vivo. The excellent stability characteristics of the collagen–asparaginase membrane that was used for perfusion trials are shown in Figure 8. Each point represents an *in vitro* assay between perfusions. During a 4-month period this membrane was used in 13 perfusions. After each perfusion, the reactor was washed in cold tap water, placed in a broad-spectrum penicillin solution, and stored at 4°C. Periodic *in vitro* assays during the time of study showed no apparent loss of activity.

c. Biocompatibility of the Collagen–Asparaginase Extracorporeal Reactor. The effects of extracorporeal perfusions on the dogs' blood are shown in Figure 9. Red blood cell counts, hemoglobin, hematocrit, and fibrinogen levels were essentially unchanged during and following the perfusions. In addition, osmotic fragility and filterability of red blood cells remained unaltered. Platelet counts were decreased on the average of 24% over a 2-hr perfusion, but returned to initial values within 24 hr following termination of treatment.

There was approximately a 48% drop in white blood cells observed after 1 hr of perfusion. These levels began to increase during the second hour of the perfusion, and after 24 hr following the treatment generally rebounded to values that were equal to or as much as 50% greater than initial readings. Follow-up studies indicate that white blood cell counts rebound to initial levels within 3 hr

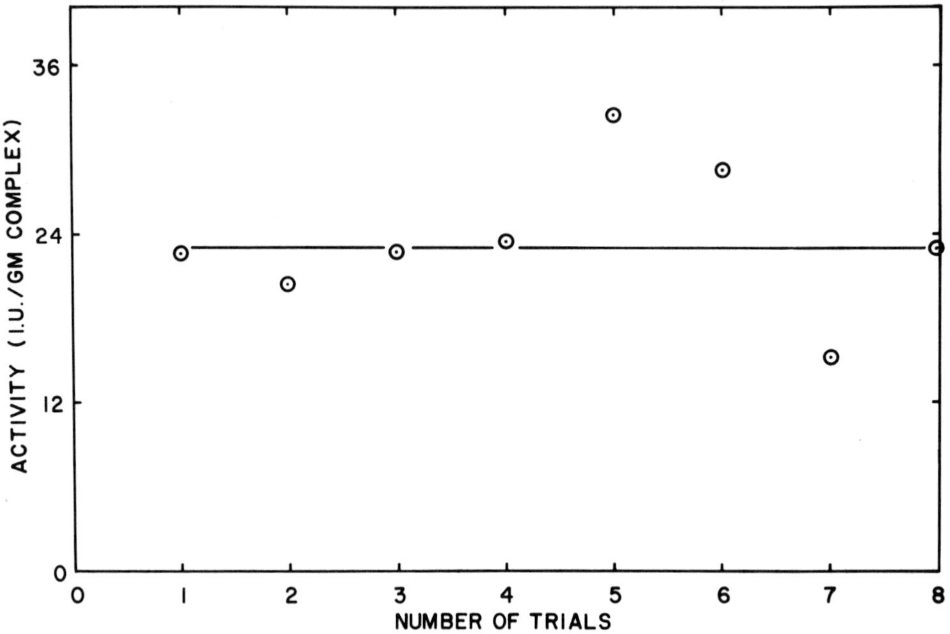

FIGURE 8. *In vivo* stability characteristics of the collagen–asparaginase reactor. Points represent *in vitro* assays between perfusions. Thirteen perfusions and 8 *in vitro* assays were performed over a period of 4 months.

following a perfusion. In some cases these levels are maintained during the following 44-hr period; in others, the counts continue to increase during an 8- to 12-hr period following treatment, remain constant for a period of 12 hr or so, and then decrease to original values during the next 24 hr. A differential count showed that lymphocytes were unaffected, and that the drop was due primarily to loss of polymorphonuclear leukocytes. This drop was almost the same when the collagen membrane was removed, indicating that white blood cells are being adsorbed on other components of the system, such as the Vexar spacing element, Tygon tubing, or Lexan reactor shell.

Complications due to the perfusion procedure were minimal. Vomiting and nausea were observed following approximately 30% of the perfusions. Dogs suffered signs of chills in two cases and moderately sized hematomas resulted in four instances. Also, weakness was detected in the dogs during four runs. All complications were temporary in nature, and follow-up observations detected no further signs of these disorders or any other antigenic responses after the treatments. Any substantial blood loss suffered by the animals was compensated for by the addition of an equal amount of lactate Ringer's solution. Clotting complications and mechanical difficulties were minimal and were usually attributable to insufficient levels of heparinization, a malfunction of the flowmeter, the leakage of air into the flow circuit, or improper cannulation of the veins.

The only maintenance necessary for the reactor was flushing with cold tap water to remove any residual blood deposits, and the use of penicillin infusions to prevent bacterial contamination during storage. When removed from the housing after 10 consecutive perfusions with four different animals, the membrane was completely intact. No stable clot formations, discoloring, detrimental protein deposits, or wear flaws were noticeable at any point on the membrane.

d. Preparation of an Implantable Collagen–Asparaginase Vascular Prosthesis. To develop a system that would enable continous long-term enzyme treatment and would serve as a model for enzyme replacement, we initiated studies aimed at preparing an implantable collagen–asparaginase vascular prosthesis. Bovine arterial heterografts (Artegraft) were generously supplied by Johnson and Johnson, New Brunswick, N.J. The Artegraft, which is essentially a highly cross-linked collagen tube, was chosen as a prosthesis because of its successful use in hemodialysis patients. We obtained the grafts before they were subjected to cross-linking, which is the final step in their preparation. The un-cross-linked Artegraft was immersed in a solution of L-asparaginase at pH 8.5 and 4°C for a period of 3 days, dried completely, and then immersed in 1% glutaraldehyde solution at pH 7.0 and room temperature for a period of 30 sec. Following the cross-linking step, the

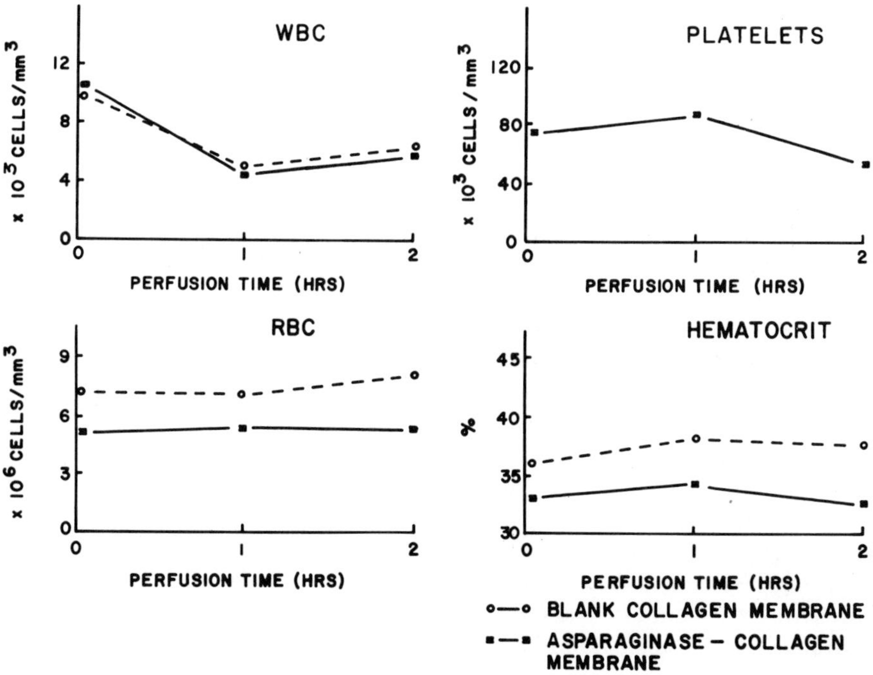

FIGURE 9. Blood damage characteristics of collagen–asparaginase and blank collagen (control) membranes for a 2-hr perfusion. Typical results. Average decrease in WBC and platelets for 13 perfusions was 48% and 24%, respectively.

graft was washed in cold tap water to remove unreacted glutaraldehyde. The method of immobilization described here was originally developed to bind enzymes to preformed collagen membranes and has been used by our group for a number of years (Bernath and Vieth, 1974).

The collagen–asparaginase heterograft was assayed in the recycle mode as described above, using 0.02 M asparagine in pH 8.5 Tris buffer at 37°C as the substrate solution. To date, the complex has been subjected to four batch assays during the period of approximately 1 week. During that time it has maintained an essentially constant activity of approximately 30 IU. This level of activity is quite substantial for a tube that measures approximately 9 mm by 26 cm, and as a result the system appears to be an excellent candidate for implant studies. Initially, the collagen–asparaginase graft will be used in extracorporeal perfusions to assess its asparagine-lowering and blood-damage characteristics. Following these studies, the graft will be implanted in a healthy dog and serum samples will be taken periodically to monitor asparagine concentrations and blood characteristics.

5.4. *Discussion*

a. *Activity Characteristics*

In Vitro. The stable enzymatic activities of collagen–asparaginase membranes prepared to date have measured between 25 and 40 IU/g of collagen. The binding mechanism has been discussed elsewhere (Bernath and Vieth, 1974). Briefly, we postulate that collagen binds large amounts of enzyme primarily by multiple, cooperative, noncovalent interactions between its own large array of reactive groups and those of the enzyme. These hydrogen bonds, electrostatic bonds, and hydrophobic interactions are of the same type that lead to stable protein–protein interactions between antibodies and proteins and also between protein subunits. Complete drying of the membrane after immobilization is necessary to allow collagen fibrils and enzyme molecules to approach very intimate intermolecular contacts and thereby form a very tightly knit stable matrix. High ionic strength decreases the activity of the immobilized enzyme preparation by interfering with electrostatic interactions and also, possibly, by causing dissociation of asparaginase subunits.

Glutaraldehyde treatment is used primarily to augment the mechanical strength of the membrane. It tends to reduce the initial activity of the immobilized enzyme preparation, probably due to unfavorable interactions with individual molecules such as binding near the active site or inducing inactivating conformational changes. However, it also acts to covalently tie together the entire matrix by introducing collagen–collagen and collagen–enzyme cross-links. As a result, glutaraldehyde treatment increases the stable activity of the complex (i.e., the activity measured after all loosely bound enzyme has been washed out of the matrix). It is extremely important to determine the optimum conditions of glutaraldehyde treatment (concentration, time, pH, and temperature) which will produce maximum stable activity while providing adequate mechanical and biocompatibility characteristics.

The apparent K_m of the complex (8.33×10^{-4} M) is approximately 80 times the K_m for soluble *E. coli* L-asparaginase (1×10^{-5}). According to Wriston and Yellin (1973), this parameter is extremely important in determining the oncolytic potency of an asparaginase preparation; high K_m values are related to low affinities for serum asparagine and therefore to low clearance rates at physiological substrate concentrations (10^{-5} M or lower). This may be true for soluble enzyme therapy where the rate of clearance of the injected enzyme from the systemic circulation is competing with the rate of asparagine reduction, but the potency of immobilized forms of the enzyme should not be so sensitive to the K_m. Strictly speaking, the K_m is not an affinity constant in most cases but only a measure of the substrate concentration which produces one-half of the maximum rate of reaction possible for a given amount of enzyme. For treatment by an immobilized enzyme, the total amount of enzyme activity in the reactor is the most important parameter. Maintenance of that activity over a long period of treatment is, of course, equally important.

In Vivo. The data from perfusion experiments indicate that collagen–asparaginase complexes are extremely efficient at reducing serum asparagine and maintaining it at low levels (<0.5 nmole/ml) for a significant time following treatment. There is no evidence of serum proteins depositing on the membrane and lowering enzyme activity. This was true even for preliminary runs using collagen–enzyme reactors that contained only 75 IU. In these cases pseudo-zero-order behavior was observed through 2 hr and approximately 45% conversion, indicating no decrease in reaction rate due to protein buildup or other complications during this extended reaction time. The kinetics of asparagine recovery following a perfusion are currently being studied in an attempt to correlate recovery time and levels to the length of treatment, initial rate of asparagine reduction, and other properties of the system.

b. Stability Characteristics. Collagen–asparaginase membranes are capable of maintaining their high levels of activity over long periods of intermittent storage and use, both *in vivo* and *in vitro.* No evidence of enzyme elution has been observed. Also, bacterial contamination and mechanical disintegration do not appear to occur during long periods of extended use. These characteristics are especially valuable in light of the observations that the same membrane can be reused many times with the same animal or with different animals. In addition to reducing the cost of enzyme therapy, this system should be very useful for studies involving new experimental enzymes which have limited availability and high cost.

c. Biocompatibility Characteristics. Extracorporeal perfusion through a circuit containing the collagen–asparaginase reactor appears to have a relatively mild effect on blood. The observed decrease in white blood cell and platelet counts, while not serious in healthy subjects, might be significant if this system were used to treat leukemic patients. White blood cells do not appear to be adsorbing to collagen, but the platelets do. Currently, we are studying the effects of various cross-linking techniques on reducing platelet adhesion to collagen. We are also investigating various sources of collagen, particularly more-purified forms, since there is probably a certain amount of platelet binding to impurities such as lipids, polysaccharides, and calcium embedded in the reconstituted collagen surface.

Complications due to the treatment were not serious and were often due to factors not related to the collagen–enzyme complex. Chills, weakness, vomiting have been considerably reduced by preheating the returning blood which had cooled significantly while traveling through the circuit, and by generally refining the techniques involved in the treatment procedure.

d. Reactor Design. The reactor configuration that was used (i.e., spiral-wound membrane on a protective backing material) is the same type that we have used in much of our work with other enzymes. The advantages of this type of reactor are as follows: (1) Good contact efficiency (i.e., low external mass-transfer resistance characteristics within the capillaric channels formed between contiguous layers of the membrane); this also serves to keep the collagen surface free from protein deposits, (2) Low pressure drop across the length of the reactor; this helps to reduce blood damage, (3) A high level of catalytic surface area per unit volume, resulting from packing a thin membrane; this characteristic lowers the effect of mass-transfer limitations internal to the enzyme membrane and increases the amount of effective enzymatic activity in the reactor.

e. Collagen–Asparaginase Bovine Arterial Heterograft. Our results indicate that we can bind a substantial amount of asparaginase to a standard Artegraft. The procedures involving implantation and utilization of the heterografts have been developed by numerous preclinical and clinical studies and should not provide serious problems in our experiments. The question that must be answered, however, is how long will the implanted enzyme prosthesis retain its activity. This will be the major focus of our initial investigation, which will involve long-term extracorporeal perfusions and implantation studies.

6. *Conclusion*

There is currently considerable interest in utilizing enzymes to treat a variety of diseases. In a few cases recent research has established the biochemical basis for enzyme therapy, and treatment strongly depends on the specificity of the enzyme to selectively act on a key substrate in the body. Lipid storage disorders and various types of cancers are two groups of diseases in which this type of selective enzyme therapy shows great promise.

The major factors limiting the development of enzyme therapy today are (1) the availability of sufficient amounts of purified enzyme for long-term treatment, and (2) methods for safely and efficiently delivering the enzyme to its site of action. These problems could be greatly alleviated by utilizing an immobilized enzyme reactor in an extracorporeal perfusion or by implanting an enzyme–carrier complex as a vascular prosthesis. In addition to preventing the free circulation of the enzyme, and therby reducing toxic and antigenic effects, immobilized enzyme therapy allows continuous long-term treatment with a relatively small amount of enzyme. This advantage results in lowering the cost of treatment and improving the availability of new experimental enzymes. The important factors determining the success of immobilized enzyme therapy are the activity of the carrier–enzyme complex, the stability or the retention of activity of the complex during long periods of storage and use, and the biocompatibility of the system. Each of these properties depends strongly on the carrier material that is used.

A number of different materials have been investigated as carriers for therapeutic enzymes. We have initiated studies utilizing reconstituted bovine collagen because of its excellent properties as an enzyme carrier and as a biomaterial. Both of these properties have been studied extensively in the past. Our studies with collagen-immobilized L-asparaginase demonstrate excellent activity, stability, and biocompatibility characteristics when administered to nonleukemic dogs via extracorporeal perfusion. As a result, this system shows great potential as a method for treating asparaginase-sensitive leukemias. At this point a project is being initiated to evaluate the use of collagen–asparaginase reactors in treating lymphocytic leukemia in humans. Collagen's outstanding characteristics also warrant experiments with other therapeutic enzymes.

The ultimate in enzyme replacement therapy or in continuous, long-term enzyme treatment would be the development of an implantable immobilized enzyme complex. The bovine collagen arterial heterograft, because of its proved success with hemodialysis patients and its ability to bind enzyme, is an excellent candidate for investigation as an immobilized enzyme vascular prosthesis. We have prepared heterografts with L-asparaginase activity and are currently evaluating them in *in vitro* and extracorporeal studies. The main focus of our work is aimed at determining the stability of this enzyme complex when used continuously for long periods of time. The second phase of this study will involve implantation in healthy dogs prior to its use in treating leukemic animals.

Immobilized enzyme reactors will be extremely valuable for the administration of therapeutic enzymes. In all probability, however, they will not be useful in treating all enzyme-sensitive disorders. For some diseases other methods of delivery, such as the injection of modified enzymes or enzymes contained within liposomes, erythrocyte ghosts, or synthetic microcapsules may prove to be successful alternatives to parenteral administration of soluble, unmodified enzymes. At this time research involving all of these novel methods of delivery is important. This work will provide the enzyme therapist with a variety of different vehicles to meet the specific requirements of a particular disease, and as a result will contribute significantly to improving the overall value and effectiveness of enzyme therapy.

ACKNOWLEDGMENTS

We thank Dr. Rene Joyeuse of the Department of Surgery, Rutgers Medical School, CMDNJ, for his general guidance and for providing his surgical expertise. The assistance of Dr. Parvin Saidi and Dr. H. C. Kim of the Division of Hematology, Rutgers Medical School, CMDNJ, in analyzing blood damage data is also gratefully acknowledged. We appreciate the assistance of Dr. R. Poretz of the Department of Biochemistry, Rutgers University, for providing serum asparagine assay results. Also, we would like to express our gratitude to Mr. Steven Topaz and to Mr. John Kehoe and his staff of the Department of Surgery, Rutgers Medical School, for their advice and help in the perfusion studies. We acknowledge with gratitude the financial support of the Charles and Johanna Busch Memorial Fund for Biological Research. Finally, we offer our appreciation to Steven Jefferies, Kevin McBride, and John Richards for their excellent technical assistance.

7. *References*

Abderhalden, R., 1961, *Clinical Enzymology,* D. Van Nostrand Company, Inc., Princeton, N.J.

Abuchowski, A., van Es, T., Palczuk, N. C., and Davis, F. F., 1974, Preparation and properties of nonimmunogenic catalase, *Federation Proc.* **33**:1317.

Allison, J. P., Davidson, L., Gutierrez-Hartman, A., and Kitto, G. B., 1972, Insolubilization of L-asparaginase by covalent attachment to nylon tubing, *Biochem. Biophy. Res. Commun.* **47**:66.

Balaji, S., 1975, Stability and activity characteristics of collagen-immobilized hesperidinase, M.S. thesis, Rutgers University.

Benson, J. B., Gordon, M. S., and Patterson, J. A., 1967, Accelerated chromatographic analysis of amino acids in physiological fluids, *Anal. Biochem.* **18**:228.

Bernath, F. R., and Vieth, W. R., 1974, Collagen as a carrier for enzymes: materials science and process engineering aspects of chemical engineering, in: *Immobilized Enzymes in Food and Microbial Chemistry* (A. Olson and C. Cooney, eds.), pp. 157–185, Plenum Press, New York.

Brady, R. O., Pentchev, P. G., and Gal, A. E., 1975, Investigations in enzyme replacement therapy in lipid storage diseases, *Federation Proc.* **34**:1310.

Broun, G., 1974, Trends in the use of immobilized enzymes and proteins in human therapeutics, in: *Enzyme Engineering* (E. K. Pye and L. B. Wingard, eds.), pp. 433–438, Plenum Press, New York.

Capizzi, R. L., Bertino, J. R., and Handschumacher, R. E., 1970, L-Asparaginase, *Ann. Rev. Med.* **21**:433.

Chang, T. M. S., 1974, Effects of different routes of *in vivo* administration of microencapsulated enzymes, in *Enzyme Engineering* (E. K. Pye and L. B. Wingard, eds.), pp. 419–424, Plenum Press, New York.

Chinitz, J., Yokoyama, T., Bower, R., and Schwartz, C., 1972, Self-sealing prosthesis for arteriovenous fistula in man, *Trans. Amer. Soc. Artificial Internal Organs* **18**:452.

Chvapil, M., Kronenthal, R. L., and VanWinkle, W., 1973, Medical and surgical applications of collagen, in *International Review of Connective Tissue Research,* Vol. VI (D. A. Hall and D. S. Jackson, eds.), pp. 1–61, Academic Press, Inc., New York.

Constantinides, A., Vieth, W. R., and Fernandes, P. M., 1973, Characterization of glucose oxidase immobilized on collagen, *Mol. Cellular Biochem.* **1**:127.

Cooney, D. A., Weetall, H. H., and Long, E., 1975, Biochemical and pharmacologic properties of L-asparaginase bonded to dacron vascular prostheses, *Biochem. Pharmacol.* **24**:503.

Gregoriadis, G., 1975, Enzyme carrier potential of liposomes in enzyme replacement therapy, *N. Engl. J. Med.* **292**:215.

Hadden, J. W., 1974, Report from Sloan-Kettering: enzyme therapy and nutritional depletion of tumors, *Clin. Bull.* **4**(2):72–74.

Heller, K. S., and Siskind, G. W., 1973, Effect of tolerance and of antibody mediated immune suppression on the avidity of the cellular and humoral immune response, *Cellular Immunol.* **6**:59.

Horvath, C., Sardi, A., and Woods, J. S., 1973, L-asparaginase tubes: kinetic behavior and applications in physiological studies, *J. Appl. Physiol.* **34**:181.

Ihler, G. M., Glew, R. H., and Schnure, F. W., 1973, Enzyme loading of erythrocytes, *Proc. Natl. Acad. Sci. U.S.* **70**:2663.

Innerfield, I., 1960, *Enzymes in Clinical Medicine,* McGraw-Hill Book Company, New York.
Katzman, H. E., Schild, A. F., and Vanderwerf, B. A., 1975, Bovine artegraft arteriovenous fistulas for hemodialysis in one hundred patients after conventional arteriovenous fistulas failed, *J. Thoracic Cardiovascular Surg.,* submitted for publication.
Miyata, T., Sohde, T., Rubin, A. L., and Stanzel, K. H., 1971, Effects of ultraviolet irradiation on native and telopeptide-poor collagen, *Biochim. Biophys. Acta* **229**:672.
Nishihara, T., Rubin, A. L., and Stanzel, K. H., 1967, Biologically derived collagen membranes, *Trans. Amer. Soc. Artificial Internal Organs* **13**:243.
Oettgen, H. F., 1970, Clinical evaluation of L-asparaginase, in: *Oncology,* Vol. II: *Experimental Cancer Therapy* (R. L. Clark, R. W. Cumley, J. E. McCay, and M. M. Copeland, eds.), p. 159, Year Book Medical Publishers, Inc., Chicago.
Paillot, B., Remy, M. H., Thomas, D., and Broun, G., 1974, Soluble crosslinked enzyme polymers: physicochemical properties, *Pathol.- Biol.* **22**(6):491.
Pentchev, P. G., Brady, R. O., Hibbert, S. R., Gal, A. E., and Shapiro, D., 1973, Isolation and characterization of glucocerebrosidase from human placental tissue, *J. Biol. Chem.* **248**:5256.
Rosenberg, N., Gaughran, E. R. L., Henderson, J., Lord, G. H., and Douglas, J. F., 1955, Use of segmental arterial implants prepared by enzymatic modification of heterologous blood vessels, *Surg. Forum* **6**:242.
Rubin, A. L. Riggio, R. R., Nachman, R. L., Schwartz, G. H., Miyata, T., and Stenzel, K. H., 1968, Collagen materials in dialysis and implantation, *Trans. Amer. Soc. Artificial Internal Organs* **14**:169.
Salzman, E. W., 1971, Role of platelets in blood-surface interactions, *Federation Proc.* **30**:1503.
Sampson, D., Hersh, L. S., Cooney, D., and Murphy, G. P., 1972, A new technique employing extracorporeal chemotherapy, *Trans. Amer. Soc. Artificial Internal Organs* **18**:54.
Schmeck, H. M., Jr., 1974, An enzyme used for rare disease, *The New York Times,* November 7, 1974.
Shier, W. T., and Trotter, J. T., III, 1975, Lectin–enzyme conjugates: potential site-specific agents for enzyme therapy, presented at 170th American Chemical Society National Meeting, Chicago, August 28, 1975.
Stenzel, K. H., Miyata, and Rubin, A. L., 1974, Collagen as a biomaterial, *Ann. Rev. Biophy. Bioeng.* **3**:231.
Venkatasubramanian, K., Vieth, W. R., and Bernath, F. R., 1974, Use of immobilized enzymes in blood treatment, in: *Enzyme Engineering 2* (E. K. Pye and L. B. Wingard, eds.), pp. 439–445, Plenum Press, New York.
Vieth, W. R., and Venkatasubramanian, K., 1976, Collagen-immobilized enzyme systems, in: *Methods of Enzymology: Immobilized Enzymes* (K. Mosbach, eds.), Academic Press, Inc., New York, in press.
Wilner, G. D., Nossel, H. L., and LeRoy, E. C., 1968, Aggregation of platelets by collagen, *J. Clin. Inves.* **47**:2616.
Wolf, M., and Ransberger, K., 1972, *Enzyme Therapy,* Vantage Press, New York.
Wolnak, B., 1972, *Present and Future Technological and Commercial Status of Enzymes, Report PB-219636, National Technical Information Service,* U.S. Department of Commerce, Springfield, Va.
Wriston, J. C., and Yellin, T. O., 1973, L-Asparaginase: a review, *Adv. Enzymol.* **39**:185.

24

Possible Roles of Enzymes in Development of a Fuel Cell Power Source for the Cardiac Pacemaker

SIDNEY K. WOLFSON, JR., LEMUEL B. WINGARD, JR., CHUNG C. LIU, AND SHANG J. YAO

1. *Introduction*

The Heart . . . moves of itself and does not stop unless for ever.
Leonardo da Vinci
Dell'Anatonia, ca. 1489 A.D.

Many types of "permanent" pacemaker devices and techniques have evolved since successful clinical application was first reported (Senning, 1959; Chardack *et al.,* 1960; Zoll *et al.,* 1961). These efforts have had an enormous impact upon the treatment of cardiac arrhythmias; however, they also have brought a host of new problems for the management of patients receiving pacers (Grendahl *et al.,* 1969; Goldstein *et al.,* 1970; Barold, 1973). Among the problems is that of providing a suitable power source for long-term trouble-free pacemaker function. This chapter will be devoted largely to discussion of one type of device potentially capable of solving this problem, namely, the biofuel or, more specifically, bioautofuel cell. Particular emphasis will be given to the role that enzyme catalysts might play. The term "bioautofuel cell" refers to a biofuel cell which can be implanted in the host and which then relies on the host for the delivery of fuel and the removal of wastes. Such an implantable fuel cell may incorporate enzymes to catalyze one or more of the reactions or to produce the fuel. A general introduction to cardiac pacing and a brief description of alternate power sources is included to place the enzyme-containing systems in perspective.

SIDNEY K. WOLFSON, JR. AND SHANG J. YAO • University of Pittsburgh School of Medicine, Pittsburgh, Pa. 15213 LEMUEL B. WINGARD, JR. • University of Pittsburgh Schools of Medicine and Engineering, Pittsburgh, Pa. CHUNG C. LIU • University of Pittsburgh School of Engineering, Pittsburgh, Pa. 15261

2. *Clinical Requirements*

The implanted electrical pacemaker is now the treatment of choice for heart block. This condition may be defined simply as an interference of propagation of a normal cardiac beat from the natural pacemaker, located in the right atrium to the ventricular muscle, which is responsible for the effective pumping action of the heart. The artificial pacemaker acts by providing a series of electrical stimuli or shocks directly to the ventricular muscle (myocardium). The ventricle then contracts in response to the stimulus.

The ideal pacemaker is small and light enough to be attached directly to the heart without leads. It has a source of energy which will last for its intended functional life and has the ability to permit the heart to respond to physiologic needs for increased cardiac output (as in the case of exercise). Of course, all its exposed components must be biocompatible, permitting it to exist within the body without interfering with normal physiologic processes. It must also be immune to attack by body tissues and fluids.

While the "ideal" pacemaker has not been developed, a number of variations in design and theory of operation have been tried. These variations can be subdivided into those relating to electrodes, mode of stimulation, circuit mode, and power sources.

2.1. *Electrodes*

The ultimate power requirement depends upon circuit design, lead losses, and electrode efficiency. The two major classes of electrode design are epicardial: directly attached to the exterior of the heart, and endocardial: located within the right ventricular cavity on the end of a catheter or semirigid lead. Epicardial electrodes usually are fixed to the cardiac surface by suturing. In fact, many electrodes are themselves metallic sutures which are insulated except for an intramuscular segment through which current passes to the myocardium. In other cases, small noble-metal electrodes protrude into the myocardium from a pad of reinforced silicone rubber or other plastic material that is sutured in place. Flexible leads connect the electrodes to the power unit located at a remote site (usually subcutaneous). Endocardial electrodes are frequently noble-metal rings or collars, placed near the tip of the catheter, that contact the endocardium by impingement. This is aided by the springiness of the catheter and by the fact that the ventricle is lined by a complex network of muscular trabeculae which tend to engage or entrap the end of a semirigid catheter. This method of contact is less constant than direct sutures and could result in variations of adequate stimulus threshold if the electrode position were to change over the millions of cardiac contraction cycles to be encountered (4×10^7 cycles) each year. The catheter or lead must exit from the vascular system through a peripheral branch vein and then be connected to the remote power unit located in a subcutaneous fat pad.

There are advantages and disadvantages of both electrode systems. Because of the direct access, epicardial leads can be affixed to the left ventricle, where the most efficient contraction is required. This is accomplished using thoracotomy, a serious and painful operative procedure. Recently, a "limited thoracotomy" has been described which is less involved but also less certain (Varriale and Naclerio, 1975). Endocardial electrodes avoid major surgery but must stimulate the right ventricle, with left ventricular contraction depending upon retrograde conduction through myocardium and conduction fibers that may not be healthy. The smaller size and direct placement of the epicardial leads should, in theory, lead to lower stimulus intensity and to lower power requirements. The endocardial type has the advantages of ease of placement from a peripheral access point, less risk to the patient, and of deployment under local anesthesia. A number of innovative electrode designs have been studied with varying success. Numerous advantages have been attributed to these, not the least of which is reduced polarization and greater power efficiency (Lewin *et al.*, 1967; Parsonnet *et al.*, 1969; Rasor *et al.*, 1972; Tyers *et al.*, 1974).

2.2. *Stimulus Characteristics*

While much work has gone into the definition of the proper stimulus waveform and pulse frequency, most systems are designed to provide a single square-wave pulse of 1–3 msec duration which delivers between 5 and 100 μJ to the cardiac muscle. Numerous combinations of voltage amplitude (4–8 V) and current (3–14 mA) have been employed. There are too many variables relating to interactions between specific pacemakers and implant conditions to attempt to define the stimulus intensity in more specific terms. The reader may explore this area further in the literature (Preston *et al.*, 1966; Fontaine *et al.*, 1968).

2.3. *Electrical Circuits and Power Units*

The pacemaker generator or power unit is composed of an electrical circuit and a power source or battery housed in a single case. The greatest volume and mass of the unit traditionally has been occupied by the power source. There are two primary circuit types: (1) asynchronous or fixed rate, and (2) synchronous or noncompetitive. Fixed-rate pacemakers are simplest in construction and operation and are most useful for temporary service. They are also employed when it is unlikely that a reasonably normal natural rhythm will return either spontaneously or as the result of therapeutic efforts. Because of its simplicity, the fixed-rate pacemaker has the advantage of being less prone to technical problems and failures.

Synchronous pacemakers are applicable to about one-third of the patients who require pacemakers. In this group, a reasonably normal rhythm is present between attacks of heart block (Adams-Stokes attacks). These patients usually have a normal natural atrial pacemaker with intermittent episodes of conduction blockade.

3. *Energy Sources and Needs*

Two major problems exist with respect to totally implantable pacemaker power sources. The first, and perhaps most important, is longevity. The other is that of size and weight. Although great strides have been made in reducing pacemaker mass, there is, as yet, no device small and light enough to be attached directly to the heart. Thus, long leads are needed and the possibility of lead fracture or distraction is introduced. Since it is now possible to produce microminiature circuitry, the development of a miniature power source could result in a completely self-contained pacemaker that *could* be attached directly to the epicardium or placed completely within the ventricle. A number of efforts have been made by numerous investigators to provide smaller and/or more long-lived pacemakers. Two categories of energy sources have been tried: (1) completely implanted systems with power stored in cells or derived from either nuclear or biologic energy sources; and (2) external power sources mated with implanted electrodes, via secondary cells or receiving coils and using electromagnetic induction or radiofrequency transmission (Lown and Kosowsky, 1970). Although mainly of historical interest when considering the permanent pacemaker, the direct-wire (percutaneous lead) pacemaker (Lillehei *et al.,* 1960) is still used in temporary or emergency situations.

It is difficult to define precisely the actual power needs for the cardiac pacemaker. Requirements vary with the type of pacemaker, circuit design, and type of electrodes. The typical 10 μJ/pulse pacemaker that has been widely used would require 250–500 μW average continuous power after allowing for inefficiencies of electronic circuits. The use of newer, low polarization (Rasor *et al.,* 1972) or nonpolarizable (Lewin *et al.,* 1967) electrodes and the most up-to-date microelectronics could lower this to the order of 10–100 μW continuous power. The presently used and potentially more promising pacemaker power sources are described below.

3.1. *Electrochemical Cells*

The most commonly used energy source has been a battery consisting of four to six zinc-mercuric oxide cells (mercury battery). This cell has a predicted life of about 5 years and is capable of a reasonably constant output until more than 75% of its energy has been consumed. Moreover, a reasonably sized battery of four to six cells provides 5–8 V, enough to activate moderately efficient transistorized circuitry. In practice, surgical intervention for failure of this type of pacemaker has usually been required after about 2 years. More recently, effective lithium cells have been perfected and pacemakers which incorporate them are being implanted in patients. They have the advantage of longer life (5–10 years) and less weight (Griffith, 1975).

3.2. *Electromagnetic Induction with Rechargeable Power Cells*

One method of increasing the life of a pacemaker would be the use of rechargeable batteries with an implanted recharging circuit inductively coupled across the skin to a recharging coil. Several investigators have demonstrated the feasibility of this approach (Tyers *et al.,* 1974; Griffith, 1975, personal communication). However, it has been disappointing for several reasons. The available, rechargeable cells, such as Ni/Cd, can take only a limited number of charges before failure, which has limited the life of this cell to within the range of the mercury cell and certainly the newer lithium cells. Moreover, patients are frequently psychologically compromised by dependence on the external recharging device. In general, they have not received good patient acceptance (Griffith, 1975, personal communication). Other problems with this type include need for frequent recharging, production of heat, and damage from overcharging (Lown and Kosowsky, 1970).

3.3. *Radiofrequency Transmission*

Clinical experimentation with this type of pacemaker began before 1960 (Glenn *et al.,* 1959; Eisenberg *et al.,* 1961; Glenn *et al.,* 1966). A set of epicardial electrodes is connected to a receiving coil located subcutaneously on the chest wall. A transmitting coil must be continuously maintained in a precise location over the receiving coil on the skin surface. This external coil is then attached to a pacing unit that is worn by the patient. Here again we have a large measure of remote apparatus, some implanted and some to be carried. Early implants were hampered by electrode lead breakage (Widmann *et al.,* 1964).

3.4. *Radionuclide Sources*

a. Thermonuclear and Thermionic Reactors. These have been used clinically, initially in Europe but more recently in the United States. ^{238}Pu was chosen as the power source because of its ready availability and its long half-life (89.5 years). An alpha emitter, this source functions by supplying heat to an adjacent thermopile or thermionic generator. It has the advantage of long life, theoretically approaching 10 years, and adequate power levels. The ^{238}Pu reactor must be sturdy to protect the bearer and the environment. Thus, a hermetically sealed case of great strength must surround the reactor and its shielding. All this adds to the mass and requires remote placement with long leads. Moreover, the patient will need to be licensed and his comings and goings may be subject to regulation. Finally, this device is expensive: $5000 per unit at this writing. In spite of this, there are several groups presently undertaking clinical investigation of the devices produced by a number of manufacturers, including Medtronic and Arco Nuclear (Laurens, 1970; Hixson and Laurens, 1972; Purdy *et al.,* 1975; Parsonnet *et al.,* 1975). Nonetheless, this is an exciting development and the first clinically successful major innovation in pacemaker power supplies to be seen for some time.

b. Beta-Voltaic Cell. This nuclear device was originally of even greater interest than the ^{238}Pu power source because it uses a beta emitter (e.g., ^{147}Pm), which in its pure form is not dangerous and should not require special shielding and stringent containment provisions. It utilizes direct energy conversion of radiation to electricity by coupling the emitter to a semiconductor material. These units are under active investigation and have been tested clinically (Huffman and Norman, 1974; Griffith, 1975, personal communication; Martinis, 1975; Parsonnet *et al.,*

1975). There are several drawbacks, however. ^{147}Pm as obtained from the reactor is contaminated with ^{146}Pm, a high-energy gamma emitter. Heavy shielding would thus be required for the impurity. Purification is possible but very expensive. Moreover, ^{147}Pm has a short (2.6 years) half-life, providing a maximum implant life of 7–8 years. Finally, the advent of the newer lithium and even the more advanced Zn–Hg cells has extended their life span to a range competitive with ^{147}Pm but without any radiation hazard.

3.5. *Biological Energy Sources*

These may be either mechanical or electrochemical and have remained mainly in the investigative realm, although several types have been tried clinically.

a. Mechanical. Attachment to anatomical structures which normally are in motion permits these sources to convert the physiologic motion of muscle contraction to electrical energy using piezoelectric or electromagnetic transducers (Lewin *et al.*, 1968; Parsonnet *et al.*, 1969). Diaphragmatic motion, aortic wall pulsation, and pericardial motion all have been coupled to transducers for this purpose. In addition to relatively low output, this arrangement is usually defeated by gradual encasement and immobilization by tissue reaction such as fibrosis or calcification.

b. Electrochemical. Biogalvanic cells, fuel cells, and combinations of these have been studied. Since the fuel cell will be discussed in depth below, we shall here describe only the biogalvanic cell.

L. W. Reynolds and J. J. Konikoff (1963, personal communication), and later Reynolds (1964) and Konikoff (1966), reported minute electrical outputs from platinum black and high-speed steel electrodes implanted subcutaneously in rats. These currents were sufficient to operate a transmitter and produce a signal that could be detected by a telemetry receiver across the room. Other workers soon experimented with larger implants and other corrodable metals, including Zn, Al, Mg, and AgCl. Racine and Massie (1966) and Massie *et al.* (1968) implanted Al and Pt black electrodes in rabbits and dogs. These were attached to external pacemaker circuits for demonstrative purposes. Roy and Wehnert (1966) and Roy (1971) did extensive testing of systems employing Zn and Pt black. More recently, Tseung *et al.* (1971), Wan *et al.* (1972), and Wan and Tseung (1974) worked with a variety of galvanic, mixed-electrodes, and fuel cell systems. A relatively large number of patients received pacemakers powered by Zn–AgCl galvanic cells (Schaldach, 1969; Schaldach, 1970).

Most of these cells have in common either two galvanic electrodes or a galvanic anode with an O_2 cathode. They derive their energy from corrosion of the metallic electrode under conditions where the reaction products, including metallic ions such as Al^{3+}, Zn^{2+}, and Mg^{+} are liberated into the body milieu at a rate slow enough that toxicity is not a problem. The size of the corrodable electrode determines the life span of the cell. In general, voltages are low, requiring converters and transformers for pacemaker circuitry. One variation is where the corrosion products are contained permanently, for the life of the implant, while O_2 and electrolyte (tissue fluid) are permitted to diffuse in from the body extracellular space (Wan *et al.*, 1972). Problems of all these systems are related to toxicity, premature failure due to the uneven corrosion of electrodes, and low voltage.

4. *Theory of Biofuel Cells*

The fuel cell represents a second electrochemical approach for providing energy for operation of a cardiac pacemaker. Strictly speaking, a biofuel cell defines any fuel cell that utilizes living organisms, such as an implant host (patient); bacterial cultures; or products of living organisms, such as enzymes, to provide catalysis or some unique component. The biofuel cell need not be implanted but may actually contain the biosubsystem within it, as in the case of a culture of microorganisms. In this text we shall discuss only that type of biofuel cell which we call a bioautofuel cell (Wolfson *et al.*, 1968; Yao *et al.*, 1972), designed to be implanted within a living host and to rely upon that host for supplies of fuel and oxidant and for reaction-product removal. Moreover, the electrical output is for the benefit of that host and to be consumed within the host. Emphasis will be given to the application of enzymes in the design of bioautofuel cell prototypes. The bioautofuel cell affords the potential advantage of *in vivo* utilization of compounds from the blood or other tissues as a source of chemical energy for direct conversion to electrical energy.

The fuel cell operates by oxidizing a substrate to a product and making available the released chemical energy (Gibbs free energy) associated with the electrons that are transferred during the reaction. The electrons, in turn, can be passed through an external circuit and used to perform work prior to their uptake by a second substrate. The efficiency of energy transfer for a fuel cell can be very high. The theoretical efficiency for a fuel cell depends only on the change in free energy in going from substrate to product. It can be shown, using the definitions of Gibbs free energy G, enthalpy H, and the first law of thermodynamics, that for constant temperature T and pressure P,

$$\Delta G = Q - W + P\,\Delta V - T\,\Delta S \tag{1}$$

where Q is the amount of heat transferred and W is the sum of the useful work W_u and the work required to change the volume against pressure P. For a reversible system, Q and $T\,\Delta S$ are equal, so Eq. (1) reduces to

$$-\Delta G = W_u \tag{2}$$

Equation (2) states that the useful work depends only on the change in free energy that the substrate undergoes on conversion to product. Actually W_u is the maximum obtainable work; the practical work will be about 90% of W_u. This results from the definition of G, which at constant temperature and pressure substitutes into Eq. (2) to give

$$W_u = -\Delta G = \Delta H - T\,\Delta S \tag{3}$$

The term $T\,\Delta S$ often is about 10% of the value of ΔH, thus giving 90% practical efficiencies under reversible operating conditions. Since the latter are not always attainable, owing to concentration polarization and to heat effects and solution resistances, actual efficiencies of 70–80% of ΔG are more realistic. Some nonbiological fuel cells have been operated at efficiencies up to 70% (Mitchell, 1963); this is in contrast to practical efficiencies of only 20–45% for existing thermal or mechanical energy transfer processes.

Enzymes can be incorporated into a bioautofuel cell in one of two ways: (1) they can serve as the catalyst either at the oxidation or the reduction electrode, and 2) they can catalyze the conversion of a raw material into a suitable substrate for a conventional nonenzymatic fuel cell. In theory, either approach is applicable to the development of an *in vivo* pacemaker power supply; and work on each approach is described in later sections. So far, most of the work with enzyme fuel cell electrodes has been with oxidative enzymes as part of the anodic electrode. However, oxidation–reduction enzymes also could be employed as a catalyst at the cathodic electrode to enhance the rate of reduction of a substrate, although essentially no work has been done in this direction.

Most oxidoreductase enzymes require cofactors that act as electrochemical mediators. A potential problem in the use of these enzymes as electrode catalysts is the necessity for regeneration or for continuous supplementation of the necessary cofactors. However, the study of enzyme electrodes has not progressed to the point where the seriousness of this problem can be clearly defined. A less critical problem, that of converting the relatively constant output of a fuel cell into an intermittent electrical pulse of the proper form, appears relatively easy to solve using state-of-the-art microelectronics.

5. *Pacemaker Biofuel Cell Studies*

5.1. *General Considerations*

The idea of developing a bioautofuel cell for powering a pacemaker is related to the concurrent efforts toward an implantable fuel cell power source for the artificial heart. While a large effort was directed toward the artificial heart fuel cell (Batzold and Beltzer, 1969; Fishman and Henry, 1969; Giner and Malachesky, 1969; Bocciarelli, 1969; Drake, 1969), it was ultimately recognized that this was not yet a feasible approach for the generation of the 5–20 W demanded by any total heart replacement system (Appleby *et al.*, 1969).

If power requirements are much lower, as in the 10–500 μW needs of the cardiac pacemaker, the limitations become less significant. However, the problems still exist of extracting fuel and oxidant from the blood without causing clotting, hemolysis, or other damage to that tissue while at the same time protecting sensitive catalysts and other fuel cell components from deleterious effects of exposure to blood. One of the great advantages of the reduced power requirement of the pacemaker should be the ability to remove the cell from the flowing blood and thus bypass all problems related to bloodstream invasion. As to fuel and oxidant requirements, this would certainly be possible. The extracellular fluid compartments are in equilibrium with capillary blood and yet separated by biological membranes that exclude many of the unwanted blood-borne substances—especially macromolecules such as protein and particulate formed elements such as leukocytes and platelets. Removal of a foreign implant to the extravascular, intratissue spaces, however, creates a number of new problems. Among these are development of fibrotic sacs which tend to isolate the cell from its source of nourishment and route of product removal, and the likelihood of abscess around the implant. However, a number of workers, including the authors, have continued to study this approach. The realization of a high-efficiency power source depends upon the selection of suitable catalysts and the arrangement of effective and tissue-compatible electrochemical mechanics for prolonged life. In theory, sufficient power can be produced by the oxidation of a naturally occurring fuel (e.g., glucose) in exceedingly small quantities. According to Faraday's law, a system yielding 100 μW from glucose would consume less than 0.001% of the normal caloric intake of the adult man. One hundred microwatts is of the order of magnitude for powering a miniaturized pacemaker circuit even using conventional pacing electrodes. In its homeostatic processes, the body would be the ever-faithful servant in providing fuel and oxidant (e.g., glucose and O_2 at their physiologic concentrations in extracellular fluid). Because of the low fuel and oxidant requirements, this cell could continue producing power regardless of the state of health of the host. With biocompatible inert materials used for construction, the cell could continue to function for the life of the patient, even in the case of implantation in childhood. This would be much longer than the present state-of-the-art pacemaker with mercury or other primary or secondary cells. In theory, an entire pacemaker–fuel cell unit could be of such small size as to attach directly to the heart wall.

5.2. *Nonenzymatic Bioautofuel Cell*

Experiments to date have shown that a bioautofuel cell is feasible to supply the low levels of power required for a pacemaker. The cell is intended to be placed in the extracellular space, either in a dissected pocket within a tissue mass (e.g., subcutaneous) or within a coelomic cavity (e.g., pericardium, pleura, or peritoneum). Although these locations for implantation do involve contact with extravasated blood from the operative dissection, such contact is transient. If the initial effects of this blood can be overcome, then the cell may be considered to operate in a plasma ultrafiltrate containing a relatively low concentration of proteins and polysaccharides. The movement of fuel and oxidant to the bioautofuel cell ultimately depends upon their concentrations in the particular fluid bathing the cell and the degree to which this fluid remains in dynamic equilibrium with the capillary blood after operative intervention and in the presence of the foreign implant.

The very limited data for the glucose, carbohydrate, and O_2 concentrations of pericardial fluid indicate similarities with pleural and peritoneal fluid. "Sugar" levels have been reported as 86–131 mg% for human peritoneal fluid and 70–122 mg% for pleural fluid (Foord *et al.*, 1929). Glucose in ascites fluid (congestive heart failure—baboon) was reported at 120 mg% (Wolfson, 1969, unpublished observations). The authors have made some measurements of glucose and O_2 in canine pericardial fluid. These animals were sampled during a hypoxia protocol when pericardial fluid had glucose levels of 5.6–7.2 mM and $P_{O_2} < 50$ mm Hg (Wolfson, 1971, unpublished observations). The P_{O_2} was low (and glucose possibly high) as a result of the hypoxemia and hypotension of those experiments.

Utilizing experience gained in research toward the artificial heart fuel cell, Drake (1969) and Drake *et al.* (1970) reported a steady power output density of 17 $\mu W/cm^2$ of electrode area *in vitro* from an implantable glucose–O_2 cell. However, the *in vivo* result was only 0.4 $\mu W/cm^2$. This cell was implanted subcutaneously and became encased in a dense fibrous scar membrane without access to fluid with a high plasma ultrafiltrate turnover rate. The collagen sac was an outgrowth of the dermal layers in contact with the cell, which indicates that the subcutaneous location was probably a poor choice for the implantation site. These cells always decayed to very low levels after prolonged implantation.

Wan and Tseung (1974) studied an O_2–glucose fuel cell intended for implantation "inside body fluids." They used a platinum black-on-platinum mesh cathode and a platinum black-on-porous graphite anode. Linear sweep studies produced 3 mA/cm^2 after addition of glucose (25 mM). This cell produced 20 $\mu W/cm^2$ *in vitro* and 3.3 $\mu W/cm^2$ when implanted in a rat. Rao and Richter (1974) used a graphite cathode and platinized Pt anode to demonstrate *in vitro* power levels of 200 μW, under a 2.7-kΩ passive load. However, this cell was provided with O_2 from air-saturated electrolyte (P_{O_2} approximately 150 mm Hg) and 100 mM glucose (2 and 20 times the physiologic concentration of these substances, respectively). When glucose was lowered to 50 mM, output fell to 100–150 μW. However, P_{O_2} and glucose concentrations were still maintained artificially high.

We have carried out extensive experiments that have led to the development of a single electrolyte bioautofuel cell utilizing permselective membranes and embracing a dry cathode concept (Strohl *et al.*, 1966; Wolfson, *et al.*, 1968; Yao *et al.*, 1969; Wolfson, *et al.*, 1970; Dohan *et al.*, 1971; Wolfson and Yao, 1972; Ahn *et al.*, 1974a,b; Yao *et al.*, 1972). The anode was made of hydrophilic Pt black and the cathode of Au or Ag black (Figure 1). Such a cell worked for 8 months in simulated extracellular fluid at a constant voltage of 0.52 V under 10-kΩ loading, with a total power output of 27 μW. The power density was about 4 $\mu W/cm^2$ at a current density of 7 $\mu A/cm^2$. Resterilization of the cell by gamma irradiation (1.25 Mrad) at the end of the first 3-month testing period did not damage the cell. By increasing the catalyst surface area or by connecting several cells in parallel, the goal of a 100-μW implantable fuel cell energy source may be achieved.

5.3. *Enzyme-Catalyzed Preparation of Fuel*

a. Review. Del Duca (1963) may be considered the first to report the development of an enzyme-catalyzed preparation of substrate for a fuel cell. He attempted to employ urea from urine as the fuel source. The enzyme, urease, was utilized to catalyze the decomposition of urea into NH_4^+ and HCO_3^+. The NH_4^+ was the direct source of fuel for the NH_4^+–O_2 fuel cell. The NH_4^+ was anodically oxidized to N_2 and water. Since NH_4^+ might easily reach toxic levels in the blood, it is undesirable

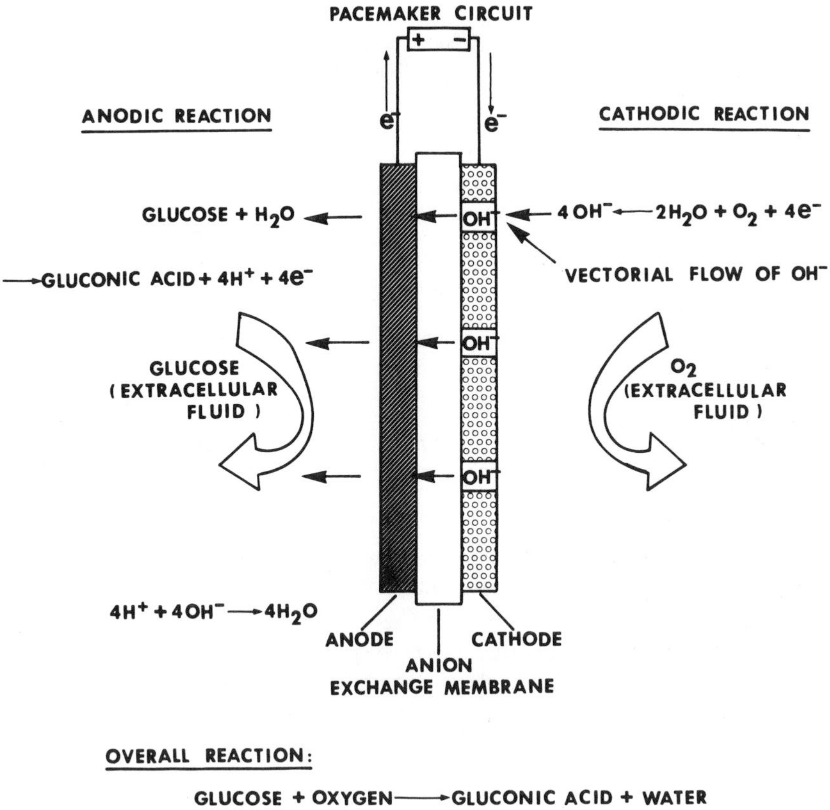

FIGURE 1. Schematic representation of the typical bioautofuel cell assembly. Implantable models would have, additionally, electrode covering membranes (e.g., cellulose for anode, silicone rubber for cathode) and a metal or plastic housing.

to generate NH_4^+ within the body. While glucose and other free carbohydrates are obvious fuel sources, many other carbonaceous compounds are present in blood plasma, mainly in combined form as polysaccharides and mucoproteins (West and Clarke, 1938); and these compounds could be made available as anodic fuels by enzymatic preparation. The presence of multiple fuels would enhance the fuel cell performance. Table 1 shows some anodic data on blood carbonaceous compounds (Yao *et al.*, 1972). Glucosamine was found to be the best fuel of those tested for a Pt anode; however, glucosamine is not present in any body fluid as a free compound but is found in polysaccharide macromolecules.

One of the most widespread polysaccharides present in interstitial fluid is hyaluronic acid, a polymer of D-glucuronic acid and *N*-acetyl glucosamine. Complete enzymatic hydrolysis of hyaluronic acid to its monomeric constituents is theoretically possible. Other polysaccharides that could yield glucosamine-like compounds upon hydrolysis are chondroitins and heparin; however, these are of limited distribution and availability (White *et al.*, 1959). Other macromolecular

TABLE 1
Anodic Data: Blood Carbohydrates and Related Compounds

Fuel 5 mM in pH 7.4 buffer	Open-circuit voltage (vs. standard hydrogen electrode)	Anode voltage (vs. standard hydrogen electrode)	
		At 50 μA/cm²	At 100 μA/cm²
Ethanol	−0.18	−0.13	−0.10
Fructose	−0.22	−0.14	−0.10
Gluconic acid	−0.11	−0.05	−0.02
Glucose	−0.22	−0.07	−0.04
Glucosamine	−0.42	−0.41	−0.40
Glucuronic acid	−0.22	−0.07	−0.04
Mannose	−0.30	−0.24	−0.16
Sorbitol	−0.25	−0.20	−0.15

sources of carbohydrate fuel exist where polysaccharides or oligosaccharides are linked to proteins. An example is mucin, a mucoprotein of saliva, which contains numerous moities of the disaccharide 6-(*N*-acetyl-*d*-neuraminyl)-*N*-acetyl-D-galactosamine in each protein molecule. Hydrolysis by neuraminidase liberates all the sialic acid (polysaccharide) and some hexosamine from the mucin polymer. This hexosamine could be another new fuel for the anode.

Another nonelectrode application for enzymes is the use of proteolytic, thrombolytic enzymes to prevent the formation of coagulum around implanted devices. When fuel cells are implanted in the extracellular space, including mesothelial lined spaces such as the peritoneum or pericardium, a coagulum of operative blood and protein transudate initially coats the implant. While this may disappear over a period of days or weeks, it can form a matrix for the ingrowth of a covering membrane or scar. This has been observed experimentally with fuel cells implanted in the subcutaneous space and peritoneal cavity. Attempts to evaluate fuel cells implanted in tissue have been hampered by the formation of these occlusive membranes (Konikoff, 1966; Drake *et al.*, 1970). Enzymes such as trypsin and chymotrypsin prepared in immobilized form possibly could be utilized to prevent the formation of the coagulum.

b. Experiments with Hyaluronidase. The finding that glucosamine gave good performance as a fuel cell substrate (Table 1) led Dohan *et al.* (1971) and Ahn *et al.* (1974a, 1974b) to study the generation of this compound from polysaccharides such as hyaluronic acid present in interstitial fluid. In the initial studies, Dohan *et al.* (1971) attempted to immobilize hyaluronidase in porous silicone sponge to hydrolyze hyaluronic acid into its lower-molecular-weight constituents. This new fuel and the increased mass transfer due to reduced viscosity enhanced the performance of a glucose fuel cell.

The concentration and the structure of hyaluronic acid in pericardial, pleural, or peritoneal fluid is not precisely known. Only synovial fluid has been investigated, where the hyaluronic acid concentration was found to be 3–5 mg/ml (Schultz and Heremans, 1966). The molecular weight of this substance is about 10^6 (Brimacombe, 1964). A value of 1 mg/ml was chosen for the simulation of a pericardial or peritoneal fluid by Dohan *et al.* (1971). Hyaluronic acid can be hydrolyzed by a group of hyaluronidases. The testicular hyaluronidase, extracted from ground bull testis, has been well investigated by Weissmann *et al.* (1954, 1964) and Linker *et al.* (1955). Highly purified hyaluronidase yields 80% tetrasaccharide and 20% disaccharide upon hydrolysis of hyaluronic acid. Complete hydrolysis to monomers is possible through the additional sequential and combined action of glucosaminidase and glucuronidase. Since glucosaminidase was not commercially available, only the hyaluronidase and β-glucuronidase were employed in that study. HSE testicular hyaluronidase (Worthington Co.) was physically entrapped in silicone rubber foam using the technique of Pennington *et al.* (1968). Liquid medical Silastic elastomer (Dow-Corning RTV 5370-S) was mixed with hyaluronidase. The mixture was cross-linked by stannous octoate to produce enzyme-bound foam. Two different types of β-glucuronoidases (Sigma Co., Types 1 and B-3) were also entrapped by the same technique. The entrapped hyaluronidase was assayed by measuring the rate of *N*-acetyl-D-glucosamine equivalent released from hyaluronic acid solution at 37°C. A modified Morgan–Elson test (Morgan and Elson, 1934; Dohan *et al.*, 1971) was adopted for the assay of both hyaluronidase- and β-glucuronidase-bound foams.

The entrapped hyaluronidase alone did not hydrolyze hyaluronic acid into monomers. The activity data showed that the average molecular weight was 820 after 48 hr of incubation. This indicated that most of the products were tetrasaccharide with some disaccharide and higher-order saccharides. It was assumed that the entrapped hyaluronidase yielded the same end products as did the free enzyme. After dry storage for 3 months at 0°C, the entrapped hyaluronidase retained about 90% of its initial activity. However, when stored dry at 37°C, no activity of the entrapped enzyme was observed. No evidence of activity from the entrapped β-glucuronidases was found. Since the mechanism of action of glucuronidase was unknown, no proper explanation was given.

Fuel cell tests of hydrolysate of hyaluronic acid were performed in a Pt/A310/Pt sandwich immersible cell (Figure 1). The anode was covered with a cellulose membrane and the cathode was coated with silicone rubber. A small disc of the hyaluronidase-bound foam was placed against the anode. Only a very low, 3 μW, but constant power output was observed for a week. This low value could have been the direct consequence of the high molecular weight of the tetrasaccharide, since this tetramer is at the limit of permissible diffusion through the cellulose membrane. On the other hand, the hyaluronidase-bound membrane did improve mass transfer when glucose was present as an alternative fuel. Figure 2 demonstrates the effect of macromolecules upon the performance of a fuel cell. Another set of experiments was carried out in the same way but with a 350-mg disc of hyaluronidase-loaded foam placed near the anode after the experiment was begun (Figure 3). All the curves were measured at 37°C.

The initial data on the performance of hyaluronidase immobilized by physical entrapment seemed promising. The stored dry enzyme did retain most of its

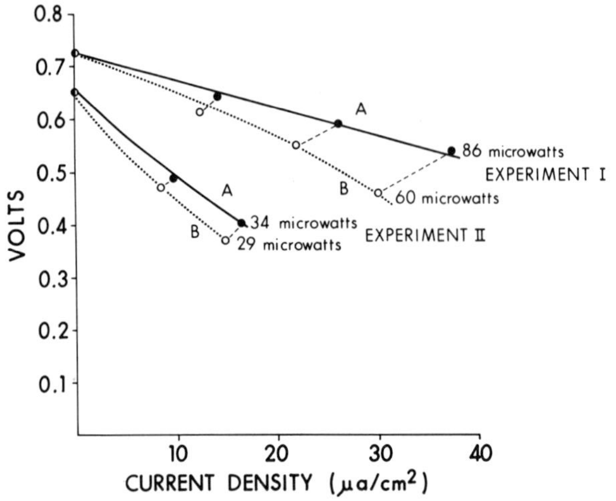

FIGURE 2. Effect of added macromolecules upon performance of glucose fuel cells. Sandwich-type cells were immersed in Krebs–Ringer bicarbonate containing 5 mM glucose. Experiment I, curve A: glucose alone; curve B: glucose plus hyaluronic acid, 1 mg/ml. Experiment II, curve A: glucose alone; curve B: glucose plus bovine serum albumin, 30 mg/ml. Since different cells were utilized for experiments I and II, overall performance was not comparable.

initial activity. However, enzymes immobilized in this way may not remain in the trapsites indefinitely, since they could be leached out by the body fluid. A more desirable approach is chemical coupling.

Ahn *et al.* (1974a,b; 1975) employed the cyanogen bromide (BrCN) technique (Porath *et al.,* 1967) to covalently bind hyaluronidase to Sepharose (Sigma Company, 4B-200). The bound enzyme had about 14% of the activity initial-

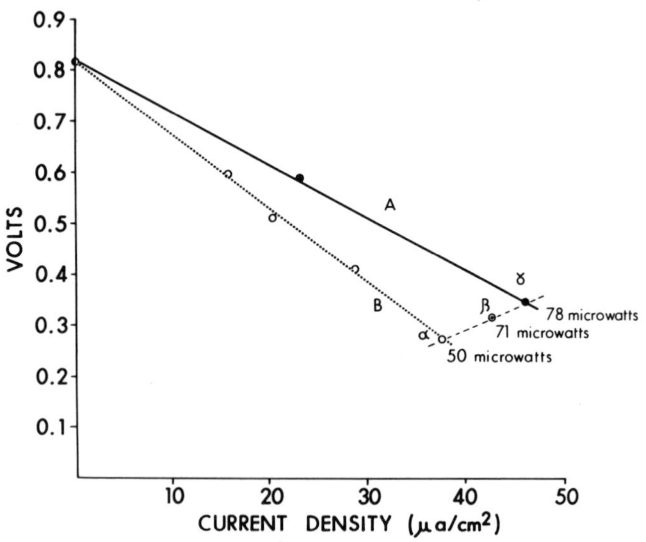

FIGURE 3. Improved performance of glucose cell containing hyaluronic acid after addition of hyaluronidase-loaded sponge. Curve A: glucose, 5 mM alone; curve B; glucose plus hyaluronic acid, 1 mg/ml. The enzyme-loaded sponge was introduced at α. Improved performance was recorded at β.

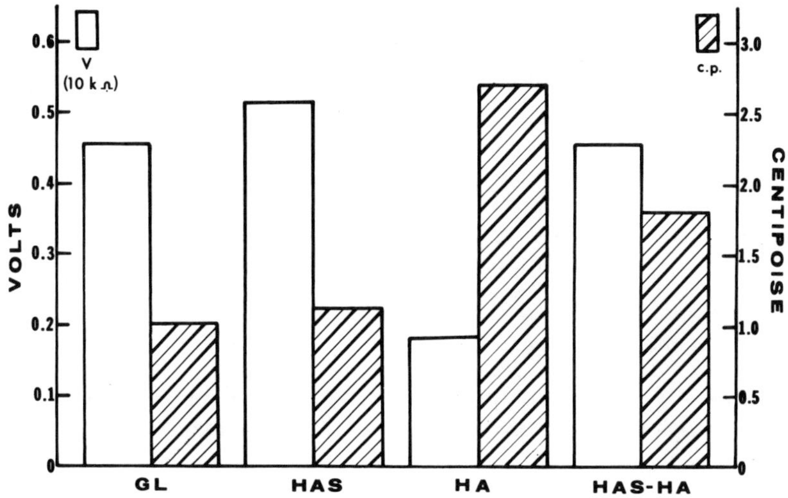

FIGURE 4. Effect of hyaluronidase-bound membrane on viscosity and on fuel cell performance (37°C). The four buffer mixtures all contain glucose 5mM. *GL:* glucose with blank membrane; *HAS:* glucose with hyaluronidase-bound membrane; *HA:* glucose plus hyaluronic acid with blank membrane; *HAS-HA:* glucose plus hyaluronic acid with hyaluronidase-bound membrane.

ly present before binding. This reduction in activity was likely due to the following: (1) the enzyme was partially inactivated during binding, and (2) only a portion of the bound enzyme was available for the reaction with substrate. Beside Sepharose, other matrix materials, such as cellulose, polyacrylamide, and cross-linked poly (vinyl alcohol) were tested. None was found to be successful upon binding by either BrCN activation or by glutaraldehyde cross-linking. Insufficient —OH groups, steric hindrance, and/or inactivation of hyaluronidase by glutaraldehyde treatment may be the cause of these failures. The activity of the bound enzyme was stable for more than 6 months at 0°C. After 30 days at 37°C, dissolved hyaluronidase was completely inactivated while the gel-bound enzyme retained almost 100% of the initial activity. However, the gel-bound enzyme stored at 37°C rapidly decreased in activity after 30 days. This could be due to bacterial degradation at that temperature.

For fuel cell evaluation the Sepharose gel was converted to a thermoplastic membrane and hyaluronidase was then bound to it by the BrCN method. A 1-mm-thick disc was attached against the anode. The cell employed was similar to the one employed by Dohan *et al.* (1971), except that the cathode was made of Ag black. The system was, thus, a Pt/A310/Ag sandwich with the cellulose anode covering membrane replaced by the enzyme-bound gel membrane. The results of adding hyaluronic acid and hyaluronidase-bound membrane to these fuel cells are illustrated in Figure 4. The presence of hyaluronic acid (1 mg/ml) increased the buffer electrolyte viscosity from 1.1 to 2.5–2.8 cP and reduced the voltage output under 10-kΩ load by 60% in 3 hr (0.45–0.51 to 0.19 V). When the hyaluronidase-bound membrane was placed on the anode, the viscosity of the glucose–hyaluronic acid solution was lowered to 1.8 cP and the cell output increased to the original level of a glucose-fueled cell (0.46 V, 10 kΩ) in 3 hr.

The glucosamine equivalent in the glucose–hyaluronic acid cell mixture with added hyaluronidase-membrane reached 1.24 mg% after 22 hr. The restoration of cell output was likely a combination of enhanced transfer of fuel, O_2, reaction products, and/or the appearance of a new fuel: glucosamine equivalent molecules.

The works of Dohan *et al.* (1971) and Ahn *et al.* (1974a,b; 1975) have indicated that the selection and immobilization of a suitable enzyme should enhance the performance of a fuel cell working in a body fluid environment.

5.4. *Enzyme-Catalyzed Fuel Cell Electrodes*

a. Review. The early attempts to use enzymes in solution to catalyze the direct release and transfer of electrons from a substrate to an external circuit have been summarized by Austin (1967) and Cenek (1969). Urease, amino acid oxidase, nitrate and sulfite reductases, arginase, glutamic decarboxylase, glucose oxidase, and other enzymes were tested in solution with an appropriate substrate and a platinum electrode with little success. Only when a redox mediator, such as methylene blue or $K_3Fe(CN)_6$, was present with an oxidative enzyme at the anode did activity result. With 1 mg/ml of glucose oxidase in solution in 0.1 mole of glucose and pH 5.7 phosphate buffer at 25°C, the maximum obtainable current was 900 $\mu A/cm^2$ at 0.08 V, as measured against a calomel electrode with a Pt black cathode (Cenek, 1969). In another study Yahiro *et al.* (1964) dissolved glucose oxidase, containing significant catalase activity, in pH 6.7 phosphate-buffered glucose (0.11 M) solution on the anode side of a cell. With Pt-foil electrodes connected through a resistive load and the addition of ferrous iron, current densities of 10 $\mu A/cm^2$ at 400 mV were obtained. The iron presumably served to regenerate the flavin cofactor associated with glucose oxidase. However, since oxygen was present at the anode, it is not clear which reaction was responsible for generating the electrons. Still poorer performance was observed with D-amino acid oxidase and D-alanine as substrate. In addition to the low current densities, this approach of using enzymes free in solution for a fuel cell electrode has the practical disadvantage of requiring a continuous supply of enzyme as well as substrate. Immobilization of the enzyme to keep it in the vicinity of the electrode and have it available for reuse is a promising approach for overcoming this practical problem.

The first major study of immobilized enzyme electrodes for energy production was by Drake (1968) as part of a project to utilize the oxidation of glucose to power an artificial heart. In that work glucose oxidase was coupled covalently to polyacrylic acid, polyglutamic acid, ethylene–maleic acid polymers, and carboxymethyl cellulose with retention of 80–90% enzyme activity. Electrodes were fabricated by (1) adsorbing the enzyme–polyacrylic acid material on carbon powder followed by compression of the mixture, (2) covalent attachment of the enzyme to carboxylic acid groups on the carbon surface (one test), (3) covalent coupling to various materials adsorbed on the carbon surface, and (4) adsorption of the enzyme–polymer material plus the addition of redox couplers such as FAD, NAD,

$K_3Fe(CN)_6$, methylene blue, and quinone, as well as the addition of peroxidase and a charge-transfer polymer. The best unit sustained a current density of 0.5 mA/cm^2 for at least 50 min when tested at about pH 5.5 in 0.5 M glucose at 37°C and with about 0.5 V of polarization. This represented a significant advance in current density when compared with the earlier studies using glucose oxidase in solution; however, the work was discontinued because other catalysts for glucose oxidation looked more promising for the near future (Drake, 1968). Similar electrochemical performance was obtained for the oxidation of glucose-6-phosphate in the presence of the corresponding dehydrogenase with NADP as the electron acceptor (Takahashi *et al.,* 1970).

Additional work has been done on the use of immobilized oxidative enzymes for the construction of analytical chemistry electrodes (Guilbault, 1972; Clark, 1972; Weibel *et al.,* 1973; Bessman and Schultz, 1973); however, tests of these additional electrodes under fuel cell or electrochemical production modes of operation have not been reported.

b. Glucose Oxidase Anode. Much of the work on enzyme-catalyzed fuel cell electrodes, particularly for pacemaker power supplies, has been done using glucose oxidase. The rationale, of course, has been to try to utilize glucose and O_2 from the circulatory system for the anodic and cathodic substrates, respectively. The early work with glucose oxidase free in solution and the initial studies by Drake (1968), using immobilized glucose oxidase, are described in the previous section. The following discussion is a description of our studies with immobilized glucose oxidase and a listing of the major problems still to be resolved.

Glucose oxidase (EC 1.1.3.4) catalyzes the oxidation of β-D-glucopyranose to δ-D-gluconolactone. The latter hydrolyzes rapidly to D-gluconic acid, which is readily metabolized by the body. This enzyme has its maximum activity at a pH of about 5.6 and at 35–40°C and requires FAD as a cofactor (Keilin and Hartree, 1948; Bentley, 1963). A condensed reaction scheme according to Bright and Gibson (1967) is as follows:

$$\begin{aligned} \text{E–FAD} + \text{G} &\rightleftharpoons \text{I}_1 \\ \text{I}_1 &\rightleftharpoons \text{I}_2 + \text{L} \\ \text{I}_2 + \text{O}_2 &\rightleftharpoons \text{E–FAD} + \text{H}_2\text{O}_2 \end{aligned}$$

where G is glucose, L is lactone, I_1 and I_2 are intermediates, and E–FAD is the enzyme–cofactor complex.

If O_2 is excluded at the anode, then only the oxidation of glucose to lactone proceeds. The electrons may go directly onto an external circuit. Or, the electrons may be taken up by FAD to produce $FADH_2$, which in turn can release electrons to an external circuit upon regeneration of FAD. The current produced corresponds to the velocity of the reaction. Using the above simplified reaction sequence and assuming the following concentrations: glucose oxidase 6×10^{-9} M, glucose 5.5×10^{-4} M (1/10 normal blood glucose), O_2 8.6×10^{-4} M (1/10 normal blood O_2), we calculated a steady-state current of 0.27 mA for a working electrode volume of 10 ml (Wingard and Liu, 1969). The maximum calculated open-circuit potential for the conversion of glucose to the lactone is +0.63 V. Thus, even with considerable concentration polarization, a few milliwatts of power could be expected, which would be more than ample to power the newer types of pacemakers. However, in order to minimize the amount of glucose consumed and to ensure essentially error-free operation over a 5- to 10-year period, considerable research remains to be done before a practical glucose oxidase pacemaker fuel cell can be expected.

Our studies have been oriented primarily to investigating the factors that affect the efficiency of electron transfer from substrate to an external circuit, particularly in regard to the methods for immobilization of the glucose oxidase or the cofactor. In our work (Wingard *et al.,* 1971); Lahoda *et al.,* 1975) glucose oxidase was immobilized on platinum screens at about pH 7 and tested for use as the anode in a fuel cell. Three different immobilization techniques were emloyed. In one the enzyme was trapped in a polyacrylamide gel matrix that was formed around the platinum screen. In a second method the enzyme was cross-linked with glutaraldehyde in the presence of the platinum screen. And, finally, the glutaraldehyde–enzyme–platinum preparation was formed with +0.04 V of direct current applied to the platinum screen. The details of the electrode preparations are described in the references cited above.

The glucose oxidase electrodes were tested in an H cell under constant-current voltametry and load (fuel cell) modes of operation. The measured open-circuit potentials with an equimolar mixture of phosphate-buffered glucose and gluconolactone charged to the cell were about −700 mV with respect to a saturated calomel reference electrode, and as expected were independent of the amount of enzyme. The open-circuit potential varied some with the immobilization technique, suggesting that there may have been competing reactions present which were influenced to different degrees by the microenvironment of the different enzyme–support preparations. The open-circuit potential represents the maximum driving force; in practice part of this driving force is used to perform useful work, while the remainder is wasted in overcoming concentration polarization and solution or junction resistances. In our studies we made only modest attempts to reduce polarization through mild stirring within the cell; however, polarization of 80–120 mV was observed.

Operation under load or in a fuel cell mode was achieved by connecting a 10-Ω resistor between the anodic enzyme electrode and a platinum- or silver-impregnated cathode. The cathodic reaction was the reduction of O_2 to water. With phosphate-buffered equimolar glucose and gluconolactone charged to the cell, steady currents of 5–60 μA were obtained, as shown in Figure 5. This corresponded to a current density of 1.6–19 $\mu A/cm^2$, respectively, based on the gross area of the platinum screen at the enzyme electrodes. This was less than the current densities obtained in some of Drake's work (1968). The differences could have been caused by a number of factors, including less enzyme, greater diffusional resistances within the matrix, greater resistance to electron transfer, less effective utilization of cofactors, and possibly other effects in our electrodes as compared to those studied by Drake. The choice of a suitable area upon which to base the current density calculations also made it difficult to compare our platinum screen electrodes with Drake's solid or porous solid electrodes. The greater

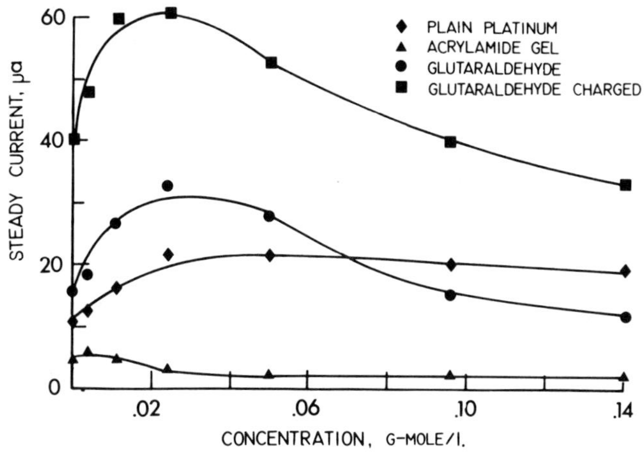

Figure 5. Steady-state current with glucose oxidase anode.

steady current with the charged glutaraldehyde enzyme electrode (Figure 5) could be attributed to a variety of effects including a larger amount of immobilized enzyme or differences in the microenvironment of the immobilized enzyme as compared to the other electrodes. Our immobilized glucose oxidase retained at least 10–20% of its original activity on immobilization up to pH 8. There was no evidence of loss of enzyme activity with repeated use over a period of 2–10 days.

Considerable work remains to define the reasons for the very small current densities obtained not only in our work but with all studies to date with enzyme electrodes. The interrelationships among the enzyme protein, the cofactor, and the external support need to be defined, particularly in regard to electron transfer, cofactor regeneration, and different immobilization schemes. The enhancement of enzyme stability and the reduction of concentration polarization are other practical problems that will require considerable effort for use of enzyme electrodes in a pacemaker. However, at present the principal problem remains the improvement of electron transfer from substrate to external circuit to give greater capacity and more reliable fuel cell performance.

6. *Outlook*

It seems possible that the continuous power requirements for cardiac pacemakers may be in the vicinity of only 10–50 μW in the near future. In theory such small amounts of power can be produced in a variety of ways. Thus, the size, mass, biocompatibility, and long-term reliability of the complete pacemaker unit will ultimately become the deciding factors.

The bioautofuel cell is particularly appealing as a potential pacemaker power source because it probably can be made quite small and of low mass, possibly sufficiently so to allow attachment directly to the heart (epicardium). The major problem of the bioautofuel cell is reliability, particularly in the long-term, of the electrode catalysts. There is ample evidence that sufficient fuel and oxidant are available. Enzymes, of course, are logical choices for the electrode catalysts because of their high selectivity and their ability to function efficiently under physiological conditions. Yet, the general understanding of enzyme-catalyzed electron transfer is quite poor; and until this process is better understood, it is doubtful if enzyme-catalyzed fuel cell electrodes will have sufficient long-term reliability for use in the pacemaker power unit. Although it may be practical to obtain the needed reliability with nonenzymatic electrode catalysts, enzyme electrodes retain their ultimate attractiveness but must await a scientific breakthrough sufficient to permit capture of enzyme-catalyzed electron transfer for electronic circuitry. In order to provide reliable power in the long term, enzymes utilized for this purpose would need to be quite stable. This again requires greater understanding of enzyme structure and functional relationships. This area, however, is receiving considerable study.

In summary, the bioautofuel cell is a promising approach for the development of a small, lightweight, implantable pacemaker power supply. It is one capable of being combined with the microcircuitry elements for application directly on or in the heart. At present, the long-term reliability and biocompatibility still require considerable work. The use of enzymes in such a fuel cell, *in vivo,* in the area of fuel production or fuel cell electrode catalysis is appealing but requires one or more significant scientific advances for practical development.

ACKNOWLEDGMENTS

This research was supported in part by grants from the Delaware Heart Association and the American Heart Association and by the following grants from the National Institute of Health: Career Development Award 5K3HE21,871 (Dr. Wolfson) and research grants HE12555, GM16654, and GM19419.

7. *References*

Ahn, B. K., Wolfson, S. K., Jr., Yao, S. J., and Liu, C. C., 1974a, Hyaluronidase-membrane for an implantable fuel cell, *Amer. Soc. Artificial Internal Organs, Abstr.* **3**:2.

Ahn, B. K., Wolfson, S. K., Jr., Yao, S. J., and Liu, C. C., 1974b, A sepharose membrane bound hyaluronidase to process extracellular fluid for the implantable bioautofuel cell, *Proc. 27th Ann. Conf. Eng. Med. Biol.,* Philadelphia, p. 200.

Ahn, B. K., Wolfson, S. K., Jr., Yao, S. J., Liu, C. C., Todd, R. C., and Weiner, S. B., 1976, Hyaluronidase-bound membrane as a biomaterial for implantable fuel cells, *J. Biomed. Mater. Res.* **10**:283.

Appleby, A. J., Ng, D. Y. C., Wolfson, S. K., Jr., and Weinstein, H., 1969, An implantable biological fuel cell with an air-breathing cathode, *Proc. 4th Intersoc. Energy Conversion Eng. Conf.,* Washington, D.C., pp. 22–26.

Austin, L. G., 1967, *Fuel Cells: A Review of Government-Sponsored Research 1950–1964, National Aeronautics and Space Administration Rept SP-120,* Government Printing Office, Washington, D.C.

Barold, S. S., 1973, Modern concepts of cardiac pacing, *Heart & Lung* **2**:238–252.

Batzold, J. S., and Beltzer, M., 1969, Feasibility studies—biological fuel cell, *Proc. Artificial Heart Program Conf.,* Washington, D.C., pp. 817–824.

Bentley, R., 1963, Glucose oxidase, in: *The Enzymes* (P. D. Boyer, H. Lardy, K. Myrback, eds.), 2nd ed., Vol. 7, pp. 567–586, Academic Press, Inc., New York.

Bessman, S. P., and Schultz, R. D., 1973, Prototype glucose oxygen sensor for the artificial pancreas, *Trans. Amer. Soc. Artificial Internal Organs* **19**:361.

Bocciarelli, C. V., 1969, On the design of catalysts for biological fuel cells, *Proc. Artificial Heart Program Conf.*, Washington, D.C., pp. 861–868.

Bright, H. J., and Gibson, Q. H., 1967, The oxidation of 1-deuterated glucose by glucose oxidase, *J. Biol. Chem.* **242**:994.

Brimacombe, J. S., 1964, Hyaluronic acid, in: *Mucopolysaccharides,* pp. 43–63, American Elsevier Publishing Co., Inc., New York.

Cenek, M., 1969, *Biochemical Fuel Cells, U.S. Air Force Rept. AD694072* (translated from *Chem. Listy* **62**:927–974, 1968).

Chardack, W. M., Gage, A. A., and Greatbatch, W., 1960, A transistorized, self-contained, implantable pacemaker for the long-term correction of complete heart block, *Surgery* **48**:643.

Clark, L. C., Jr., 1972, A family of polarographic enzyme electrodes and the measurement of alcohol, in: *Enzyme Engineering* (L. B. Wingard, Jr., ed.), pp. 377–394, John Wiley & Sons, Inc., New York.

Del Duca, M. G., 1963, Direct and indirect bioelectrochemical energy conversion system, in: *Develop. Ind. Microbiol.,* pp. 81–91, Plenum Press, New York.

Dohan, L. A., Yao, S. J., and Wolfson, S. K., Jr., 1971, An enzymatic converter for the implanted fuel cell, *Trans. Amer. Soc. Artificial Internal Organs* **17**:411–414.

Drake, R. F., 1968, *Implantable Fuel Cell for an Artificial Heart, U.S. Government Publication PB177695* (February).

Drake, R. F., 1969, Implantable fuel cell for an artificial heart, *Proc. Artificial Heart Program Conf.,* Washington, D.C., pp. 869–880.

Drake, R. F., Kusserow, B. K., Messinger, S., and Matsuda, S., 1970, A tissue implantable fuel cell power supply, *Trans. Amer. Soc. Artificial Internal Organs* **16**:199–205.

Eisenberg, L., Mauro, A., and Glenn, W. W. L., 1961, Transistorized pacemaker for remote stimulation of the heart by radio-frequency transmission, *IREE Trans. Bio-Med. Electron.* **8**:253.

Fishman, J. H., and Henry, J. F., 1969, Oxygen reduction on gold-palladium alloys in neutral media, *Proc. Artificial Heart Program Conf.,* Washington, D.C., pp. 825–838.

Fontaine, G., Kevorkian, M., and Welti, J., 1968, Thresholds for electrical stimulation, *Ann. Cardiol. Angiol.* **17**:251.

Foord, A. G., Youngberg, G. E., and Wetmore, V., 1929, The chemistry and cytology of serous fluids, *J. Lab. Clin. Med.* **14**:417–428.

Giner, J., and Malachesky, P., 1969, Anodic oxidation of glucose, *Proc. Artificial Heart Program Conf.* Washington, D.C., pp. 839–848.

Glenn, W. W. L., Mauro, A., Longo, E., Lavietes, P. H., and MacKay, F. J., 1959, Remote stimulation of the heart by radiofrequency transmission, *New Engl. J. Med.* **261**:948.

Glenn, W. W. L., Furman, S., Gordon, A. J., Escher, D. J. W., and van Heeckeren, D. W., 1966, Radio-frequency-controlled catheter pacemaker, *New Engl. J. Med.* **275**:137–140.

Goldstein, S., Moss, A. J., Rivers, R. J., Jr., and Weiner, R. S., 1970, Transthoracic and transvenous pacemakers: A comparative clinical experience with 131 implantable units, *Brit. Heart J.* **32**:35–45.

Grendahl, H., Sivertssen, E., Bay, G., and Bergan, F., 1969, Permanent cardiac pacing, *Acta Med. Scand.* **185**:139–143.

Guilbault, G. G., 1972, Analytical uses of immobilized enzymes, in *Enzyme Engineering* (L. B. Wingard, Jr., ed.), pp. 361–376, John Wiley & Sons, Inc., New York.

Hixson, J. D., and Laurens, P., 1972, Design criteria and two-year clinical results of Pu-238 fueled demand pacemaker, *Proc. 7th Intersoc. Energy Conversion Eng. Conf.,* Washington, D.C., pp. 765–770.

Huffman, F. N., and Norman, J. C., 1974, Nuclear fueled cardiac pacemakers, *Chest* **65**:667–672.

Keilin, D., and Hartree, E. F., 1948, Properties of glucose oxidase (Notatin), *Biochem. J.* **42**:221.

Konikoff, J. J., 1966, In-vivo experiments with bioelectric potentials, *Aerospace Med.* **37**:824–828.

Lahoda, E. J., Liu, C. C., and Wingard, L. B., Jr., 1975, Electrochemical evaluation of glucose oxidase immobilized by different methods, *Biotechnol. Bioeng.* **17**:413.

Laurens, P., 1970, French nuclear-powered pacemaker program, *Trans. Amer. Nucl. Soc.* **13**:508.

Lewin, G., Myers, G., Parsonnet, V., and Zucher, I. R., 1967, A non-polarizing electrode for physiological stimulation, *Trans. Amer. Soc. Artificial Internal Organs* **13**:345–349.

Lewin, G., Myers, G. H., Parsonnet, V., and Raman, K. V., 1968, An improved piezoelectric biological power source for cardiac pacemakers, *Trans. Amer. Soc. Artificial Internal Organs* **14**:215–217.

Lillehei, C. W., Gott, V. L., Hodges, P. C., Jr., Long, D. M., and Bakken, E. E., 1960, Transistor pacemaker for treatment of complete atrioventricular dissociation, *J. Amer. Med. Assoc.* **172**:2006–2010.

Linker, A., Meyer, K., and Weissman, B., 1955, Enzymatic formation of monosaccharides from hyaluronate, *J. Biol. Chem.* **213**:237.

Lown, B., and Kosowsky, B. D., 1970, Artificial cardiac pacemakers, *New Engl. J. Med.* **283**:907–916.

Martinis, A. J., 1975, Initial U.S. experience with promethium-147 fueled cardiac pacemakers, in: *Advances in Pacemaker Technology* (M. Schaldach and S. Furman, eds.), Springer-Verlag, Berlin.

Massie, H. L., Racine, P. J., Pasker, R., Hahn, A. W., and Sun, H. H., 1968, Study of power generating implantable electrodes, *Med. Biol. Eng.* **6**:503–516.

Mitchell, W., Jr., 1963, *Fuel Cell,* Academic Press, Inc., New York.

Morgan, W. Y. J., and Elson, L. A., 1934, A colorimetric method for the determination of *N*-acetylglucosamine and *N*-acetylchondrosamine, *Biochem. J.* **28**:988.

Parsonnet, V., Zucker, I. R., Gilbert, G., Lewin, G., and Myers, G., 1969, Clinical use of a new transvenous electrode, *Ann. N.Y. Acad. Sci.* **167**:756.

Parsonnet, V., Myers, G. H., Gilbert, L., and Zucker, I. R., 1975, Clinical experience with the nuclear pacemaker, *Surgery* **78**:776–786.

Pennington, S. N., Brown, H. D., Patel, A. B., and Chattopadhyay, S. K., 1968, Silastic entrapment of glucose oxidase-peroxidase and acetylcholine esterase, *J. Biomed. Mater. Research* **2**:443.

Porath, J., Axen, R., and Ernback, S., 1967, Chemical coupling of proteins to agarose, *Nature* **215**:1491–1492.

Preston, T., Judge, R., Lacchesi, B. and Bowers, D., 1966, Myocardial threshold in patients with artificial pacemakers, *Amer. J. Cardiol.* **18**:83.

Purdy, D. L., McGovern, G. J., and Smyth, N., 1975, A new radioisotope-powered cardiac pacer, *J. Thoracic Cardiovascular Surg.* **69**:82–91.

Racine, P. J., and Massie, H., 1966, An experimental internally powered cardiac pacemaker, *Med. Res. Eng.* **5**:24–27.

Rao, J. R., and Richter, G., 1974, Implantable bioelectrochemical power sources, *Naturwissenschaften* **61**:200.

Rasor, N. S., Spickler, J. W., and Clabaugh, T. L., 1972, Comparison of power sources for advanced pacemaker applications, *Proc. 7th Intersoc. Energy Conversion Eng. Conf.,* Washington, D.C., pp. 757–760.

Reynolds, L. W., 1963, *Utilization of Bioelectric Potentials, Ames Research Center, NASA, Quarterly Report* (September and November 1963).

Reynolds, L. W., 1964, *Utilization of Bioelectric Potentials, Ames Research Center, NASA, Quarterly Report* (February 1964).

Roy, O. Z., 1971, Biological energy sources: a review, *Biomed. Eng.* **6**:250–256.

Roy, O. Z., and Wehnert, R. W., 1966, Keeping the heart alive with a biological battery, *Electronics* **105**:107.

Schaldach, M., 1969, Bioelectric energy sources for cardiac pacing, *Ann. N.Y. Acad. Sci.* **167**:1016–1024.

Schaldach, M., 1970, Klinische Erfahrungen und experimentelle Ergelnisse über korpereigene electrochemical Energiequellen, *Z. Exptl. Chirurgie* **3**:200–216.

Schultz, H. E., and Heremans, J. F., 1966, *Molecular Biology and Human Proteins,* Vol. I: *Nature and Metabolism of Extracellular Protein,* Sec. IV: Proteins of Extravascular Fluids, Chap. I, American Elsevier Publishing Co., Inc., New York.

Senning, A., 1959, Discussion of Goot, B., Miller, F. A.: Prevention of posthypercapneic ventricular fibrillation in dogs, *J. Thoracic Cardiovascular Surg.* **38**:630–642.

Strohl, C. L., Scott, R. D., Frezel, W. J. and Wolfson, S. K., Jr., 1966, Studies of bioelectric power sources for cardiac pacemakers, *Trans. Amer. Soc. Artificial Internal Organs* **12**:318–326.

Takahashi, F., Aizawa, M., Mizuguchi, J., and Suzuki, S., 1970, Cell with NADP–NADPH Redox system, *Kogyo Kagaku Zasshi* **73**:908.

Tseung, A. C. C., King, W. J., and Wan, B. Y. C., 1971, An encapsulated, implantable metal–oxygen cell as a long-term power source for medical and biological applications, *Med. Biol. Eng.* **9**:175–184.

Tyers, G. F. O., Torman, H. A., Hughes, H. C., 1974, Comparative studies of "state of the art" and presently used clinical pacemaker electrodes, *J. Thoracic Cardiovascular Surg.* **67**:849–856.

Varriale, P., and Naclerio, E. A., 1975, Sutureless electrode for permanent ventricular pacing—observations and results of a three year follow-up study, *Circulation,* Suppl. II, **52**:252.

Wan, B. Y. C., and Tseung, A. C. C., 1974, Some studies related to electricity generation from biological fuel cells and galvanic cells, in vitro and in vivo, *Med. Biol. Eng.* **12**:14–28.

Wan, B. Y. C., Tseung, A. C. C., Kenny, J., and Wilds, A., 1972, The development and long-term implantation studies of an encapsulated, implantable aluminum/oxygen cell as an *in vivo* power source, *Proc. 7th Intersoc. Energy Conversion Eng. Conf.*, San Diego, pp. 745–751.

Weibel, M. K., Dritschilo, W., Bright, H. J., and Humphrey, A. E., 1973, Immobilized enzymes: a prototype apparatus for oxidase enzymes in chemical analysis utilizing covalently bound glucose oxidase, *Anal. Biochem.* **52**:402.

Weissmann, B., Meyer, K., Sampson, P., and Linker, A., 1954, Isolation of oligosaccharides enzymatically produced from hyaluronic acid, *J. Biol. Chem.* **208**:417.

Weissmann, B., Hadjiioannou, S., and Tornheim, J., 1964, Oligosaccharase activity of β-*N*-acetyl-*D*-glucosaminidase of beef liver, *J. Biol Chem.* **239**:59.

West, R., and Clarke, D. H., 1938, The concentration of glucosamine in normal and pathological sera, *J. Clin. Invest.* **17**:173.

White, A., Handler, P., and Smith, E. L., 1959, *Principles of Biochemistry,* pp. 54–57, McGraw-Hill Book Company, New York.

Widmann, W. D., Glenn, W. W. L., Eisenberg, L., and Mauro, A., 1964, Radiofrequency cardiac pacemaker, *Ann. N.Y. Acad. Sci.* **111**:992–1006.

Wingard, L. B., Jr., and Liu, C. C., 1969, Development of a glucose oxidase fuel cell, *Proc. 8th Int. Conf. Med. Biol. Eng.,* Chicago, p. 26.

Wingard, L. B., Jr., Liu, C. C., and Nagda, N. L., 1971, Electrochemical measurements with glucose oxidase immobilized in polyacrylamide gel: constant current voltametry, *Biotechnol. Bioeng.,* **13**:629.

Wolfson, S. K., Jr., and Yao, S. J., 1972, Implantable fuel cells: effect of added endogenous dialyzable materials, *Proc. 7th Intersoc. Energy Conversion Eng. Conf.,* San Diego, pp. 733–739.

Wolfson, S. K., Jr., Gofberg, S. L., Prusiner, P., and Nanis, L., 1968, The bioautofuel cell: a device for pacemaker power from direct energy conversion consuming autogenous fuel, *Trans. Amer. Soc. Artificial Internal Organs* **14**:198–203.

Wolfson, S. K., Jr., Yao, S. J., Geisel, A., and Cash, H. R., Jr., 1970, A single electrolyte fuel cell utilizing permselective membranes, *Trans. Amer. Soc. Artificial Internal Organs* **16**:193–198.

Yahiro, A. T., Lee, S. M., and Kimble, D. O., 1964, Bioelectrochemistry: I. Enzyme utilizing bio-fuel cell studies, *Biochim. Biophys. Acta* **88**:375.

Yao, S. J., Appleby, A. J., Geisel, A., Cash, H. R., Jr., Wolfson, S. K., Jr., 1969, Anodic oxidation of carbohydrates and their derivatives in neutral saline solution, *Nature* **224**:921–922.

Yao, S. J., Michuda, M., Markley, F., and Wolfson, S. K., Jr., 1972, A bioautofuel cell for pacemaker power, in: *From Electrocatalysis to Fuel Cells* (G. Sandstede, ed.), pp. 291–299, University of Washington Press, Seattle, Washington.

Zoll, P. M., Frank, H. A., Zarsky, L. R. N., Linenthal, A. J., and Belgard, A. H., 1961, Long-term electric stimulation of the heart for Stokes-Adams diseases, *Ann. Surg.* **154**:330.

25

The Use of Enzymes for Oxygenator Membranes

GEORGES B. BROUN AND CANH TRAN MINH

1. *Introduction*

The trend in internal organ prosthesis is toward the construction of portable organs, easily adaptable to the changing conditions of current life, whose performance it is possible to control. Since the oxygenating device is permanently in contact with the blood, good hemocompatibility is important. The present oxygenators are clearly in need of improvement in terms of these criteria.

In the case of membrane oxygenators, the area and permeability of artificial membranes have a definite influence on the performance of the prosthesis. Important improvements were achieved recently in membrane gas permeation and blood distributions. But the slow diffusion of gases through the blood in the unstirred layer close to the membrane limits the diffusion of oxygen and CO_2 in most oxygenators, which necessitates the use of large active areas, giving rise to cumbersome devices. Consequently, efforts aimed at reducing the thickness of blood films in oxygenators raises the question of the relative importance of hydrodynamic and hemocompatibility factors (Buckles, 1966).

Analysis of the limitations of these improved oxygenators brought us to the opinion that the next step should be to accelerate transfer in the unstirred level of the oxygenating unit.

Such an improvement seemed to be attainable only through a new concept of the whole system. Our basic approach to the search for better oxygenators has been to mimic natural processes by using enzyme-bound membranes.

New techniques of chemical aggregation of enzymes with human albumin onto permeable supports have been developed in our laboratory (Broun *et al.*, 1973). These led to the attempt to develop catalytic membranes to facilitate and increase gas transfer at the point where it bottlenecked, that is, in the diffusion layer at the interface between blood flow and membrane.

GEORGES B. BROUN • Département de Génie Biologique, Université de Technologie, Compiègne France. CANH TRAN MINH • Ecole Nationale des Mines, Saint Etienne, France. (Reprints are available from École Nationale des Mines.)

Two applications of this approach will be described here. One of them deals with the coating of carbonic anhydrase on a silastic membrane, to obtain more efficient transfer of CO_2. The other describes an oxygen-generating membrane, made of catalase coated on cellophane, which produces oxygen in the blood itself, starting with a concentrated solution of hydrogen peroxide.

Carbonic anhydride is weakly soluble in water, so its solubility in water at physiological pH's is obtained primarily as hydrogen carbonate ions. Consequently, transfer of CO_2 across a water film between two phases is facilitated by the intermediary synthesis of hydrogen carbonate ions.

Biologically specific-facilitated transport was first studied by Scholander (1960), then again by several authors for the transport of oxygen and carbon monoxide across aqueous solutions of hemoglobin or myoglobin (La Force, 1966; Wittenberg, 1970). The same phenomenon was shown to be observable with CO_2 when using solutions of carbonic anhydrase. Enzyme-facilitated transport of CO_2 across a film of buffered saline (Longmuir *et al.*, 1966; Enns, 1964) was studied in a system in which it was sandwiched between two gas-permeable membranes. Application of this enzyme close to the surface of the membrane appeared to be an especially effective way of activating CO_2 transfer either between two buffer solutions or between one buffered solution and one gas phase.

The use of hydrogen peroxide as a donor of oxygen to the blood was recently assayed *in vivo* (White and Teasdale, 1966). Several authors tried to provide it either directly by venous injections or in extracorporeal shunts. Boisseau (1967) injected intravenously 0.05 M hydrogen peroxide in saline solution, and showed that it could aid the patient in acute anoxia, without harm. Awad and Caron (1969) developed a method of extracorporeal blood oxygenation of dogs using dialyzer tubing immersed in a solution of hydrogen peroxide, which liberated oxygen in the blood under the effect of erythrocyte catalase. To obtain complete degradation of H_2O_2, they supplied the blood of the dog with 300–500,000 IU of catalase in saline solution.

In the system developed by us, catalase was bound to the H_2O_2-permeable membrane, so hydrogen peroxide was completely destroyed by a permanent catalase activity which remained stable for a relatively long period and gave rise to controllable oxygen production from a concentrated solution of hydrogen peroxide (10 M).

2. *Methods*

A binding method which keeps a major part of bound active sites effective (Broun *et al.*, 1973) has been adapted to the chemical binding of carbonic anhydrase and catalase.

2.1. *Binding Carbonic Anhydrase on Hydrophobic Supports*

The enzyme was bound onto a thin silicone membrane very permeable to CO_2. A "single-backed" 30-μm-thick General Electric membrane was used. The enzyme was bound to the silicon-coated side facing the blood. A very thin layer of enzyme was cross-linked close to the silicon, to reduce the resistance of the enzyme layer to diffusion. A 50-μl solution containing 2,000 IU/ml of carbonic anhydrase (Sigma) with 0.8% glutaraldehyde in a 0.02 M, pH 6.8 phosphate buffer was layered onto the silicone sheet. The cross-linking of the enzyme was continued at +4°C until there was complete evaporation of the water. The membrane was rinsed with distilled water, with a solution of glycine 0.05 M, then again with water (Broun *et al.,* 1970).

The enzyme activity of the membrane was assayed according to Waygood (1965), by the rate of pH shift of a standard solution from 7.95 to 6.3 after the addition of a saturated CO_2 solution at +1°C, as compared to a reference without enzyme.

2.2. *Coating a Hydrophilic Support with Catalase*

Catalase was bound to the cellophane membrane which separated the blood from the hydrogen peroxide solution. A glycerol-free 30-μm-thick sheet of cellophane (Rhône-Poullenc) was used as a support. The membrane was coated with enzyme on the side facing the blood. The immobilization of catalase on cellophane was difficult because of the size of the catalase molecule. Only a slight penetration of the cellophane matrix by the catalase was obtained. Cross-linking of a mixture of catalase and plasma albumin helped to anchor the catalase, since albumin penetrated the cellophane more readily. Molecules of albumin were cross-linked by glutaraldehyde in the cellophane matrix, anchoring the cross-linked material to the support. Then, catalase was bound to the albumin by the same agent.

To establish complete enzymatic conversion of the hydrogen peroxide which diffuses through the cellophane membrane, a large excess of catalase was bound to the membrane. For each square centimeter, 20 μl of solution containing (per ml) 18,000 IU of catalase (Sigma), 50 mg of plasma albumin, and 5 mg of glutaraldehyde in a 0.05 M pH 6.8 phosphate buffer were spread on the cellophane sheet. The resulting activity of catalase was higher than 350 IU/m^2. This value was 10 times higher than the theoretically required value for the complete degradation of H_2O_2 which diffuses through cellophane per time unit. Crosslinking was performed at +4°C. The membrane was rinsed with water, with glycine 0.05 M, then again with water (Broun *et al.,* 1971). Catalase activity was determined using the decrease of absorbance of a 0.02 M H_2O_2 solution in a 0.1 M, pH 7 phosphate buffer at 240 nm.

2.3. *CO_2 Transfer Determinations*

The transfer of CO_2 was measured according to Lautier *et al.* (1970) under the conditions used for the assay of blood oxygenation membranes (Figure 1). The device consisted of a membrane (A), which provided CO_2; a diffusion cell (B), where CO_2 is taken out of solution across a membrane in contact with a current of oxygen devoid of CO_2; an absorber (C) of CO_2; a pH stat (D) permanently maintaining (C) at pH 10 by the addition of 0.05 M barium hydroxide; and a recorder (E), giving the volume of added barium hydroxide as a function of time.

The CO_2 supplied to the cell is obtained by a flow of gas in which CO_2

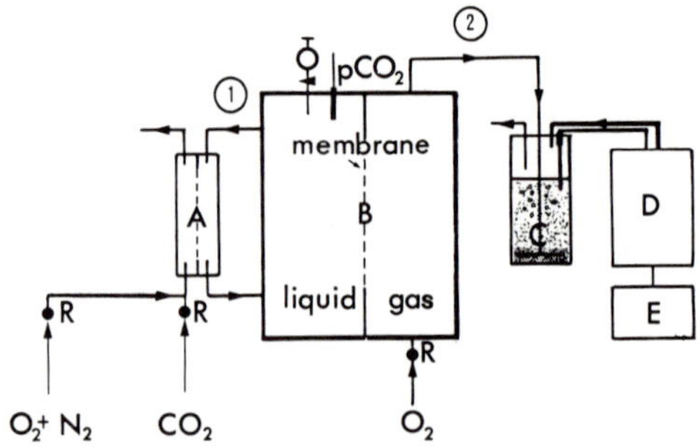

FIGURE 1. Determination of CO_2 transport through a membrane using a gas flow. A, Supplying device; B, transfer of CO_2 through the membrane; C, CO_2 absorber; D, pH stat; E, recorder; R, rotameter.

concentration is adjusted to the desired level by a mixture of N_2, O_2, and CO_2. The diffusion cell consisted of a plexiglass unit divided by a diffusion membrane of known area. The buffer solutions or heparinized blood were moderately stirred by a magnetic stirrer. The outward-diffusing CO_2 was absorbed and measured by a pH stat. At stationary state, the pH-stat response to CO_2 became linear, so CO_2 transfer could be measured precisely.

2.4. *Determination of Oxygen Production by Catalase-Bearing Membranes*

The applied device was a Technicon dialyzer made of two plates with semicircular grooves separated by the active membrane. On one side of the membrane, stabilized, concentrated hydrogen peroxide solution was kept in constant circulation by means of a peristaltic pump to obtain a flow of 0.33 ml/sec. On the other side of the membrane, the oxygen acceptor solution flowed at the rate of 2 ml/sec. For the preliminary assays, phosphate buffer 0.1 M, pH 7.4 was used (Figure 2).

The partial pressure of oxygen (p_{O_2}) was measured polarographically by dipping two P_{O_2} electrodes in blood near the inlet and outlet of the unit. A difference in the responses of the electrodes resulted from the activity of the

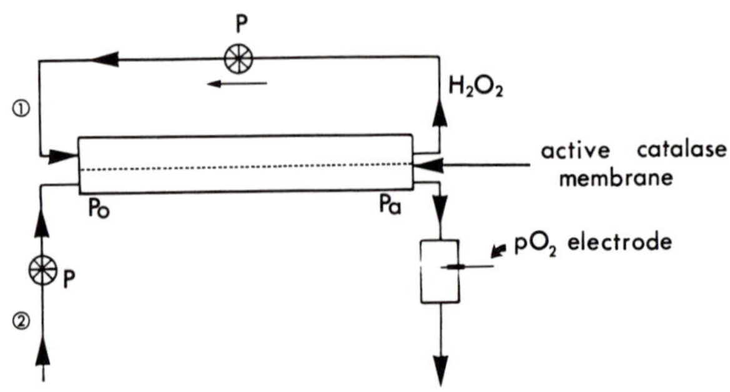

FIGURE 2. Oxygen production by a catalase-bearing membrane. P. peristaltic pump; p_{O_2} determination of blood p_{O_2}; 1, H_2O_2 circulation; 2, blood circulation.

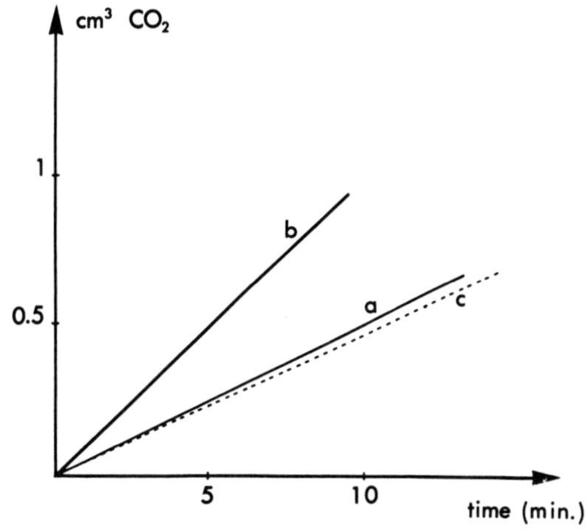

FIGURE 3. Facilitated transport of CO_2 through membranes of dimethylsilicone (General Electric). a, Noncoated membranes; b, carbonic anhydrase–coated membranes on the blood side; c, carbonic anhydrase–coated membranes in the presence of an inhibitor (acetazolamide).

membrane. When the membrane was active, the p_{O_2} of blood was quite a bit higher after contact with the catalase membrane. Knowing the blood flow and the difference of p_{O_2} the flux of oxygen produced from hydrogen peroxide was easily determined.

3. *Results and Discussion*

3.1. *Acceleration of CO_2 Transfer*

The rate of diffusion of CO_2 across the diffusion membranes determined by the quantities of barium hydroxide added is given in Figure 3. The CO_2 flux across the silicon membrane (curve a) was increased when the membrane was coated with carbonic anhydrase (curve b). The addition of 0.01 mg/ml of acetazolamide—an inhibitor of the enzyme (Enns, 1967)—reduces the flux of CO_2 across the enzyme but not across the control membrane. In these experiments, the flux of CO_2 was doubled by the presence of the enzyme. The flux of CO_2 across the acetazolamide-inhibited enzyme membrane gave an insight into the resistance of the enzyme layer to CO_2 diffusion (curve c).

3.2. *Oxygen Generation*

Results in terms of partial pressure of O_2 are given in Figure 4. The increase of oxygen partial pressure $p_a - p_0$ is controlled by the buffer (or blood) flow along the membrane to avoid bubbling in the acceptor solution. It remains permanently below the saturation limit.

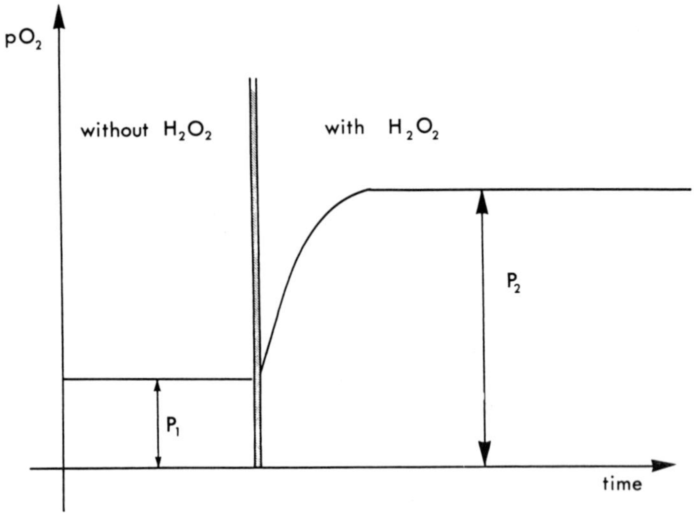

FIGURE 4. Recording of p_{O_2} in the blood before (p_{O_2}) and after (p_a) introduction of H_2O_2 in the donor compartment.

The flux of oxygen can be calculated from the following data:

$$J = \frac{dn}{S\,dt}$$

where S = membrane area

$\frac{dn}{dt}$ = flow of oxygenated solution

$$dn = d(cV) = \sigma(p_a - p_0)\,dV$$

where c = gas concentration in the solution

V = volume of solution

σ = solubility constant of oxygen

$$J = \frac{\sigma(p_a - p_0)}{S}\frac{dV}{dt}$$

where dV/dt = liquid flow. Since $\sigma = 3 \times 10^{\times 4}$ cm^3/cm^2-cm Hg, if p_0 is 15 cm Hg and p_a is 70 cm Hg, S is 35 cm^2 (the area of the assayed membrane). A flow of $J = 10^{-3}$ cm^3 of O_2/sec-cm^2 is obtained with a liquid flow of $dV/dt = 2$ cm^3/sec.

4. *Discussion*

4.1. *Analysis of CO_2-Facilitated Transport Mechanism in Solution*

a. Evidence of Diffusion Boundary Layers. The presence of a liquid phase in contact with a membrane reduces gas exchange across this membrane (Yasuda,

1967). The resulting apparent decrease of membrane permeability is due to the diffusion barrier which appears in the liquid phase close to the membrane. This diffusion boundary layer controls gas transfer across the membrane, mostly when it is very permeable. The same phenomenon cannot be seen with a membrane in contact with a gas phase because of the higher mobility of the gas phase.

Considering that the ratio e/p of the thickness of the membrane to its permeability is comparable to a resistance, the law of association of two resistances in series can be applied. The global resistance, which is the reciprocal of the global permeability, is the sum of the resistances of the membrane and its boundary layer (Tran Minh, 1971).

$$\frac{1}{G} = \frac{e_{liq}}{p_{liq}} + \frac{e_{mem}}{p_{mem}}$$

The membrane resistance is quite low in recently available membranes (e_{mem} low and p_{mem} high). The global resistance is then nearly equal to $e$liq/$p$liq, which controls the diffusion rate of the system.

b. Facilitated Transport. If transfer of dissolved gas through a membrane is accompanied by a reversible reaction where the transfered species reacts, transport is facilitated. Transfer of a combined molecule in the same direction is associated with the transfer of the free gas. This happens with carbon dioxide in solution, which is reversibly changed into hydrogen carbonic ions, more soluble than gas itself.

$$CO_2 + OH^- \underset{k_{-1}}{\overset{k_{+1}}{\rightleftarrows}} HCO_3^-$$

These ions move together with CO_2. These results show that the transfer rate is directly dependent on the reaction rate.

Several authors have described the influence of dissolved carbonic anhydrase on facilitated transport of CO_2. An analogous effect was obtained at the interface between a membrane and a solution when the boundary layer controls the

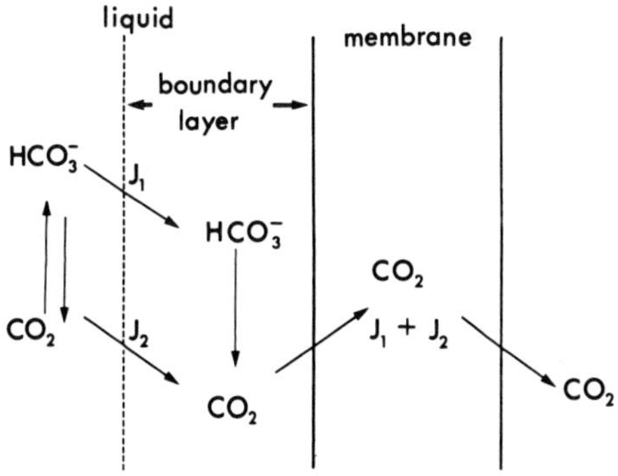

FIGURE 5. Schematic figuration of CO_2-facilitated transport through a hydrophobic membrane. J_1, hydrogen carbonate flux in the boundary layer; J_2, CO_2 flux in the boundary layer; $J_1 + J_2$, total flux of CO_2 diffusing through the membrane.

transfer rate. The selective extraction of CO_2 from the membrane displaced the reaction from the equilibrium, and the lack of CO_2 in the boundary layer gave rise to the hydrolysis of HCO_3^-.

Three kinetic processes can be analyzed in the boundary layer (Figure 5): (1) the diffusion of CO_2 through the boundary layer toward the membrane; (2) the degradation of HCO_3^- into CO_2 in order to get closer to the equilibrium ratio HCO_3^-/CO_2 at the given pH of the solution; and (3) the diffusion of HCO_3^- from the solution toward the boundary layer to replace the degraded HCO_3^-.

In the boundary layer, the fluxes J_1 of HCO_3^- and J_2 of CO_2 are to be studied separately. The flux of CO_2 crossing the membrane is the sum $J_1 + J_2$. Without anhydrase, transformation of HCO_3^- in CO_2 is slow (Magid and Turbeck, 1968). The maximum value of J_1 cannot be reached under these conditions. When carbonic anhydrase is added to the system, the reaction rate is several thousand times higher (Gibbons and Edsall, 1963; McIntosh, 1968), so equilibrium is established almost immediately. As a result, there is better access to the membrane and a steeper gradient of CO_2 concentration in the membrane (Figure 6).

The importance of flux J_1 is a direct function of the ratio of concentrations of HCO_3^- to CO_2. The value of the ratio is pH-dependent according to the Henderson–Hasselbach equation:

$$\mathrm{pH} = \mathrm{p}K_a + \log \frac{[\mathrm{HCO_3^-}]}{[\mathrm{CO_2}]}$$

At pH 7.4 and 37°C, there are 20 times more hydrogen carbonic ions than CO_2 molecules.

c. Analysis of Fluxes by Concentration Profiles. The kinetics of this facilitated transport can be deduced from the analysis of concentration profiles in the

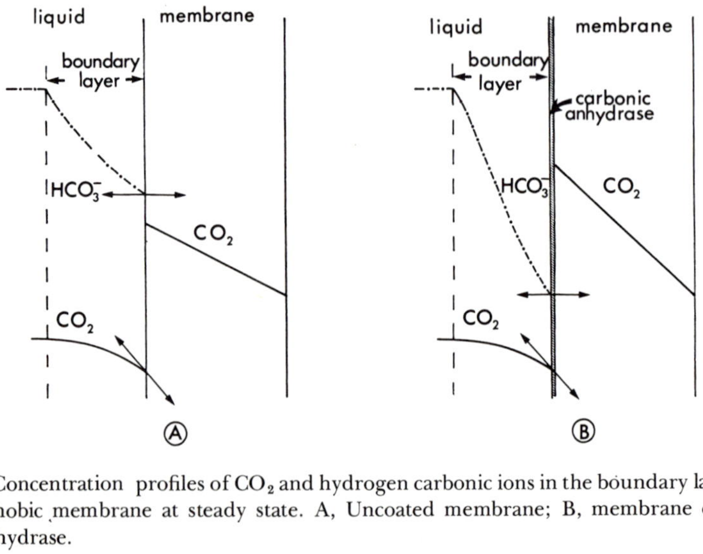

FIGURE 6. Concentration profiles of CO_2 and hydrogen carbonic ions in the boundary layer and in the hydrophobic membrane at steady state. A, Uncoated membrane; B, membrane coated with carbonic anhydrase.

combined system of the diffusion membrane with the enzymatically catalyzed unstirred layer. In the unstirred layer, the concentration profiles of CO_2 and HCO_3^- are obtained from the differential equations of diffusion with reaction (Tran Minh, 1971).

$$\frac{\partial[CO_2]}{\partial t} = k_{+1}[HCO_3^-] - k'_{-1}[CO_2] + D_{CO_2}\frac{\partial^2[CO_2]}{\partial x^2}$$

$$\frac{\partial[HCO_3^-]}{\partial t} = k'_{-1}[CO_2] - k_{+1}[HCO_3^-] + D_{HCO_3}\frac{\partial^2[HCO_3^-]}{\partial x^2}$$

$$\text{with } k'_{-1} = k_{-1}[OH^-].$$

Addition of the previous equations gives a relation between CO_2 and HCO_3^-:

$$\frac{\partial[CO_2] + \partial[HCO_3^-]}{\partial t} = D_{CO_2}\frac{\partial^2[CO_2]}{\partial x^2} + D_{HCO_3^-}\frac{\partial^2[HCO_3^-]}{\partial x^2}$$

Analytical solutions of these equations were obtained at steady state. For the "evolutive" state, the profiles were given by computerized numerical analysis (Kernevez and Thomas, 1975). Concentration profiles with and without carbonic anhydrase are given in Figure 6.

4.2. *Oxygen-Generating Membrane Model*

The oxygen-generating membrane model consists of a hydrophilic membrane between an oxygen donor and an oxygen acceptor compartment. In the first a hydrogen peroxide solution is circulated; in the second, blood or buffer. Catalase molecules are bound to the surface of the membrane turned toward the second compartment. The enzyme degrades H_2O_2 and provides oxygen to blood.

The hydrophilic support has a regulating effect: hydrogen peroxide diffuses through the support according to its concentration gradient. It is very actively destroyed at the level of catalase (the turnover of the enzyme is 3.7×10^7 molecules/sec), so the concentration of H_2O_2 is very close to nil in the blood compartment (Ogura, 1955).

Oxygen production depends only on the diffusion of hydrogen peroxide, that is, the initial concentration of H_2O_2 solution and the diffusivity of the support.

a. Calculation of the Theoretically Maximum Flux of O_2. The flux across the membrane is

$$J_{H_2O_2} = -D_{H_2O_2}\frac{\delta[H_2O_2]}{\delta x}$$

It can be expressed in O_2 equivalent flux as

$$J_{O_2} = -D_{H_2O_2}\frac{\delta[O_2]}{\delta x}$$

With a 10 M hydrogen peroxide solution and a cellophane membrane 3×10^{-3}cm thick, considering that the diffusion coefficient of H_2O_2 in cellophane is $D = 10^{-5}$ cm²/sec, a theoretical maximum flux of oxygen can be calculated:

$$J_{O_2} = 0.4 \text{ cm of } O_2\text{/sec-cm}$$

To obtain the same flux of oxygen through a silicone membrane 30 μm thick (General Electric), a pressure difference of more than 600 bars would be needed.

b. Theoretical Amount of Enzyme To Be Provided. One condition of complete clearance of H_2O_2 after its diffusion through the cellophane membrane is adequate catalase activity. The flux J_{O_2} previously determined is issued from a flux of hydrogen peroxide:

$$J_{H_2O_2} = \frac{0.40 \times 2}{22{,}400} = 0.36 \times 10^{-4} \text{ mole of } H_2O_2\text{/sec-cm}^2$$

Using a purified enzyme (Sigma) rating 3000 IU/mg, the minimum amount of enzyme preparation should be 1.2×10^{-2} mg/cm². Even with a 10% ratio of bound activity, this amount of protein is easily grafted to cellophane without a notable change in the membrane diffusivity.

c. Control of the System In a membrane oxygenator (gas–membrane–liquid), the oxygen flux is directly related to the oxygen concentration gradient. With a constant flow of gaseous oxygen along the membrane, oxygenation depends on oxygen concentration of the blood near the membrane, which is a function of the blood flow in the oxygenation unit.

In the enzyme device, the oxygenation is independent of blood flow but depends on hydrogen peroxide flow. Acting on the flow of the donor solution controls oxygenation without exerting noxious effects on the blood components (Figure 7). Different thicknesses of membranes can also help to adapt the oxygenation device to different needs of oxygen supply.

d. Solution of Some Practical Problems

1. How to protect the cellophane membrane from degradation by peroxides: The commercially available hydrogen peroxide is stabilized by acidification by H_2SO_4 at pH 4. The neutralization by sodium hydroxide results in a quick degradation of the membrane by H_2O_2. Protection of the membrane at neutral pH by a nontoxic reagent is possible. The addition of 50 mg/liter of uric acid fulfills this aim.

2. How to protect catalase against concentrated hydrogen peroxide: Catalase in solution is immediately destroyed by concentrated H_2O_2. Catalase bound to cellophane is protected against the concentrated solution by binding to the support and by the screening effect of the support itself. The concentration profile of H_2O_2 in the membrane is such that the local concentration near the enzyme is quite low, since it is controlled at the same time by membrane diffusion and enzyme reaction.

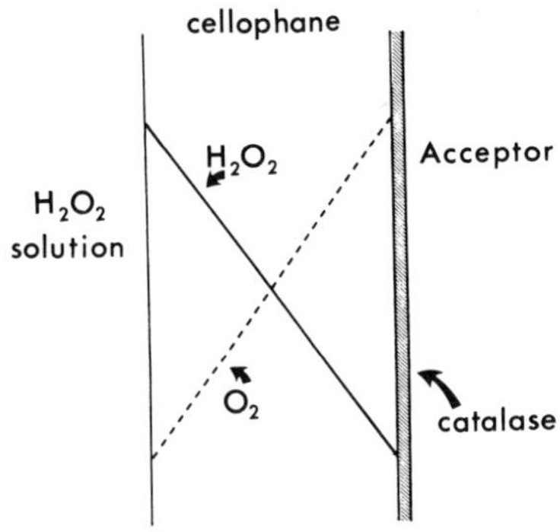

FIGURE 7. Model of an oxygen-generating membrane. Approximated concentration profiles of $[H_2O_2]$ and of $[O_2]$ in the cellophane sheet at steady state.

3. Degradation of the remaining H_2O_2: Previous experiments performed by injecting H_2O_2 solutions in animals (Boisseau, 1967) have shown that, as long as small quantities were used, catalase disappeared from the blood without any harm to the animal. The cellular catalase, quite active in erythrocytes, probably quickly destroyed the circulating H_2O_2. With our active membrane, only small quantities of H_2O_2, which were always analytically undetectable, could reach the blood.

4. Simultaneous extraction of CO_2: An enzyme-oxygenating membrane could at the same time extract CO_2 in an extracorporeal prosthesis. Preliminary assays showed that cellophane is highly permeable to dissolved CO_2. The bubbling of nitrogen in the solution could clear the H_2O_2 solution from the CO_2. Alternatively, carbonic anhydride can be absorbed on ion exchangers.

e. Performance of an Enzyme Oxygenator as Compared with a Passive Membrane Oxygenator. As mentioned above, oxygen flux can be controlled by the flow of H_2O_2 solution. In practice, the flow is adjusted to obtain the maximum oxygenation without bubbling of oxygen in the blood. With this procedure, for each square centimeter of catalase-bound cellophane, the resulting blood oxygenation under the conditions described was 8–10 times higher that oxygenation obtained using a specially designed nonenzymatic silicone membrane (General Electric).

5. *Conclusions*

To increase exchange in membrane oxygenators, specific solutions were developed, applying facilitated transport of CO_2 and generation of oxygen from H_2O_2. The efficiency of CO_2 extraction with carbonic anhydrase was maximum when located at the interface between the blood and the hydrophobic membrane. This effect could be explained by the catalytic transformation of CO_3H^- into OH^- + CO_2 which diffused through the membrane. The importance of this step increases with the permeability of the membrane to CO_2. At the present time, the specially designed membranes are more permeable to CO_2 than blood itself, so the permeability of the membrane to CO_2 coming from blood has become the rate-limiting step.

As concerns oxygen, the described procedure takes advantage of the local production of O_2 from hydrogen peroxide without the risky direct injection tentatively performed in some acute respiratory failures (Lorincz *et al.*, 1948). This system avoids the risk of methemoglobinization of hemoglobin (Paniker and Iyer, 1965). The same membrane could also help in extraction of CO_2 from the blood. The proteic nature of the coating could help to obtain better hemocompatibility of the membrane. The versatility of its control permits adequate adjustments to varying conditions. According to our calculations, a surface of 0.8 m^2 of active membrane would theoretically suffice to supply oxygen to an organism completely deprived of respiratory capacity.

A great many problems remain to be solved concerning hemocompatibility, membrane stability on storage, membrane lifetime in operation, and safety. The resulting blood oxygenation seemed promising enough to perform the multiple tests necessary for an application to oxygenation in man. Recently, our group has dealt mainly with biocompatibility problems, to reduce the risk of accidents secondary to immunologic or cellular reactions that could result from contact of the enzyme-grafted membrane with the blood.

6. *Summary*

Two applications of immobilized enzymes to the artificial oxygenation of blood are described. The first concerns facilitated transport of CO_2 moving from the blood through a hydrophobic membrane. It is based on activation of the rate-limiting step, which is the dehydration of hydrogen carbonic ions to CO_2 at the interface between the blood and the membrane. The addition of grafted carbonic anhydrase may hypothetically increase the diffusion rate 50 times. Multiplication by 2–8 of the CO_2 flux across a silicone membrane was obtained experimentally.

The second application is the use of a catalase-bound cellophane membrane which creates oxygen in the blood stream, starting from a concentrated solution of hydrogen peroxide. Experimentally, a flux of 0.4 $cm^3/sec\text{-}cm^2$ could be obtained without any bubbling and without any analytically detectable hydrogen peroxide in the acceptor solution. Oxygen production by catalase is controlled by the movement of H_2O_2 in the donor compartment. Catalase remains active for 48 hr in such a system. The technical problems linked to the presence of the stabilizer and to the protection of catalase against concentrated H_2O_2 were successfully solved. One constraining factor in the use of the catalase oxygenator was the attack on the hydrophilic support of the enzyme by the hydrogen peroxide solution.

A system using both enzymes in the same device was also tentatively designed.

7. *References*

Awad, J. A., and Caron, W. M., 1969, Extracorporeal oxygenation with hydrogen peroxide, *J. Surg. Res.* **9**:339.

Boisseau, R., 1967, De l'emploi du peroxyde d'hydrogène dans les anoxies, *Acad. Nat. Med.* **151**:134.

Broun, G., Selegny, E., Tran Minh, C., and Thomas, D., 1970, Facilitated transport of CO_2 across a membrane bearing carbonic anhydrase, *FEBS Letters,* **7**:223.

Broun, G., Tran Minh, C., Thomas, D., Domurado, D., and Selegny, E., 1971, The use of proteic and enzymatic coatings and membranes for oxygenators (trans.), *Amer. Soc. Artificial Internal Organs.* **17**:341.

Broun, G., Thomas, D., Gellf, G., Domurado, D., Berjonneau, A. M., and Guillon, C., 1973, New methods for binding enzyme molecules into a water insoluble matrix, *Biotech. Bioeng.* **15**:359.

Buckles, R. G., 1966, Analysis of gas exchange in a membrane oxygenator, Thesis, Massachusetts Institute of Technology.

Enns, T., 1964, Molecular collision exchange: transport of oxygen by hemoglobin, *Proc. Natl. Acad. Sci., U.S.* **51**:247.

Enns, T., 1967, Facilitation by carbonic anhydrase of CO_2 transport, *Science* **155**:44.

Gibbons, B. H., and Edsall, J. T., 1963, Rate of hydration of carbon dioxide and dehydration of carbonic acid at 25°, *J. Biol. Chem.* **238**:3502.

Kernevez, J. P., and Thomas, D., 1975, Numerical analysis of immobilized enzyme systems, *J. Appl. Math. Optimiz.* **1**:222.

La Force, R. C., 1966, Steady-state diffusion in the carbon monoxide + oxygen + hemoglobin system, *Trans. Faraday Soc.* **62**:1458.

Lautier, A., Laurent, D., Granger, A., and Sausse, A., 1970, A testing device for evaluation of gas transfer through synthetic membranes, *J. Biomed. Mater. Res.* **4**:189.

Longmuir, I. S., Forster, R. E., and Chi-Yuan, W., 1966. Diffusion of carbon dioxide through thin layers of solution, *Nature* **209**:393.

Lorincz, A. L., Jacoby, J. J., and Livingstone, H. M., 1948, Studies on the parenteral administration of hydrogen peroxide, *Anaesthesiology* **9**:168.

Magid, E., and Turbeck, B. O., 1968, The rates of the spontaneous hydration of CO_2 and the reciprocal reaction in neutral aqueous solutions between 0° and 38°, *Biochim. Biophys. Acta* **165**:515.

McIntosh, J. E. A., 1968, Assay of carbonic anhydrase by titration at constant pH, *Biochem. J.* **109**:203.

Ogura, Y., 1955, Catalase activity at high concentration of hydrogen peroxide, *Arch. Biochem. Biophys.* **57**:288.

Paniker, N. V., and Iyer, G. Y. N., 1965, Erythrocyte catalase and detoxification of hydrogen peroxide, *Can. J. Biochem.* **43**:1029.

Scholander, P. F., 1960, Oxygen transport through hemoglobin solutions, *Science* **131**:585.

Tran Minh, C., 1971, Transfert et génération des gaz par les membranes: application aux oxygénateurs et aux capteurs, Thèse AO 6262 C.N.R.S., Rouen, France.

Waygood, E. R., 1965, Carbonic anhydrase (plant and animal), in: *Methods in Enzymology* Vol. 2 (S. P. Colowick and N. O. Kaplan, eds.), p. 836, Academic Press Inc., New York.

White, D. C., and Teasdale, P. R., 1966, The oxygenation of blood by hydrogen peroxide: in vitro studies, *Brit. J. Anaesthesia* **38**:339.

Wittenberg, J. B., 1970, Myoglobin facilitated oxygen diffusion; role of myoglobin in oxygen entry into muscle, *Physiol. Rev.* **50**:559.

Yasuda, H., 1967, Basic consideration of permeability of polymer membrane to dissolved oxygen, *J. Polymer Sci.* **5**:2952.

Index